Principles and Techniques of Biochemistry and Molecular Biology

Principles and Techniques of Biochemistry and Molecular Biology

Edited by William O'Brien

Syrawood
PUBLISHING HOUSE

New York

Published by Syrawood Publishing House,
750 Third Avenue, 9th Floor,
New York, NY 10017, USA
www.syrawoodpublishinghouse.com

Principles and Techniques of Biochemistry and Molecular Biology
Edited by William O'Brien

International Standard Book Number: 978-1-64740-096-5 (Hardback)

Cataloging-in-Publication Data

Principles and techniques of biochemistry and molecular biology / edited by William O'Brien.
    p. cm.
Includes bibliographical references and index.
ISBN 978-1-64740-096-5
1. Biochemistry. 2. Molecular biology. 3. Biochemistry--Technique. 4. Molecular biology--Technique. I. O'Brien, William.
QH345 .P75 2022
572--dc23

TABLE OF CONTENTS

PREFACE

Biochemistry is a sub-discipline of both chemistry and biology that delves into the study of chemical processes within and relating to living organisms. It deals with the structures, interactions and functions of biological macromolecules such as carbohydrates, proteins, lipids and nucleic acids. It is divided into metabolism, molecular genetics and protein science. Biochemistry and molecular biology are closely related to each other. The field of study that is concerned with the molecular basis of biological activity between biomolecules in the various systems of a cell is called molecular biology. This includes the interactions between RNA, DNA, proteins and their biosynthesis, as well as the regulation of these interactions. Microarrays, gel electrophoresis, macromolecule blotting and probing, polymerase chain reaction and molecular cloning are some of the techniques used in this field. This book is compiled in such a manner, that it will provide in-depth knowledge about the principles and techniques of biochemistry and molecular biology. From theories to research to practical applications, case studies related to all contemporary topics of relevance have been included in it. This book includes contributions of experts and scientists, which will provide the readers with innovative insights into these fields.

After months of intensive research and writing, this book is the end result of all who devoted their time and efforts in the initiation and progress of this book. It will surely be a source of reference in enhancing the required knowledge of the new developments in the area. During the course of developing this book, certain measures such as accuracy, authenticity and research focused analytical studies were given preference in order to produce a comprehensive book in the area of study.

This book would not have been possible without the efforts of the authors and the publisher. I extend my sincere thanks to them. Secondly, I express my gratitude to my family and well-wishers. And most importantly, I thank my students for constantly expressing their willingness and curiosity in enhancing their knowledge in the field, which encourages me to take up further research projects for the advancement of the area.

Editor

The organization of metabolic genotype space facilitates adaptive evolution in nitrogen metabolism

Andreas Wagner[1,2,3], Vardan Andriasyan[4] and Aditya Barve[1,2]

[1]Institute of Evolutionary Biology and Environmental Studies, University of Zurich, Zurich, 8057, Switzerland
[2]The Swiss Institute of Bioinformatics, Basel, Switzerland
[3]The Santa Fe Institute, 1399 Hyde Park Road, Santa Fe, New Mexico 87501, USA
[4]Institute of Molecular Life Sciences, University of Zurich, Zurich, 8057, Switzerland

Correspondence should be addressed to Andreas Wagner; Email: andreas.wagner@ieu.uzh.ch

Abstract

A metabolism is a complex chemical reaction system, whose metabolic genotype – the DNA encoding the enzymes catalyzing these reactions – can be compactly represented by its complement of metabolic reactions. Here, we analyze a space of such metabolic genotypes. Specifically, we study nitrogen metabolism and focus on nitrogen utilization phenotypes that are defined through the viability of a metabolism – its ability to synthesize all essential small biomass precursors – on a given combination of sole nitrogen sources. We randomly sample metabolisms with known phenotypes from metabolic genotype space with the aid of a method based on Markov Chain Monte Carlo sampling. We find that metabolisms viable on a given nitrogen source or a combination of nitrogen sources can differ in as much as 80 percent of their reactions, but can form networks of genotypes that are connected to one another through sequences of single reaction changes. The reactions that cannot vary in any one metabolism differ among metabolisms, and include a small core of "absolutely superessential" reactions that are required in all metabolisms we study. Only a small number of reaction changes are needed to reach the genotype network of one metabolic phenotype from the genotype network of another metabolic phenotype. Our observations indicate deep similarities between the genotype spaces of macromolecules, regulatory circuits, and metabolism that can facilitate the origin of novel phenotypes in evolution.

Introduction

Nitrogen is among the top five chemical elements occurring in living systems, comprising of the order of 10 percent of biomass in bacteria, for example (Fagerbakke *et al.* 1996, Heldal *et al.* 1985). Most of this nitrogen occurs in the form of amino acids, but some of it also as RNA and DNA nucleotides, as well as cofactors such as NAD and heme (Neidhardt 1996). The biomass of a free-living heterotrophic organism such as *E. coli* is built from approximately sixty small molecule biomass precursors, of which 48 contain nitrogen (Table S1).

Highly abundant but chemically inert atmospheric molecular nitrogen gas can only be converted into biomass by a select few organisms (Sadava *et al.* 2006). Many other nitrogen sources are less abundant and can limit an organism's rate of growth or reproduction. Organisms can circumvent such limitations by using more than one nitrogen source. For free-living heterotrophic organisms like the bacterium *Escherichia coli*, three nitrogen-containing molecules play an especially important role as nitrogen sources, because the biosynthesis pathways leading to nitrogen-containing biomass precursors require one or more of them. These are ammonia, glutamine, and glutamate. Among them, ammonium supports the fastest growth in *E. coli* and is thus considered a preferred carbon source. Glutamine and glutamate are not only potentially important nitrogen sources; they also serve as precursors for the biosynthesis of amino acids, and of purine and pyrimidine nucleotides (Merrick & Edwards 1995, Neidhardt 1996, Reitzer 2003).

Metabolic generalists like *E. coli* can use dozens of nitrogen sources, including many amino acids, but also compounds such as nitrate and urea (Neidhardt 1996). They can also vary considerably in their ability to use any one nitrogen source. For example, while the proteobacterium *Klebsiella aerogenes* can use histidine (Neidhardt 1996) as a sole nitrogen source, its relative *E. coli* cannot. Some strains of *E.coli* can use agmatine, an intermediate in the degradation of arginine, as a sole nitrogen source, but *Salmonella typhimurium* cannot (Neidhardt 1996). Variation also exists in the biochemical pathways that metabolize or synthesize nitrogen-containing molecules. For example, arginine can be metabolized by multiple different pathways, two of which occur in enteric bacteria (Neidhardt 1996). The first uses arginine decarboxylase to degrade arginine to γ-aminobutyric acid in multiple steps, which then serves as a nitrogen source. In the second pathway arginine is succinylated and then metabolized further to produce succinate and glutamate. Similarly, L-alanine can be synthesized from pyruvate by glutamic-pyruvic transaminase using glutamate as an amino donor, or by transaminase C with valine as an amino donor (Neidhardt 1996). L-γ-glutamic semialdehyde, a precursor to the amino acid proline, can be synthesized from two different compounds, N-acetylglutamic γ-semialdehyde and L-γ-glutamyl phosphate.

Observations like these suggest that the pathways leading from any one nitrogen source to nitrogen-containing biomass precursors are flexible. We aimed to understand the extent of this flexibility, not just at the level of individual pathways, but on the level of the entire complex metabolic reaction network that is needed to synthesize all biomass precursors. More generally, we wanted to understand the basic organizational features of the space of possible metabolisms that can utilize different nitrogen sources. To this end, we used a recently developed approach to study large ensembles of metabolic networks that share the same biosynthetic abilities, but contain an otherwise random complement of biochemical reactions. We next introduce some necessary terminology and sketch the method behind this approach, which has been described in greater detail elsewhere (Rodrigues & Wagner 2009, Samal *et al.* 2010).

Methods

Metabolic genotypes, phenotypes and viability

A metabolism is a complex network of chemical reactions catalyzed by enzymes that are encoded by genes. The metabolic *genotype* of an organism is the part of a genome's DNA sequence that encodes these enzymes. This DNA-level representation of a metabolic genotype is too unwieldy to study qualitative and large-scale differences in the complement of enzyme-catalyzed reactions that specifies a metabolism. A more suitable and more compact representation is based on the observation that our current knowledge of metabolism comprises more than 5,000 enzyme-catalyzed reactions with known stoichiometry that occur in some organism (Goto *et al.* 1998, Kanehisa & Goto 2000). One can write these reactions as a long list, as indicated in Figure 1a. If the genome of an organism, such as that of a human, a fruit fly, or of *E. coli* encodes an enzyme that can catalyze a specific reaction, write "1" next to the reaction, otherwise write "0" (Figure 1a). The result is a representation of a metabolic genotype as a binary vector that completely specifies the reaction complement of a metabolism. This representation also makes clear that any one metabolic genotype is a member of a giant space of genotypes, a metabolic *genotype space* or a space of possible metabolisms. Since the universe of metabolic reactions comprises more than 5000 reactions, this space comprises more than 2^{5000} possible genotypes, many more than could be realized in the history of life on earth. Two metabolisms are *neighbors* in this space if they differ in a single reaction. A metabolism's *neighborhood* comprises all its neighbors. The *genotype distance* of two metabolisms can be defined through a variety of distance measures. We here use the fraction of reactions in which two metabolisms differ (in the representation of Figure 1a) as a distance measure. Specifically, if n_{12} denotes the reactions that two metabolic genotypes G_1 and G_2 have in common, and n_i denotes the number of reactions in genotype G_i, then this distance measure can be written as $1 - (n_{12} / (n_1 + n_2 - n_{12}))$.

The metabolic genotype of any one organism encodes its metabolic *phenotype*. There are many ways to define a metabolic phenotype, but the best-suited for the purpose of this paper is described hereafter. It starts from the observation that the most fundamental task of any one metabolism is to sustain life; that is, to synthesize all major biomass molecules that an organism needs for growth and reproduction, which include all amino acids, nucleotide precursors, lipids, and several co-factors (Feist *et al.* 2007, 2009, Feist & Palsson 2010). An organism whose metabolism is able to do that in a given chemical environment can survive in this environment - we refer to it as *viable*. Clearly, the potential nutrient molecules that occur in a given environment strongly influence whether a metabolism is viable. We will here consider minimal chemical environments that contain a single source of carbon (D-glucose), phosphorus (inorganic phosphate), sulfur (sulfate), oxygen, iron (Fe^{2+}, Fe^{3+}), as well as a single one among multiple possible nitrogen sources. One can write these potential nitrogen sources as a list, as shown in Figure 1b and associate a "1" with a nitrogen source if a metabolism is viable on it, that is, if it can synthesize all nitrogen-containing biomass precursors from it, and a "0" otherwise. In this way, a metabolic (nitrogen utilization) phenotype can be represented as a binary vector that indicates the spectrum of nitrogen

a) Genotype

b) Phenotype

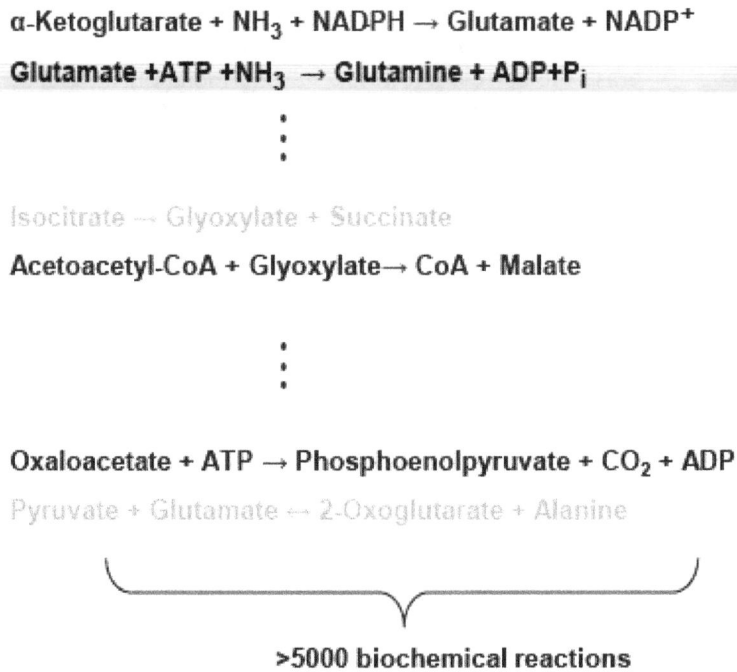

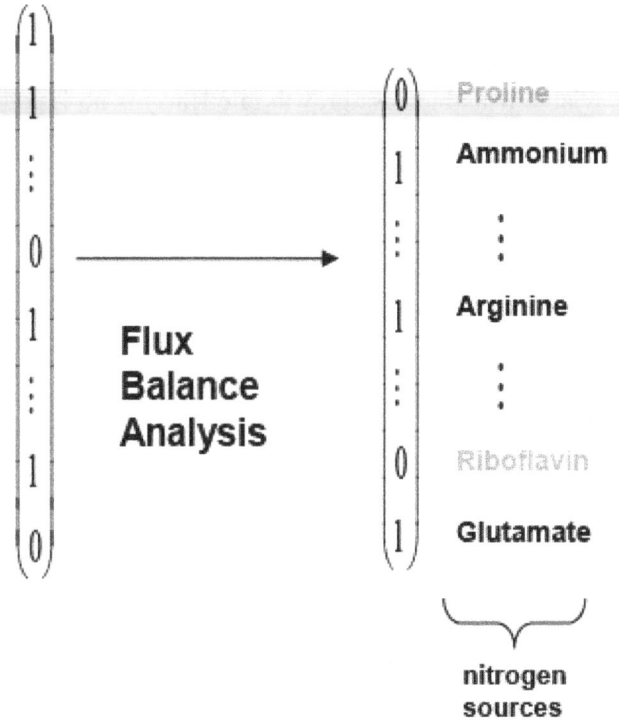

Figure 1. Metabolic genotypes and phenotypes. See text for details.

sources on which a metabolism is viable. In this paper, we consider 50 different nitrogen sources (Table S2).

To characterize those metabolic genotypes within the metabolic genotype space that are viable on a given number of nitrogen sources, we need to study many different metabolic genotypes and their phenotypes. It is possible to determine the metabolic phenotype of any one organism and its metabolic network experimentally on multiple different sources of chemical elements, such as through large scale metabolic phenotyping (Bochner 2009). However, it is currently not yet possible to experimentally manipulate metabolic genotypes on the large scale necessary to create many metabolic networks that are very different from each other. Fortunately, during the last 15 years computational approaches have been developed that allow us to predict metabolic phenotypes (Figure 1b) from qualitative information about metabolic genotypes, such as the stoichiometric equations shown in Figure 1a (Becker *et al.* 2007, Edwards & Palsson 2000, Feist & Palsson 2008, Heinrich & Schuster 1996, Schilling *et al.* 1999). Most notable among such approaches is the constraint-based modeling framework called flux balance analysis (Becker *et al.* 2007, Schilling *et al.* 1999). For a network that operates in a metabolic steady-state, such as would occur in a constant chemi-

cal environment under a steady nutrient supply, flux balance analysis predicts the maximal rate of biomass synthesis that a network can achieve in this chemical environment. Importantly, flux balance analysis requires only information about the stoichiometry of a metabolic reaction, and not about its kinetics or the concentrations of the enzyme catalyzing it. For metabolic networks with a well-studied genotype, the predictions of flux balance analysis are in good qualitative agreement with experimental data, for example on the viability of gene deletion mutations (Feist *et al.* 2007, Wang & Zhang 2009). The most important limitation of flux balance analysis is that it can incorporate regulatory constraints on biomass only with difficulty. Aside from the fact that many such constrains are easily broken in laboratory evolution experiments (Fong *et al.* 2005, Fong *et al.* 2003), such constraints are not of central importance for our purpose, because we are concerned mainly with the more fundamental constraints on viability caused by the complete absence of a reaction (enzyme-coding gene) from a metabolic genotype.

In our analysis, we constrained uptake rates of each nitrogen source to a maximum of 10 mmol/gdw/hr, and that of oxygen to a maximum of 20 mmol/gdw/hr. All other nutrients, including glucose as the sole carbon source in the minimal environment were effectively unconstrained in their uptake rate ($<10^{20}$ mmol/gdw/hr). As we are studying the metabolism of nitrogen sources, choosing s low uptake rate for the nitrogen source makes it the rate-limiting nutrient. This is especially relevant because we define a network as viable if its biomass growth rate (flux) is no less than one percent of the starting *E. coli* network in the same minimal environment. Not having the nitrogen source, but another nutrient as rate-limiting could result in a false estimation of viability. Moreover, our definition of viability also takes into account that many microbes survive in the wild even though they grow slowly (Vieira-Silva *et al.* 2011).

Sampling of random viable metabolisms
In bioengineering, flux balance analysis is often applied to a single metabolism, to help understand the role that individual reactions play in the metabolism and to improve incomplete knowledge about its metabolic genotype (Figure 1a) (Becker *et al.* 2007, Feist *et al.* 2007, 2009, Herrgard *et al.* 2008, Jamshidi & Palsson 2007). In contrast, we will characterize many different metabolisms in metabolic genotype space, as well as their viability on different nitrogen sources. To this end, we employ a variant of Markov Chain Monte Carlo (MCMC) sampling in network space that we have already described previously (Rodrigues & Wagner 2009, 2011, Samal *et al.* 2010). This technique can produce uniform samples of metabolisms with a given phenotype. Briefly, it relies on random walks through genotype space that start with a metabolism of a given number of reactions and a given metabolic phenotype, for example viability on glutamine as a sole nitrogen source. Each step of this random walk consists of a so-called reaction swap, where one reaction chosen at random from the known reaction universe is added to a metabolism, whereas another randomly chosen reaction is deleted from the metabolism. This procedure ensures that each step leaves the number of reactions in the metabolism constant. In addition, each step is required to preserve viability on the chosen nitrogen source. If a step does not fulfill this requirement it is rejected, and another step is tried until one is found that preserves viability. In this way, one can perform long random walks through metabolic genotype space and sample networks at some steps during this random walk.

During a random walk using MCMC sampling, each metabolism in the random walk differs by only a reaction pair from the preceding metabolism. In other words, successive metabolisms in the random walk are "correlated" in their genotype and thus also in their phenotypic properties. As the number of steps

between two metabolisms along the random walk increases, this autocorrelation decreases. Earlier work has shown that for metabolisms comprising about 1400 reactions, similar to the number of reactions in the *E. coli* metabolic network and the metabolisms studied here (Feist *et al.* 2007), 3×10^3 steps are sufficient to erase the correlation to the starting metabolism (Barve & Wagner 2013, Rodrigues & Wagner 2009, Samal *et al.* 2010). We thus sampled the first network after 5,000 steps, a number ensuring that the autocorrelation between the starting and the sampled metabolism was minimal. After this "burn-in" period, we sampled a metabolic genotype every 5,000 steps until we had obtained a sample of 1,000 genotypes that are viable on one or more given nitrogen sources, but contain an otherwise random complement of reactions (Rodrigues & Wagner 2009, Samal *et al.* 2010). In other words, our random walks proceeded for at least 5×10^6 steps, unless otherwise mentioned. We refer to the metabolisms in the samples we thus generated as random viable metabolisms.

Variants of random sampling used in different analyses
Different analyses required us to use different variants of the sampling procedure. To estimate maximal genotype distances of metabolisms viable on a given, single nitrogen source, we started each random walk from the *E. coli* metabolic network (Feist *et al.* 2007), which comprises 1397 metabolic reactions, and required that none of the 5000 viability-preserving steps in the random walk reduce the distance to the starting network. In this way, we generated 1000 metabolisms required to be viable on a sole nitrogen source for each of the 50 nitrogen sources, that is, a total of 5×10^4 (50 x 1000) metabolisms. We also used these samples to quantify the superessentiality of reactions for each nitrogen source (Figure 3).

To understand if the maximal genotype distances we observed for metabolisms viable on a single nitrogen source changed when metabolisms were required to be viable on multiple nitrogen sources, we needed metabolisms viable on a randomly chosen *n*-tuples (5, 10, 15, 20 and so on) of nitrogen sources. To create them, we first generated a random metabolism viable on all 50 nitrogen sources starting from the *E. coli* metabolism (which itself is viable on all 50 sources) after 5000 viability-preserving steps. We then randomly chose n nitrogen sources and continued the random walk for another 5000 steps while ensuring that the metabolism was viable on these n nitrogen sources. We then generated 100 random metabolisms from this starting metabolism through another 5000 steps of the random walk, with the constraint that none

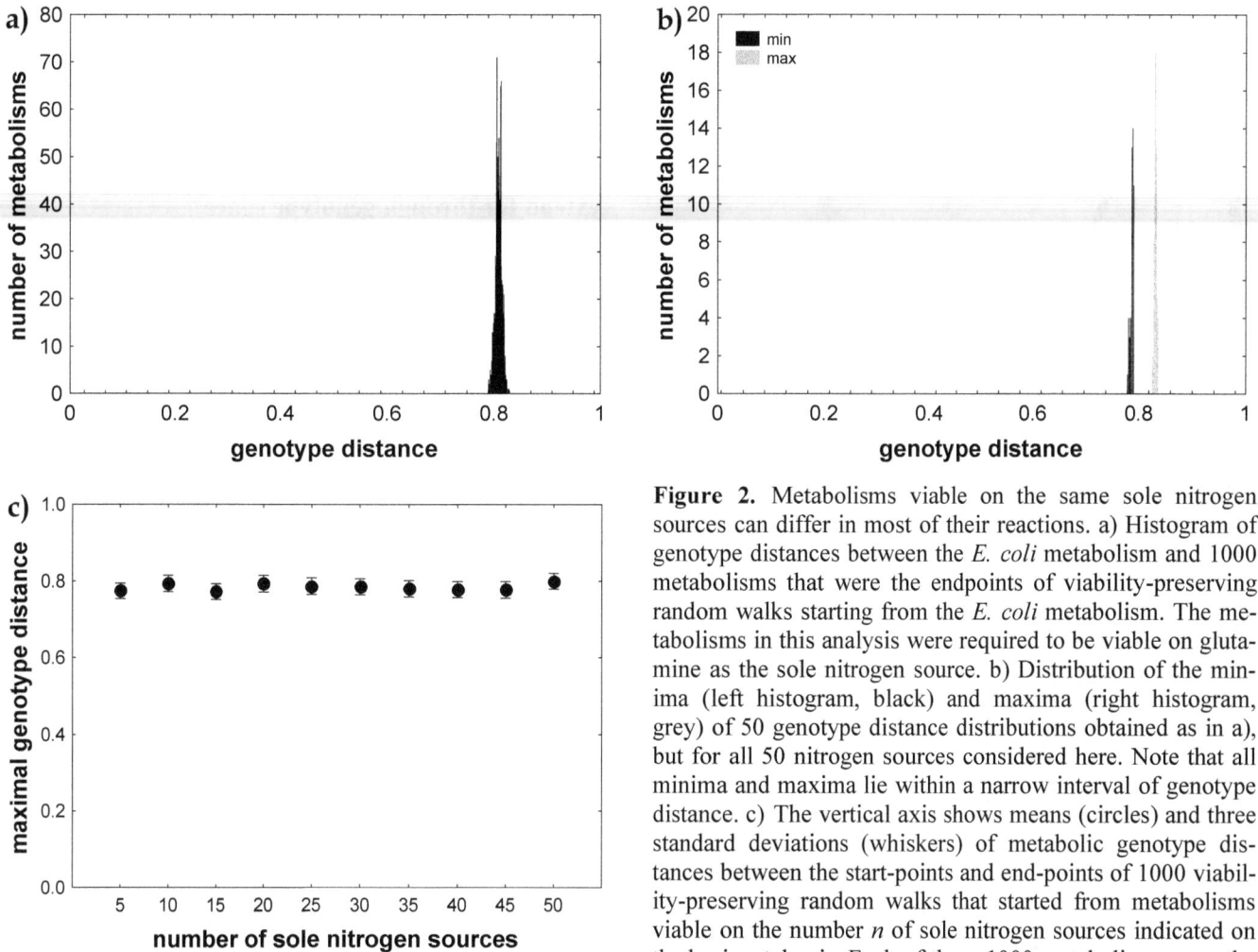

Figure 2. Metabolisms viable on the same sole nitrogen sources can differ in most of their reactions. a) Histogram of genotype distances between the *E. coli* metabolism and 1000 metabolisms that were the endpoints of viability-preserving random walks starting from the *E. coli* metabolism. The metabolisms in this analysis were required to be viable on glutamine as the sole nitrogen source. b) Distribution of the minima (left histogram, black) and maxima (right histogram, grey) of 50 genotype distance distributions obtained as in a), but for all 50 nitrogen sources considered here. Note that all minima and maxima lie within a narrow interval of genotype distance. c) The vertical axis shows means (circles) and three standard deviations (whiskers) of metabolic genotype distances between the start-points and end-points of 1000 viability-preserving random walks that started from metabolisms viable on the number *n* of sole nitrogen sources indicated on the horizontal axis. Each of these 1000 metabolisms were the starting points of a random walk where each step (i) had to preserve viability on the *n*-tuple of nitrogen sources, and (ii) was not allowed to decrease the distance to the starting metabolism (see Methods).

of the 5000 steps reduced the distance from the starting metabolism. We repeated this procedure 9 more times, with a different, randomly chosen n-tuple. In other words, we used this procedure to generate 1000 metabolisms viable on a given number *n* of nitrogen sources (100 metabolisms for each of 10 n-tuples).

As a starting point of our analysis of genotype network closeness, we required metabolisms that were viable on a specific nitrogen source, but not on other nitrogen sources. To generate such metabolisms, we returned to our sample of 1000 metabolisms viable on a sole nitrogen source. All of them were viable on one nitrogen source, but each may also be viable on other nitrogen sources (Barve & Wagner 2013). We chose an arbitrary metabolism among them, that was viable only on the focal nitrogen source (at least one of such metabolisms happened to exist in all of our samples). We used this metabolism as the starting point for random walks in which each reaction-swapping step was required to retain viability only on the focal nitrogen source. That is, if a step created viability on additional nitrogen sources, we discarded it. Through such random walks, we generated 10 random viable metabo-

lisms that were strictly viable only on the focal nitrogen source. We repeated this approach for all 50 nitrogen sources, thus creating a total of 500 metabolisms, in groups of 10, where each group contained metabolisms viable on a specific nitrogen source. To estimate how close two genotype networks of metabolisms viable on two nitrogen sources (termed source 1 and 2) are in genotype space, we chose, with uniform probability, one metabolism G_1 among the ten metabolisms viable on source 1 and another metabolism G_2 among the ten metabolisms viable on source 2. We then used G_1 as a starting point for a phenotype-preserving random walk towards G_2, where each step was required to preserve viability only on the focal nitrogen source, and was not allowed to create viability on a new nitrogen source. After 5000 reaction swaps, we recorded the remaining distance between the random walker and G_2. We note that this distance is an upper bound for the point of closest proximity between genotype networks.

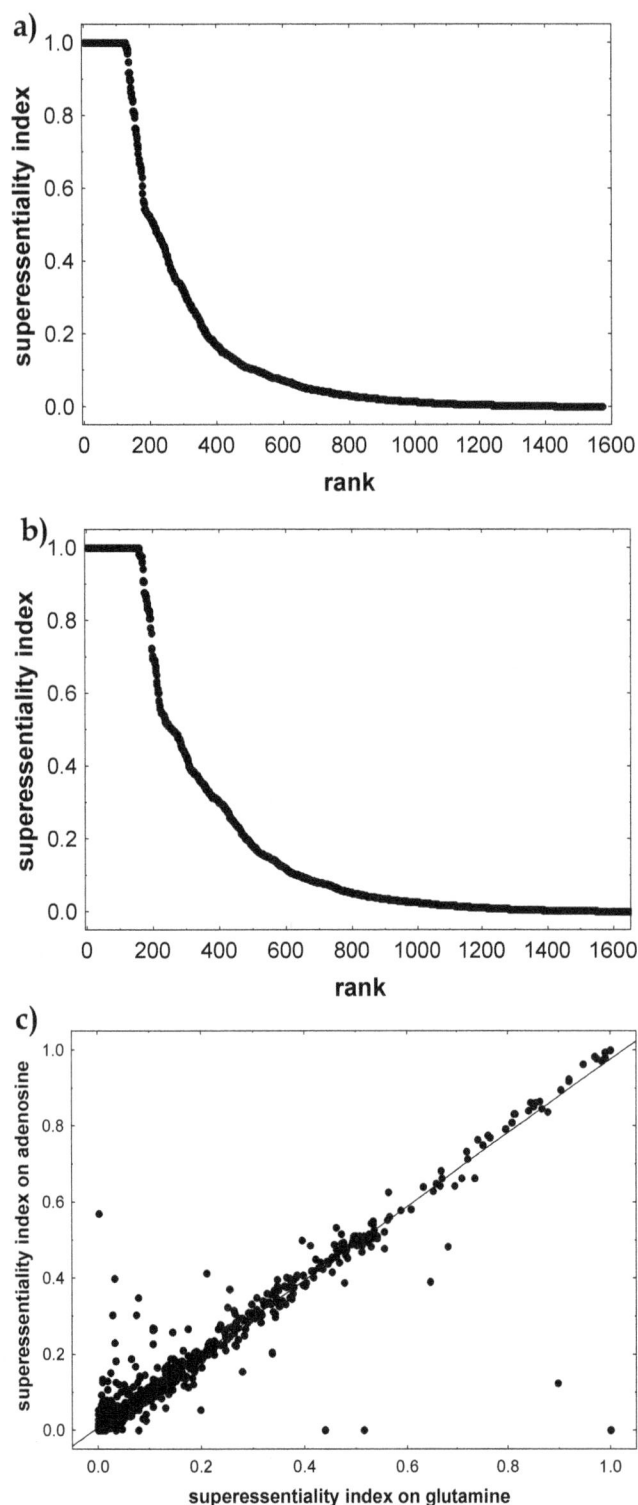

Figure 3. Reaction superessentiality in nitrogen metabolism. Rank plots of superessentiality indices (vertical axis) I_{SE} based on 1000 random metabolisms viable on a) glutamine, b) all 50 nitrogen sources considered here (when each is provided as the sole nitrogen source). c) Superessentiality indices of reactions where $I_{SE} > 0$, for metabolisms viable on glutamine (horizontal axis) or adenosine (vertical axis) as sole nitrogen sources. Data are based on a sample of 1000 random viable metabolic metabolisms generated as described in methods.

We repeated this procedure 100 times, i.e., for 100 randomly chosen pairs of nitrogen sources.

Results and Discussion

Connected networks of viable nitrogen metabolisms extend far through genotype space

We first inquired how different two metabolisms (metabolic genotypes) can be while preserving their viability on a given spectrum of nitrogen sources. To answer this question, we performed the following analysis. Starting from the *E. coli* metabolic network, we performed a random walk of 5000 viability-preserving steps, where none of these steps was allowed to reduce the distance to the starting network. At the end of this walk, we recorded the genotype distance between the random walker and the starting network. We repeated this random walk 1000 times. Figure 2a shows a histogram of the genotype distance from *E. coli*, for 1000 networks viable on glutamine as the sole nitrogen source. The distribution of genotype distances is sharply peaked around a mean of 0.81, with a standard deviation of 0.006, a minimum of 0.79 and a maximum of 0.83. This means that two networks can differ in the vast majority of their reactions – approximately 80 percent – and still retain viability on glutamine as a sole nitrogen source. In addition, the networks that we used in this analysis can be connected in genotype space through long sequences of single reaction changes, none of which eliminates viability. In other words, they form part of a single connected network of genotype networks with the same viability phenotype, a genotype network (Rodrigues & Wagner 2009, Samal *et al.* 2010).

This observation is not a peculiarity of glutamine as a nitrogen source. To show this, we performed 1000 additional random walks for each of the 49 remaining nitrogen sources N, such that each random walk had to preserve viability on N. The results were 49 further genotype distance distributions like the one shown in Figure 2a. Figure 2b shows a histogram of the minima (black) and the maxima (grey) of all 50 distributions. It demonstrates that these distributions are confined within a narrow interval. Specifically, the smallest minimum genotype distance for all 50 nitrogen sources is 0.78 and the largest maximal genotype distance for all 50 nitrogen sources is 0.83.

Next, we asked whether these observations change fundamentally if we require networks to be viable on multiple different nitrogen sources, when each of these sources is provided as the sole nitrogen source. The answer is shown in Figure 2c, for 1000 end-points of viability-preserving random walks starting from networks viable on different numbers of *sole* nitrogen sources, as shown on the horizontal axis. Clearly, the large genotype distances we observed for metabolisms required to be viable on only one nitrogen source change little when we consider multiple nitrogen sources. Taken together, these observations mean that metabolisms viable on one or more nitrogen source can differ greatly in the complement of metabolic reactions they harbor. Regardless of their specific nitrogen metabolism phenotype, they form connected networks of metabolic genotypes that range far through genotype space. In other words, they show great internal flexibility in their reaction complements.

Reactions vary widely in their superessentiality

The observation that 20 percent of metabolic reactions cannot change if viability on specific nitrogen sources is to be preserved raises the question of what these unchangeable reactions are, and whether they are the same for each of the 1000 metabolisms we studied during any one random walk. In other words, are some reactions more important than others in this sense? In previous works on carbon metabolism, we had shown that this is indeed the case, and that one can quantify this importance, as described hereafter (Barve *et al.* 2012). In any one metabolism, a reaction can be *essential* to synthesize biomass, that is, its removal will abolish the metabolism's viability. In a random sample of viable metabolisms, a reaction may be essential in some metabolisms but not in others. We call a reaction that is essential in more than one metabolism *superessential* – it is more than just essential. We introduced a *superessentiality index* I_{SE} that denotes the fraction of metabolic networks in which this reaction is essential. This index can range from zero (the reaction is never essential) to one (the reaction is essential in all metabolisms). A reaction with a superessentiality index of one is special, because it cannot be by-passed through an alternative sequence of reactions. We previously showed that assessing superessentiality based on random samples of at least 500 viable networks gives results that are in good agreement with a complementary approach that estimates superessentiality independently of random sampling (Barve *et al.* 2012). We thus proceeded to analyze the distribution of reaction superessentiality in randomly sampled metabolisms.

Figure 3a shows a rank plot of the superessentiality index of those reactions that were essential in at least one metabolism in a sample of 1000 random metabolisms viable on glutamine as a sole nitrogen source. The graph clearly shows that a small number of reactions are essential in all metabolisms – they are absolutely superessential and have a superessentiality index of one. Specifically, there are 126 such reactions, 102 of which involve nitrogen-containing mole-

cules. The vast majority of reactions whose superessentiality index is shown are not always essential and rank from being essential in most metabolisms (left) to being essential only in few metabolisms (right). Figure 3b shows an analogous rank plot, but for metabolisms viable on all 50 nitrogen sources shown here. The overall shape of this plot is very similar, except that the number of absolutely superessential reactions is somewhat larger (157 reactions, 114 of which involve nitrogen-containing compounds). Table S3 shows a list of these reactions.

The absolutely superessential reactions include riboflavin synthase, the last step in the biosynthesis of riboflavin, a component of the cofactors flavin adenine dinucleotide (FAD) and flavin adenine mononucleotide (FMN). Another example is the reaction catalyzed by the enzyme phosphomethylpyrimidine kinase, encoded by the gene thiD (Blattner number b2103), which participates in the biosynthesis of thiamine diphosphate (also known as thiamine pyrophosphate TPP). TPP is an essential cofactor in enzymes such as pyruvate dehydrogenase (Nemeria *et al.* 2010). Yet another example concerns the enzyme histidinol phosphatase, encoded by the gene hisB (Blattner number b2022). Histidinol-phosphatase is essential for the biosynthesis of the amino acid histidine, while the same gene product also catalyzes a further reaction essential for histidine synthesis, that of imidazoleglycerol-phosphate dehydratase. The superessentiality indices of reactions on different nitrogen sources are statistically associated with one another. For example, Figure 3c shows a scatterplot of superessentiality indices of reactions where this index is greater than zero, for metabolisms viable on glutamine (horizontal axis) and adenosine (vertical axis) as sole nitrogen sources (Spearman's r=0.94, P<<10^{-17}, n=1480).

Though most reactions have very similar superessentiality indices for growth on glutamine and adenosine, it is instructive to discuss those outlier reactions whose superessentiality differs greatly on these nitrogen sources. One of them is the reaction catalyzed by the enzyme adenylosuccinate lyase (encoded by the gene purB). It is essential in all metabolisms viable on glutamine, but only in 0.2 percent of metabolisms viable on adenosine. The reaction converts adenylosuccinate to adenosine monophosphate. Whenever glutamine is the sole nitrogen source, this reaction is essential for the synthesis of the DNA precursor deoxyadenosine-triphosphate (dATP) (Baba *et al.* 2006). However, when adenosine is provided as a nutrient, this reaction can be by-passed, because dATP can be synthesized directly from adenosine. Indeed, *E. coli* strains lacking the gene purB are able to grow only when adenosine or adenine is supplied to a minimal

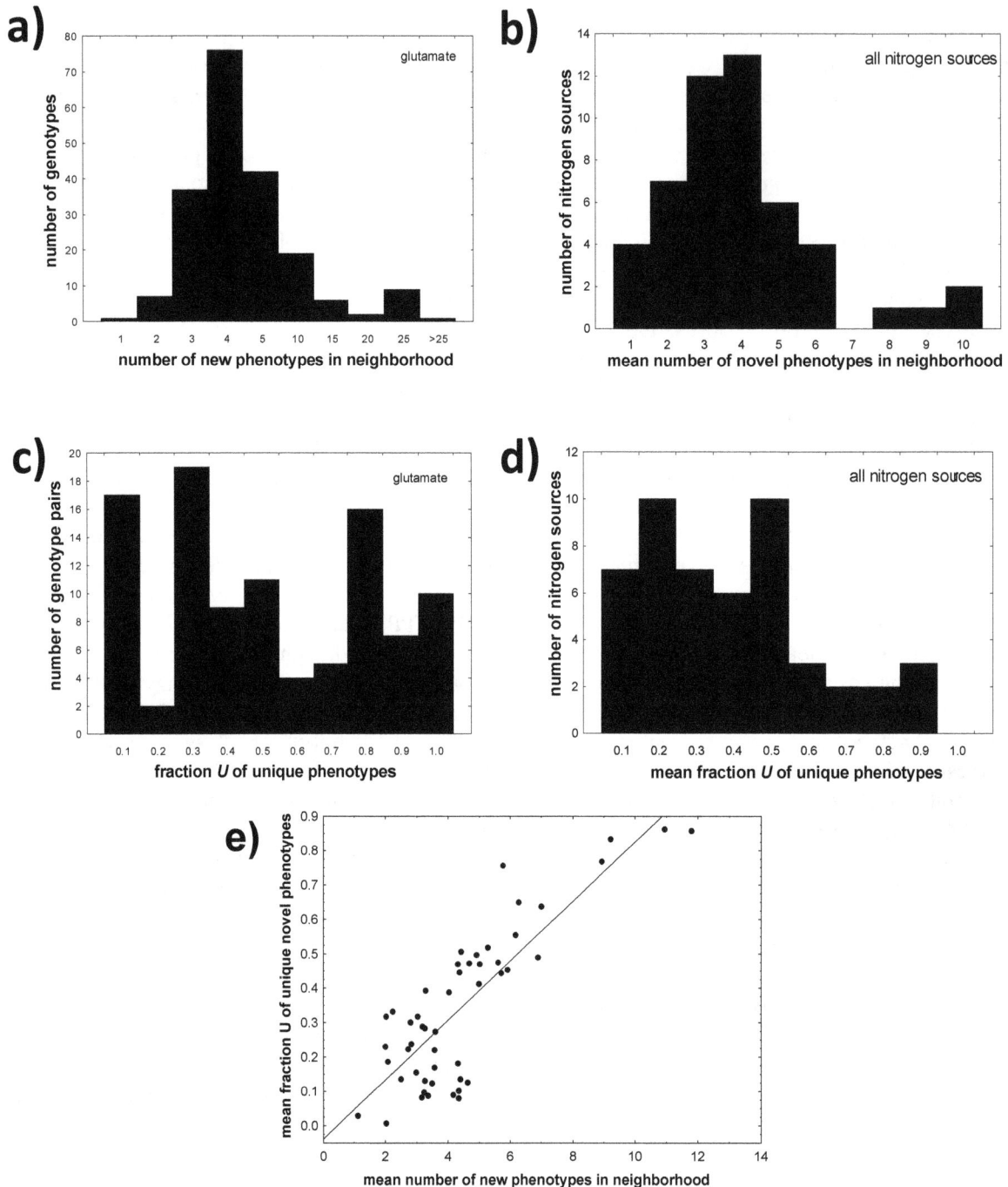

Figure 4. Genotypes contain novel phenotypes in their neighborhoods and some fractions of these novel phenotypes are unique. a) Distribution of the number of novel phenotypes in the neighborhood of 200 genotypes viable on glutamate. b) Distribution of the *mean* number of novel phenotypes in the neighborhood of genotypes for all 50 nitrogen sources considered here. Each data item is based on a sample of 200 genotypes (and each genotype's neighborhoods) for each of 50 nitrogen sources. Thus, the histogram is based on 50 samples of 200 genotypes each. c) For genotypes G_1 and G_2 sampled from the same genotype network, that is, they are viable on the same nitrogen source, and for P_i the set of all phenotypes that are found among genotypes in the neighborhood of G_i, the figure shows the distribution of $U=(|P_1| - |P_1 \cap P_2|)/|P_1|$. This is the fraction of phenotypes unique to one neighborhood, i.e. without occurring in the other genotype's neighborhood. Specifically, the vertical axis shows the number of genotype pairs whose value of U is shown on the horizontal axis. The data is based on 100 random genotype pairs viable on glutamate. d) Histogram of the mean value of U (horizontal axis) for genotype pairs viable on each of the 50 nitrogen sources considered here. The vertical axis shows the number of nitrogen sources for which genotype pairs have the mean value of U shown on the horizontal axis. The data is based on 100 random genotype pairs (and their neighborhoods) for each of 50 nitrogen sources. Panels b) and d) are based on the mean as a summary statistic, because it is the simplest such statistic for distributions that are not extremely right- or left-skewed.

growth medium (Sun *et al.* 2011). Another example involves citrate synthase, which is essential in 0.3 and 56.9 percent of metabolisms on glutamine and adenosine, respectively. The enzyme citrate synthase is encoded by the gene gltA, which participates in the tricarboxylic acid cycle and produces citrate, which is in turn necessary for the synthesis of important biomass precursors such as 2-oxoglutarate and succinyl-CoA (generated from 2-oxoglutarate) (Noor *et al.* 2010). On glutamine as the sole nitrogen source, enzymes such as glutaminase can convert glutamine to glutamate (Brown *et al.* 2008), which can be further metabolized to 2-oxoglutarate via other enzymes such as aspartate-transaminase (encoded by the gene aspC) (Marcus & Halpern 1969). These biochemical pathways make the enzyme citrate synthase dispensable on glutamine as the sole nitrogen source, because they allow 2-oxoglutate to be directly synthesized from glutamine without the need for citrate synthesis. In contrast, growth on adenosine does not easily allow this bypass and thus renders citrate synthase essential in the majority of metabolisms (56.9 percent).

Different neighborhoods in metabolic genotype space do not contain the same novel phenotypes

In a population of evolving organisms, metabolism would evolve through alteration of an organism's metabolic genotypes. Especially in microbes, such evolution can occur rapidly by adding individual enzyme-coding genes to a genome through horizontal gene transfer, as well as by deleting individual genes. Even in populations that evolve under stabilizing selection for an existing, well-adapted phenotype, genotypic change can occur, because of the flexibility afforded by genotype networks. Such populations will explore the genotype network associated with a well-adapted phenotype, and its member genotypes will also explore local *neighborhoods* around them and around their genotype network. In general, the neighborhood of a genotype is important from an evolutionary perspective, because it comprises all those genotypes – with potentially novel phenotypes – that are easily reached via little genotypic change. Some metabolic genotypes in this neighborhood may have novel metabolic phenotypes, i.e., they may be able to survive on novel combinations of nitrogen sources. Genotype networks would be especially important for evolutionary adaptation, if different neighborhoods contained a different spectrum of novel phenotypes: Because genotype networks allow the exploration of different regions of genotype space, they also allow the exploration of different neighborhoods, and thus the exploration of novel phenotypes that would not be accessible otherwise. We thus wished to find out whether the neighborhood of different genotypes on a genotype network contained different phenotypes. To this end, we carried out the following quantitative analysis.

Consider two arbitrary genotypes G_1 and G_2 that are sampled from the same genotype network, that is, they are viable on the same nitrogen source. Denote as P_1 the set of all phenotypes that are found among genotypes in the neighborhood of G_1, that is, among all those metabolisms that differ in a single reaction from G_1. Define P_2 analogously as the set of all phenotypes found in the neighborhood of G_2. We wished to quantify the fraction of U of phenotypes that are contained in P_1 but not in P_2, i.e., the number of phenotypes that are in this sense unique to P_1. To this end we computed the quantity $U = (|P_1| - |P_1 \cap P_2|)/|P_1|$, where $|X|$ denotes the number of elements in a set. For example, if $|P_1| = |P_2| = 10$ and $|P_1 \cap P_2| = 5$ (5 phenotypes are common to both sets P_1 and P_2), then $U = (10 - 5)/10) = 0.5$; that is, 50 percent of phenotypes are unique to the neighborhood of G_1. More specifically, we computed this quantity for 100 random genotype pairs viable on the same nitrogen source. We obtained these genotype pairs by randomly choosing two different metabolisms from our sample of 1000 metabolisms viable on a given nitrogen source.

Figure 4a shows the distribution of the number of novel phenotypes that occur in the neighborhood of 200 genotypes (100 pairs) viable on glutamate, illustrating that most neighborhoods contain some phenotypes viable on novel combinations of nitrogen sources. Figure 4b shows a histogram summarizing the same data for all 50 nitrogen sources. Specifically, the figure shows the distribution of the *mean* number of novel phenotypes in the neighborhood of genotypes, where each data item is the mean value of U based on a sample of 200 genotypes (and each genotype's neighborhoods) for each of 50 nitrogen source In other words, the histogram is based on 50 samples of 200 genotypes and their neighborhood. The data illustrates that the number of novel phenotypes in a genotype's neighborhood varies broadly between one and ten, depending on the nitrogen source. Figure 4c shows, as an example, the distribution of the fraction U of unique phenotypes for 200 pairs of metabolic genotypes viable on glutamate, and Figure 4d shows the distribution of the mean value of U for genotype pairs viable on each of the 50 nitrogen sources considered here. The panel is again based on 200 genotype pairs for each of the 50 nitrogen sources, i.e., on a total of 50x200 genotype pairs. The panels show that some fraction of novel phenotypes are unique to individual neighborhoods, otherwise U would be equal to zero for all genotype pairs and nitrogen sources. They also demonstrate that U varies broadly among both geno-

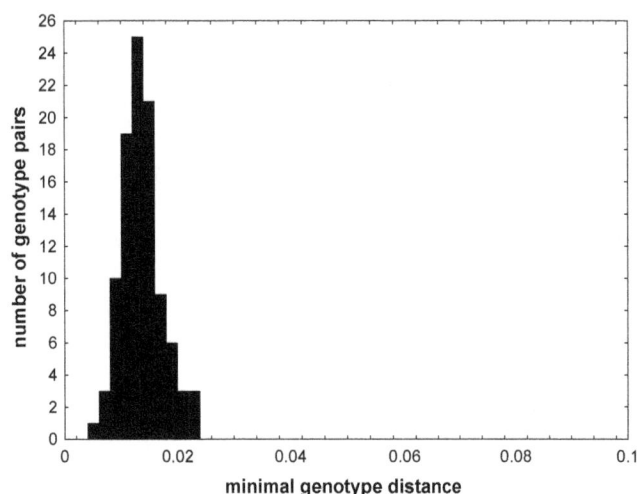

Figure 5. Metabolisms with different nitrogen utilization phenotypes can be very close together in genotype space. The figure shows the distribution of the minimal genotype distance for 100 pairs of metabolisms with different phenotypes, where each member of a pair was required to be viable on a different sole nitrogen source randomly and equiprobably chosen from the 50 nitrogen sources in Table S2 (See Methods for procedures).

types (Figure 4c) and nitrogen sources (Figure 4d). The mean U tends to be lowest for those nitrogen sources where genotypes have, on average, the smallest number of novel phenotypes in their neighborhood (Figure 4e, Spearman's r=0.71, p=7.35x10^{-9}).

Some genotypes on two different genotype networks are close to each other in genotype space

Earlier analyses on metabolic and other systems showed that two genotypes with arbitrary different phenotypes can often be found close together in genotype space (Ciliberti *et al.* 2007, Rodrigues & Wagner 2009, Schuster *et al.* 1994). In the context of metabolism, this means that few reaction changes may be necessary for a transition from one phenotype to another, unrelated phenotype. We wished to explore whether this also holds true for our nitrogen utilization phenotypes. In this regard, we conducted an analysis that starts with two metabolic genotypes, G_1 and G_2, each of which is viable only on one nitrogen source, but where these nitrogen sources are different.

We then asked how similar we can make the reaction complement of G_1 to that of G_2 without altering its phenotype. To this end, we carried out reaction-swapping and phenotype-preserving random walks that started from G_1 and approached G_2, i.e., each step in such a random walk was not allowed to increase the distance to the target G_2. After 5000 steps we recorded the distance remaining between G_1 and G_2. We emphasize that our estimates of minimal distances are upper bounds, since our procedure does not exclude the pos-

sibility that metabolisms with different phenotypes differ in even fewer reactions. Figure 5 shows the results of this approach for 100 different pairs of metabolisms viable on different nitrogen sources. The figure shows that the minimum genotype distance of networks with different phenotypes is very small, and comprises less than 2% of the total diameter (maximal distance) of genotype space. In other words, metabolism pairs that are viable on different nitrogen sources can share 98% or more of their chemical reactions. Only very few reaction changes are minimally needed to produce one nitrogen utilization phenotype from another such phenotype.

Conclusion

To summarise, our analysis has shown that metabolic genotypes can differ in most of the biochemical reactions they encode, yet share the same nitrogen utilization phenotype. In addition, our Markov chain Monte Carlo approach shows that even very different genotypes with the same phenotype can be transformed into one another through a series of single reaction changes. In other words, such genotypes form large connected networks – genotype networks – that extend far through metabolic genotype space. A second qualitative feature we observed is that different neighborhoods of genotypes on the same genotype network usually do not contain the same novel phenotypes. Together, these properties can facilitate the exploration of novel phenotypes by a population whose metabolism evolves through the addition and deletion of enzyme-coding genes in a genome. Specifically, the individuals in such a population can preserve existing, well adapted phenotypes, while at the same time altering their genotypes in a step-by-step manner, thus exploring different regions and neighborhoods of a genotype network. Because different neighborhoods contain different novel phenotypes, the existence of genotype networks can help in the exploration of novel phenotypes. Any evolutionary search for novel and adaptive phenotypes may be further facilitated by the observation that different genotype networks tend to be highly interwoven and close together in genotype space (Figure 5). These observations are in qualitative agreement with earlier ones on carbon and sulfur metabolism (Rodrigues & Wagner 2009, 2011, Samal *et al.* 2010), and genotype spaces with these features also exist in proteins, RNA, as well as in regulatory circuits, where they help facilitate evolutionary adaptation (Wagner 2011). Although some 80% of chemical reactions in a genotype network may change without affecting nitrogen utilization phenotype, not all reactions are equally changeable. In particular, there is a small core of super essential reactions that cannot be altered without abolishing viability on any one nitrogen source, at least based on current biochemical knowledge. Reactions like these are potential targets for antimetabolic drugs whose action cannot be easily circumvented through the evolution of alternative metabolic pathways in pathogens targeted by these drugs (Barve *et al.* 2012).

Acknowledgements

We acknowledge support through Swiss National Science Foundation grants 315230-129708, as well as through the YeastX project of SystemsX.ch, and the University Priority Research Program in Systems Biology at the University of Zurich.

References

Baba T, Ara T, Hasegawa M, Takai Y, Okumura Y, Baba M, Datsenko KA, Tomita M, Wanner BL & Mori H 2006 Construction of Escherichia coli K-12 in-frame, single-gene knockout mutants: the Keio collection. *Mol Syst Biol* **2** 2006.0008

Barve A, Rodrigues JFM & Wagner A 2012 Superessential reactions in metabolic networks. *Proc Natl Acad Sci U S A* **118** E1121-E1130

Barve A & Wagner A 2013 A latent capacity for evolutionary innovation through exaptation in metabolic systems. *Nature* **500** 203-206

Becker SA, Feist AM, Mo ML, Hannum G, Palsson BO & Herrgard MJ 2007 Quantitative prediction of cellular metabolism with constraint-based models: the COBRA Toolbox. *Nature Prot* **2** 727-738

Bochner BR 2009 Global phenotypic characterization of bacteria. *FEMS Microbiol Rev* **33** 191-205

Brown G, Singer A, Proudfoot M, Skarina T, Kim Y, Chang C, Dementieva I, Kuznetsova E, Gonzalez CF, Joachimiak A, Savchenko A & Yakunin AF 2008 Functional and structural characterization of four glutaminases from Escherichia coli and Bacillus subtilis. *Biochemistry* **47** 5724-5735

Ciliberti S, Martin OC & Wagner A 2007 Innovation and robustness in complex regulatory gene networks. *Proc Natl Acad Sci U S A* **104** 13591-13596

Edwards JS & Palsson BO 2000 The Escherichia coli MG1655 in silico metabolic genotype: Its definition, characteristics, and capabilities. Proc Natl Acad Sci U S A **97** 5528-5533

Fagerbakke KM, Heldal M & Norland S 1996 Content of carbon, nitrogen, oxygen, sulfur and phosphorus in native aquatic and cultured bacteria. *Aquat Microb Ecol* **10** 15-27

Feist AM, Henry CS, Reed JL, Krummenacker M, Joyce AR, Karp PD, Broadbelt LJ, Hatzimanikatis V & Palsson BØ 2007 A genome-scale metabolic reconstruction for Escherichia coli K-12 MG1655 that accounts for 1260 ORFs and thermodynamic information. *Mol Syst Biol* **3** 121

Feist AM, Herrgard MJ, Thiele I, Reed JL & Palsson BO 2009 Reconstruction of biochemical networks in microorganisms. *Nat Rev Microbiol* **7** 129-143

Feist AM & Palsson BO 2008 The growing scope of applications of genome-scale metabolic reconstructions using Escherichia coli. *Nat Biotechnol* **26** 659-667

Feist AM & Palsson BO 2010 The biomass objective function. *Curr Opin Microbiol* **13** 344-349

Fong SS, Joyce AR & Palsson BO 2005 Parallel adaptive evolution cultures of *Escherichia coli* lead to convergent growth phenotypes with different gene expression states. *Genome Res* **15** 1365-1372

Fong SS, Marciniak JY & Palsson BO 2003 Description and interpretation of adaptive evolution of Escherichia coli K-12 MG1655 by using a genome-scale in silico metabolic model. *J Bacteriol* **185** 6400-6408

Goto S, Nishioka T & Kanehisa M 1998 LIGAND: chemical database for enzyme reactions. *Bioinformatics* **14** 591-599

Heinrich R & Schuster S 1996 *The regulation of cellular systems*. New York: Chapman and Hall

Heldal M, Norland S & Tumyr O 1985 X-ray microanalytic method for measurement of dry-matter and elemental content of individual bacteria. *Appl Environment Microbiol* **50** 1251-1257

Herrgård MJ, Swainston N, Dobson P, Dunn WB, Arga KY, Arvas M, Blüthgen N, Borger S, Costenoble R, Heinemann M, Hucka M, Le Novère N, Li P, Liebermeister W, Mo ML, Oliveira AP, Petranovic D, Pettifer S, Simeonidis E, Smallbone K, Spasić I, Weichart D, Brent R, Broomhead DS, Westerhoff HV, Kirdar B, Penttilä M, Klipp E, Palsson BØ, Sauer U, Oliver SG, Mendes P, Nielsen J & Kell DB 2008 A consensus yeast metabolic network reconstruction obtained from a community approach to systems biology. *Nat Biotechnol* **26** 1155-1160

Jamshidi N & Palsson BO 2007 Investigating the metabolic capabilities of Mycobacterium tuberculosis H37Rv using the in silico strain iNJ661 and proposing alternative drug targets. *BMC Systems Biol* **1** 26

Kanehisa M & Goto S 2000 KEGG: Kyoto Encyclopedia of Genes and Genomes. *Nucleic Acids Res* **28** 27-30

Marcus M & Halpern YS 1969 The metabolic pathway

of glutamate in Escherichia coli K-12. *Biochim Biophys Acta* **177** 314-320

Merrick MJ & Edwards RA 1995 Nitrogen control in bacteria. *Microbiologic Rev* **59** 604-622

Neidhardt FC (Ed.) 1996 *Escherichia coli and Salmonella*. Washington, DC: ASM Press.

Nemeria NS, Arjunan P, Chandrasekhar K, Mossad M, Tittmann K, Furey W & Jordan F 2010 Communication between thiamin cofactors in the Escherichia coli pyruvate dehydrogenase complex E1 component active centers: evidence for a "direct pathway" between the 4'-aminopyrimidine N1' atoms. *J Biol Chem* **285** 11197-11209

Noor E, Eden E, Milo R & Alon U 2010 Central carbon metabolism as a minimal biochemical walk between precursors for biomass and energy. *Mol Cell* **39** 809-820

Reitzer L 2003 Nitrogen assimilation and global regulation in Escherichia coli. *Ann Rev Microbiol* **57** 155-176

Rodrigues JFM & Wagner A 2009 Evolutionary plasticity and innovations in complex metabolic reaction networks. *PLoS Comput Biol* **5** e1000613

Rodrigues JFM & Wagner A 2011 Genotype networks, innovation, and robustness in sulfur metabolism. *BMC Systems Biol* **5** 39

Sadava D, Heller C, Orians G, Purves W & Hillis D 2006 *Life: The science of biology*. (8th ed.). New York: WH Freeman.

Samal A, Rodrigues JFM, Jost J, Martin OC & Wagner A 2010 Genotype networks in metabolic reaction spaces. *BMC Systems Biol* **4** 30.

Schilling CH, Edwards JS & Palsson BO 1999 Toward metabolic phenomics: Analysis of genomic data using flux balances. *Biotechnol Prog* **15** 288-295

Schuster P, Fontana W, Stadler P & Hofacker I 1994 From sequences to shapes and back - a case-study in RNA secondary structures. *Proc Royal Soc Lon Series B* **255** 279-284

Sun YR, Fukamachi T, Saito H & Kobayashi H 2011 ATP requirement for acidic resistance in Escherichia coli. *J Bacteriol* **193** 3072-3077

Vieira-Silva S, Touchon M, Abby SS & Rocha EPC 2011 Investment in rapid growth shapes the evolutionary rates of essential proteins. *Proc Natl Acad Sci U S A* **108** 20030-20035

Wagner A 2011 The molecular origins of evolutionary innovations. *Trends Genet* **27** 397-410

Wang Z & Zhang JZ 2009 Abundant indispensable redundancies in cellular metabolic networks. *Genome Biol Evol* **1** 23-33

2

MicroRNA-365 is a negative regulator of endothelial cell proliferation

Xiao Wu and Fan Jiang

Department of Pathology and Pathophysiology, School of Basic Medicine, Shandong University, Jinan, Shandong Province, China

Correspondence should be addressed to Fan Jiang; E-mail: fjiang@sdu.edu.cn

Abstract

Recent studies have shown that human microRNA miR-365 has significant inhibitory effects on proliferation of transformed cancer cells and vascular smooth muscle cells. However, the effects of miR-365 on proliferation and migration of vascular endothelial cells remain unknown. Using human umbilical vein endothelial cells and in vitro assays, we demonstrated that miR-365 was a suppressor of endothelial cell proliferation, whereas cell migration was not affected by miR-365. We also identified that the expression level of the serine/threonine protein kinase serum- and glucocorticoid-regulated kinase-1 (SGK-1) was regulated by miR-365. The cytostatic effect of miR-365 was mimicked by the specific SGK-1 inhibitor GSK 650394. We further demonstrated that microvesicles isolated from plasma of patients with intracerebral hemorrhage, in which the level of miR-365 was elevated, decreased the expression level of SGK-1, and this effect was abolished in cells pretreated with miR-365 antagomir. However, we did not observe a significant effect of the microvesicles on cell proliferation. It is suggested that miR-365 may have important roles in vascular physiology and/or pathophysiology by modulating endothelial cell proliferation.

Introduction

Mounting evidence has suggested that microRNAs (miRNAs) are important epigenetic regulators of physiology and pathophysiology of the cardiovascular system (Condorelli *et al.* 2014, Hata 2013). The expression of a number of miRNA species has been found to be enriched in vascular endothelial (Santoro & Nicoli 2013) and smooth muscle cells (Bonauer *et al.* 2010), and cardiac muscles (Thum *et al.* 2008). Not surprisingly, aberrant expression or function of different miRNAs has been linked to multiple cardiovascular pathologies, including cardiac hypertrophy and remodeling, myocardial infarction, heart failure, arrhythmia, hypertension, atherosclerosis, aneurysm, and stroke (Condorelli *et al.* 2014, Romaine *et al.* 2015), at least in animal models. Inside the cell, miRNA molecules regulate gene expression primarily via sequence-specific interaction with target mRNAs, leading to mRNA degradation or translational repression. Individual miRNAs may regulate the expression of multiple genes, while the expression of an individual gene may be regulated by multiple miRNAs (Romaine *et al.* 2015).

In addition to intracellular miRNA molecules, miRNAs can also exist in the extracellular fluids such as plasma or serum. Recent studies have suggested that circulating miRNAs may be used as new biomarkers for cardiovascular disease (Wang *et al.* 2014). Moreover, it has been shown that microvesicle-encapsulated miRNAs secreted by cells can be internalized by other cells and modulate biological functions in the recipient cells (Muller *et al.* 2011, Valadi *et al.* 2007, Zhang *et al.* 2010). Hence it is thought that microvesicle-encapsulated miRNAs in the circulation are biologically functional and may be involved in long-distance cell-cell communications, raising the possibility that circulating miRNAs may represent novel therapeutic targets for human diseases (Wang *et al.* 2014).

Genomic profiling studies have revealed that changes of the circulating miRNA profile are associated with different cardiovascular diseases in humans, such as coronary arterial disease (Fichtlscherer *et al.* 2010), myocardial infarction (Wang *et al.* 2010), heart

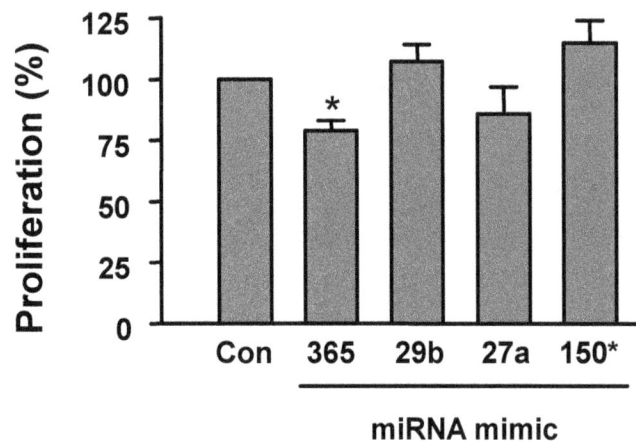

Figure 1. Effects of various miRNA mimics on proliferation of human umbilical vein endothelial cells (HUVECs). Cells transfected with a non-targeting miRNA mimic were used as control (Con). Cell proliferation was assayed 48 hours after transfection. Data represent mean ± SEM. * $P < 0.05$ vs control, one-way ANOVA followed by Tukey's test, $n = 3$ experiments.

failure (Goren et al. 2012, Tijsen et al. 2010), aortic aneurysms (Kin et al. 2012), and hypertension (Li et al. 2011). In addition, altered circulating miRNA levels have also been reported in patients with cerebral vascular diseases including intracranial aneurysms (Li et al. 2014) and stroke (Guo et al. 2013, Li et al. 2015). In our previous study in intracerebral hemorrhage patients, we identified a group of 30 circulating miRNAs that were selectively upregulated in both male and female patients as compared to healthy controls (Guo et al. 2013). Bioinformatic analysis revealed that these miRNAs were overrepresented in biological processes associated with inflammation (Guo et al. 2013). One of these specifically changed miRNAs is miR-365, which has been shown to have an important role in regulating cell proliferation in transformed cancer cells (Chen et al. 2015, Kang et al. 2013, Nie et al. 2012). Currently, however, the effects of miR-365 on functions of the vascular endothelial cell remain unknown. Therefore, in the present study, we investigated the impacts of miR-365 on endothelial cell proliferation.

Materials and Methods

Reagents
Synthetic miRNA mimics and antagomirs were purchased from GenePharma (Shanghai, China). GSK 650394 was obtained from Tocris Bioscience (Bristol, UK).

Cell culture
Human umbilical vein endothelial cells (HUVECs) were purchased from American Type Culture Collection. HUVECs were cultured in Endothelial Cell Medium (ECM) supplemented with 1% endothelial cell growth supplement (ECGS) (ScienCell Research Laboratories, Carlsbad, CA, USA), penicillin/streptomycin (Invitrogen), and 5% fetal bovine serum (FBS) (Invitrogen). Cells were cultured in 95% air/5% CO_2 at 37°C. For experimentation, cells within passage 10 were used.

Transfection of miRNA mimics and antagomirs
Twenty four hours before transfection, cells were subcultured into 24-well plates at a density of 1.5×10^5 cells per well. Cells were incubated with miRNA (final concentration 30 nM) mixed with Lipofectamine RNAiMAX Reagent (Invitrogen) for 6 hours and then changed to fresh medium. In our preliminary experiments, we characterized the gene silencing efficacy of transfected miR-365 mimic at different time points. We found that the optimal effect was obtained at 24 hours after transfection, whereas the effects at 48 or 72 hours were much smaller. We therefore carried out western blot and qPCR experiments 24 hours post transfection. Cell proliferation was assayed 48 hours after transfection.

Cell proliferation
Cell proliferation was assessed with a tetrazolium-based non-radioactive assay kit (CellTiter 96 Aqueous from Promega, Madison, WI, USA) as used previously (Datla et al. 2007).

Cell wound healing assay
Cell migratory activity was tested with the wound healing assay. Cells were cultured in 24-well plates to 100% confluency. Wound healing assay was performed by scratching the cell monolayer with a 200 μl pipette tip (Datla et al. 2007), and the cells were further cultured in reduced serum (1%) ECM for 6 - 24 hr. The recovery rate of the denuded area was monitored by taking digital pictures at different time points, and the rate was quantified as percentage recovery of the total denuded area.

Quantitative real-time PCR (qPCR)
Total RNA was isolated using Trizol (Invitrogen) according to the manufacturer's instructions. cDNA was reverse transcribed from 1 μg of total RNA using random hexamer primer and TaqMan Reverse Transcription Reagents (Applied Biosystems, Carlsbad, CA, USA). Thermal cycling parameter settings for reverse transcription were: 25°C for 10 min; 37°C for 60 min;

95°C for 5 min. Real-time PCR was performed using the Taqman Gene Expression Assay primer-probe set for human SGK-1, and the TaqMan Gene Expression Master Mix (all from Applied Biosystems). An ABI Prism 7500 system (Applied Biosystems) was used for real-time PCR amplification. Thermal cycling parameters were as follows: 50°C for 2 min; 95°C for 10 min; 40 cycles of 95°C for 15 s plus 60°C for 1 min. 18S rRNA was used as the house-keeping gene. The $2^{-\Delta\Delta CT}$ method was used to assess the relative mRNA expression level. The results were expressed as fold of control.

Western blot
Total proteins were separated by 10% SDS-PAGE and transferred to Immobilon-P PVDF membranes (from Millipore, Billerica, MA, USA). The membrane was blocked with 5% non-fat milk and then incubated with primary antibodies at 4°C overnight. The blots were developed with an ECL-Plus (enhanced chemiluminescence) reagent (GE Healthcare, Giles, UK). All primary antibodies were purchased from Cell Signaling Technology (Danvers, MA, USA). HRP-conjugated secondary antibodies were from Jackson ImmunoResearch (West Grove, PA, USA).

miRNA target gene prediction
Potential miRNA target genes were predicted using five independent bioinformatics tools: miRanda, MirTarget2, PITA, RNAhybrid and TargetScan.

Isolation of microvesicles from human plasma

Collection of human samples was approved by the hospital Human Ethics Committee, and informed consents were obtained from all subjects. Peripheral blood was collected from patients with intracerebral hemorrhage (diagnosed by multi-detector-row computed tomography scanning). Isolation of microvesicles from plasma was performed as described (Caby et al. 2005). Briefly, fresh plasma was centrifuged at 500 g for 30 min, followed by 12,000 g for 45 min. The supernatant was then transferred to a fresh tube and microvesicles pelleted by ultra-centrifugation at 110,000 g for 2 hours. For microvesicle treatment in cultured cells, microvesicles isolated from 4 ml of plasma were washed and resuspended in 1 ml of culture medium, and cells were incubated for 24 hrs. Equivalent amount of bovine serum albumin was used as control.

Statistical analysis
Data are expressed as mean ± standard error of the mean (SEM). Differences between groups were analyzed with unpaired t-test (two groups) or one-way analysis of variance (ANOVA) followed by post hoc Tukey's test (for multiple groups) unless stated otherwise. Statistical analysis was performed with GraphPad Prism. A value of $P < 0.05$ was regarded as statistically significant.

Results

To determine whether miR-365 had any effect on proliferation of vascular endothelial cells, we transfected HUVEC cells with a miR-365 mimic construct and

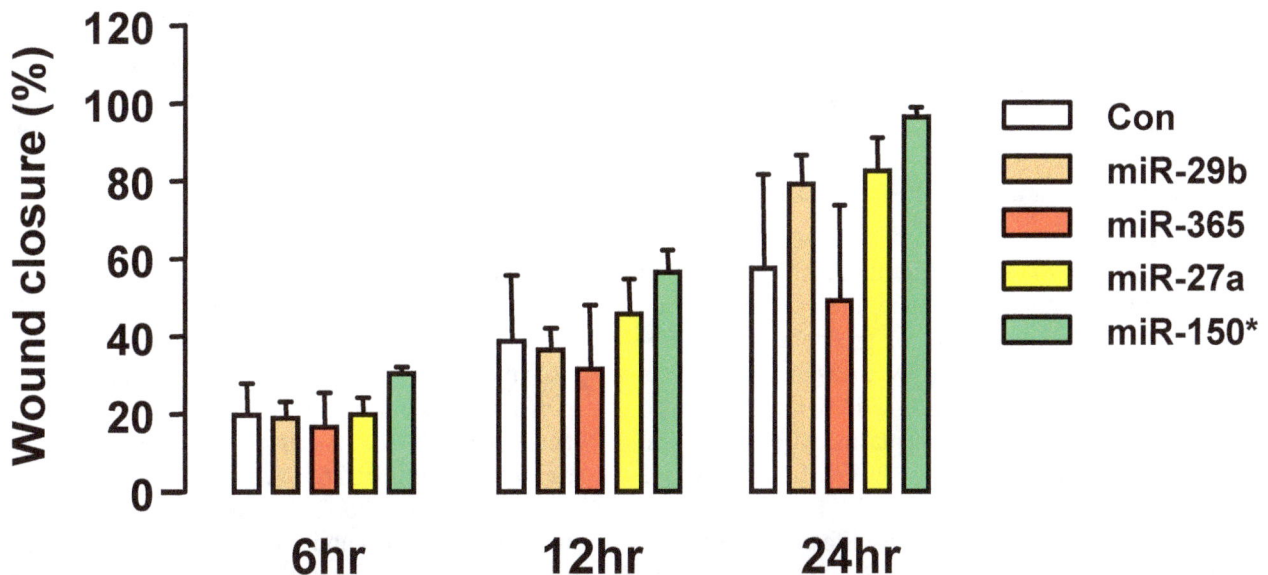

Figure 2. Effects of various miRNA mimics on migration of HUVECs assessed with the wound healing assay. Wound healing response was expressed as % recovery of the total denuded area. Data were mean ± SEM. * $P < 0.05$ vs control (Con), one-way ANOVA followed by Tukey's test, $n = 3\text{-}4$.

(A) **Human**

3' uauuccuaaaaaUCCCCGUAAu 5' hsa-miR-365

| |||||||||

754:5' gucaccuguaaaAUGGGCAUUa 3' Sgk1

Mouse

3' uauuccuaaaaaUCCCCGUAAu 5' mmu-miR-365

| |||||||||

799:5' gucaccuauaaaACGGGCAUUa 3' Sgk1

Rat

3' uauuccuaaaaaUCCCCGUAAu 5' rno-miR-365

| |||||||||

811:5' gucaccugcaaaAUGGGCAUUa 3' Sgk1

(B)

(C)

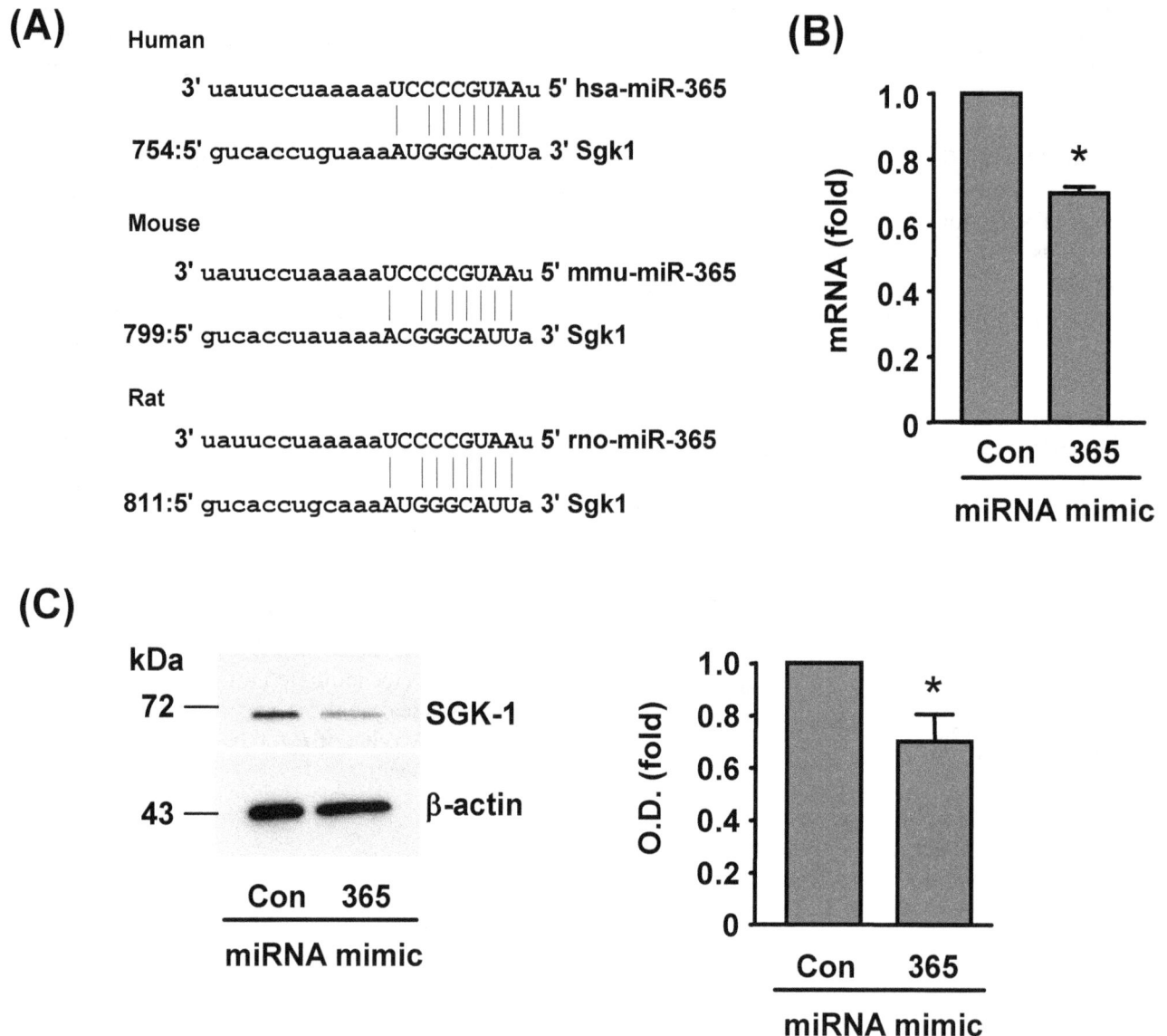

Figure 3. Regulation of the expression of serum- and glucocorticoid-regulated kinase-1 (SGK-1) by miR-365. (A) Conserved miR-365 targeting sequence in SGK-1 genes of human, mouse and rat. (B) Transfection with the miR-365 mimic in HUVECs decreased the mRNA level of SGK-1 as measured by qPCR, which was carried out at 24 hours after transfection. (C) Western blot and densitometry data showing the effect of miR-365 mimic on protein expression of SGK-1 24 hours after transfection. Data were mean ± SEM. * $P < 0.05$ vs control (Con), unpaired t-test, $n = 3 - 6$.

measured the rate of cell proliferation. As compared to the non-specific control sequence, the miR-365 mimic significantly reduced the rate of proliferation of HU-VECs (Figure 1). In addition, we measured the effects of miR-29b, miR-27a and miR-150*, which were also specifically upregulated in the plasma of intracerebral hemorrhage patients (Guo et al. 2013). In contrast, we found that miR-29b, miR-27a or miR-150* had no significant effects on HUVEC proliferation. To test whether miR-365, as well as miR-29b, miR-27a and miR-150* had any effects on HUVEC migratory activity, we transfected mimics of these miRNAs and measured the cell monolayer wound healing response. We

found that these miRNAs showed no significant effects on HUVEC migration (Figure 2).

To understand the mechanism by which miR-365 modulated endothelial cell proliferation, we first performed bioinformatics analysis to predict target genes for miR-365. As listed in Table 1, several potential targets for miR-365 were identified by 5 independent bioinformatic tools. Among these target genes, we focused on serum- and glucocorticoid-regulated kinase-1 (SGK-1), a serine/threonine protein kinase with high homology in amino acid sequence and biological functions to Akt (Webster et al. 1993). Sequence analysis revealed that the target sequence for miR-365

Table 1. Predicted target genes for miR-365

Gene Symbol	Description	MFE (kcal/mol)
CSK	c-src tyrosine kinase	-24.2
ANKRD11	ankyrin repeat domain 11	-23.1
LAMP2	lysosomal-associated membrane protein 2	-20.7
SOCS5	suppressor of cytokine signaling 5	-24.9
EFEMP1	EGF-containing fibulin-like extracellular matrix protein 1	-24.5
SGK3	serum/glucocorticoid regulated kinase family, member 3	-21.3
TMOD3	tropomodulin 3 (ubiquitous)	-19.7
DTNA	dystrobrevin, alpha	-23.3
ARRB2	arrestin, beta 2	-21.2
MYLK	myosin light chain kinase	-22.7
SGK1	serum/glucocorticoid regulated kinase 1	-27.0
OAZ2	ornithine decarboxylase antizyme 2	-24.2
GALNT4	UDP-N-acetyl-alpha-D-galactosamine:polypeptide N-acetylgalactosaminyltransferase 4 (GalNAc-T4)	-22.5
MAPK1IP1L	mitogen-activated protein kinase 1 interacting protein 1-like	-23.3
PAX6	paired box 6	-23.4
HMGCR	3-hydroxy-3-methylglutaryl-Coenzyme A reductase	-21.9
NR3C2	nuclear receptor subfamily 3, group C, member 2	-24.9
CREB5	cAMP responsive element binding protein 5	-22.5
USP48	ubiquitin specific peptidase 48	-24.0

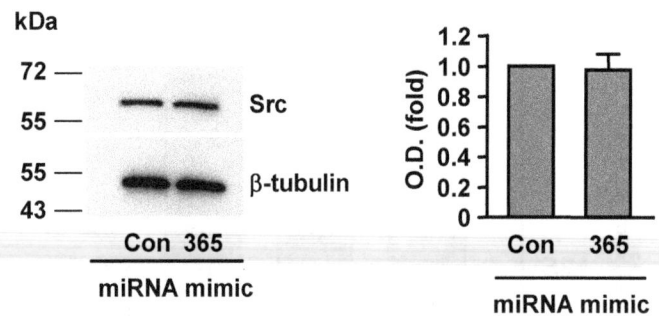

Figure 4. Western blot and densitometry data showing that the miR-365 mimic had no effect on the protein expression level of Src in HUVECs. Data were mean ± SEM, $n = 3$. Con, control.

in the SGK-1 gene was totally conserved in human, mouse and rat (Figure 3A). We next performed qPCR analysis and showed that treatment with the miR-365 mimic significantly decreased the mRNA level of SGK-1 (Figure 3B). To confirm the effect of miR-365 on the expression level of SGK-1, we did western blot experiments and demonstrated that the miR-365 mimic also significantly decreased the SGK-1 protein in HUVECs (Figure 3C). We noted that the tyrosine kinase c-Src was also identified as a potential target of miR-365 (see Table 1).

Since c-Src is tightly involved in modulating cell proliferation (Roskoski 2004), we tested whether miR-365 could also modulate the expression level of c-Src. As shown in Figure 4, however, we found that transfection with miR-365 mimic did not show any effect on Src expression. To confirm the role of SGK-1 in modulating endothelial cell proliferation, we treated

HUVECs with a specific SGK-1 inhibitor GSK 650394. We showed that GSK 650394 at 100 mM significantly suppressed HUVEC proliferation (Figure 5). To further establish that modulating SGK-1 expression had a causal role in mediating the miR-365 effect, we tested the effect of miR-365 mimic in the presence of GSK 650394. We found that in cells pretreated with GSK 650394, miR-365 mimic had no significant effect on proliferation ($101.6 \pm 5.5\%$ versus control miRNA, $P > 0.05$, $n = 3$), supporting a causal role of SGK-1 in mediating the miR-365 effect.

Previous studies have shown that miRNA-rich microvesicles can be captured by endothelial cells, while the encapsulated miRNA molecules may produce biological effects once they are internalized into the cells (Zhang et al. 2010). To test this possibility for miR-365, we treated HUVECs with concentrated native microvesicles isolated from the plasma of intracerebral hemorrhage patients in the presence or absence of the miR-365 antagomir. As shown in Figure 6A, microvesicles slightly but significantly ($P < 0.05$ with unpaired t-test) decreased the expression level of SGK-1, and this effect was abolished in cells pretreated with miR-365 antagomir. However, we failed to detect an effect of the microvesicles on endothelial cell proliferation either with or without miR-365 antagomir (Figure 6B).

Discussion

Several recent studies have shown that increased miR-365 expression has inhibitory effects on the proliferation of various cancer cells (Chen et al. 2015, Kang et al. 2013, Nie et al. 2012). In the present study, we have identified miR-365 as a suppressor of cell proliferation in primary human vascular endothelial cells, indicating that miR-365 may be involved in modulating cell proliferation not only in transformed cells, but also in normal cells. Our results are in line with two recent reports showing that miR-365 inhibits prolifera-

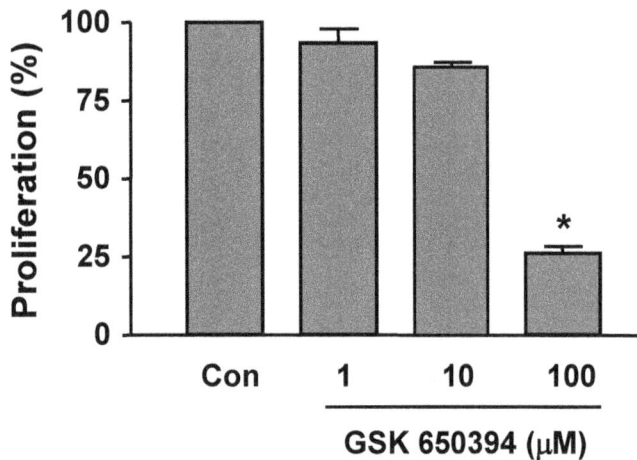

Figure 5. Effects of the SGK-1 inhibitor GSK 650394 on HUVEC proliferation. DMSO was used as vehicle control (Con). Data were mean ± SEM. * $P < 0.05$ *vs* Con, one-way ANOVA followed by Tukey's test, $n = 4$.

tion of vascular smooth muscle cells (Kim *et al.* 2014; Zhang *et al.* 2014). In one study, the authors found that the expression level of miR-365 was significantly reduced in the rat carotid artery after balloon injury, a pathological process characterized by aberrant proliferation of the medial smooth muscle cells (Zhang *et al.* 2014). In the other study, it was demonstrated that mitogens such as platelet-derived growth factor, angiotensin II and serum all downregulated miR-365 expression in vascular smooth muscle cells, whereas miR-

365 overexpression induced cell cycle arrest (Kim *et al.* 2014). Moreover, Qin et al. provided evidence showing that treatment of endothelial cells with oxidized low-density lipoproteins increased miR-365 expression, which in turn mediated an apoptotic response by repressing Bcl-2 expression (Qin *et al.* 2011). These data suggest that miR-365 may have important roles in regulating vascular physiology and may be involved in the pathogenesis of vascular disease.

Evidence suggests that miR-365 may target multiple molecular targets to regulate proliferation in a cell-specific manner. For example, in human colon cancer cells, the cytostatic effect of miR-365 appears to be mediated by targeting and repressing the expressions of Cyclin D1 and Bcl-2 (Nie *et al.* 2012). On the other hand, it was shown that the transcription factor NKX2-1 was the target of miR-365 in suppressing proliferation of lung cancer cells (Kang *et al.* 2013). In vascular smooth muscle cells, both groups identified that miR-365 exerted its anti-proliferative actions by inhibiting the expression of cyclin D1 (Kim *et al.* 2014, Zhang *et al.* 2014).

In the present study, we showed that the serine/threonine protein kinase SGK-1 appeared to be a target of miR-365. This notion has also been reported by others (Xu *et al.* 2011). We clarified that miR-365 mimic significantly decreased SGK-1 expression at both mRNA and protein levels in endothelial cells. We further demonstrated that the SGK-1 inhibitor GSK

(A)

(B)

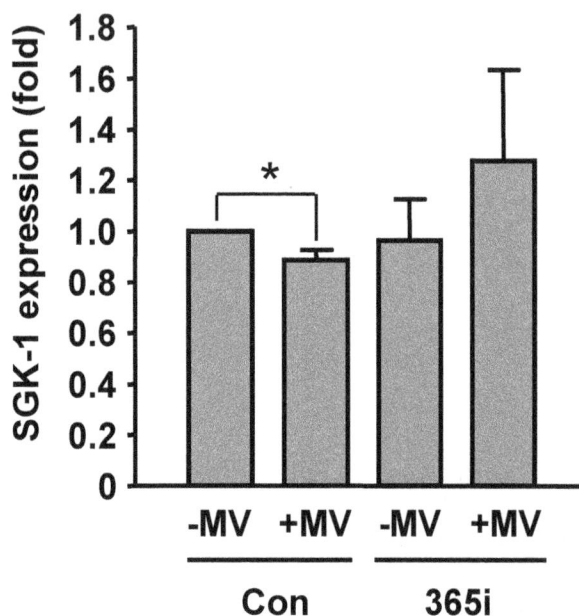

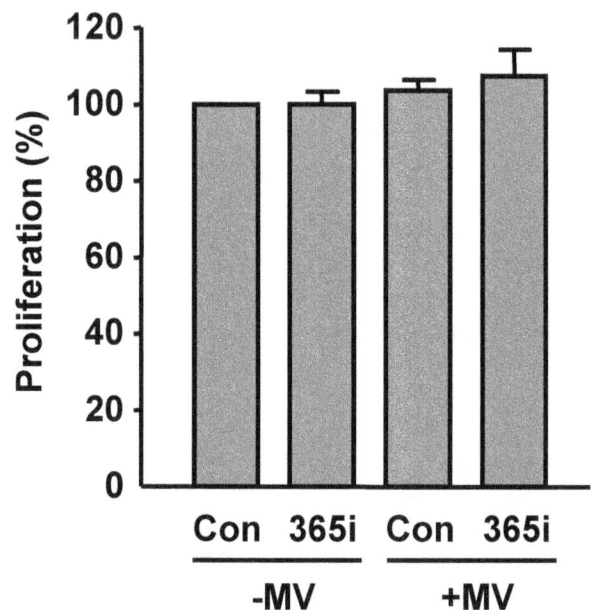

Figure 6. Effects of microvesicles (MV) isolated from the plasma of intracerebral hemorrhage patients, with or without miR-365 antagomir (365i) on (A) SGK-1 expression (measured by qPCR) and (B) proliferation in cultured HUVECs. A nonspecific antagomir molecule was used as control (Con). Data were mean ± SEM. * $P < 0.05$ (only with unpaired *t*-test); $n = 3 - 5$.

650394 mimicked the effect of miR-365 expression on endothelial cell proliferation, whereas in the presence of GSK 650394, miR-365 mimic had no significant effect, supporting a causal role of SGK-1 in mediating the miR-365 effect. We noted, however, that GSK 650394 was reported to have potential off-target effects other than inhibiting SGK-1 (Burgon *et al.* 2014). Upon stimulation with mitogens such as insulin or insulin-like growth factor-1, SGK-1 is activated by phosphorylation by phosphoinositide-3-kinase (PI3K) or mammalian target of rapamycin complex (mTORC), and the activated SGK-1 shares common downstream substrates with the prosurvival kinase Akt (Bruhn *et al.* 2010). Interestingly, a recent study has shown that SGK-1 can promote vascular smooth muscle cell proliferation by inducing phosphorylation of glycogen synthase kinase-3β and β-catenin activation (Zhong *et al.* 2014). Taken together, these findings support the notion that SGK-1 is a critical regulator of cell proliferation, including vascular cells (Bruhn *et al.* 2010). Nevertheless, a limitation of the present study was that we could not exclude an involvement of other miR-365 targets in the anti-proliferative effects.

It is thought that circulating miRNAs in the plasma are encapsulated in membranous microvesicles (Reid *et al.* 2010), which can be internalized by other cells and subsequently modulate the functions of recipient cells (Zhang *et al.* 2010). We treated endothelial cells with native microvesicles isolated from the plasma of intracerebral hemorrhage patients. We found that microvesicle treatment decreased the expression level of SGK-1, although the effect was relatively small. Moreover, we showed that inhibiting the miR-365 function with an antagomir abrogated this effect. These results indicate that miR-365 contained in the microvesicles can exert biological functions in vascular endothelial cells, supporting the notion that microvesicle-mediated microRNA transportation may represent a novel means of remote cell-cell communications (Valadi *et al.* 2007; Zhang *et al.* 2010). However, although we showed that microvesicle treatment reduced SGK-1 expression, it failed to affect the rate of proliferation of endothelial cells. A possible reason to explain this observation is that microvesicles contain numerous biologically active molecules, while the effect of miR-365 may be masked by those of other molecules.

Given that the levels of multiple miRNAs are significantly elevated in the plasma of intracerebral hemorrhage patients (Guo *et al.* 2013), we were inspired to explore whether these altered miRNAs have any effects on the regenerative functions (e.g. proliferation and migration) of vascular endothelial cells. Based on the present experiemnts, however, we could not directly link miR-365 with complications or outcomes following intracerebral hemorrhage. Instead, our results have provided new clues for molecular mechanisms of vascular diseases with impaired endothelial proliferative functions. Although our results and those from other groups indicate that miR-365 may be actively involved in modulating vascular cell functions, the pathophysiological significance of this miRNA in influencing vascular disease progression and/or the vascular healing process in vivo warrants more investigations.

In summary, we have identified that miR-365 is a suppressor of endothelial cell proliferation, and this effect is related to inhibition of SGK-1 expression by miR-365. Together with the data from previous studies (Qin *et al.* 2011), it is suggested that miR-365 may have important roles in vascular physiology and/or pathophysiology by modulating endothelial cell functions.

Authors' contributions

Xiao Wu performed the experiments. Fan Jiang conceived the study and wrote the manuscript.

Acknowledgements

This work was supported by the Natural Science Foundation China (Grant #81271269).

References

Bonauer A, Boon RA & Dimmeler S 2010 Vascular microRNAs. *Curr Drug Targets* **11** 943-949

Bruhn MA, Pearson RB, Hannan RD & Sheppard KE 2010 Second AKT: the rise of SGK in cancer signalling. *Growth Factors* **28** 394-408

Burgon J, Robertson AL, Sadiku P, Wang X, Hooper-Greenhill E, Prince LR, Walker P, Hoggett EE, Ward JR, Farrow SN, Zuercher WJ, Jeffrey P, Savage CO, Ingham PW, Hurlstone AF, Whyte MK & Renshaw SA 2014 Serum and glucocorticoid-regulated kinase 1 regulates neutrophil clearance during inflammation resolution. *J Immunol* **192** 1796-805

Caby MP, Lankar D, Vincendeau-Scherrer C, Raposo G & Bonnerot C 2005 Exosomal-like vesicles are present in human blood plasma. *Int Immunol* **17** 879-887

Chen Z, Huang Z, Ye Q, Ming Y, Zhang S, Zhao Y, Liu L, Wang Q & Cheng K 2015 Prognostic significance and anti-proliferation effect of microRNA-365 in hepatocellular carcinoma. *Int J Clin Exp Pathol* **8** 1705-1711

Condorelli G, Latronico MV & Cavarretta E 2014 microRNAs in cardiovascular diseases: current knowledge and the road ahead. *J Am Coll Cardiol* **63** 2177-2187

Datla SR, Peshavariya H, Dusting GJ, Mahadev K, Goldstein BJ & Jiang F 2007 Important role of Nox4 type NADPH oxidase in angiogenic responses in human microvascular endothelial cells in vitro. *Arterioscler Thromb Vasc Biol* **27** 2319-2324

Fichtlscherer S, De Rosa S, Fox H, Schwietz T, Fischer A, Liebetrau C, Weber M, Hamm CW, Roxe T, Muller-Ardogan M, Bonauer A, Zeiher AM & Dimmeler S 2010 Circulating microRNAs in patients with coronary artery disease. *Circ Res* **107** 677-684

Goren Y, Kushnir M, Zafrir B, Tabak S, Lewis BS & Amir O 2012 Serum levels of microRNAs in patients with heart failure. *Eur J Heart Fail* **14** 147-154

Guo D, Liu J, Wang W, Hao F, Sun X, Wu X, Bu P, Zhang Y, Liu Y, Liu F, Zhang Q & Jiang F 2013 Alteration in Abundance and Compartmentalization of Inflammation-Related miRNAs in Plasma After Intracerebral Hemorrhage. *Stroke* **44** 1739-1742

Hata A 2013 Functions of microRNAs in cardiovascular biology and disease. *Annu Rev Physiol* **75** 69-93

Kang SM, Lee HJ & Cho JY 2013 MicroRNA-365 regulates NKX2-1, a key mediator of lung cancer. *Cancer Lett* **335** 487-494

Kim MH, Ham O, Lee SY, Choi E, Lee CY, Park JH, Lee J, Seo HH, Seung M, Min PK & Hwang KC 2014 MicroRNA-365 inhibits the proliferation of vascular smooth muscle cells by targeting cyclin D1. *J Cell Biochem* **115** 1752-1761

Kin K, Miyagawa S, Fukushima S, Shirakawa Y, Torikai K, Shimamura K, Daimon T, Kawahara Y, Kuratani T & Sawa Y 2012 Tissue- and plasma-specific MicroRNA signatures for atherosclerotic abdominal aortic aneurysm. *J Am Heart Assoc* **1** e000745

Li P, Teng F, Gao F, Zhang M, Wu J & Zhang C 2015 Identification of circulating microRNAs as potential biomarkers for detecting acute ischemic stroke. *Cell Mol Neurobiol* **35** 433-447

Li P, Zhang Q, Wu X, Yang X, Zhang Y, Li Y & Jiang F 2014 Circulating microRNAs serve as novel biological markers for intracranial aneurysms. *J Am Heart Assoc* **3** e000972

Li S, Zhu J, Zhang W, Chen Y, Zhang K, Popescu LM, Ma X, Lau WB, Rong R, Yu X, Wang B, Li Y, Xiao C, Zhang M, Wang S, Yu L, Chen AF, Yang X & Cai J 2011 Signature microRNA expression profile of essential hypertension and its novel link to human cytomegalovirus infection. *Circulation* **124** 175-184

Muller G, Schneider M, Biemer-Daub G & Wied S 2011 Microvesicles released from rat adipocytes and harboring glycosylphosphatidylinositol-anchored proteins transfer RNA stimulating lipid synthesis. *Cell Signal* **23** 1207-1223

Nie J, Liu L, Zheng W, Chen L, Wu X, Xu Y, Du X & Han W 2012 microRNA-365, down-regulated in colon cancer, inhibits cell cycle progression and promotes apoptosis of colon cancer cells by probably targeting Cyclin D1 and Bcl-2. *Carcinogenesis* **33** 220-225

Qin B, Xiao B, Liang D, Xia J, Li Y & Yang H 2011 MicroRNAs expression in ox-LDL treated HUVECs: MiR-365 modulates apoptosis and Bcl-2 expression. *Biochem Biophys Res Commun* **410** 127-133

Reid G, Kirschner MB & van Zandwijk N 2010 Circulating microRNAs: Association with disease and potential use as biomarkers. *Crit Rev Oncol Hematol* **80** 193-208

Romaine SP, Tomaszewski M, Condorelli G & Samani NJ 2015 MicroRNAs in cardiovascular disease: an introduction for clinicians. *Heart* **101** 921-928

Roskoski R, Jr. 2004 Src protein-tyrosine kinase structure and regulation. *Biochem Biophys Res Commun* **324** 1155-1164

Santoro MM & Nicoli S 2013 miRNAs in endothelial cell signaling: the endomiRNAs. *Exp Cell Res* **319** 1324-1330

Thum T, Catalucci D & Bauersachs J 2008 MicroRNAs: novel regulators in cardiac development and disease. *Cardiovasc Res* **79** 562-570

Tijsen AJ, Creemers EE, Moerland PD, de Windt LJ, van der Wal AC, Kok WE & Pinto YM 2010 MiR423-5p as a circulating biomarker for heart failure. *Circ Res* **106** 1035-1039

Valadi H, Ekstrom K, Bossios A, Sjostrand M, Lee JJ & Lotvall JO 2007 Exosome-mediated transfer of mRNAs and microRNAs is a novel mechanism of genetic exchange between cells. *Nat Cell Biol* **9** 654-659

Wang F, Chen C & Wang D 2014 Circulating microRNAs in cardiovascular diseases: from biomarkers to therapeutic targets. *Front Med* **8** 404-418

Wang GK, Zhu JQ, Zhang JT, Li Q, Li Y, He J, Qin YW & Jing Q 2010 Circulating microRNA: a novel potential biomarker for early diagnosis of acute myocardial infarction in humans. *Eur Heart J* **31** 659-666

Webster MK, Goya L, Ge Y, Maiyar AC & Firestone GL 1993 Characterization of sgk, a novel member of the serine/threonine protein kinase gene family which is transcriptionally induced by glucocorticoids and serum. *Mol Cell Biol* **13** 2031-2040

Xu Z, Xiao SB, Xu P, Xie Q, Cao L, Wang D, Luo R, Zhong Y, Chen HC & Fang LR 2011 miR-365, a novel negative regulator of interleukin-6 gene expression, is cooperatively regulated by Sp1 and NF-kappaB. *J Biol Chem* **286** 21401-21412

Zhang P, Zheng C, Ye H, Teng Y, Zheng B, Yang X & Zhang J 2014 MicroRNA-365 inhibits vascular smooth muscle cell proliferation through targeting cyclin D1. *Int J Med Sci* **11** 765-770

Zhang Y, Liu D, Chen X, Li J, Li L, Bian Z, Sun F, Lu J, Yin Y, Cai X, Sun Q, Wang K, Ba Y, Wang Q, Wang D, Yang J, Liu P, Xu T, Yan Q, Zhang J, Zen K & Zhang CY 2010 Secreted monocytic miR-150 enhances targeted endothelial cell migration. *Mol Cell* **39** 133-144

Zhong W, Oguljahan B, Xiao Y, Nelson J, Hernandez L, Garcia-Barrio M & Francis SC 2014 Serum and glucocorticoid-regulated kinase 1 promotes vascular smooth muscle cell proliferation via regulation of beta-catenin dynamics. *Cell Signal* **26** 2765-2772

3

A series of Notch3 mutations in CADASIL; insights from 3D molecular modelling and evolutionary analyses

Dimitrios Vlachakis[1#], Spyridon Champeris Tsaniras[2#], Katerina Ioannidou[3], Louis Papageorgiou[1], Marc Baumann[4] and Sophia Kossida[1]

[#] These authors contributed equally to this study

[1] Bioinformatics & Medical Informatics Team, Biomedical Research Foundation, Academy of Athens, Athens, Greece

[2] Department of Physiology, Medical School, University of Patras, Rio, 26504 Patras, Greece

[3] School of Electrical and Computer Engineering, National Technical University of Athens, Greece

[4] Protein Chemistry/Proteomics Unit, Biomedicum Helsinki, Institute of Biomedicine, University of Helsinki, P.O. Box 63, Finland

Correspondence should be addressed to Dimitrios Vlachakis (dvlachakis@bioacademy.gr) and Sophia Kossida (skossida@bioacademy.gr)

Abstract

CADASIL disease belongs to the group of rare diseases. It is well established that the Notch3 protein is primarily responsible for the development of CADASIL syndrome. Herein, we attempt to shed light to the actual molecular mechanism underlying CADASIL via insights that we have from preliminary *in silico* and proteomics studies on the Notch3 protein. At the moment, we are aware of a series of Notch3 point mutations that promote CADASIL. In this direction, we investigate the nature, extent, physicochemical and structural significance of the mutant species in an effort to identify the underlying mechanism of Notch3 role and implications in cell signal transduction. Overall, our *in silico* study has revealed a rather complex molecular mechanism of Notch3 on the structural level; depending of the nature and position of each mutation, a consensus significant loss of beta-sheet structure is observed throughout all *in silico* modeled mutant/wild type biological systems.

Introduction

CADASIL stands for Cerebral Autosomal Dominant Arteriopathy Subcortical Infarcts Leukoencephalopathy. In other words, cerebral relates to the brain, autosomal dominant to the inheritance pattern, arteriopathy to arterial disease (in this case within the brain), subcortical to the area of the brain involved in higher functioning, infarcts to tissue necrosis due to a lack of oxygen and leukoencephalopathy to myelin damage in the brain (Joutel *et al.* 1996). CADASIL is also known as hereditary multi-infarct dementia, familial disorder with subcortical ischemic strokes and familial Binswanger's disease.

This disease is the most prevalent type of hereditary cerebral angiopathy (Sourander & Walinder 1977) or stroke disorder, and mutations of the NOTCH3 gene located on chromosome 19 are regarded as its main cause. It belongs to a family of disorders called leukodystrophies (Chabriat *et al.* 1995, Joutel *et al.* 1996). More than 30 years have passed since the first time the condition was described in a Swedish family. However, the acronym CADASIL did not come out until the early 1990s (Sourander & Walinder 1977, Tournier-Lasserve *et al.* 1993). Most patients with CADASIL have a family member with the disorder. However, there have been cases reported where a patient has no first-degree relative with CADASIL symptoms (Joutel *et al.* 2000). Nevertheless, as genetic testing for this disease was developed after 2000, many cases were misdiagnosed as different neurodegenerative disorders (Razvi *et al.* 2005).

This inherited condition mainly affects the small arteries in the white matter of the brain, causing thickening of their walls that, consequently, blocks the blood flow. The poorer blood supply can cause areas of tissue death; for example, multiple lacunar and subcortical white matter infarctions. The vascular smooth muscle (VSM) cells surrounding the arteries become abnormal and gradually die. Disproportionate cortical hypometabolism has also been reported (Tatsch *et al.* 2003).

The most common first sign of CADASIL is migraines, often followed by visual sensations or auras, usually at the age of 20-30. However, the most usual symptom, which has also been reported as a first sign in some cases, is the ischemic episodes, which include Transient Ischemic Attacks (TIAs) and strokes. Strokes appear for the first time between the ages of 30 and 50, while TIAs usually occur prior to them. Individuals with CADASIL may suffer recurrent strokes (subcortical cerebral infarctions) (Sourander & Walinder 1977) throughout their lifetime that damage the brain as time goes by. Strokes can cause paralysis like pseudobulbar palsy (Chabriat *et al.* 1995, Dalkas *et al.* 2013), loss of sensation, walking problems or slurred speech, while recurrent ischemic episodes may lead to urinary incontinence or severe disability. The strokes that appear in the subcortical area of the brain, associated with reasoning and memory, are also responsible for cognitive function problems. Cognitive deterioration progresses to difficulties with concentration, attention or loss of intellectual function (Toni-Uebari 2013). Approximately 70% of the patients over 65 years old suffer from subcortical dementia, which appears with slowing of motor function, loss of memory or apathy. Furthermore, almost 10-20% of the patients demonstrate psychiatric disorders, sometimes even as a first symptom. Such disorders include mood or perception changes (mania or depression), changes in personality, hallucinations, delusions, anxiety and panic disorders (Valenti *et al.* 2011). Rarely, CADASIL also causes recurrent epileptic seizures (5-10% of patients) (Buffon *et al.* 2006, Dichgans *et al.* 1998, Velizarova *et al.* 2011). Leukoencephalopathy is diagnosed through magnetic resonance imaging (MRI) as diffuse white matter lesions (Ueda *et al.* 2009).

The onset of cerebrovascular disease usually occurs in the mid-adult (30s-60s), while recurrent transient ischemic attacks (TIAs) or strokes in multiple vascular territories appear between 30-50 years of age. Symptoms and disease onset can vary widely (Chabriat et al. 1995).

The Notch3 Protein

In general, Notch genes encode receptors that arbitrate short-range signalling events. A typical Notch gene encodes a single-pass transmembrane receptor protein (Fiuza & Arias 2007, Sourander & Walinder 1977).

In particular, the NOTCH3 gene was identified in the early 90s and its expression was first identified in proliferating neuroepithelium (Bellavia *et al.* 2008). It provides instructions for generating the Notch3 receptor protein, found on the surface of VSM cells (Tikka *et al.* 2009). Upon Notch3 receptor activation, a signaling cascade is initiated resulting in the regulation of specific genes. In the case of VSM cells, Notch3 receptors play a pivotal role in the optimal functionality, survival and maintenance of muscle cells in the arterial network of the brain.

The NOTCH3 gene is located on the short (p) arm of chromosome 19 between positions 13.2 and 13.1 (cytogenetic location), as retrieved from NCBI database (Gene ID: 4854). More precisely, the molecular location of NOTCH3 gene is between the base pairs 15,270,443 to 15,311,791 on chromosome 19.

Methods

Bioinformatics
A Blast search, with the Notch 3 protein sequence was performed at the European Bioinformatics Institute (EBI), to find similar sequences in the Uniprot database (Consortium 2011). All sequences with an E-value lower than 7.0E-4 were selected to perform an analysis with the Multiple EM for Motif Elicitation (MEME) software (Papageorgiou *et al.* 2014, Papangelopoulos *et al.* 2014). The analysis was performed at http://meme.sdsc.edu/meme4_6_1/ using default values. Logos for the found motifs were generated using the WebLogo software (Crooks *et al.* 2004).

Medical Data
A list of point mutations of the Notch3 protein was gracefully provided by Prof. Marc Baumann, the Biomedicum Helsinki Unit. Table 1 summarizes the point mutations examined in this study.

Table 1. Point mutations examined in the current study.

A	Arg107Trp	Canadian case, previously unreported (data provided by Prof. Baumann)
B	Ser497Leu	Reported in the ESP database (NHLBI GO Exome Sequencing Project) (data provided by Prof. Baumann)
C	Glu813Lys	Reported in the ESP database (NHLBI GO Exome Sequencing Project) (data provided by Prof. Baumann)
D	Ala1020Pro	Scheid *et al.* 2008
E	Ser978Arg	Ferreira *et al.* 2007
F	His1133Glu	Canadian case (data provided by Prof. Baumann)

Homology Modelling and Energy Minimization
Homology modelling for the Notch3 protein was carried out using Modeller (Sali 1995). The structure of the Human Notch1 EGFS structure was used as template for this study (PDB entry: 2VJ3) (Cordle *et al.* 2008). Subsequent energy minimization was performed using the Gromacs-implemented, Charmm27 forcefield (Vlachakis *et al.* 2013a, b). Models were structurally evaluated using the Procheck utility (Laskowski 1996, Vlachakis *et al.* 2013c). Energy minimizations were used to remove any residual geometrical strain in each molecular system, using the Charmm forcefield as it is implemented into the Gromacs suite, version 4.5.5 (Hess *et al.* 2008, Pronk *et al.* 2013). All Gromacs-related simulations were performed through our previously developed graphical interface (Sellis *et al.* 2009, Vlachakis *et al.* 2013d). An implicit Generalized Born (GB) solvation was chosen at this stage, in an attempt to speed up the energy

minimization process (Vlachakis & Kossida 2014).

Molecular Dynamics Simulations

Molecular systems of the Notch3 protein were subjected to unrestrained Molecular Dynamics simulations (MDs) using the Gromacs suite, version 4.5.5 (Hess *et al.* 2008, Pronk *et al.* 2013, Vlachakis *et al.* 2014). MDs took place in a SPC water-solvated, periodic environment. Water molecules were added using the truncated octahedron box extending 7Å from each atom. Molecular systems were neutralized with counter-ions as required. For the purposes of this study all MDs were performed using the NVT ensemble in a canonical environment, at 300 K, 1 atm and a step size equal to 2 femtoseconds for a total 500 nanoseconds simulation time (Balatsos *et al.* 2012, Vlachakis *et al.* 2012b, 2013f, g). An NVT ensemble requires that the number of atoms, volume and temperature remain constant throughout the simulation (Vlachakis *et al.* 2012a, 2013e).

Results & Discussion

Molecular Mechanism

More than 190 mutations in the NOTCH3 gene have been found to cause CADASIL (Tournier-Lasserve *et al.* 1993). The gene encodes a large transmembrane (Notch3) receptor protein (Joutel *et al.* 1996), which has a crucial role within the vascular smooth muscle signaling pathways. It is comprised of 33 exons (Monet-Leprêtre 2009). At the molecular level, all pathogenic mutations have been found to result in an odd number of cysteine residues (which stay unpaired, due to addition or deletion of a cysteine molecule) in one of the 34 epidermal growth factor (EGF)-like repeats that are present in the extracellular domain (N3ECD). This results in misfolding of the Notch3 protein. NOTCH3 is expressed in VSM cells and this misfolding leads to a gradual accumulation of N3ECD on their surface, resulting in their degeneration. Apart for N3ECD accumulation, there is also granular osmiophilic material (GOM) deposited in the walls of the vessels. These two features can only be found in CADASIL (Ishiko *et al.* 2006, Lewandowska *et al.* 2011, Valenti *et al.* 2011).

Various neurodegenerative diseases have presented the key pathological feature of aggregation and accumulation of misfolded proteins. It was recently reported that Notch3 multimerization is enhanced by the presence of CADASIL-associated mutations. In the case that GOM deposits contain N3ECD, the mutated Nothc3 protein oligomers that are formed may be the lodgements of GOM. Degradation does not affect mutant Notch3 aggregates and their formation may cause

cellular dysfunction and/or cell death. However, VSM cell degeneration and loss may not be the only cause of blood vessel wall deterioration in CADASIL patients. Other factors contributing to this process may include hyalinization of the vessels, fibrosis and an abnormal increase in the intercellular space at the vessel wall (Ishiko *et al.* 2006, Lewandowska *et al.* 2011, Valenti *et al.* 2011).

However, recently some non-Cys mutations have been reported. Those mutations do not follow the characteristic pathology and pattern of the disease. Furthermore, it has been identified that some of the non-Cys mutations cause GOM whereas others do not (Bersano *et al.* 2012).

AAR Pattern

Amino Acid Repeats (AARs) can be frequently found within protein sequences and play various roles in protein function and evolution. Therefore, the identification of AAR patterns usually serves as the first step for further investigation. Various algorithms have been developed to assist in this identification. However, in this case, a pattern was identified solely through observation.

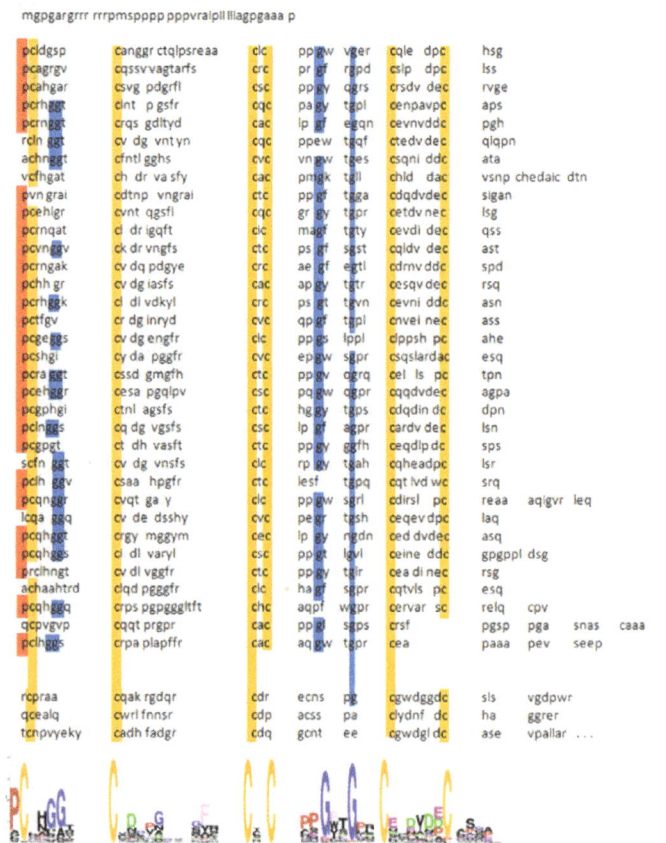

Figure 1. Repeating pattern of the Notch3 protein sequence, divided into repeats, based on the highlighted conserved residues, followed by a Weblogo representing the repeating motif.

The whole Notch3 protein sequences as retrieved from the NCBI protein database is comprised of 2321 amino acids. However, the repeating pattern appears among 1461 amino acids (Figure 1). The latter sequence was divided into 34 most similar repeats and another 3 that demonstrate only partial similarity, as shown. From the last repeat to the end, no pattern is observed.

The alignment was based on the residues highlighted on the sequence. The repeating motif is obvious. The conserved positions which supported the forming of the pattern are apparent on the logo following the sequence (Figure 1).

After having mapped all the point mutations on the sequence of Notch3, as they have been reported from cases examined by Prof. Baumann, the resulting secondary structure is shown in Figure 2. The selected residues represent the mutation positions. The repeating pattern of the mutation positions is obvious in this representation as well. The red arrows indicate the 3 disulphide bonds that are followed by 3 hydrogen bonds. The pattern is repeated 3 times.

As the mutations appear at specific locations on the secondary structure, it is assumable that these locations serve specific purposes, structure-wise, for the Notch3 protein function.

Description of Pattern

As shown in Figure 3A (color distribution of residues based on classification according to structure and chemical characteristics) and 3B (color distribution of residues based on side-chains, charge and hydrophobicity), the repeating pattern starts with a highly conserved proline (pos. 1) which appears on a coil-coil

conformation on the secondary structure, followed by a cysteine (pos. 2) almost conserved. Cys residues are responsible for the formation of disulphide bonds, which affect the proper folding and biological function of a protein. The high appearance of mutations on Cys residues, which are mostly conserved, in Notch3 sequence accounts for misfolding and loss of function that may lead to CADASIL disease. The succession of two contiguous higly conserved Gly residues (pos. 5 and 6) constitutes a flexible linker between the two conserved Cys residues (pos. 2 and 8) providing protein stability at that point (Robinson & Sauer 1998, Yan & Sun 1997). The position of the second Gly (pos. 6) seems to be an aliphatic conserved residue, as it is mainly either Gly or Ala. However, Ala does not appear so frequently due to its bulky side chain. On positions 13-17, although not higly conserved, Gly residues are mostly observed, forming a flexible linker once again. The appearance of aromatic residues (F, Y, W) (pos. 18 and 26) right after Gly (pos. 17 and 25) on the strand of the beta sheet on the secondary structure is also noticable. Aromatic residues provide favorable interaction energy through cross-strand pairing with glycine in order to reduce glycine's destabilizing effect (Merkel & Regan 1998). These positions appear conserved by hydrophobic residues (F, Y, A, W, Y) as

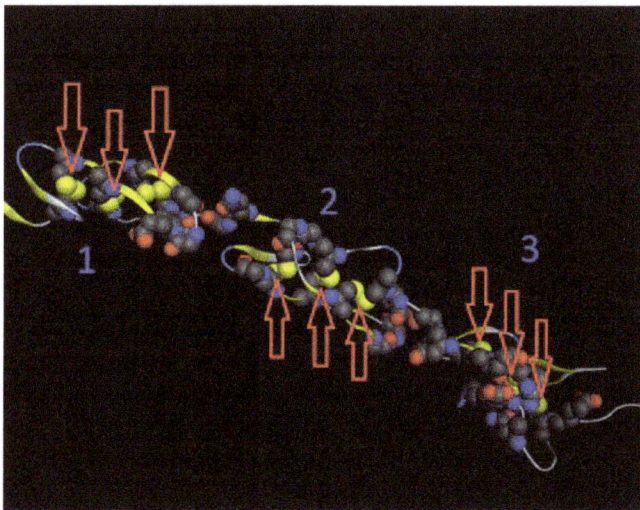

Figure 2. 3D structure of Notch3 protein. The selected residues represent all the point mutations reported by Prof. Baumann. The pattern is obvious in the secondary structure and is repeated 3 times (number above).

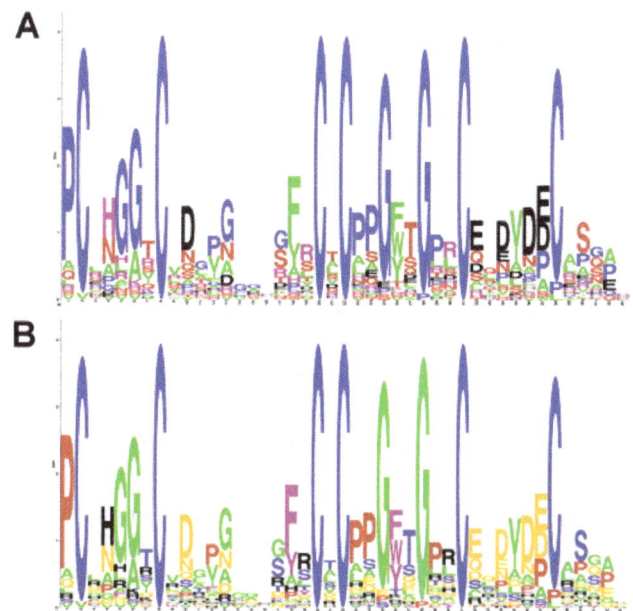

Figure 3. Weblogo representing conserved Notch3 protein repeats. **(A)** Color distribution based on structure and chemical characteristics of residues (green: aliphatic, blue: hydroxyl or sulfur-containing, red: cyclic, black: basic, purple: aromatic, yellow: acidic & their amide). **(B)** Color distribution based on side-chains, charge and hydrophobicity of residues (purple: positively charged, black: negatively charged, red: uncharged, blue: special cases, green: hydrophobic).

well. The Cys-X-Cys motif (pos. 20-22) that follows is completely conservative and is a strong pattern responsible for catalysis of redox reactions (Zhang *et al.* 2008). The two contiguous prolines right afterwards (pos. 23, 24), due to their cyclic character, seem to aid in the formation of the beta turn on the secondary structure. The amino acid sequence that follows contains two quite conservative glycines (pos. 25 and 28), which substitute the smallest amino acid with the simplest side chain of a hydrogen molecule. These two glycines are separated by aromatic residues (F, Y, W) (pos. 26) that have bulky side chains and others with hydroxyl-containing or acidic side chains (T, S, Q, D, E) (pos. 27) that are likely to participate in H-bonding. Position 27 seems to be mostly conserved by uncharged residues (T, S, Q). On positions 31-38, the sequence starts and ends with conserved cysteines with hydrophilic residues (E, D, N) in between, interrupted only by an aliphatic hydrophobic residue (V, I, A) on position 35. This region seems to be exposed to the solvent due to its hydrophilic character. Residues Asp and Glu seem equally replaced by one another at this region, as they are both negatively charged. The high content of Gly and Pro in Notch3, as well as their conservation score in the sequence, serve critical roles for the protein's formation into elastomeric or amyloid fibrils. This finding stands of great significance as amyloid fibrils are inextricably connected with tissue-degenerative diseases, such as CADASIL.

Protein Structure

The Notch extracellular domain is composed of a conserved array of up to 36 epidermal growth factor (EGF)-like repeats, that contribute to ligand interaction, as well as three Lin-12-Notch (LN) repeats, which are juxtamembrane repeats responsible for interactions between the extracellular and intracellular domains. The intracellular domain of Notch comprises of seven ankyrin repeats together with a PEST (proline, glutamine, serine, threonine-rich) and a TAD (transactivation) domain. The C-terminal heterodimerization domain is a hydrophobic region of extracellular Notch, able to create a stable complex with the extracellular domain of transmembrane Notch. The cleavage of the Notch receptor occurs at the S1 site within the trans-Golgi network, during the secretion process. The process of cleavage as well as the resulting structure are vital for Notch function in mammals.

At the surface of the cell, Notch can interact with its ligands, either Delta or Serrate. This interaction sheds the ectodomain and makes the Notch protein vulnerable to further cleavages. Following 3 different cleavages (S2-S4), the intracellular domain of the remaining Notch fragment is released and translocates to

the nucleus where it modulates gene expression by acting as a co-activator of the transcription factor suppressor of hairless (Fiuza & Arias 2007).

The negative regulatory region of the Notch receptor, which can be found between the ligand-binding and transmembrane domains, is essential for protecting the Notch receptor against cleavage, when no ligand is present. This region includes three cysteine-rich LIN-12–Notch repeats and a 'heterodimerization domain', which comes right before the membrane and contains both the S1 and S2 cleavage sites. Several studies, over a long time, drove to the conclusion that the negative regulatory region functions as the regulatory switch to activate the receptor. Receptors missing the EGF-like repeats remain functionally inactive. However, deleting the LIN12-Notch repeats or point mutations in this area have been reported to result in gain-of function phenotypes as well as ligand-independent metalloprotease cleavage (Gordon *et al.* 2008).

The NOTCH proteins, *i.e.* NOTCH1-4, consist of 34 EGF-like repeats and work as receptor for Jagged and Delta ligands. They can function as both transmembrane receptors and transcription factors. Upon activation and subsequent release of the notch intracellular domain (NICD), they effect cellular differentiation and proliferation, during development and adult life, including homeostasis in adult tissues (Boucher *et al.* 2012, Fiuza & Arias 2007, Gordon *et al.* 2008).

Molecular Modelling of Notch3 3D Structure

In order to understand the consequences for the NOTCH3 protein when introducing amino acid changes, we took a homology modelling approach since no structure has been determined for this protein. Using Modeller, Notch3 was modelled and its *in silico* 3D structure was established (Sali 1995). Moreover, the consequence of the introduced mutations could also be predicted via molecular dynamics simulations using Gromacs (Hess *et al.* 2008). Modeller predicted a set of two anti-parallel beta-sheet network in arranged repeats across the Notch3 protein. One pair is short, made up of 5 to 6 residues, while the other one is longer as it consists of almost 10 to 12 residues. The beta-sheets are held together by beta turns and coil fragments that contain mainly non-hydrophobic amino acids. A set of strategically positioned cysteine residues is responsible for the structural integrity of the Notch3 structure via disulphide bridges (Figure 4).

The induced R107W mutation provoked the partial loss of structure of one of the beta sheets in the long pair antiparallel conformation of Notch3. The position of the 107 residue was most likely crucial for the proper formation of the first antiparallel set of sheet-loop-sheet, as it was modelled to be located right

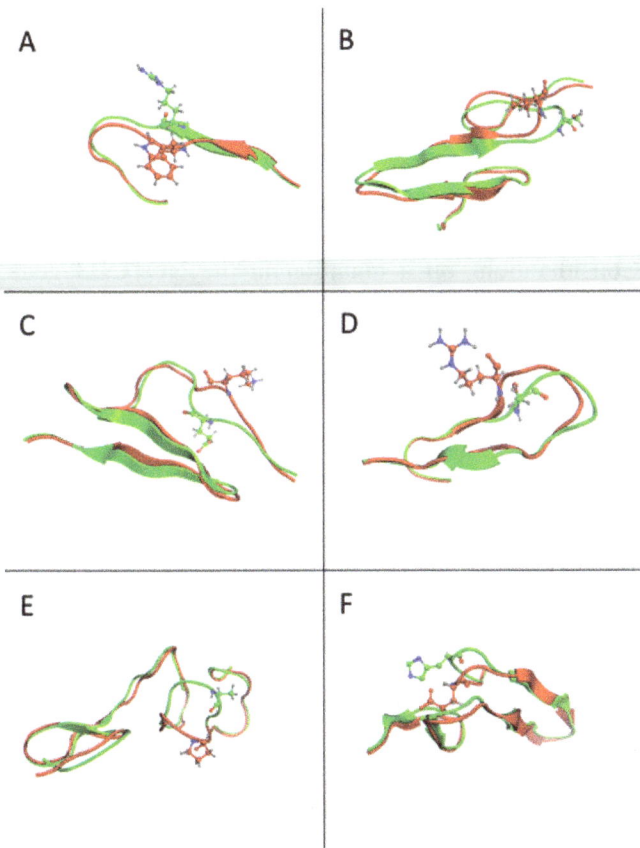

Figure 4. Structural superposition of the wild type and mutant type of Notch3 protein. Color coding is as follows: green for wild type and red for mutant, all shown in ribbon representation. **(A)** Arg107Trp, **(B)** Ser497Leu, **(C)** Glu813Lys, **(D)** Ser978Arg, **(E)** Ala1020Pro, **(F)** His1133Glu.

in the middle of the connecting loop. The bulky and conjugated side chain of the Tryptophan is pushed away from the core of the protein. Therefore it is found to claim the available space in the same plane that the two beta sheets define. As a result, the protein backbone in the proximity of the 107 position is twisted and converted to a coil conformation, having lost its original beta-sheet planarity (Figure 4A).

The S497L mutation was predicted to lead to a partial loss of the first long pair of beta-sheet formations, as it significantly changes the unstructured coil conformation it belongs to, to an almost perpendicular to the beta-sheet plane structure. As a result the network of hydrogen bonds in the nearby anti-parallel beta sheet pair is lost, the gap between the secondary elements becomes slightly bigger and inevitably one of the beta-sheets collapses and loses more than half of its structure (Figure 4B).

The same structural principle applies to the E813K mutation. The negatively charged glutamic acid residue was predicted to actively establish hydrogen bonds to the residues in close proximity that stabi-

lise the nearby pair of beta sheets. The induced E813K mutation introduces a bulky lysine residue in the original glutamic acid position that was modelled to induce breakage of the contact between the two upper beta-sheets. The obvious reason behind this change is attributed to the physicochemical change of the side chain from negatively to positively charged, in the case of the lysine, and to the fact that the lysine is slightly larger in size than the glutamic acid amino acid. It seems the available room is not big enough to accommodate an amino acid larger than glutamic acid (Figure 4C).

The A1020P mutation leads to a significant loss of structure of the nearby beta-sheet arrangement of the Notch3 structure. Even though the 1020P residue is found on a coil-coil structural conformation, the phi/psi dihedral angles of the amino acids involved are very close to a-helical Ramachandran plot regions. Proline cannot establish a hydrogen bond via its backbone amine residue since it has no amide hydrogen. However, even though proline has very poor alpha-helix-forming tendency, it is quite often seen as the first or last residue of an alpha-helix, presumably due to its structural rigidity as its side chain forms a ring. Alanine on the other hand belongs to a group of amino acid residues with low helix forming propensity. In this case, it seems that the induced proline is trying to stabilize an almost alpha-helical local conformation for the small stretch of loop it sits on. Consequently, the structural shift is too great for the nearby beta-sheets to sustain and they collapse, thus losing most of their structure (Figure 4E)

The S978R mutation also leads to loss of beta-sheet structure from Notch3. The Arginine 978 was predicted to be outside of the beta-sheet formation since there is no space available to accommodate it. However, the tiny side chain of the original serine residue was small enough to allow the loop to further bend backwards in favor of the beta-sheet formation. As a result of the induced arginine residue the phi/psi dihedral angle geometry of the nearby residues changes and some residues of the first beta-sheet formation are converted to a coil structure (Figure 4D).

The His1133E mutation is located on a beta-turn conformation that stabilizes the short pair of the antiparallel beta sheets. The negatively charged glutamic acid is drawn to the inner part of the Notch3 core via newly, randomly established hydrogen bonds to nearby residues. The positive charge of the Histidine's imidazole ring was positioned in a neutral rotamer that allowed the two beta sheets to interact through hydrogen bonding. The tilt of the glutamic acid's side chain and the consecutive rearrangement of the backbone of the protein locally, prohibits the formation of one of the two beta sheets completely. As a result, the Notch3 protein is very much disorganized and locally almost denatured as it completely loses a vital secondary structural element (Figure 4F).

Conclusions

Herein, we presented an *in silico* study of a series of Notch3 mutations that have been recently identified. In most cases these mutations refer to a Cys residue that leads to another unpaired Cys residue. In some cases of this disease, we see accumulation of Granular Osmiophilic Material (GOM), which has been a hallmark for the final diagnosis based on electron microcopy. However, recently some non-Cys mutations have also been identified. These mutations do not follow the characteristic pathology and pattern of the disease. Furthermore, it has been established that some of the non-Cys mutations cause GOM whereas some others do not. The ultimate aim of the *in silico* study was to establish, via the 3D structural analysis, an explanation of what actually happens in these cases. In this direction, exhaustive molecular dynamics simulations showed that these non-Cys mutations trigger significant loss of structure in the Notch3 protein, compared to the wild type. Even though these are mainly point mutations, we have established that the effect of each one of them on the three dimensional structure of the Notch3 protein is significant.

Acknowledgements

This work was partially supported by: (1) The BIOEXPLORE research project. BIOEXPLORE research project falls under the Operational Program "Education and Lifelong Learning" and is co-financed by the European Social Fund (ESF) and National Resources. (2) European Union (European Social Fund - ESF) and Greek national funds through the Operational Program "Education and Lifelong Learning" of the National Strategic Reference Framework (NSRF) - Research Funding Program: Thales. Investing in knowledge society through the European Social Fund.

References

Balatsos N, Vlachakis D, Chatzigeorgiou V, Manta S, Komiotis D, Vlassi M & Stathopoulos C 2012 Kinetic and *in silico* analysis of the slow-binding inhibition of human poly (A)-specific ribonuclease (PARN) by novel nucleoside analogues. *Biochimie* 94 **1** 214-221

Bellavia D, Checquolo S, Campese AF, Felli MP, Gulino A & Screpanti I 2008 Notch3: from subtle structural differences to functional diversity. *Oncogene* **27** 5092-5098

Bersano A, Ranieri M, Ciammola A, Cinnante C, Lanfranconi S, Dotti MT, Candelise L, Baschirotto C, Ghione I, Ballabio E, Bresolin N & Bassi MT 2012 Considerations on a mutation in the NOTCH3 gene sparing a cysteine residue: a rare polymorphism rather than a CADASIL variant. *Funct Neurol* **27** 247-252

Boucher J, Gridley T & Liaw L 2012 Molecular Pathways of Notch Signaling in Vascular Smooth Muscle Cells. *Front Physiol* **3** 81

Buffon F, Porcher R, Hernandez K, Kurtz A, Pointeau S, Vahedi K, Bousser MG & Chabriat H 2006 Cognitive profile in CADASIL. *J Neurol Neurosurg Psychiatry* **77** 175-180

Chabriat H, Vahedi K, Iba-Zizen MT, Joutel A, Nibbio A, Nagy TG, Krebs MO, Julien J, Dubois B, Ducrocq X *et al.* 1995 Clinical spectrum of CADASIL: a study of 7 families. Cerebral autosomal dominant arteriopathy with subcortical infarcts and leukoencephalopathy. *Lancet* **346** 934-939

Consortium TU 2011 Ongoing and future developments at the Universal Protein Resource. *Nucleic Acids Res* **39** D214-D219

Cordle J1, Johnson S, Tay JZ, Roversi P, Wilkin MB, de Madrid BH, Shimizu H, Jensen S, Whiteman P, Jin B, Redfield C, Baron M, Lea SM & Handford PA 2008 A conserved face of the Jagged/Serrate DSL domain is involved in Notch trans-activation and cis-inhibition. *Nat Struct Mol Biol* **15** 849-857

Crooks GE, Hon G, Chandonia JM & Brenner SE 2004 WebLogo: a sequence logo generator. *Genome Res* **14** 1188-1190

Dalkas GA, Vlachakis D, Tsagkrasoulis D, Kastania A & Kossida S 2013 State-of-the-art technology in modern computer-aided drug design. *Brief Bioinform* **14** 745-752

Dichgans M, Mayer M, Uttner I, Brüning R, Müller-Höcker J, Rungger G, Ebke M, Klockgether T & Gasser T 1998 The phenotypic spectrum of CADASIL: clinical findings in 102 cases. *Ann Neurol* **44** 731-739

Ferreira S, Costa C & Oliveira JP 2007 Novel human pathological mutations. Gene symbol: NOTCH3. Disease: cerebral autosomal dominant arteriopathy with subcortical infarcts and leucoencephalopathy (CADASIL). *Hum Genet* **121** 649

Fiuza UM & Arias AM 2007 Cell and molecular biology of Notch. *J Endocrinol* **194** 459-474

Gordon WR, Arnett KL & Blacklow SC 2008 The molecular logic of Notch signaling – a structural and biochemical perspective. *J Cell Sci* **121** 3109-3119

Hess B, Kutzner C, van der Spoel D & Lindahl E 2008 GROMACS 4: Algorithms for Highly Efficient, Load-Balanced, and Scalable Molecular Simulation. *J Chem Theor Comput* **4** 435-447

Ishiko A, Shimizu A, Nagata E, Takahashi K, Tabira T & Suzuki N 2006 Notch3 ectodomain is a major component of granular osmiophilic material (GOM) in CADASIL. *Acta Neuropathol* **112** 333-339

Joutel A, Corpechot C, Ducros A, Vahedi K, Chabriat H, Mouton P, Alamowitch S, Domenga V, Cécillion M, Marechal E, Maciazek J, Vayssiere C, Cruaud C, Cabanis EA, Ruchoux MM, Weissenbach J, Bach JF, Bousser MG & Tournier-Lasserve E 1996 Notch3 mutations in CADASIL, a hereditary adult-onset condition causing stroke and dementia. *Nature* **383** 707-710

Joutel A, Dodick DD, Parisi JE, Cecillon M, Tournier-Lasserve E & Bousser MG 2000 De novo mutation in the NOTCH3 gene causing CADASIL. *Ann Neurol* **47** 388-391

Laskowski RA, Rullmannn JA, MacArthur MW, Kaptein R & Thornton JM 1996 AQUA and PROCHECK-NMR: programs for checking the quality of protein structures solved by NMR. *J Biomol NMR* **8** 477-486

Lewandowska E, Dziewulska D, Parys M & Pasennik E 2011 Ultrastructure of granular osmiophilic material deposits (GOM) in arterioles of CADASIL patients. *Folia Neuropathol* **49** 174-180

Loukatou S, Papageorgiou L, Fakourelis P, Filtisi A, Polychronidou E, Bassis I, Megalooikonomou V, Makalowski W, Vlachakis D & Kossida S 2014 Molecular dynamics simulations through GPU video games technologies. *J Mol Biochem* **3** 64-71

Merkel JS & Regan L 1998 Aromatic rescue of glycine in beta sheets. *Fold Des* **3** 449-455

Monet-Leprêtre M 2009 Distinct phenotypic and functional features of CADASIL mutations in the Notch3 ligand binding domain. *Brain* **132** 1601-1612

Papagelopoulos N, Vlachakis D, Filntisi A, Fakourelis P, Papageorgiou L, Megalooikonomou V & Kossida S 2014 State of the art GPGPU applications in bioinformatics. *Int J Sys Biol Biomed Technol* **2** 24-48

Papageorgiou L, Vlachakis D, Koumandou VL, Papagelopoulos N & Kossida S 2014 Computer-Aided Drug Design and Biological Evaluation of Novel Anti-Greek Goat Encephalitis Agents. *Int J Sys Biol Biomed Technol* **2** 1-16

Pronk S, Páll S, Schulz R, Larsson P, Bjelkmar P, Apostolov R, Shirts MR, Smith JC, Kasson PM, van der Spoel D, Hess B & Lindahl E 2013 GROMACS 4.5: a high-throughput and highly parallel open source molecular simulation toolkit. *Bioinformatics* **29** 845-854

Razvi SS, Davidson R, Bone I & Muir KW 2005 The prevalence of cerebral autosomal dominant arteriopathy with subcortical infarcts and leucoencephalopathy (CADASIL) in the west of Scotland. *J Neurol Neurosurg Psychiatry* **76** 739-341

Robinson CR & Sauer RT 1998 Optimizing the stability of single-chain proteins by linker length and composition mutagenesis. *Proc Natl Acad Sci U S A* **95** 5929-5934

Sali A, Potterton L, Yuan F, van Vlijmen H & Karplus M 1995 Evaluation of comparative protein modeling by MODELLER. *Proteins* **23** 318-326

Scheid R, Heinritz W, Leyhe T, Thal DR, Schober R, Strenge S, von Cramon DY & Froster UG 2008 Cysteine-sparing notch3 mutations: cadasil or cadasil variants? *Neurology* **71** 774-776

Sellis D, Vlachakis D & Vlassi M 2009 Gromita: a fully integrated graphical user interface to gromacs 4. *Bioinform Biol Ins* **3** 99-102

Sourander P & Walinder J 1977 Hereditary multi-infarct dementia. Morphological and clinical studies of a new disease. *Acta Neuropathol* **39** 247-254

Tatsch K, Koch W, Linke R, Poepperl G, Peters N, Holtmannspoetter M & Dichgans M 2003 Cortical hypometabolism and crossed cerebellar diaschisis suggest subcortically induced disconnection in CADASIL: an 18F-FDG PET study. *J Nucl Med* **44** 862-869

Tikka S, Mykkänen K, Ruchoux MM, Bergholm R, Junna M, Pöyhönen M, Yki-Järvinen H, Joutel A, Viitanen M, Baumann M & Kalimo H 2009 Congruence between NOTCH3 mutations and GOM in 131 CADASIL patients. *Brain* **132** 933-939

Toni-Uebari TK 2013 Cerebral autosomal dominant arteriopathy with subcortical infarcts and leucoencephalopathy (CADASIL): a rare cause of dementia. *BMJ Case Rep* bcr2012007285

Tournier-Lasserve E1, Joutel A, Melki J, Weissenbach J, Lathrop GM, Chabriat H, Mas JL, Cabanis EA, Baudrimont M, Maciazek J, Bach MA & Bousser MG 1993 Cerebral autosomal dominant arteriopathy with subcortical infarcts and leukoencephalopathy maps to chromosome 19q12. *Nat Genet* **3** 256-259

Ueda M, Nakaguma R & Ando Y 2009 Cerebral autosomal dominant arteriopathy with subcortical infarcts and leukoencephalopathy (CADASIL)]. [Article in Japanese]. *Rinsho Byori* **57** 242-251

Valenti R, Pescini F, Antonini S, Castellini G, Poggesi A, Bianchi S, Inzitari D, Pallanti S & Pantoni L 2011 Major depression and bipolar disorders in CADASIL: a study using the DSM-IV semi-structured interview. *Acta Neurol. Scand* **124** 390-395

Velizarova R, Mourand I, Serafini A, Crespel A & Gelisse P 2011 Focal epilepsy as first symptom in CADASIL. *Seizure* **20** 502-504

Vlachakis D, Argiro A & Kossida S 2013a An update on virology and emerging viral epidemics. *J Mol Biochem* **2** 80-84

Vlachakis D, Bencurova E, Papangelopoulos N & Kossida S 2014 Current State-of-the-Art Molecular Dynamics Methods and Applications. *Adv Protein Chem Struct Biol* **94** 269-313

Vlachakis D, Karozou A & Kossida S 2013b 3D Molecular Modelling Study of the H7N9 RNA-Dependent RNA Polymerase as an Emerging Pharmacological Target. *Influenza Res Treat* **2013** 645348

Vlachakis D, Kontopoulos DG & Kossida S 2013c. Space constrained homology modelling: the paradigm of the RNA-dependent RNA polymerase of dengue (type II) virus. *Comput Math Methods Med* **2013** 108910

Vlachakis D & Kossida S 2013 Molecular modeling and pharmacophore elucidation study of the Classical Swine Fever virus helicase as a promising pharmacological target. *PeerJ* **1** e85

Vlachakis D, Koumandou VL & Kossida S 2013d A holistic evolutionary and structural study of flaviviridae provides insights into the function and inhibition of HCV helicase. *PeerJ* **1** e74

Vlachakis D, Pavlopoulou A, Tsiliki G, Komiotis D, Stathopoulos C, Balatsos NA & Kossida S 2012a An integrated in silico approach to design specific inhibitors targeting human poly(a)-specific ribonuclease. *PLoS One* **7** e51113

Vlachakis D, Tsaniras SC, Feidakis C & Kossida S 2013e Molecular modelling study of the 3D structure of the biglycan core protein, using homology modelling techniques. *J Mol Biochem* **2** 85-93

Vlachakis D, Tsaniras SC & Kossida S 2012b Current viral infections and epidemics of flaviviridae; lots of grief but also some hope. *J Mol Biochem* **1** 144-149

Vlachakis D, Tsiliki G & Kossida S 2013f 3D Molecular Modelling of the Helicase Enzyme of the Endemic, Zoonotic Greek Goat Encephalitis Virus. *Comm Com Inf Sci* **383** 165-171

Vlachakis D, Tsiliki G, Pavlopoulou A, Roubelakis MG, Tsaniras SC & Kossida S 2013g Antiviral Stratagems Against HIV-1 Using RNA Interference (RNAi) Technology. *Evol Bioinform Online* **9** 203-213

Yan BX & Sun YQ 1997 Glycine residues provide flexibility for enzyme active sites. *J Biol Chem* **272** 3190-3194

Zhang T, Zhang H, Chen K, Shen S, Ruan J & Kurgan L 2008 Accurate sequence-based prediction of catalytic residues. *Bioinformatics* **24** 2329-2338

4

Antimicrobial activities of caffeic acid phenethyl ester

Sivan Meyuhas[1], Morad Assali[2], Mahmoud Huleihil[3] and Mahmoud Huleihel[1]

[1]Department of Microbiology, Immunology and Genetics, Faculty of Health Sciences, Ben-Gurion University of the Negev, Beer-Sheva, Israel
[2]Department of Orology, Soroka Medical Center, Beer-Sheva, Israel
[3]Academic Institute for Training Arab Teachers (AITAT), Beit Berl College, Israel

Correspondence should be addressed to Mahmoud Huleihel; E-mail: mahmoudh@bgumail.bgu.ac.il

Abstract

Caffeic acid phenethyl ester (CAPE) is considered as one of the most active components of Propolis extract (PE), a natural product obtained from beehives. PE comprises a complex of chemicals and has been found to have various biological activities. The aim of the present study is to assess the antibacterial activities CAPE has against various gram positive [gram (+)] and gram negative [gram (-)] bacteria and try to elucidate its mechanism of action. Bacteria were grown in the presence of various doses of CAPE and examined at different periods of time for their growth, both by absorbance (OD) measurement and colony assay. The results show that CAPE significantly inhibited the growth of most examined gram (+) bacteria while having only a slight inhibitory effect on most tested gram (-) bacteria. Our results also show that continuous treatment of gram (+) bacteria with CAPE for at least 6h caused irreversible inhibition of the bacterial growth (bacteriocidic effect); however, treatment for shorter periods of time caused only a stopping of bacterial growth (bacteriostatic effect). It seems that these effects were caused, at least partially, as a result of disruptions of the treated bacterial outer and plasma membranes. There is no significant synergistic effect between CAPE and ampicillin, although an additive effect has been found.

Introduction

The use of natural products as medicines for the treatment of various diseases has had a long history. Traditional healers, throughout history, have acquired detailed knowledge regarding the use of medicinal plants (Abel & Busia 2005).

It has been estimated that at least 25% of the active compounds present in currently prescribed synthetic drugs were first identified in natural sources. In this regard, the investigation of natural products and their potential therapeutic properties is essential (Halberstein 2005). Increasing efforts are currently being devoted towards novel applications for natural products and their derivatives for treating human diseases.

PE is obtained from beehives and based on resins produced by honeybees from certain trees and plants, bee's wax and their secretions (Onlen et al. 2007). Scientific research has revealed various biological activities of PE such as antibacterial (Huang et al. 2006, Velazquez et al. 2007), antiviral (Drago et al. 2007, Huleihel & Eshanu 2002, Orsolić & Basić 2005, Salomão et al. 2008, Viuda-Martos et al. 2008, Yao et al. 2004), antioxidant (Ahn et al. 2007, Kerem et al. 2006), antifungal (Quiroga et al. 2006, Silici & Koc 2006), antiinflamatory (Harris et al. 2006, Wu et al. 2006), antitumor (Akao et al. 2003) and other activities (de Rezende et al. 2008, Kosalec et al. 2005, Viuda-Martos et al. 2008). During the last few years, PE has also been used in the food industry as an additive as well as in beverages and nutritional supplements to enhance health and prevent diseases (Velazquez et al. 2007). In addition, previous studies have reported antibacterial activity of PE mainly against gram (+) but not gram (-) bacteria (Velazquez et al. 2007).

The chemical composition of PE is complex and has not been completely elucidated. Mainly, differences in the composition of PE have been found to depend on its origin but also among topical samples, depending on the local flora at the site of collection

(Salomão *et al.* 2008). Nevertheless, it is known that the most important group of compounds - in terms of amount and biochemical activity - is the flavonoids, which are thought to play a significant role in its biological activities (Viuda-Martos *et al.* 2008).

The most active and studied component of PE is caffeic acid phenethyl ester (CAPE), which is known to be a potent inhibitor of activation of NF-kB (Natarajan *et al.* 1996). A major part of CAPE bioactivities is thought to be related to NF-kB inhibition (Marquez *et al.* 2004, Song *et al.* 2002, Yang *et al.* 2005). CAPE has also been reported to be a potent anti-inflammatory and antioxidant agent and possess several antiviral, antibacterial and antifungal properties (Alici *et al.* 2015, Ilhan *et al.* 1999, Tolba *et al.* 2013, Velazquez *et al.* 2007).

In this work, we tested the inhibitory effects of an aqueous extract of CAPE *in vitro* against different gram (+) and gram (-) bacteria, trying to elucidate its antibacterial mechanism of action.

Materials and Methods

CAPE
CAPE was purchased as a powder from Sigma-Aldrich Corporation, USA. A stock solution of this product was prepared by dissolving it in dimethyl sulfoxide (DMSO) and then making the appropriate concentrations for examining its activity by dilution with bacterial growth medium (LB medium).

Bacteria
In the present study, the following gram (-) bacteria were used: *Escherichia (E.) coli, Serratia (S.) marcescens, Pseudomonas (P.) aeruginosa, Haemophilus (H.) influenza, Pseudomonas, Shegella, Salmonella (S.) entenidis, Neisseria* and *Klebsiella.* The gram (+) bacteria used were the following: *Staphylococcus (Staph.) aureus, Micrococcus, Streptococcus (S.) olysgalactiae, Streptococcus (S.) mitis, Bacillus (B.) subtilis, B. cereus, B. megaterium* and *B. thuringiensis.*

All the above bacterial strains were supplied by Dr Valentina Pavlov from the microbiology department in our institute. All bacteria were grown on Nutrient Agar (Difco) at 37 °C.

Measurement of bacterial amount
The amounts of bacteria were evaluated by 2 different methods. Their optical density (OD) was examined by spectrophotometer at a wavelength of 620nm. This method gives an evaluation of the live and dead bacteria. The other method involved counting bacterial colonies, by plating raising dilutions of each bacteria

on LB agar plates for 24h at 37°C and counting the number of the obtained colonies. This method only gives the number of live bacteria.

Antibacterial activity
In order to examine the antibacterial activity of the tested products, we grew the appropriate bacteria overnight and diluted them with LB medium to obtain about 10^4 colony-forming units (cfu)/mL, as estimated by optical density (OD_{630nm}) and plating on agar examinations. These suspensions were used as inoculates in the antimicrobial activity tests. Various concentrations of the tested material (CAPE) were added to the appropriate bacterial suspension with a final volume of 10ml. Following this, treated and untreated bacterial suspensions were incubated at 37°C in a shaking incubator. Thereafter, the following steps were undertaken at different time-points post-treatment: (a) the OD_{630nm} of these samples was measured and corrected by subtraction of the OD_{630nm} of PE or CAPE alone in sterile LB; (b) 0.5ml from each suspension was centrifuged at 2000g for 15 minutes, washed twice with sterile distilled water (ddH$_2$O), resuspended in 0.2ml LB, plated on agar plates, incubated at 37°C for 24h and the number of colonies was counted. The lowest concentrate of CAPE that prevented bacterial growth was considered to be the minimum inhibitory concentration (MIC).

Table 1. Minimum inhibitory concentration (MIC) of CAPE on different species of gram (+) and gram (-) bacteria.

Bacterial species	Gram (+/-)	MIC (mM)
S. aureus	+	48±2.5
S. olysgalactiae	+	48±3.1
S. mitis	+	55±2.8
B. thuringiensi	+	250±10.5
(B.) subtilis	+	70±3.2
B. cereus	+	96±2.8
B. megaterium	+	50±2.6
Micrococcus	+	55±2.6
E. coli	-	254±15.1
S. marcescens	-	400±23.7
P. aeruginosa	-	500±25.1
H. influenza	-	200±12.1
Pseudomonas	-	400±26.1
S. entenidis	-	60±2.70
Neisseria	-	500±24.3
Klebsiella	-	50±3.10

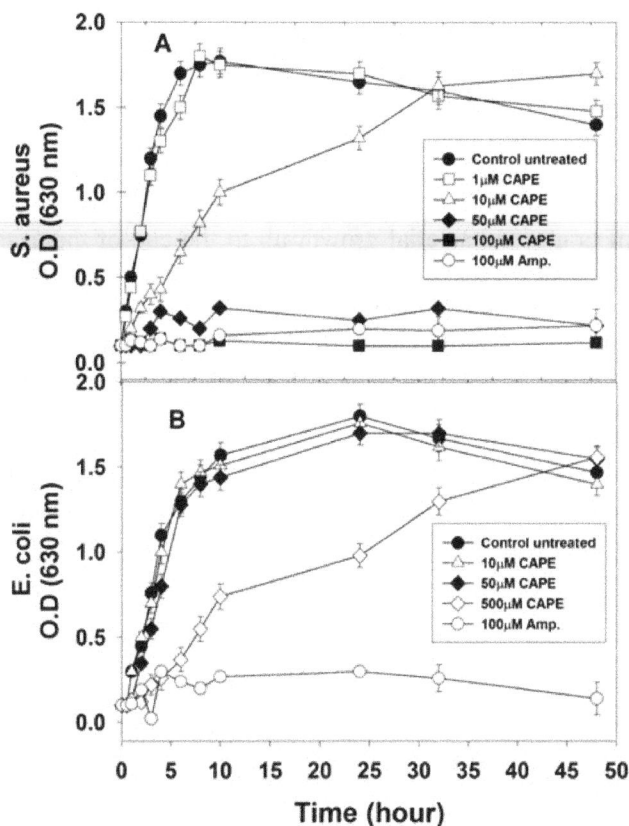

Figure 1. Effect of CAPE on Gram (+) and Gram (-) bacteria. Various concentrations of CAPE were added to 0.1 OD suspensions of *S. aureus* (A) and *E. coli* (B) and incubated at 37°C with shaking. In parallel, the appropriate suspensions were treated with 100 µg/ml of ampicillin as a positive control. At different time-points post-treatment, the amount of the bacteria was determined by measuring their OD_{630nm}. The results are means ± SD (n=5).

Combined antibacterial activity with ampicillin

In the present study we examined possible synergistic or additive activity between CAPE and ampicillin, as a representative antibiotic against different gram (+) and gram (-) bacteria, according to Bonapace *et al.* (2002). Briefly, overnight working bacterial broth cultures were diluted in LB medium together with 10% serial dilutions of CAPE and different ampicillin combinations, so as to give about 10^4 cfu/mL. Samples without treatment were used as controls. The cultures were incubated for 24h at a 37°C incubator with shaking. MICs were determined for each tested bacterial strain and the interaction between CAPE and ampicillin was calculated by the fractional inhibitory concentration (FIC) index of the combination. The FIC for each antibacterial agent was calculated as the ratio of the MIC of the agent in combination (with the other tested agent) to its MIC alone. In addition, the FIC index equals the FIC of CAPE plus the FIC of ampicillin. The interaction between the two agents was considered

as synergy when the FIC index was ≤ 0.5, additivity/no interaction when the FIC index was between 0.5 to 4 and antagonism when the FIC index >4.

Gentian violet uptake by gram (+) and gram (-) bacteria

CAPE treated bacteria (1 ml) were centrifuged at 2000 rpm for 15 minutes at room temperature, washed twice with ddH_2O, resuspended in 1 ml of ddH_2O containing gentian violet (10µg/ml) and incubated for 10 min at 37°C under vigorous shaking. The cells were removed by centrifugation and the amount of gentian violet remaining in the supernatant was measured at 590 nm in a spectrophotometer.

Loss of 260nm absorbing material

The release of UV-absorbing materials was measured by a UV–VIS spectrophotometer. Overnight cultures of gram (+) and gram (-) bacteria were adjusted to an OD600 of 1.0. Bacterial cells were centrifuged at 2000 rpm for 15 minutes, the supernatant was discarded, and the resulting pellet was washed twice and then resuspended in 1 ml of PBS (pH 7.4). Different concentrations of CAPE were added to the bacterial suspensions. All samples were incubated at 37°C for different periods of time (2 or 6h) and centrifuged at 13,000rpm for 15 min, before the OD260 value of the supernatant was measured.

Fourier-transform infrared (FTIR) microscopy

The alteration in structural features of gram (+) and gram (-) bacteria at the molecular level upon treatment with PE or CAPE was analyzed by FTIR spectroscopy. Different doses of CAPE were added to the cell suspensions of the examined bacteria, from overnight cultures with an OD600 of 1.0. Treatment was performed for 6h at 37°C. The cells were then pelleted by centrifugation at 3000 rpm for 5 min and washed three times with distilled water. The bacterial pellet was resuspended in 100µl of distilled water and 1µl of the obtained suspension was placed on a certain area on a zinc sellenide crystal, air dried for 15min at room temperature (or for 5 min by air drying in a laminar flow) and examined by FTIR microscopy.

FTIR measurements were performed in the transmission mode with a liquid-nitrogen-cooled MCT detector of the FTIR microscope (Bruker IRScope II) coupled to an FTIR spectrometer (BRUKER EQUINOX model 55/S, OPUS software). The spectra were obtained in the wave number range of 600-4000 cm-1. Spectral resolution was set at 4 cm-1. Baseline correction by the rubber band method and vector normalization was obtained for all the spectra by OPUS soft-

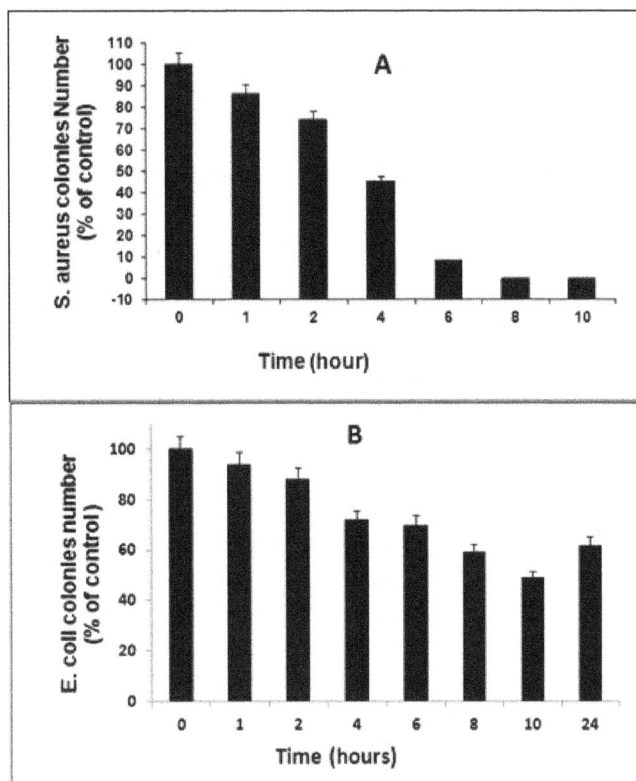

Figure 2. Effect of CAPE on bacteria colony growth. Suspensions (0.1 OD) of *S. aureus* (A) and *E. coli* (B) were treated with 50mM of CAPE and incubated at 37°C with shaking. At different time-points post-treatment, the amount of the living bacteria was determined by counting bacterial colonies. The results are means $\pm$ SD (n=5).

ware. Peak positions were determined by means of a second derivation method by OPUS software.

Results

Effect of CAPE on Gram (+) and Gram (-) bacteria
To determine the antibacterial activity of CAPE, we evaluated its effect on Gram (+) (*S. aureus, Micrococcus, S. olysgalactiae, S. mitis, B. subtilis, B. cereus, B. megaterium* and *B. thuringiensis*) and Gram (-) (*E. coli, S. marcescens, P. aeruginosa, H. influenza, Pseudomonas, Shegella, S. entenidis, Neisseria* and *Klebsiella*) bacteria.

Various concentrations of the tested material (CAPE) were added to the appropriate bacterial suspension (0.1 OD) and incubated at 37°C with shaking. At various time-points post-treatment the amount of the bacteria was determined by both measuring their OD_{630nm} and counting growing colonies.

CAPE showed strong antibacterial activity against most of the gram (+) bacteria examined, in a concentration dependent manner (Table 1 and Figure 1A, C), while it only had a slight and weak antibacterial activity against most of the gram (-) bacteria exam-

ined (Table 1 and Figure 1B, D). However, CAPE showed weak antibacterial activity against the gram (+) bacteria *B. Thuringiensi* and a strong activity against the gram (-) bacteria *S. Entenidis* and *Klebsiella* (Table 1).

It can also be seen that continuous treatment with 50mM of CAPE caused a complete inhibition of all gram (+) bacterial growth up to the end of the experiment (48h post-treatment), while continuous treatment with lower doses (10mM of CAPE) caused a slower growth of gram (+) bacteria compared to the untreated control bacteria, as determined by measuring their OD_{630nm} (Figure 1A, C). However, continuous treatment with even 500mM of CAPE only caused a reduction in the growth rate of all gram (-) bacteria examined, as can be seen in the representative results in Figure 1B and D.

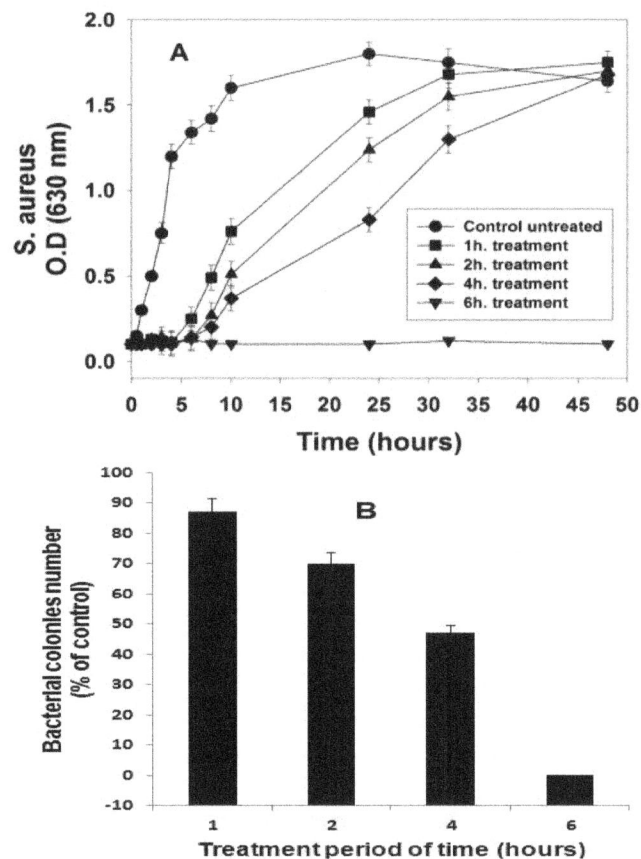

Figure 3. Effect of CAPE treatment termination on bacterial growth. Suspensions (0.1 OD) of *S. aureus* were treated with 50 mM CAPE and at different time-points treatment was terminated by exchanging the bacterial culture medium with a fresh medium free of the examined products. These bacterial cultures were incubated at 37°C with shaking and their growth was examined at different time-points post-treatment by measuring their absorbance (OD) (A) and by counting growing colonies (colony assay) (B). The results are means $\pm$ SD (n=5).

When the amount of living bacteria was determined by counting bacterial colonies, at different time-points post-treatment with 50mM of CAPE, it was found that treatment for up to 4h caused a significant decrease in the number of the gram (+) bacteria colonies, compared to the untreated control (Figure 2A). This result is mainly attributed to the increase in the number of the control bacteria, while the number of the treated bacteria did not change significantly over this

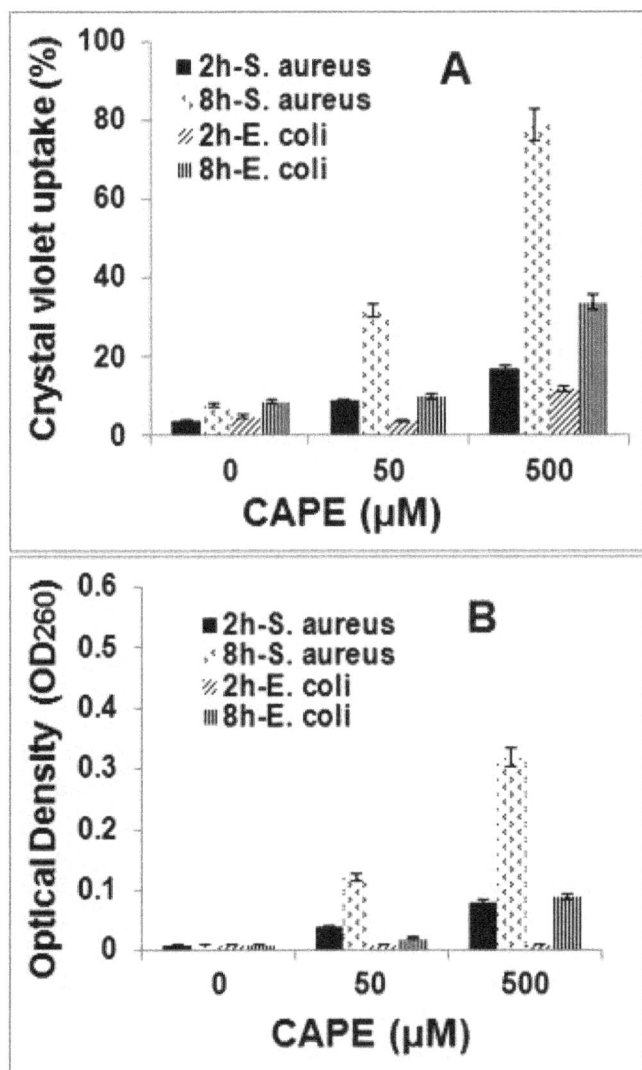

period of time. In addition, the size of the treated bacterial colonies was significantly smaller than that of the controls. However, treatment for 6h or more caused complete death of all bacteria treated (Figure 2A). Treatment of the gram (-) bacteria with 500mM of CAPE reduced the number of bacterial colonies as a function of treatment time (Figure 2B).

Effect of CAPE treatment termination on bacterial growth

Trying to examine the antibacterial mechanism of CAPE, different gram (+) bacteria were treated with 50 mM CAPE, and at different time-points the treatment was terminated by exchanging the bacterial culture medium with a fresh medium, free of the examined products. These bacterial cultures were incubated at 37°C in a shaking incubator and their growth was examined at different time-points. The results obtained, either by OD measurement (Figure 3A and B) or colonies counting (Figure 3C) showed that continuous treatment of the bacterial culture with PE or CAPE for up to 4h only caused a delay of about 5h in the growth of the bacteria, which was followed by a moderate growth compared to the untreated control cultures. However, continuous treatment for 6h caused complete death of the bacteria without any recovery, even 2 days after the termination of the treatment.

Effect of CAPE treatment on bacterial outer membrane (OM) permeability

The effect of CAPE treatment on OM permeability of gram (+) and gram (-) bacteria to crystal violet was examined. It is known that gentian violet poorly penetrates the intact bacterial OM (Devi *et al.* 2010); therefore, it is possible to examine the effect of CAPE treatment on OM permeability by examining the effect on gentian violet uptake by the treated bacteria. Different gram (+) and gram (-) bacteria were treated with various doses of CAPE for 2 or 8h; the bacteria were the centrifuged, resuspended in 1 ml of ddH$_2$O containing gentian violet (10µg/ml) and incubated for 10 min at 37°C, under vigorous shaking as described in the "Materials and Methods" section. The gentian violet uptake was determined by measuring the remaining amount in the supernatant. The gentian violet uptake of the control untreated gram (+) and gram (-) bacteria was 4-10% of the input; however, it increased dramatically in gram (+) bacteria in a dose dependent mater, reaching over 95% uptake at the higher doses after 8h of treatment with CAPE (Figure 4A). Also, in the case of gram (-) bacteria there was a gradual and significant increase in the gentian violet uptake at the higher doses of CAPE (Figure 4A), although this increase was profoundly lower compared to the gram (+) bacteria. It

Figure 4. Effect of CAPE treatment on gentian violet uptake and on 260nm absorbing material loss by gram (+) and gram (-) bacteria. 1ml suspensions (0.5 OD) of *S. aureus* and *E. coli* were treated with different doses of CAPE for different periods of time (2 or 8h), then: (A) the bacteria were centrifuged at 2000 g for 15 minutes, bacterial pellets were washed twice with ddH2O, resuspended in 1ml of ddH2O containing gentian violet (10µg/ml) and incubated for 10 min at 37°C under vigorous shaking. The bacteria were removed by centrifugation and the amount of gentian violet remaining in the supernatant was measured at 590 nm in a spectrophotometer; (B) the bacteria were pelleted as above and the OD260 value of the supernatant was measured. The results are means ± SD (n=5).

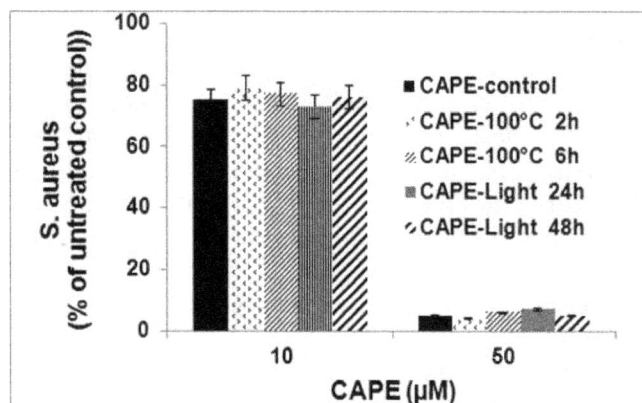

Figure 5. Effect of heat and light on CAPE antibacterial activities. Different doses of CAPE were incubated at 100C° or continuously exposed to a 100w lamp light for different periods of time. They were then added to 0.1 OD suspensions of *S. aureus*, incubated at 37°C with shaking and bacterial growth was examined at different time-points post-treatment by measuring their absorbance (OD). The results are means ± SD (n=5).

should also be noted that a short treatment of 2h with CAPE caused a moderate increase in the gentian violet uptake in gram (+) and gram (-) bacteria.

Effect of CAPE on release of 260nm absorbing materials from gram (+) and gram (-) bacteria

Different gram (+) and gram (-) bacteria (1 OD) were treated with various amounts of CAPE for 2 or 8h. The bacteria were then centrifuged and the OD260 values of the supernatants were measured. It is known that measuring the release of UV-absorbing materials is an index of cell lysis (Zhou *et al.* 2008). Our results showed that an 8h treatment of gram (+) bacteria with CAPE caused a significant increase in the OD (from 0.01 up to 0.3) while it resulted in a significantly lower increase in the case of gram (-) bacteria (from 0.01 up to 0.09) (Figure 4B).

Effect of heat treatment or light exposure on CAPE antibacterial activity

PE is rich with aromatic and flavonoid components which are, at least in part, sensitive to photo-oxidation that might affect their antibacterial activity. Two main theories have been put forward to explain the antibacterial capacity of honey and probably of PE (Viuda-Martos *et al.* 2008). The first one assumes that it is due to the action of hydrogen peroxide that is produced by glucose oxidase in the presence of light and heat (Dustmann 1979), while the other one states that it is a nonperoxide activity that is independent of light and heat (Roth *et al.* 1986).

In this regard, we sought to examine the effect of high temperatures and light exposure on the antibacterial activity of CAPE. Different doses of CAPE were exposed to a high temperature or light for different periods of time and their antibacterial activity against gram (+) bacteria was examined. The results showed that neither incubation of CAPE in 100°C for 2 or 6h nor exposure to light for 24 or 48h had an effect on its antibacterial activity (Figure 5).

Combined antibacterial activity with ampicillin

We examined possible synergistic or additive activity between CAPE and ampicillin, as a representative antibiotic against different gram (+) and gram (-) bacteria. Our results, presented in Table 2, show that the FIC indices of CAPE plus ampicillin against all examined bacteria were above 0.5. Therefore, it seems that these combinations had only additive activity against the different tested bacteria without synergistic activity.

Spectral changes in gram (+) and gram (-) bacteria treated with CAPE

FTIR spectroscopy can be used to detect and monitor characteristic changes in molecular compositions and structures of living cells, providing a wealth of qualitative and quantitative information about a given sample. The infrared spectrum of any compound is known to give a unique "finger print" (Naumann *et al.* 1991). In addition, the large information already known about spectral peaks obtained from FTIR spectra of living cells (Diem *et al.* 1999) make FTIR spectroscopy an attractive technique for detection and identification of pathogens. This technique has been previously used for the detection and characterization of cancer cells (Erukhimovitch *et al.* 2002), cells infected with viruses

Table 2. Minimum inhibitory concentration (MIC), fractional inhibitory concentration (FIC) and FIC index of CAPE and ampicillin against different gram (+) and gram (-) bacteria.

Bacteria	MIC			FIC		FIC index
	CAPE (µM)	Amp (µg/ml)	CAPE + Amp	CAPE	Amp	
S. aureus	48	64	16 + 24	0.33	0.37	0.70
B. cereus	96	64	30 + 32	0.31	0.50	0.81
E. coli	254	48	96 + 30	0.38	0.62	1.00
Neisseria	500	96	180 + 50	0.36	0.52	0.88

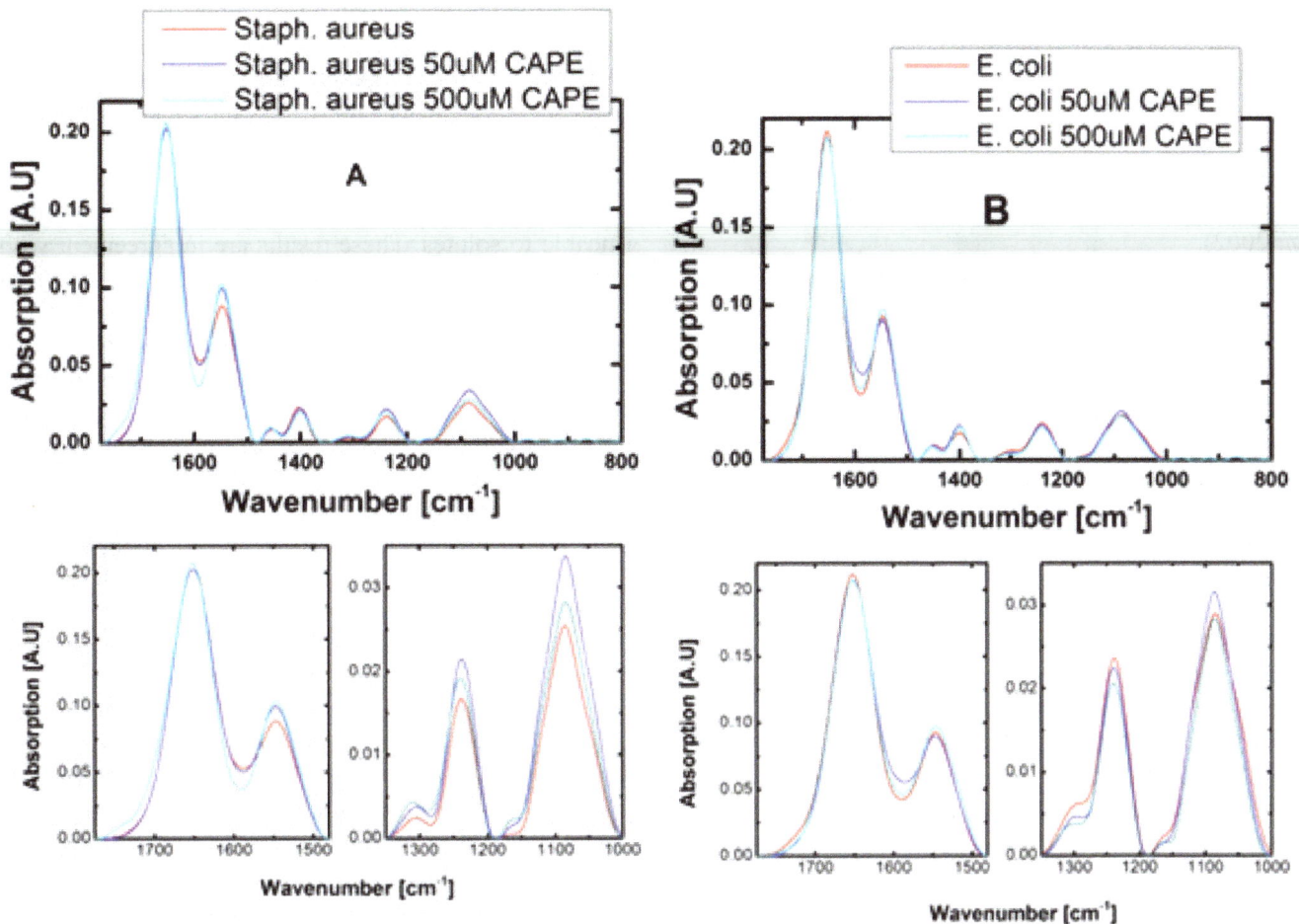

Figure 6. FTIR spectroscopy of gram (+) and gram (-) bacteria treated with CAPE. *S. aureus* and *E. coli* were treated with different doses of CAPE for 6h and examined by FTIR spectroscopy.

(Salman *et al.* 2002) and microorganisms (Erukhimovitch *et al.* 2002, Maquelin *et al.* 2003, Mariey *et al.* 2001).

In the present study, gram (+) and gram (-) bacteria were treated with different doses of PE for 6h and examined by FTIR spectroscopy. Significant spectral changes were observed in the frequency between 1700 and 1000cm^{-1} in CAPE treated samples of *S. aureus* (Figure 6A) compared to the slight spectral changes in the treated *E. coli* (Figure 6B), which may indicate an apparent deformation in the treated cells. These spectral changes were associated with changes in different components of the treated bacteria, such as proteins, polysaccharides, nucleic acids and lipids. There is a significant increase in frequency at 1100 and 1250 cm^{-1} of the CAPE-treated bacteria, which may indicate an alteration in polysaccharide and phosphate components. These differences may reflect changes in the content or distribution of the above components and possibly others.

Discussion

In the present study we examined the antibacterial activity of CAPE (one of the active components of PE) against various gram (+) and gram (-) bacteria. CAPE showed potent and impressive antibacterial activity against most tested gram (+) bacteria, while it only had moderate and partial activity against gram (-) bacteria. This observation is in complete agreement with previous studies which have shown that gram (+) bacteria are more susceptible to the antibacterial effect of PE or CAPE than gram (-) bacteria (Drago *et al.* 2007, Velazquez *et al.* 2007). In fact, various kinds of PE have been found to have multiple biological effects such as antiviral (Drago *et al.* 2007, Huleihel & Eshanu 2002, Orsolic & Basic 2005, Salomão *et al.* 2008, Viuda-Martos *et al.* 2008, Yao *et al.* 2004), antioxidant (Ahn *et al.* 2007, Kerem *et al.* 2006), antifungal (Quiroga *et al.* 2006, Silici & Koc 2006) and antibacterial (Huang *et al.* 2006, Velazquez *et al.* 2007) activities, mainly due to the presence of phenolic compounds and flavonoids (Viuda-Martos *et al.* 2008). Flavonoids, cinnamic acids and their ester derivatives have been found to be the most abundant and effective antimicrobial compounds in PE (Diem *et al.* 1999, Dustmann 1979, Erukhimovitch *et al.* 2002, Fontana *et al.* 2004, Fujiwara *et al.* 1990, Isla *et al.* 2001, Maquelin *et al.* 2003, Mariey *et al.* 2001, Martos *et al.* 1997, Naumann *et al.* 1991, Pascual *et al.* 1994, Roth *et al.* 1986, Salman *et al.* 2002).

Although our results didn't show any synergistic effect between CAPE and ampicillin, a significant additive effect between these agents can been seen (Table 2). These results may indicate that the mechanisms of action of CAPE and ampicillin are not related.

The exact mechanism of anti-bacterial activity of CAPE is still unclear. Our results showed that continuous treatment with CAPE for up to 6h caused a reversible antibacterial effect; however continuous treatment for longer periods of time caused irreversible inhibition of the bacterial growth (Figure 3). Also, the data presented in Figure 2 showed that 4h of treatment with CAPE did not affect the viability of the treated bacteria, while treatment for 6h or more caused complete bacterial death. It seems that this product has bacteriostatic effect during the first 5-6 h of treatment, while further treatment is required for a bacteriocidic effect.

Burdock (1998) attributes the antibacterial activity of PE to the aromatic acids and esters, while another study (Takaisi & Schilcher 1994) suggested it is due to flavonone pinocembrin, the flavonol galangin and CAPE, whose mechanism of action seems to rely on the inhibition of the bacterial RNA polymerase. Others have shown that other flavonoids, including galangin, can also have antibacterial activity (Cushnie & Lamb 2005). The mechanism of action involves degrading the bacterial cytoplasmic membrane, which causes a loss of potassium ions and further damage, leading to cell autolysis. Furthermore, quercetin, which is also found in PE, has been reported to increase membrane permeability and dissipate its potential, causing loss of the ATP synthesis capacity by the bacteria as well as dysfunction of membrane transport and mobility (Mirzoeva *et al.* 1997).

The effect of CAPE on the permeability of the bacterial outer membrane was assessed by the uptake of the crystal violet dye. Generally, crystal violet penetrates the outer membrane poorly, but it easily enters when the membrane is defective (Devi *et al.* 2010). Our results showed a dramatic enhancement in the uptake of crystal violet in gram (+) bacteria in a dose dependent mater, reaching over 95% uptake at the higher doses after 8h of treatment with CAPE (Figure 4A). In the case of gram (-) bacteria, there also was a signifi-

cant increase in gentian violet uptake at the higher doses of CAPE (Figure 4A), even though this increase was much lower compared to gram (+) bacteria. After a 2h short treatment with CAPE, only a slight increase in the gentian violet uptake in gram (+) and gram (-) bacteria was observed. This shows that CAPE alters membrane permeability and makes the cells hyperpermeable to solutes. These results are in agreement with our results showing significantly higher antibacterial activity of this product against gram (+), as compared to gram (-) bacteria. Taken together, these results together with our findings that CAPE treatment for over 6h caused a bacteriocidic effect may indicate a dramatic irreversible disruption of the bacterial outer membrane, leading to bacterial death.

The escape of UV-absorbing substances from the cells is an index of membrane cell disruption and probably cell lysis and nonselective pore formation (Maisnier-Patin *et al.* 1996, Zhou *et al.* 2008). As seen in Figure 4B, treatment of *S. aureus* (a gram (+) strain) with CAPE caused a high leakage of intracellular components compared to a lower leakage from the treated gram (-) bacteria. These results suggest that the effect of CAPE on gram (+) and partially on gram (-) bacteria could be the formation of pores, or even a more severe disruption of the plasma membrane.

In addition, Scazzocchio *et al.* (2006) has reported that PE suppresses the expression of different bacterial virulence factors such as lipase and coagulase, while producing an evident suffering state of bacterial cells.

In this study, CAPE effectively inhibited the growth of most of the examined gram (+) bacteria, while they only had a limited effect on part of the gram (+) and on all tested gram (-) bacteria. This selective effect may be due to the presence of a capsule around these bacteria, which can prevent the penetration of the tested products into the bacteria. Therefore, at very high doses of the products, only small amounts of these products can penetrate into the capsular bacteria and partially affect their growth.

FTIR spectroscopy is a potential method used to study spectral changes in living cells (Erukhimovitch *et al.* 2002, Fontana *et al.* 2004, Fujiwara *et al.* 1990). Infrared spectra of microbial cells reflect the biochemical structure and composition of the cellular constituents such as water, fatty acids, proteins, polysaccharides and nucleic acids. In our study, major spectral variations were observed in the frequencies between 1800 and $1000cm^{-1}$. More specifically, our results showed a significant increase in frequency at $1100cm^{-1}$ of the CAPE-treated bacteria. This may indicate an alteration in the polysaccharide content or distribution, probably of the bacterial envelope. This result supports our findings showing a disruption of the bacterial outer membrane as a result of treatment with CAPE (Figure 4A). In addition, there was another notable increase in the frequency at $1250cm^{-1}$ of the CAPE-treated bacteria. This peak, which represents phosphate vibrations (Diem *et al.* 1999) may reflect significant changes in the phosphates. This could result from deformation in membrane phospholipids, suggesting that CAPE alters the macromolecular structures in the membrane, which further results in the complete loss of its integrity. These findings are in agreement with our results pointing towards formation of pores or even more severe disruption of the plasma membrane in CAPE-treated bacteria (Figure 4B).

Although the obtained results in this study provided a significant contribution for understanding the antibacterial mechanism of CAPE action, still further research is required for elucidating its exact mechanism of action.

Acknowledgements

This study was supported the Ministry of Health, Israel. Number: 3-3931.

References

Alici O, Kavakli HS, Koca C, Altintas ND, Aydin M & Alici S 2015 Value of Caffeic Acid Phenethyl Ester Pretreatment in Experimental Sepsis Model in Rats. *Mediators of Inflammation* **2015** 810948

Abel C & Busia K 2005 An Exploratory Ethnobotanical Study of the Practice of Herbal Medicine by the Akan Peoples of Ghana. *Altern Med Rev* **10** 112-122

Ahn MR, Kumazawa S, Usui Y, Nakamura J, Matsuka M, Zhu F & Nakayama T 2007 Antioxidant activity and constituents of propolis collected in various areas of China. *Food Chem* **101** 1383-1392

Akao Y, Maruyama H, Matsumoto K, Ohguchi K, Nishizawa K, Sakamoto T, Araki Y, Mishima S & Nozawa Y 2003 Cell growth inhibitory effect of cinnamic acid derivatives from propolis on human tumor cell lines. *Biol Pharm Bull* **26** 1057-1059

Bonapace CR, Bosso JA, Friedrich LV & White RL 2002 Comparison of methods of interpretation of checkerboard synergy testing. *Diagn Microbiol Infect Dis* **44** 363-366

Burdock GA 1998 Review of the biological properties and toxicity of bee propolis. *Food Chem Toxicol* **36** 347-363

Cushnie TP & Lamb AJ 2005 Detection of galangin-induced cytoplasmic membrane damage in Staphylococcus aureus by measuring potassium loss. *J Ethnopharmacol* **101** 243-248

de Rezende GP, da Costa LR, Pimenta FC & Baroni DA 2008 In vitro Antimicrobial Activity of Endodontic Pastes with Propolis Extracts and Calcium Hydroxide: A Preliminary Study *Braz Dent J* **19** 301-305

Devi KP, Nisha SA, Sakthivel R & Pandian SK 2010 Eugenol (an essential oil of clove) acts as an antibacterial agent against Salmonella typhi by disrupting the cellular membrane. *J Ethnopharmacol* **130** 107-115

Diem M, Boydstom-White S & Chiriboga L 1999 Infrared spectroscopy of cells and tissues: shining light onto a novel subject. *Appl Spectroscopy* **53** 148-161

Drago L, De Vecchi E, Nicola L & Gismondo MR 2007 In vitro antimicrobial activity of a novel propolis formulation (Actichelated propolis). *J Appl Microbiol* **103** 1914-1912

Dustmann JH 1979 Antibacterial effect of honey. *Apiacta* **14** 7-11

Erukhimovitch V, Talyshinsky M, Souprun Y & Huleihel M 2002 Spectroscopic characterization of human and mouse primary cells, cell lines and malignant cells. *Photochem and Photobiol* **76** 446-451

Fontana R, Mendes MA, De Souza BM, Konno K, Cezar LM, Malaspina O & Palma MS 2004 Jelleines: a family of antimicrobial peptides from the Royal Jelly of honeybees (Apis mellifera) *Peptides* **25** 919-928

Fujiwara S, Imai J, Fujiwara M, Yaeshima T, Kawasgima T & Kobayasgi K 1990 A potent antibacterial protein in royal jelly. Purification and determination of the primary structure of royalisin 220. *J Biol Chem* **265** 11333-11337

Halberstein RA 2005 Medicinal Plants: Historical and Cross-Cultural Usage Patterns. *Ann Epidemiol* **15** 686-699

Harris GK, Qian Y, Leonard SS, Sbarra DC & Shi XL 2006 Luteolin and chrysin differentially inhibit cyclooxygenase-2 expression and scavenge reactive oxygen species but similarly inhibit prostaglandin-E2 formation in RAW264.7 cells. *J Nutr* **136** 1517-1521

Huang WZ, Dai XJ, Liu YQ, Zhang CF, Zhang M & Wang ZT 2006 Studies on antibacterial activity of flavonoids and diarylheptanoids from Alpinia katsumadai. *J Plant Resour Environ* **15** 37-40

Huleihel M & Eshanu V 2002 Anti-HSV effect of an aqueous extract of propolis. *Isr Med Assoc J* **4** 923-927

Ilhan A, Koltuksuz U, Ozen S, Uz E, Ciralik H & Akyol O 1999 The effects of caffeic acid phenethyl ester (CAPE) on spinal cord ischemia/reperfusion injury in rabbits. *Eur J Cardiothrac Surg* **16** 458-463.

Isla MI, Moreno MIN, Sampietro AR & Vattuone MA 2001 Antioxidant activity of Argentine propolis extracts. *J Ethnopharmacol* **76** 165-170

Kerem Z, Chetrit D, Shoseyov O & Regev-Shoshani G 2006 Protection of lipids from oxidation by epicatechin, trans-resveratrol, and gallic and caffeic acids in intestinal model systems. *J Agric Food Chem* **54** 10288-10293

Kosalec I, Pepeljnjak S, Bakmaz M & Vladimir-Knezevic S 2005 Flavonoid analysis and antimicrobial activity of commercially available propolis products. *Acta Pharm* **55** 423-430

Maisnier-Patin S, Forni E & Richard J 1996 Purification, partial characterization and mode of action of enterococcin EFS2, an anti-listerial bacteriocin produced by a strain of Enterococcus faecalis isolated from a cheese. *Inter J of Food Microbiol* **30** 255-270

Maquelin K, Kirschner C, Choo-Smith LP, Ngo-Thi NA, van Vreeswijk T, Stämmler M, Endtz HP, Bruining HA, Naumann D & Puppels GJ 2003 Prospective study of the performance of vibrational spectroscopies for rapid identification of bacterial and fungal pathogens recovered from blood cultures. *J Clin Microbiol* **41** 324-429

Mariey L, Signolle P, Amiel C & Travert J 2001 Discrimination, classification, identification of microorganisms using FTIR spectroscopy and chemometrics. *Vib Spectroscopy* **26** 151-159

Marquez N, Sancho R, Macho A, Calzado M, Fiebich B & Munoz E 2004 Caffeic acid phenethyl ester inhibits T-cell activation by targeting both nuclear factor of activated T-cells and NF-kappaB transcription factors. *J Pharmacol Exp Ther* **308** 993-1001

Martos I, Cossentini M, Ferreres F & Tomas-Barberan FA 1997 Flavonoid composition of Tunisian honeys and propolis. *J Agric Food Chem* **45** 2824-2829

Mirzoeva OK, Grisjanin RN & Calder PC 1997 Antimicrobial action of propolis and some of its components: the effects on growth, membrane potential and motility of bacteria. *Microbiol Res* **152** 239-246

Natarajan K, Singh S, Burke TJ & Grunberger D 1996 Caffeic acid phenethyl ester is a potent and specific inhibitor of activation of nuclear transcription factor NF-kappa B. *Proc Natl Acad Sci U S A* **93** 9090-9095

Naumann D, Helm D & Labischinski H 1991 Microbiological characterizations by FT-IR spectroscopy. *Nature* **351** 81-82

Onlen Y, Tamer C, Oksuz H, Duran N, Altug ME & Yakan S 2007 Comparative trial of different antibacterial combinations with propolis and ciprofloxacin on Pseudomonas keratitis in rabbits. *Microbiol Res* **62** 62-68

Orsolić N & Basić I 2005 Water-soluble derivative of propolis and its polyphenolic compounds enhance tumoricidal activity of macrophages. *J Ethnopharmacol* **102** 37-45

Pascual C, Torricella RG & Gonzalez R 1994 Scavenging action of propolis extract against oxygen radicals. *J Ethnopharmacol* **41** 9-13

Quiroga EN, Sampietro DA, Soberon JR, Sgariglia MA & Vattuone MA 2006 Propolis from the northwest of Argentina as a source of antifungal principles. *J Appl Microbiol* **101** 103-110

Roth LA, Kwan S & Sporns P 1986 Use of a disc-assay system to detect oxytetracycline residues in honey. *J Food Prot* **49** 436-441

Salman A, Erukhimovitch V, Talyshinsky M, Huleihil M & Huleihel M 2002 FTIR- spectroscopic method for detection of cells infected with herpes viruses. *Biopolymers* **67** 406-412

Salomão K, Pereira PR, Campos LC, Borba CM, Cabello PH, Marcucci MC & de Castro SL 2008 Brazilian Propolis: Correlation Between Chemical Composition and ntimicrobial Activity. *Evid Based Complement Alternat Med* **5** 317-324

Scazzocchio F, D'Auria FD, Alessandrini D & Pantanella F 2006 Multifactorial aspects of antimicrobial activity of propolis. *Microbiol Res* **161** 327-333

Silici S & Koc AN 2006 Comparative study of in vitro methods to analyse the antifungal activity of propolis against yeasts isolated from patients with superficial mycoses. *Lett Appl Microbiol* **43** 318-324

Song YS, Park EH, Hur GM, Ryu YS, Lee YS, Lee JY, Kim M & Jin C 2002 Caffeic acid phenethyl ester inhibits nitric oxide synthase gene expression and enzyme activity. *Cancer Letters* **175** 53-61

Takaisi NB & Scjoncjer H 1994 Electron microscopy and microcalorimetric investigations of the possible mechanism of the antibacterial action of a defined propole provenance. *Planta Med* **60** 222-227

Tolba MF, Azab SS, Khalifa AE, Abdel-Rahman SZ & Abdel-Naim AB 2013 Caffic acid phenethyl ester, a promising component of propolis with a plethora of biological activities: a review on its anti-inflammatory, neuroprotective, hepatoprotective, and cardioprotective effcts. *IUBMB Life* **65** 699-709

Velazquez C, Navarro M, Acosta A, Angulo A, Dominguez Z, Robles R, Robles-Zepeda R, Lugo E, Goycoolea FM, Velazquez EF, Astiazaran H & Hernandez J 2007 Antibacterial and free-radical scavenging activities of Sonoran propolis. *J Appl Microbiol* **103** 1747-1756

Viuda-Martos M, Ruiz-Navajas Y, Fernández-López J & Pérez-Alvarez JA 2008 Functional Properties of Honey, Propolis, and Royal Jelly. *J of Food Sci* **73** 1750-3841

Wu Y, Zhou C, Li X, Song L, Wu X, Lin W, Chen H, Bai H, Zhao J, Zhang R, Sun H & Zhao Y 2006 Evaluation of antiinflammatory activity of the total flavonoids of Laggera pterodonta on acute and chronic inflammation models. *Phytother Res* **20** 585-590

Yang C, Wu J, Zhang R, Zhang P, Eckard J, Yusuf R, Huang X, Rossman TG & Frenkel K 2005 Caffeic acid phenethyl ester (CAPE) prevents transformation of human cells by arsenite (As) and suppresses growth of As-transformed cells. *Toxicology* **213** 81-96

Yao L, Jiang YM, D'Arcy B, Singanusong R, Datta N, Caffin N & Raymont K 2004 Quantitative high-performance liquid chromatography analyses of flavonoids in Australian Eucalyptus honeys. *J Agric*

Food Chem **52** 210-214

Zhou K, Zhou W, Li P, Liu G, Zhang J & Dai Y 2008 Mode of action of pentocin 31-1: an antilisteria bacteriocin produced by Lactobacillus pentosus from Chinese traditional ham. *Food Control* **19** 817-822

5

Crystal structure and molecular docking studies of benzo[8]annulenes as potential inhibitors against *Mycobacterium tuberculosis*

RA Nagalakshmi[1], J Suresh[1], S Maharani[2] and R Ranjith Kumar[2]

[1]Department of Physics, The Madura College (Autonomous), Madurai, 625 011, India
[2]Department of Organic Chemistry, School of Chemistry, Madurai Kamaraj University, Madurai, 625 021, India

Correspondence should be addressed to J Suresh; E-mail: ambujasureshj@yahoo.com

Abstract

Tuberculosis is a disease caused by *Mycobacterium tuberculosis*. The bacterial cell wall has a characteristic low permeability, which essentially makes antibiotics ineffective. The cell wall material must be regulated so that its deposition does not compromise its structure. In this study, two new inhibitors, 2-amino-4-(4-cholorophenyl)-5,6,7,8,9,10-hexahydrobenzo[8] annulene-1,3,3(4H)-tricarbonitrile(Ia) and 2-amino-4-(4-bromophenyl)-5,6,7,8,9,10-hexahydrobenzo[8] annulene-1,3,3(4H)-tricarbonitrile(Ib) were synthesized. The crystal structures of the above compounds were determined by single crystal X-ray diffraction. The compounds $C_{21} H_{19} Cl N_3$ (Ia) and $C_{21} H_{19} Br N_3$ (Ib) were crystallized in the monoclinic and triclinic system. In both compounds, the cyclohexane ring was found to adopt a boat conformation. The cyclooctane ring of both compounds adopted a twisted chair-chair conformation. *In silico* analyses revealed that both compounds showed good anti-mycobacterial activities against the enoyl-acyl carrier enzyme and the N-acetyl-gamma protein, both of which are critical for bacterial survival. Synthesis, structure determination, conformation, intra, inter-molecular interactions and docking studies of both compounds are presented herein.

Introduction

Tuberculosis (TB) is an infection caused by slow-growing bacteria and remains a leading cause of human suffering and death (Tiruviluamala *et al.* 2002, Smith 2003, Schluger *et al.* 2005). *Mycobacterium tuberculosis* has a unique membrane structure composed largely of lipids that have long-chain fatty acids, called mycolic acids.

The enoyl-acyl carrier protein reductase enzyme (InhA) catalyzes the NADH-dependent reduction of unsaturated, long chain, β-branched fatty acids (mycolic acids), which are essential for bacterial cell wall synthesis. *Mycobacterium tuberculosis* InhA is a fundamental target for anti-tuberculosis intervention (Bradford *et al.* 1998, Dessen *et al.* 1995, Rozwarski *et al.* 1998). Inhibition of the enzyme leads to an increased bacterial vulnerability to external oxidative attacks and ultimately to bacterial death.

The enzyme N-acetyl-gamma-glutamyl-phosphate reductase (AGPR) catalyzes the nicotina-

mide adenine dinucleotide phosphate (NADPH)-dependent reductive dephosphorylation of N-acetyl-gamma-glutamyl-phosphate to N-acetylglutamate-gamma-semialdehyde. This reaction is the third step in the argentine-biosynthetic pathway (Cybis & Davis 1975) that is essential for some microorganisms and plants, and in particular for *Mycobacterium tuberculo-*

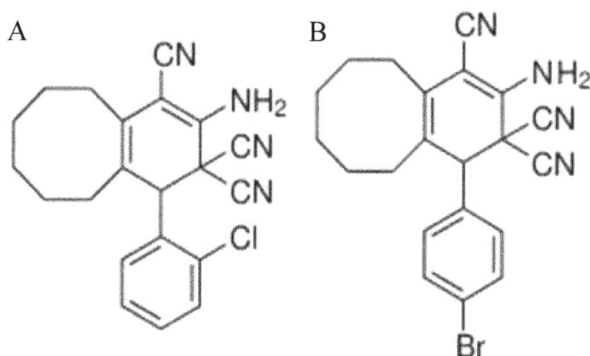

Figure 1. Chemical diagram of the molecule (Ia) (A) and (Ib) (B).

A

B

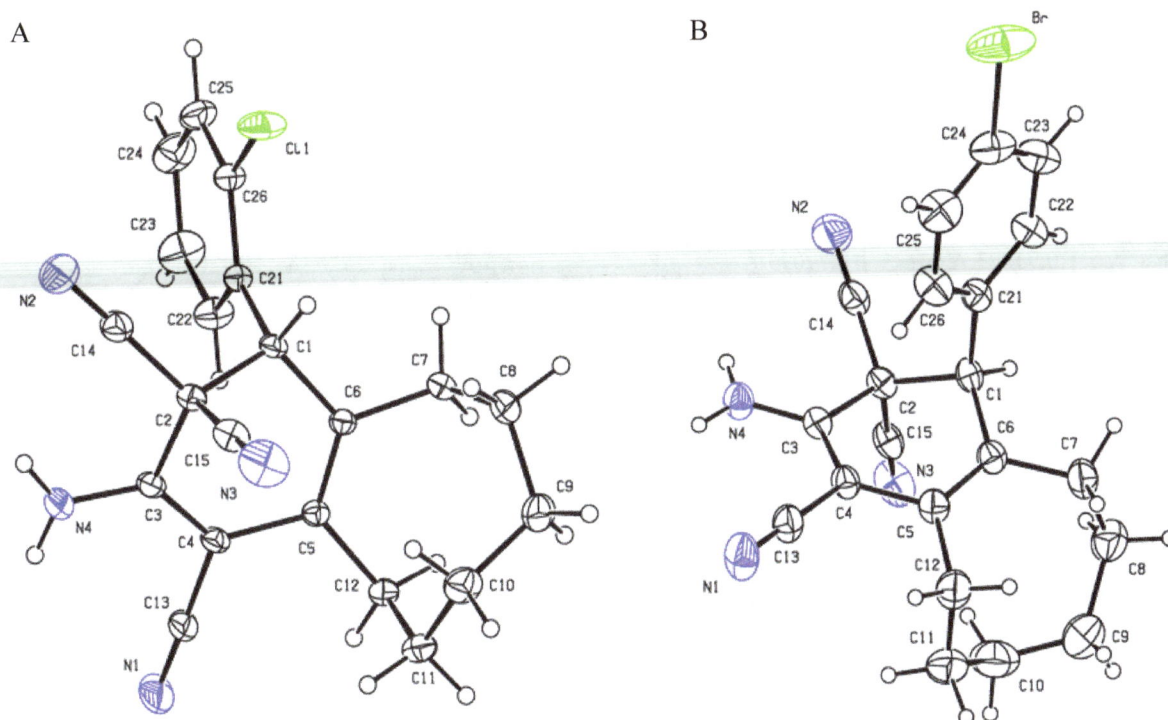

Figure 2. The molecular structure of compounds (Ia) (A) and (Ib) (B), showing the atom numbering scheme. Displacement ellipsoids are drawn at 30% probability level, using Platon (Spek 2008).

sis. Inhibition of this protein leads to inhibition of the argentine biosynthesis.

Isoniazid is a narrow spectrum anti-mycobacterial agent. Activated isoniazid causes inhibition of mycolic acid synthesis. A series of 2-benzylsulfanyl derivatives of benzooxazole and benzothiazole have shown their anti-mycobacterial activity against *Mycobacterium tuberculosis* (Koci *et al.* 2002). Sanna *et al.* (2000) synthesized a series of aryl substituted-[1H(2H)benzotriazol-1(2)-yl]acrylnitriles and have also reported anti-mycobacterial activity of the latter. The emergence of multi drug resistant TB (*MDR-TB*) and extensively drug resistance (XDR-TB) makes the treatment ineffective. The drug resistant and HIV co-infection have resulted in the need for new anti-tuberculosis drugs. Continuing our work in developing new drug-like agents for *Mycobacterium tuberculosis* (Nagalakshmi *et.al.* 2014, Suresh *et al.* 2012, Vishnupriya *et al.* 2014), two new benzo[8]annulene compounds were synthesized. The structures of both compounds were studied using single crystal X-ray diffraction. Here we report the synthesis and single crystal X-ray studies of 2-amino-4-(4-cholorophenyl)-5,6,7,8,9,10-hexahydrobenzo[8]annulene-1,3,3(4H)-tricarbonitrile (Ia) and 2-amino-4-(4-bromo phenyl)-5,6,7,8,9,10-hexahydrobenzo[8]annulene-1,3,3(4H)-tricarbonitrile (Ib) (Figure 1) together with docking studies.

Materials and Methods

Preparation of compound (Ia)

A mixture of cyclooctanone (1 mmol), 2-choloro benzaldehyde (1 mmol) and malononitrile (2 mmol) was solvated in ethanol to which NaOEt (0.5 mmol) was added. All chemicals were purchased from Sigma-Aldrich (St. Louis, MO) and were used without further purification. The reaction mixture was heated under reflux for 2–3 hours. As soon as the completion of the reaction was confirmed by thin layer chromatography, the remaining solid mater was filtered and dried. The solid was then crystallized from ethyl acetate, which yielded colorless crystals. The melting point was 470 K and yield was 85%.

Preparation of compound (Ib)

A mixture of cyclooctanone (1 mmol), 4-bromo benzaldehyde (1 mmol) and malononitrile (2 mmol) was taken in ethanol to which NaOEt (0.5 mmol) was added. The reaction mixture was heated under reflux for 2-3 hours. As soon as the completion of the reaction was confirmed by thin layer chromatography, the remaining solid mater was filtered and dried. The solid was then crystallized from ethyl acetate, which yielded colorless crystals. The melting point was 461 K and yield was 85%.

Table 1. The crystal data, intensity data collection and structure solution and structure refinement parameters of compounds (Ia) and (Ib).

Empirical formula	$C_{21} H_{19} Cl N_4$	$C_{21} H_{19} Br N_4$
Formula weight	362.85	407.31
Temperature	293(2) K	293(2) K
Wavelength	0.71073 Å	0.71073 Å
Crystal system / space group	Monoclinic/ $P2_1/c$	Triclinic/ P-1
Unit cell Dimensions	a = 13.630(11) Å	a = 7.4208(5) Å
	b = 10.475(8) Å	b = 10.160(6) Å
	c = 13.8690(4) Å	c = 13.955(8) Å
	$\alpha = 90°$	$\alpha = 70.635(3)°$
	$\beta = 107.403(2)°$	$\beta = 77.70(3)°$
	$\gamma = 90°$	$\gamma = 87.80(3)°$
Volume	1889.6(3) Å3	969.35(11) Å3
Z/ Density (calculated)	4/ 1.275 mg/m^3	2/1.395 mg/m^3
Absorption coefficient	0.214 mm^{-1}	2.132 mm^{-1}
F(000)	760	416
Crystal size	0.21x0.19 x0.18 mm^3	0.20 x 0.19 x 0.17 mm^3
Theta range for data collection	2.48 to 30.17°.	2.81 to 25.49°.
Limiting indices	-19<=h<=19, -14<=k<=14, -16<=l<=19	-8<=h<=8, -12<=k<=12, -16<=l<=16
Reflections collected	25664	19116
Independent reflections	5443 [R(int) = 0.0363]	3591 [R(int) = 0.036]
Completeness to theta = 25.50°	99.9 %	99.9 %
Refinement method	Full-matrix least-squares on F^2	Full-matrix least-squares on F^2
Data / restraints / parameters	5443 / 2 / 235	3591 / 1 / 235
Goodness-of-fit on F2	1.025	1.05
Final R indices [I>2sigma(I)]	R1 = 0.0491, wR2 = 0.1068	R1 = 0.0534, wR2 = 0.1252
R indices (all data)	R1 = 0.0898, wR2 = 0.1280	R1 = 0.0926, wR2 = 0.1425
Largest diff. peak and hole	0.214 and -0.303 e.Å-3	0.782 and -1.342 e.Å-3

Structure Determination and Refinement

X-ray diffraction intensity data were collected for compounds (Ia) and (Ib) on the Bruker Smart Apex II single crystal X-ray diffractometer, equipped with graphite mono-chromated (MoKα) λ=0.7103 Å radiation and CCD detector. Crystals were cut to a suitable size and mounted on a glass fibre using cyanoacrylate adhesive. The unit cell parameters were determined

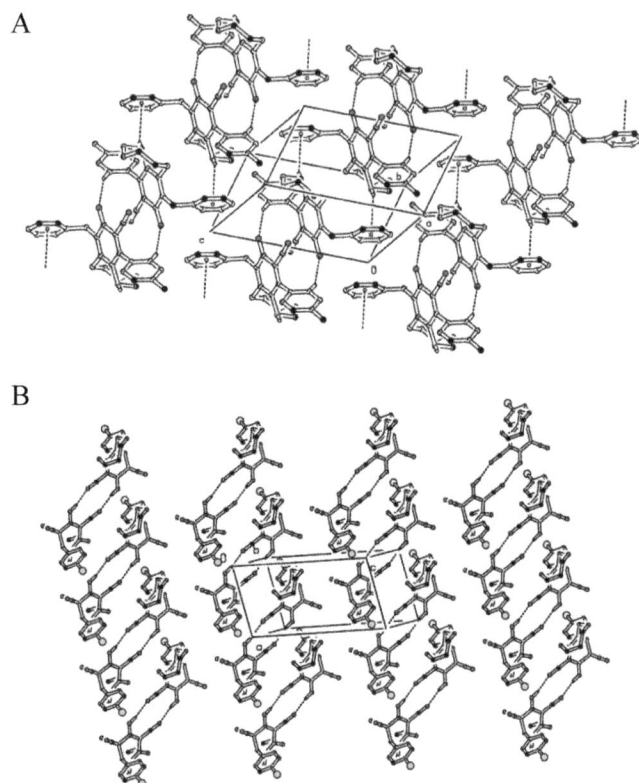

A

B

Figure 3: The inversely related molecules forms a ring motif $R_2^2(12)$ which are linked through Van der Waal's interactions. A, compound (Ia); B, compound (Ib).

from 36 frames measured (0.5° ω and φ scans) from three different crystallographic zones, using the method of difference vectors. The intensity data collection, frames integration, Lorentz-Polarization correction and decay correction were performed using *SAINT* (Bruker 2004). Empirical absorption correction multiscan was performed using the *SADABS* (Bruker 2004) program. The crystal structures of both compounds were obtained by direct methods using SHELXD2013. The structures were refined by the full-matrix least-squares method using SHELXL2013 (Sheldrick 2008). All non-hydrogen atoms were refined using anisotropic temperature factors. The final difference Fourier maps for all compounds at this stage were critically inspected and the hydrogen atoms were located and included in the refinement, only with isotropic temperature factors. The accuracy of the crystal structures was evidenced from the final residual *R*- and *wR*- factors and other parameters including estimated standard deviations in the values of bond length and bond angles, 'data-to-parameter' ratio, etc. Details of the crystal data, data collection and refinement are given in Table 1.

Docking studies

The coordinates of enoyl acyl carrier protein (2NSD)

Table 2. Hydrogen bonds [Å and °] of compound (Ia). Symmetry transformations used to generate equivalent atoms: *(-x, - y, - z)*.

D-HA	d(D-H)	...d(HA)	...d(DA)	<DHA
C1-H1-Cl1	0.98	2.54	3.087 (15)	115
N4-H4A-N1 [(i)]	0.86	2.16	3.015(2)	170

Table 3. Hydrogen bonds [Å and °] of compound (Ib). Symmetry transformation used to generate equivalent atoms: *(1 - x, 1 - y, -z)*.

D-HA	d(D-H)	...d(HA)	...d(DA)	<DHA
N4-H4A-N1 [(i)]	0.86	2.13	2.974(5)	169

www.ebi.ac.uk/pdbsum/).

with Nad and N-acetyl-gamma-glutamyl-phosphate reductase (2NQT) were retrieved from the RCSB protein data bank (http://www.pdb.org). The target protein of 2NSD contains two monomers. Only one monomer was selected for docking analysis. The protein structures were cleaned using the whatif online server (http://swift.cmbi.ru.nl/whatif/). The protein's active site pocket was identified using 'Computed Atlas of Surface Topography' (http://sts.bioe.uic.edu/castp/). Preparation of protein and ligand input structures and the definition of the binding sites were carried out under a GRID-based procedure. A rectangular grid box was constructed over the protein with grid points 90X90X90 Å^3 separated by 0.375 Å under the docking procedure. The lowest energy cluster returned by AutoDock for each compound was considered and used for further analysis. Consequent runs were set up to 100 for each inhibitor. All other parameters were maintained at their default settings. All the docking result visualizations are achieved by using the 'PDBsum Generate' online server (https://

Results

The molecular structures of compounds (Ia) and (Ib) are shown in Figure 2. The two compounds differ in the nature of the substituent at position 2 and 4 of the phenyl ring. This simple change in the structures results in the change of structure types. In both compounds, the cyclohexane ring adopts a boat conformation with puckering parameters Q = 0.445(17) Å, Θ = 65.5(2)° and Φ = 39.4(2)° in compound (Ia); Q = 0.406(4) Å, Θ = 115.6(6)° and Φ = 210.4(7)° in compound (Ib). The cyclooctane ring adopts a twisted chair-chair conformation in compounds (Ia) and (Ib) as found in related structures (Fun *et al.* 2010, Suresh *et al.* 2007). The triple-bond characters of C13 ≡ N1, C14 ≡ N2 and C15 ≡ N3 [1.138(2) Å, 1.131(2) Å, 1.129(2) Å in compound (Ia) and 1.137(5) Å, 1.126(5) Å, 1.127(5) Å in compound (Ib), respectively] as well as the bond angles of C4-C13≡N1, C2-C14≡N2 and C2-C15≡N3 [179.2(2)°, 178.5(2)°, 178.2(2)° in compound (Ia) and

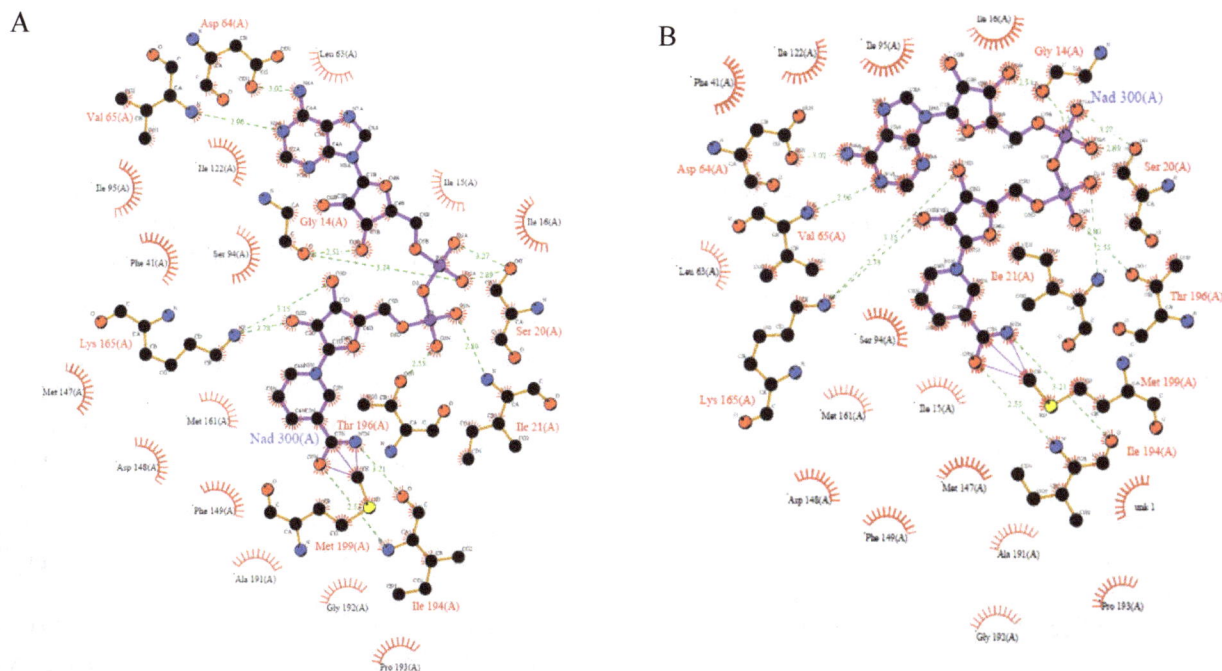

Figure 4: Enoyl acyl carrier protein with Nad interactions for compound (Ia) (A) and (Ib) (B).

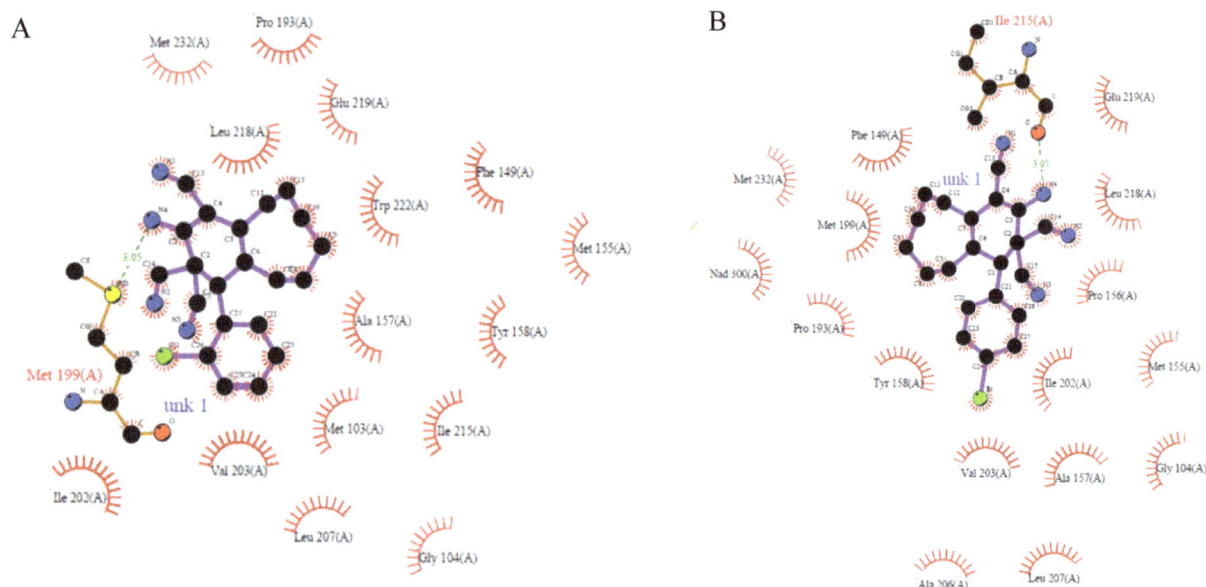

Figure 5. Interactions of compound (Ia) (A) and (Ib) (B) with protein.

179.2(5)°, 178.9(4)°, 178.7(4)° in compound (Ib), respectively] in both compounds define the linearity of the cyano carbonitrile compounds. The phenyl substituent of the cyclohexane ring at C1 has a negative synclinal conformation in compound (Ia) and negative anticlinal conformation in compound (Ib), as evidenced by the C2–C1–C21–C22 [-83.67° in compound (Ia) and -112.5° in compound (Ib) torsion angles]. The aryl ring in the structure (Ib) is not coplanar with the mean plane of the cyclohexane ring. This lack of coplanarity is caused by non-bonded interactions between one of the ortho H atoms in the aryl ring and a hydrogen atom at the C1 position of the cyclohexane ring. Steric repulsions are reduced by the expansion of the C1–C21– C26 angle in structure (Ib).

Crystal packing

Inter molecular hydrogen bond geometry of compound (Ia) and compound (Ib) are listed in Table 2. In the crystal structure of compound (Ia) the N4 atom of the cyclohexane ring is involved in the intermolecular interaction N4—H4A···N1$^{(i)}$ with the N1 atom (Table 2). This forms a $R_2^2(12)$ ring motif (Bernstein *et al.* 1995) [symmetry code: (i) (- x, - y, – z)]. These adjacent ring motifs are linked together via Van der Waal's interactions as shown in Figure 3.

    In the crystal structure of the compound (Ib) the intermolecular interaction N4—H4A···N1$^{(i)}$ (with the N1 atom) (Table 2) forms a R_2^2 ring motif (12) (Bernstein *et al.* 1995) [symmetry code: (i) (2 - x, 1 - y, – z)]. These adjacent ring motifs are linked together by van der Waal's interactions, as shown in Figure 3.

    The N—N distance is longer in compound (Ia) [3.015(2) Å] compared to (Ib) [2.974(5) Å] and the

N—H···N angle in compounds (Ia) and (Ib) are 169° and 170°, respectively. This shows that the 4-bromine substituent forms a stronger hydrogen bond than the 2-cholorine substituent.

Docking analysis

The target protein structure of 2NSD with Nad was docked with the synthesized compounds using Auto-Dock4v2 (Goodsell 1998). The synthesized compounds were found to display good binding affinity to the receptor with minimum binding energy equal to - 9.47 and -10.70 for (Ia) and (Ib), respectively. The Nad interactions of the compounds are shown in Figure 4.

    In compound (Ia), the nitrogen atom of the benzo ring is hydrogen bonded to MET[199], one of the catalytic residues in the InhA active site, as shown in Figure 4. Rozwarski *et al.* (1999) have shown that hydrophobic aminoacids of the loop are important for proper substrate binding into the cavity. Interestingly, the last few carbon atoms of the fatty acid interact with the hydrophobic amino acids Ala[198], Met[199], and Ile[202]. Our (Ia) inhibitor is able to directly interact with one of these residues (Met[199]), leads to a defined loop structure and has hydrophobic interactions to the important loop residues of InhA, resulting in a slow tight binding inhibition. In our (Ib) compound, the nitrogen atom of the benzo ring is hydrogen bonded to Ile[215], one of the catalytic residues in the InhA active site and has hydrophobic interactions with Ala[198] and Met[199], as shown in Figure 5.

    The target protein structure of 2NQT was docked with the synthesized compounds using Auto-Dock4v2 (Goodsell, 1998). The synthesized compounds were found to display good binding affinity to

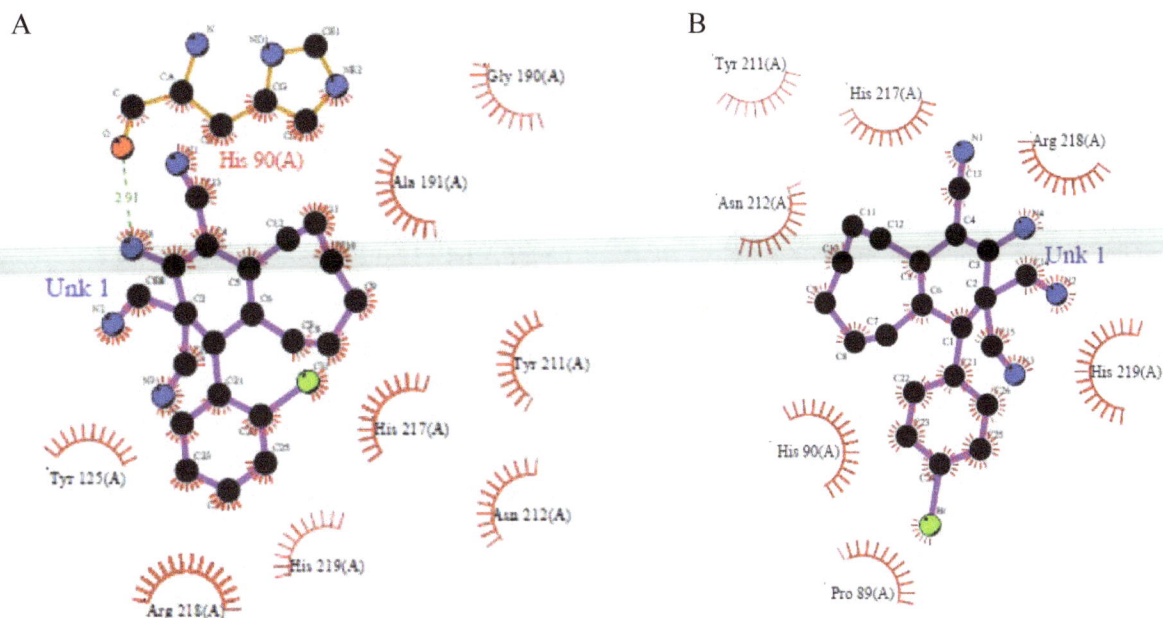

Figure 6. Interactions of compounds (Ia) (A) and (Ib) (B) with protein.

the receptor with minimum binding energy equal to -6.25 and -7.18, for compounds (Ia) and (Ib), respectively.

In compound (Ia), the nitrogen atom of the benzo ring is hydrogen bonded to His[90], where as in compound (Ib), the ligand has hydrophobic interactions with the protein (Figure 6).

Conclusion

Our goal is to propose new TB inhibitors. In this direction, we synthesized two novel compounds of benzoannulene derivatives. The compounds were docked with the enoyl-aceyl carrier protein and the N-acetyl-gamma-glutamyl-phosphate reductase. Their free binding energies were evaluated. Both compounds showed good binding affinity, which is indicative of stable, energetically viable complexes. Hence, our synthesized ligands bear promising *in silico* anti-tubercular activities. From the docking analysis of the two compounds with both protein receptors, it was found that (Ib) shows better binding with the receptors, as compared to (Ia).

Authors' Contribution

Authors 1 and 2 were involved in the data collection, crystallography and docking work. Authors 3 and 4 were involved in the synthesis and crystal growth of the title compounds. First author is the second author's student and third author is the fourth author's student.

Acknowledgements

JS and RAN thank the management of The Madura College (Autonomous), Madurai for their encouragement and support. RRK thanks the University Grants Commission, New Delhi, for funds through Major Research Project F. No. 42-242/2013 (SR). SM thanks the university grants commission, New Delhi for the fellowship under the UGC-BSR-JRF meritorious scheme.

References

Bernstein J, Lott WA, Steinberg BA & Yale HL 1952 Chemotherapy of experimental tuberculosis. *Am Rev Tuberc* **65** 357-374

Bradford WZ & Daley CL 1998 Multiple drug-resistant tuberculosis. *Infect Dis Clin North Am* **12** 157 -172

Bruker 2004 APEX2 and SAINT. Bruker AXS Inc., Madison, Wisconsin, USA.

Cybis J & Davis RH 1975 Organization and control in the arginine biosynthetic pathway of Neurospora. *J Bacteriol* **123** 196-202

Dessen A, Quémard A, Blanchard JS, Jacobs WR Jr & Sacchettini JC 1995 Crystal structure and function of the isoniazid target of Mycobacterium tuberculosis. *Science* **267** 1638-1641

Fun HK, Yeap CS, Ragavan RV, Vijayakumar V & Sarveswari S 2010 4,5,6,7,8,9-Hexahydro-2H-cycloocta[c]pyrazol-1-ium-3-olate. *Acta Crystallogr Sect E Struct Rep Online* **66** o3019

Goodsell DS, Morris GM & Olson AJ 1998 Automated docking of flexible ligands: Applications of autodock. *J Mol Recog* **9** 1-5

Koci J, Klimesova V, Waisser K, Kaustova J,Dahse H-M & Mollmann U 2002 Heterocyclic benzazole derivatives with antimycobacterial in vitro activity. *Bioorg Med Chem Lett* **12** 3275-3278

Nagalakshmi RA, Suresh J, Maharani S & Ranjith Kumar R 2014 Crystal structure and molecular docking studies of octahydrocycloocta[b]pyridine-3-carbonitriles as potential inhibitors against Mycobacterium tuberculosis. *J Mol Biochem* **3** 77-84

Rozwarski DA, Grant GA, Barton DH, Jacobs WR Jr & Sacchettini JC 1998 Modification of the NADH of the isoniazid target (InhA) from Mycobacterium tuberculosis. *Science* **279** 98-102

Rozwarski DA, Vilchèze C, Sugantino M, Bittman R & Sacchettini JC 1999 Crystal structure of the Mycobacterium tuberculosis enoyl-ACP reductase, InhA, in complex with NAD+ and a C16 fatty acyl substrate. *J Biol Chem* **274** 15582-15589

Sanna P, Carta A & Nikookar ME 2000 Synthesis and antitubercular activity of 3-aryl substituted-2-[1H(2H) benzotriazol-1(2)-yl]acrylonitriles. *Eur J Med Chem* **35** 535-543

Schluger NW, 2005 The pathogenesis of tuberculosis: the first one hundred (and twenty-three) years. *Am J Respir Cell Mol Biol* **32** 251-256

Sheldrick GM 2008 A short history of *SHELX*. *Acta Cryst* A **64** 112-122

Smith I 2003 Mycobacterium tuberculosis pathogenesis and molecular determinants of virulence. *Clin Microbiol Rev* **16** 463-496

Spek AL 2009 Structure validation in chemicalcrystallography. *Acta Cryst* D **65** 148-155

Suresh J, Vishnu Priya R, Sivakumar S & Ranjith Kumar R 2012 Spectral analysis and crystal structure of two substituted spiro acenaphthene structures. *J Mol Biochem* **1** 183-188

Tiruviluamala P & Reichman LB 2002 Tuberculosis. *Annu Rev Public Health* **23** 403-426

Vishnupriya R, Kowsalyadevi AVKM, Suresh J, Maharani S & Kumar RR 2014 Crystal structure and docking studies of hexahydrocycloocta[b]pyridine-3-carbonitriles. *J Mol Biochem* **3** 50-57

6

Molecular dynamics simulations through GPU video games technologies

Styliani Loukatou[1,†], Louis Papageorgiou[1,2,†], Paraskevas Fakourelis[1,3], Arianna Filntisi[1,4], Eleftheria Polychronidou[1,6], Ioannis Bassis[1], Vasileios Megalooikonomou[3], Wojciech Makałowski[5], Dimitrios Vlachakis[1,3,7] and Sophia Kossida[1]

[1]Computational Biology & Medicine Group, Biomedical Research Foundation, Academy of Athens, Soranou Efessiou 4, Athens 11527, Greece
[2]Department of Informatics and Telecommunications, National and Kapodistrian University of Athens, University Campus, Athens 15784, Greece
[3]Department of Computer Engineering and Informatics Faculty, University of Patras, Patras 26500, Greece
[4]School of Electrical and Computer Engineering, National Technical University of Athens, Athens, Greece
[5]Institute of Bioinformatics, University of Münster, Niels-Stensen-Straße 14, Münster 48149, Germany
[6]Department of Informatics, Ionian University, Tsirigoti Square 7, Corfu 49100, Greece
[7] Bionetwork ltd. 15234, Chalandri, Athens, Greece

Correspondence should be addressed to Sophia Kossida; E-mail: skossida@bioacademy.gr

Abstract

Bioinformatics is the scientific field that focuses on the application of computer technology to the management of biological information. Over the years, bioinformatics applications have been used to store, process and integrate biological and genetic information, using a wide range of methodologies. One of the most *de novo* techniques used to understand the physical movements of atoms and molecules is molecular dynamics (MD).

MD is an *in silico* method to simulate the physical motions of atoms and molecules under certain conditions. This has become a state strategic technique and now plays a key role in many areas of exact sciences, such as chemistry, biology, physics and medicine. Due to their complexity, MD calculations could require enormous amounts of computer memory and time and therefore their execution has been a big problem. Despite the huge computational cost, molecular dynamics have been implemented using traditional computers with a central memory unit (CPU).

A graphics processing unit (GPU) computing technology was first designed with the goal to improve video games, by rapidly creating and displaying images in a frame buffer such as screens. The hybrid GPU-CPU implementation, combined with parallel computing is a novel technology to perform a wide range of calculations. GPUs have been proposed and used to accelerate many scientific computations including MD simulations. Herein, we describe the new methodologies developed initially as video games and how they are now applied in MD simulations.

Introduction

The sharp increase in biological data, the need for its analysis and the increased interest of the scientific community to understand the structural and biological processes of biomolecules were the main reasons for the development of bioinformatics. The first milestone of this field was in 1970 when Paulien Hogewed studed information processes in biotic systems (Hesper & Hogeweg 1970). Later in 1980s, the term "bioinformatics" was used in genome analysis data to refer to *in silico* methods (Fago *et al.* 1992). Nowadays, bioinformatics is an interdisciplinary science incorporating computer science, biology, math and physics that develops novel methods applied in biological data. The biological knowledge that we can extract using *in silico* methods is significant and can be generated rapidly, as opposed to traditional biological methods (Balatsos *et al.* 2012, Palaiomylitou *et al.* 2008, Sellis *et al.* 2009, Vlachakis 2009).

In order to understand the cellular and biomolecular activities, the biological data must be combined to form a comprehensive picture of these activities (Attwood *et al.* 2011). Bioinformaticians have been developing computationally intensive techniques that can process biological data, including nucleotide sequences, amino acid sequences, 3D structures and biological signals and images. Major research efforts in the field include pattern recognition, sequence alignment, protein structure analysis (alignment – prediction), phylogenetic analysis, molecular dynamics, homology modelling, genome analysis, genome wide association studies, drug design and drug discovery. Furthermore, there are many predictive techniques specific to gene expression and protein-protein interaction. Herein, we focus on MD techniques.

The first step in MD was the study of the macroscopic dynamic spheres by theoretical physicists in the late 1950s (Alder & Wainwright 1959). Later, given the theory that atoms and molecules are allowed to interact, MD simulations were applied to describe their physical movements in the context of N-body simulations, something that would be very difficult before the invention of computers (Mesirov *et al.* 1996). MD simulations can be extracted from two parameters. The first parameter, the potential energy function is estimated from the energy that grows between the bonded and non-bonded atoms in a molecule. The second parameter is the dynamics function that is computed from the forces between the atoms. Additionally, in order to achieve better accuracy, in many *in silico* methods, the kinetic energy of atoms in a specific temperature is also calculated.

MD simulation techniques are important methods for discovering the natural 3D structure and function of biological macromolecules. During the last 10 years a great effort has been used to link the structure and the function of biomolecules. Dynamic models provide significant information about the internal motion and conformational changes of a macromolecule, which play a major role in its function. Despite the inflexible structure models constructed to understand the physical shape of macromolecules in 3D space, MD simulations are applied to get more information about their functionality. Furthermore, MD simulations models provide the structural dynamic properties of molecules and the conformational transitions that they undergo (Karplus & McCammon 2002) such as molecular prediction and drug design (Sellis *et al.* 2012, Vangelatos *et al.* 2009, Vlachakis *et al.* 2012).

The number of studies that use MD simulations has increased during the last 10 years due to the growing availability of programs and computer power. However, the design of a dynamic model still requires a large amount of computer power. The duration of a simulation must be relevant to the timescale of the natural process being studied. A simulation of (n) particles has a complexity of $O(n^2)$, if all pairwise electrostatic and van der Waals interactions are accounted for explicitly. Therefore, the simulation of a macromolecule dynamic model in a traditional computer with a CPU requires a total time spanning from days to years. In order to reduce time cost, scientists have applied electrostatic methodologies. A novel idea has been applied in the last years, incorporating video games technology such as parallel programming and GPUS on molecular dynamic simulations (Vlachakis & Kossida 2013, Vlachakis *et al.* 2012).

Many available technologies were invented to serve purposes different to which they are currently used for. Despite the fact that GPU-CPU implementation and parallel computing were developed as video games technologies, it became clear that they are the

means to perform a wide range of calculations and therefore have been applied in many scientific fields to solve computational problems. A GPU has more transistors than the average CPU. The application of parallel programming using GPUs in MD simulations has the significant benefit of a time cost significantly reduced by many times, as compared to CPUs. In the present review we describe the general information about the usage of GPU and parallel programming in the MD field.

The advent of the GPU

Assuming that x=GP; y=U; z=(x (√y))², then z equals what? By performing these arithmetic operations on x*y (GPU), the correct answer is z=GPGPU. In the early 80's, Intel created the precursor of graphics processing unit, the iSBX 275 Video Graphics Controller Multimodule Board, designed for industrial systems based on the Multibus standard. After almost twenty years, in 1999, GPU emerged, with NVIDIA marketing the GeForce 256 as "the world's first GPU". It is a circuit similar to CPU that is composed of a processor, a RAM and an I/O system, designed specifically for performing the complex mathematical and geometric calculations that are necessary for graphics performance (Kirk & Hwu 2010). The main target of this creation was the video game industry, which had a need for a rapid processor to produce images and graphics. Initially, computer graphics were in two dimensions (2D) and after several years, 3D computer games were developed. Some of the fastest GPUs have more transistors than the average CPU. Nowadays, every widely used technological innovation has a GPU installed (Alerstam *et al.* 2010, Hoang *et al.* 2013, Sharma *et al.* 2011). One picture is worth a thousand words; that is probably the phrase that Nvidia's people had in mind back in 2007, when they decided to develop CUDA, *i.e.* compute unified device architecture. CUDA is a framework for parallel computing and programming, which allows the developers to use GPUs not solely for graphics processing but for overall use, known as GPGPU, or general-purpose GPU. GPGPU was a simple idea, but the principal purpose of the construction of GPUs made this a difficult conception.

Molecular dynamics simulations: from CPU to GPU

The main apothegm of the comparison between GPU and CPU is that the architecture of the former focuses in executing a number of simultaneous threads, while the CPU executes only one thread in a heartbeat, which gives GPU a head start (Owens *et al.* 2008). As a result, it is feasible for a single GPU to replace several CPU cluster nodes and make parallel and quick calculations without waiting for shared resources. That allows for a better execution of complex algorithms and a more efficient management of big data. Furthermore, a graphics unit has less financial and electrical power requirements, so the creation of a supercomputer using GPUs instead of CPUs indemnifies the need for immense rooms filled with computers. The installation of a second GPU duplicates the parallelism of programming. Thus, it is now possible to fit a GPU supercomputer in a university lab (Stone *et al.* 2010).

Additionally, the GPU has impressive floating-point capabilities and high memory bandwidth. It gives the opportunity of optimized memory access, controlled execution configuration and direct management of the cores through a few lines of code. Towards this end, the direct connection between GPUs and motherboards leads to a more trustworthy performance (Ueng *et al.* 2008). The intelligent, innovative technology of GPUs was extended beyond the game industry so as to be utilized by the sciences. This gave computational chemistry and biology researchers the opportunity to expand the horizons of scientific exploration. Specifically, MD, quantum chemistry, visualization and docking applications run even 5 times faster on a GPU's CUDA cores in comparison to CPU's cores (Murakami *et al.* 2010).

MD simulations require a lot of complex arithmetic operations that sometimes lead to numerical errors, result bias, wrong reports and even non-polyonimic (NP) complete problems (Browna *et al.* 2012). Before the discovery of CUDA and GPGPU, GPUs have been used only for the imaging procedure of the molecular structures and the execution of MD simulations could take from hours to days to finish. The solution was provided by the GPGPU, as the GPUs have a lot of arithmetic units which can work in parallel, each doing an arithmetic operation while collaborating with each other. From the beginning of GPU coding to present, the programming rush continues to grow dramatically. The number of GPU based applications has almost reached 200, which is a 60 per cent increase in two years. The top ranked GPU applicable programs have been designed for MD, drug design, quantum chemistry, climate, weather and physics (NVIDIA Corporation, 2014).

In the field of MD, programmers and scientists have taken several steps towards this direction. Many programs based on GPGPU have been developed, supporting even multi-GPU simulations. This innovation unfolds many opportunities for the future, especially for smaller research units. It reduces the time required for the procedures and the necessary funds for research, promoting the development of new applications and scientific progress. Recently, a group from Universitat Pompeu Fabra in Barcelona developed GPUGRID, a project based on distributed computing systems, made of many GPUs all over the world, connected on a network for the implementation of biomedical research (Buch *et al.* 2010). GPUGRID is a volunteer network, running on BOINC, an open source software platform developed at the University of California, Berkeley. GPUGRID executes MD simulations

designed to run on NVIDIA's CUDA suitable GPUs.

CUDA and OpenCL share the main central idea but also have several differences. CUDA consists of a proprietary hardware architecture, an internal security assessor (ISA), a programming language, an application programming interface (API), a software development kit (SDK) and different tools. CUDA supports several operating systems, but a lot of its functionalities depend on NVidia. On the other hand, OpenCL is an OpenAPI and a language specification, which is supports multiple GPU companies such as Apple, NVidia, AMD and IBM. Most of the OpenCL capacities depend on the implemented vendor (Scarpino 2012).

Current trends in molecular dynamics

The growing amount of programs related to MD in combination with the available computing power has led to an increase in the number of MD studies. During the last 10 years the publications using MD have increased from six thousand to twenty six thousand (Figure 1). In total, the published studies that use MD have reached seventeen thousand. An extensive analysis of MD studies in a short review is therefore not achievable; however the significant current trends in this field will be reported.

MD simulations using GPU can now rapidly track processes. Many of those processes occur in less than a millisecond and can provide useful information concerning relevant biological systems in atomic resolution. Dynamic models have a significant impact on drug discovery, resolving many problems of the pharmaceutical industry. Molecular dynamics have been used in a numerous drug design studies, providing efficient methods and simulations (Papageorgiou *et al.* 2013, Vlachakis *et al.* 2014, Vlachakis & Kossida 2013, Vlachakis *et al.* 2013). Furthermore, MD simulations have increased the knowledge of molecular biology and have led to more successful clinical trials.

Dynamic models can provide significant molecular information concerning individual particle motions in time. In comparison to *in vitro* experiments, an *in silico* MD simulation is less expensive, more accurate and less time consuming. MD has many applications in a variety of studies. Recently, MD simulations were carried out to analyze the base of ligand selectivity of estrogen receptors (Hu & Wang 2014). Moreover, MD methods in combination with other methods are used to investigate the conformational behavior of peptides (Kulkarni & Ojha 2014). Additionally, the mechanical behavior of osteopontin-hydrixapatite interfaces is investigated using MD methods in order to understand the nanomechanics of bones and develop new artificial biological materials (Lai *et al.* 2014). Similarly, MD were used to simulate the cardiac troponin core domain complex, in order to understand the structural details of troponin switching (Jayasundar *et al.* 2014). Another significant aspect of such simula-

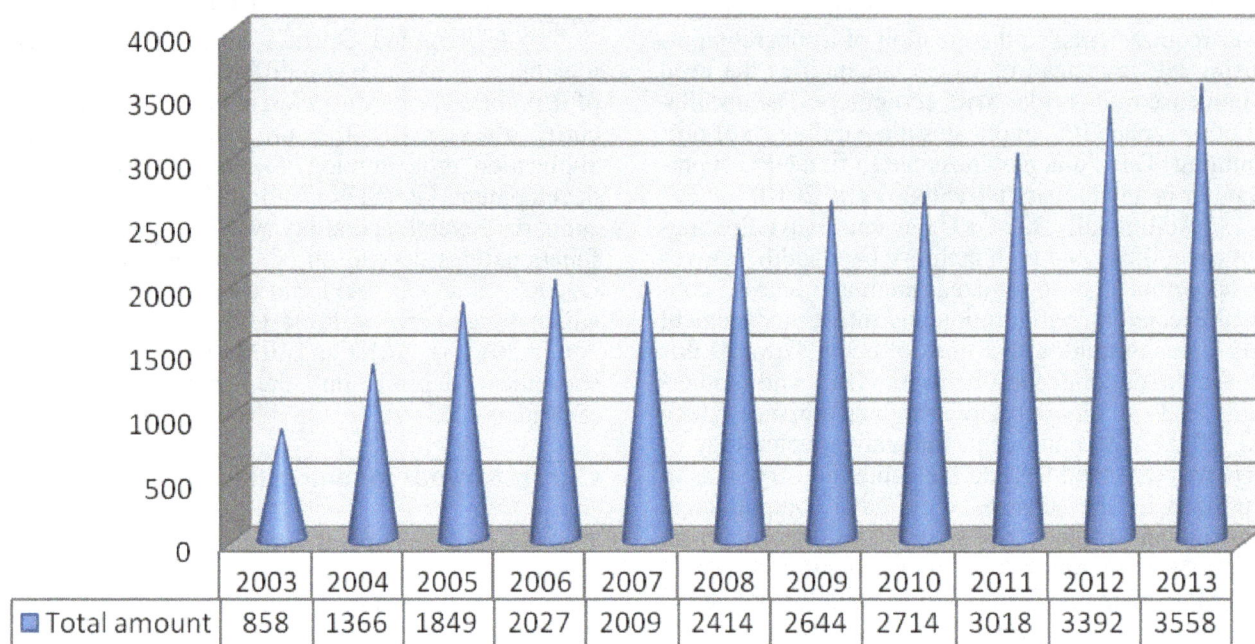

Total amount	2003	2004	2005	2006	2007	2008	2009	2010	2011	2012	2013
	858	1366	1849	2027	2009	2414	2644	2714	3018	3392	3558

Figure 1. Pubmed publications related to molecular dynamics from 2003 to 2013.

tions is that they are completely controlled by the user so by altering, elaborating and removing specific parametes their properties can be examined.

The groundbreaking idea to share complicated, unsolved scientific problems through a video game with the public sounds like a science fiction scenario. However, this extremist idea has been recently applied in practice. In a recent example, computer gamers solved a problem in HIV research that had been puzzling scientist for years, in only three weeks of continuous gaming (Lv *et al.* 2013). This radiant idea could be applied to many other areas, such as molecular dynamics.

Benefits of parallel programming using GPU

Parallel programming is based on the assumption that problems involving large volumes of data can be broken down into smaller parts, which can be resolved by parallel redundant computations in each core (Almasi & Gottlieb 1989). Parallel computing can utilize multiple processors or computers that work together to accomplish a common task. Each processor (or computer) implements a separate part of the cleaved problem. This computational method has been the cornerstone for solving problems that could not be resolved with the use of a single CPU in real-time. Parallel computing, principally in the form of multi-core processors, is the main model of programming in modern computer architecture (Asanovíc *et al.* 2006).

Recently computer engineers explored the applicability of parallel programming using GPUs (Figure 2). Algorithms running on GPUs can often accelerate the calculation speed up to 100 times compared to common CPUs, enabling the development of

several applications for neural networks, MD, physics simulations and countless other fields. However, it is essential to note that the benefits of GPU computing cannot be reached without the combinational work of the GPU with the CPU. Each unit is developed for different types of computing. CPUs have few cores that make it very effective in single-threaded serial processing, while GPUs have thousands of small cores developed for parallel computing. The reason behind such large increase in GPUs performance is their built, which is based on an escalating array of multithreaded Streaming Multiprocessors (SMs). Each one of GPU multiprocessors can carry out hundreds or even thousands of threads simultaneously. The architecture used to empower the administration of such a big number of threads is called single-instruction, multiple-thread (SIMT). This ability for parallel computing allows GPUs to fragment large and complex tasks into several small tasks, which run separately and contemporaneously on different threads of different multiprocessors (Papangelopoulos *et al.* 2014).

Available GPUs for parallel programming

Presently, there are several commercialized, CUDA-enabled GPUs adopted on desktops and notebooks, equally. The top ranked GPUs are mostly promoted by NVIDIA (Figure 3). For instance, GeForce 750, GTX 750, GTX 750 TI for desktop and GTX 850M, 840M, 830M for notebook users. These GPUs can attain the dominant computational capability, as is shown on NVidia's web page. GeForce GTX 860M has been released in two editions, the former with compute capability up to 5, and the latest up to 3, which depends on the number of CUDA cores. Addressed to the purchasing public, NVidia is promoting Quadro and Tesla relevant products (NVIDIA Corporation, 2014). Particularly, Quadro GPUs specialise in professional visualization while Tesla in high-end imaging for technical and scientific computing. However, they do not achieve the highest score in the range of GPU computing. Indicatively, Tesla K40, K20 and Quadro K6000 gather the highest score (3.5) of these families. Other GPUs companies are Intel, Qualcomm, AMD, Altera and Samsung, which are able to implement GPGPU only by OpenCL (Stone *et al.* 2010).

OpenCL is the dominant open GPGPU programming language, even though GPGPU can be accelerated by modifying most commonly used programming languages, libraries and compilers. Some of these programming languages are C and C++ which use the NVCC compiler and, together with some addendums, are the base of the CUDA platform. Almost every high level programming language can be accelerated though C wrappers and swig. Most commonly used languages linked to CUDA are Fortran CUDA, PyCUDA, JCUDA, JCUBLAS and JCUFFT, Haskell, KAPPACUDA, Ruby, Python, Perl, GPULLIB and MATLAB, each one of which uses a specific compiler and certain libraries. (Rudy *et al.* 2011). Vector is a new GPGPU programming language developed by Howard Mao and his team, who also developed a compiler that compiles Vector into CUDA (Resch 2010). Another coding language is CHESTNUT, which is a GPU programming language for non-experts as its developers characterize it (Stromme *et al.* 2012). The CUDA compute capability plays a significant role in the selection of the programming language. Namelya compute capability of 1.x can be used with C and some simple extensions while compute capability of 2.x can be used with C++. Finally, there exists Harlan; a new GPGPU programming language introduced by Eric Holk and his laboratory team (Holk *et al.* 2011).

Molecular dynamics simulators on CPU and GPU

As a result of this ever-growing trend, many programs which can implement several MD methods were developed, either using both CPU and GPU or using only GPU. Examples of this software are listed in Table 1, accompanied by their principal characteristics. As is apparent from the fifth column of the table, each program that is able to run a simulation on both GPU and CPU is superior in speed. Thus, when the simulation is running on a GPU, it requires less time to complete the task. Finally, the immediacy of each application depends on the habits and the skills of each user. If the user is not familiar with the command prompt, the best solution is to choose a program that provides a GUI interface. The program selection also depends on the operating system in which the user wants to install the program and the reasons for installing it.

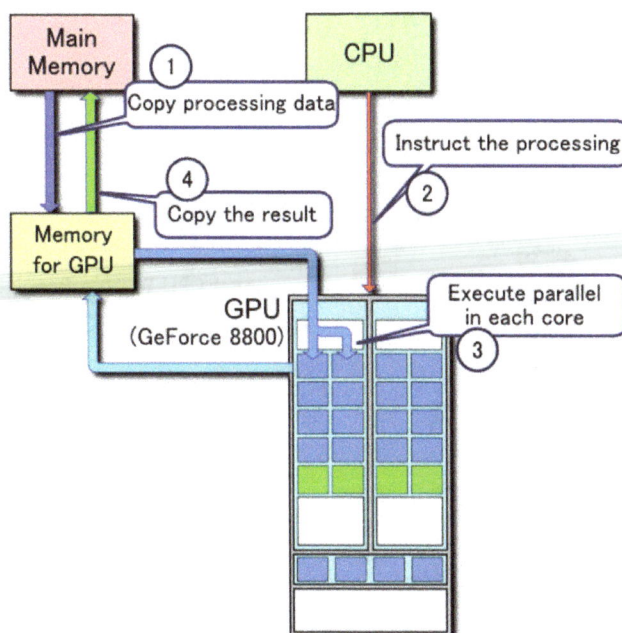

Figure 2. Diagram of Cuda parallel programming using GPU (Johnson and Vijayalakshmi 2013).

On the other hand, there are plenty of disadvantages in using CUDA. It can be accelerated only through NVIDIA's GPUs. Thus ATI's, AMD's and other users are barred from this technology. The time bottleneck which can come up at the communication between GPU and CPU still exists. Perhaps a huge progress has taken place in the area of research and development, but the number of GPU users is stable. CUDA is still under development and it is not known whether it will consolidate the marketplace or more problems will arise and CUDA will hit rock bottom (Che *et al.* 2008, Pan *et al.* 2008).

Conclusion

Video games were the starting point of GPU technology. GPUs caused a temporary retreat in the course of CPU-based computer technology. The many benefits of parallel programming and GPUs gained the interest of many scientists. GPGPU is a very promising computing technology, which holds plenty to offer in every field of science, especially in bioinformatics. Many applications running on GPUs have already been constructed and a plethora of researchers is involved in this programming method. Nowadays, there is a wide range of techniques performed with parallel programming and GPUs. The combination of GPU video game technology with molecular dynamics simulations revolutionized computational biology and drug discovery. Additionally, many computational issues have been solved such as the issue of time and computing power. In the near future a superabundance of achievements through GPU molecular dynamics is expected to occur. One of them could be the understanding of the relations between 3D structure, dynamic model and

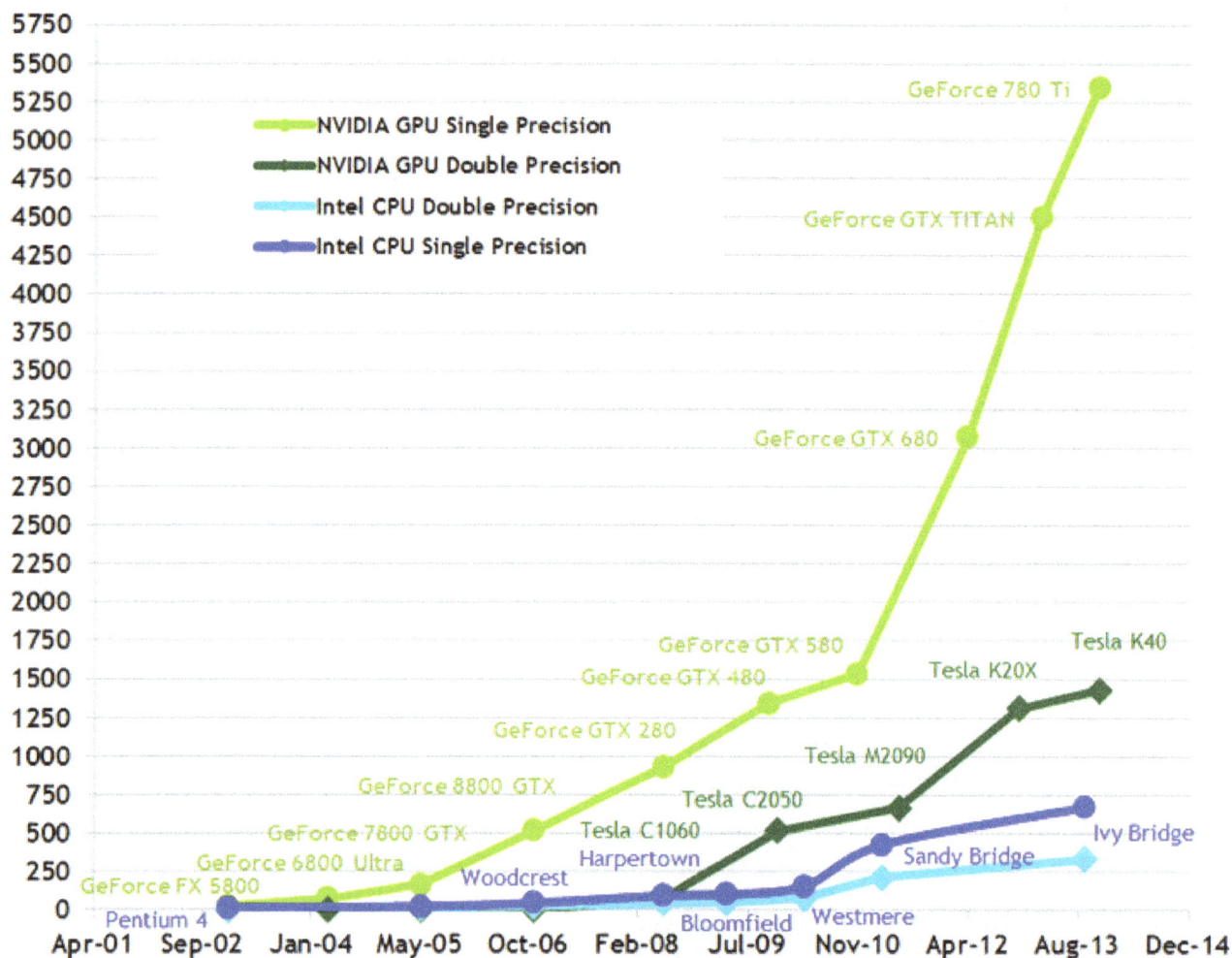

Figure 3. Memory Bandwidth for the CPU and GPU (NVIDIA, 2014).

functionality of a molecule. Drug discovery and design, genomics and computational chemistry are some of the areas most expected to adopt this programming architecture.

Funding

This work was partially supported by: (1) The BIOEXPLORE research project. BIOEXPLORE research project falls under the Operational Program "Educationand Lifelong Learning" and is co-financed by the European Social Fund (ESF) and National Resources. (2) European Union (European Social Fund - ESF) and Greek national funds through the Operational Program "Educationand Lifelong Learning" of the National Strategic Reference Framework (NSRF) - Research Funding Program: Thales. Investing in knowledge societythrough the European Social Fund.

References

Alder BJ & Wainwright TE 1959 Studies in molecular dynamics. I. General Method *Journal of Chemical Physics* **31** 459-466

Alerstam E, Lo WC, Han TD, Rose J, Andersson-Engels S & Lilge L 2010 Next-generation acceleration and code optimization for light transport in turbid media using Gpus *Biomed Opt Express* **1** 658-675

Almasi GS & Gottlieb A 1989 *Highly parallel computing* The Benjamin/Cummings series in computer science and engineering. CA, USA: Benjamin/Cummings

Asanovíc K, Bodik R, Catanzaro B, Gebis J, Husbands P, Keutzer K, Patterson D, Plishker W, Shalf J, Williams S & Yelick K 2006 The landscape of parallel computing research: a view from Berkeley *Electrical Engineering and Computer Sciences, University of California at Berkeley*

Attwood TK, Gisel A, Eriksson NE & Bongcam-Rudloff E 2011 *Concepts, historical milestones and the central place of bioinformatics* in Modern Biology: A European Perspective. Bioinformatics – Trends and Methodologies. Eds M Mahdavi. InTech

Browna WM, Kohlmeyerb A, Plimptonc SJ & Tharringtona AN 2012 Implementing molecular dynamics on hybrid high performance computers - particle–particle particle-mesh. *Comp Phys Comm* **183** 449-459

Table 1. Main softwares using GPU for MD.

Softwares for Molecular Dynamics							
Name	CPU	GPU	GUI	Speed (GPU vs CPU)	3D Graphics	Service	OS
Abalone	YES	YES	NO	4-29x	YES	Executes bio molecular simulations of proteins, DNA, ligands	Windows
ACEMD	NO	YES (multi-GPU support)	YES(Charmm -GUI)	160 ns/day	NO	Program for MD simulations running on NVidias GPUs	Windows, Linux
AMBER	YES	YES (multi-GPU support)	YES	100 ns/day	NO	Executes MD simulator on biomolecules	MacOs, Linux, Windows (through Cygwin)
Ascalaph Designer	YES	YES	YES	4-29x	YES	General purpose MD simulator	Linux, Windows
CHARMM	YES	YES (multi-GPUsupport)	YES	32-35x	NO	Program for MDs, commercial	MacOs, Linux, Windows (through Cygwin)
Discovery Studio	YES	YES	YES	x5	YES	Simulation program with many capabilities	Windows, Linux, MacOS
DL POLY	YES	YES (multi-GPU support)	YES	4x	YES	Application to facilitate MD simulations of macro-molecules, polymers, ionic systems	Linux, Windows (through cyngwin), MacOS
GROMACS	YES	YES (multi-GPUsupport)	YES (GROMITA)	165 ns/day	NO	A hybrid CPU-GPU MD program, which perform high-level simulations	Linux, Windows (through cyngwin), MacOS
GULP	YES	NO	NO	------	NO	An application which optimizes MD	Linux, Windows, MacOS
HOOMD-blue	YES	YES (multi-GPUsupport)	NO	2x	NO	Molecular dynamic simulation toolkit	Linux, MacOS
LAMMPS	YES	YES (multi-GPUsupport)	YES (Lammpsfe)	3-18x	NO	MD simulation package	Linux, Windows (through cyngwin), MacOS
MD-display	YES	NO	YES	------	YES	A platform for 3D MD	Windows, Linux
MdynaMix	YES	NO	YES	------	NO	Program for general use of MD's simulations	Linux
MOE	YES	NO	YES	------	YES	Molecular Operating Environment	Windows, Linux, MacOS
ORAC	YES	NO	NO	------	NO	MD's simulations	Linux
Namd	YES	YES (multi-GPU support)	NO (YES through VMD)	4 ns/days	NO (YES through VMD)	Parallel molecular dynamics pro-gram for Biomolecular systems	Windows, Linux, MacOS, Solaris
NWChem	YES	YES	NO	3-10x	NO	Computational chemistry and MD simulations project	Linux, Windows, MacOS, FreeBSD
Packmol	YES	NO	YES	------	YES	An application to generate MD geometries	Linux, Windows, MacOS
Protein LocalOptimization Program	YES	NO	NO	------	NO	Program for protein modeling and simulations	Linux
Protomol	NO	YES	YES	-----	YES	Object-oriented programming for MD	Linux, Windows, Solaris
RedMD	YES	NO	NO	------	YES (using aninterface)	Package for MD simulations	Linux
TeraChem	YES	YES	YES	44-650x100	NO	MD program for very large molecules' simulations	Linux
VEGA ZZ	YES	YES	YES	10x	YES	2D and 3D viewer and editor	Linux (without GUI interface), Windows
VMD	YES	YES (multi-GPU support)	YES	100-125x	YES	A program for modeling, analyzing, visualizing biomolecu-les, it is intergrated with NAMD	Windows, Linux, MacOS, Solaris
WHAT IF	YES	NO	YES	------	YES	A visualizer for MD using an interface	Linux, Windows (under conditions)

Buch I, Harvey MJ, Giorgino T, Anderson DP & De Fabritiis G 2010 High-throughput all-atom molecular dynamics simulations using distributed computing. *J Chem Inf Model* **50** 397-403

Che S, Boyer M, Meng J, Tarjan D, Sheaffer JE & Skadron K 2008 Performance study of general-purpose applications on graphics processors using Cuda *J Parall Distr Comp* **68** 1370-1380

Fago A, D'Avino R & Di Prisco G 1992 The hemoglobins of notothenia angustata, a temperate fish belonging to a family largely endemic to the antarctic ocean *Eur J Biochem* **210** 963-970

Hesper B & Hogeweg P 1970 Bioinformatica: Een Werkconcept Kameleon *Kameleon* **1** 28-29

Hoang RV, Tanna D, Jayet Bray LC, Dascalu SM & Harris FC Jr 2013 A novel Cpu/Gpu simulation environment for large-scale biologically realistic neural modeling *Front Neuroinform* **7** 19

Holk E, Byrd W, Mahajan N, Willcock J, Chauhan A & Lumsdaine A 2011 declarative parallel programming for Gpus. In *14th biennial ParCo* edited by KD Bosschere, EH D'hollander, GR Joubert, D Padua, F Peters and M Sawyer. Amsterdam, Netherlands.

Hu G & Wang J 2014 Ligand selectivity of estrogen receptors by a molecular dynamics study *Eur J Med Chem* **74** 726-735

Jayasundar JJ, Xing J, Robinson JM, Cheung HC & Dong WJ 2014 Molecular dynamics simulations of the cardiac troponin complex performed with fret distances as restraints. *PLoS One* **9** e87135

Johnson J & Vijayalakshmi G 2013 Comparison study of parallel computing with Aluand Gpu (Cuda). *Int J Sci Res (IJSR)* **2** 209-212

Karplus M & McCammon JA 2002 Molecular dynamics simulations of biomolecules *Nat Struct Biol* **9** 646-652

Kirk D & Hwu WM 2010 *Programming Massively Parallel Processors Hands-on with Cuda*. MA, USA: Morgan Kaufmann Publishers

Kulkarni AK & Ojha RP 2014 Combined 1 H-NMR and molecular dynamics studies on conformational behavior of a model heptapeptide, GRGDSPC. *Chem Biol Drug Des*

Lai ZB, Wang M, Yan C & Oloyede A 2014 Molecular dynamics simulation of mechanical behavior of osteopontin-hydroxyapatite interfaces. *J Mech Behav Biomed Mater* **36C** 12-20

Lv Z, Tek A, Da Silva F, Empereur-mot C, Chavent M & Baaden M 2013 Game on, science - how video game technology may help biologists tackle visualization challenges. *PLoS One* **8** e57990

Mesirov JP, Schulten K & Sumners DW 1996 *Mathematical Approaches to Biomolecular Structure and Dynamics*. The Ima Volumes in Mathematics and Its Applications. New York, USA: Springer

Murakami T, Kasahara R & Saito T 2010 An imple-

mentation and its evaluation of password cracking tool parallelized on Gpgpu. *IEEE* 534 - 538

NVIDIA 2014 Cuda C Programming Guide. *NVIDIA Corporation*

NVIDIA Corporation. Cuda Gpus. NVIDIA Developer. Accessed from https://developer.nvidia.com/cuda-gpus at June 6, 2014.

Owens JD, Houston M, Luebke D, Green S, Stone JE & Phillips JC 2008 Gpu computing. *Proc IEEE* **6**

Pan L, Gu L & Xu J 2008 Implementation of medical image segmentation in Cuda. *IEEE* 82-85

Papageorgiou L, Vlachakis D, Koumandou VL, Papangelopoulos N & Kossida S 2013 Computer-aided drug design and biological evaluation of novel anti-Greek goat encephalitis agents. *Int J Syst Biol Biomed Tech* **2** 1-16

Papangelopoulos N, Vlachakis D, Filntisi A, Fakourelis P, Papageorgiou L, Megalooikonomou V & Kossida S 2014 State of the art Gpgpu applications in bioinformatics. *Int J Syst Biol Biomed Tech* **2** 24-48

Resch M 2010 *High Performance Computing on Vector Systems 2009* New York, USA: Springer

Rudy G, Khan MM, Hall M, Chen C & Chame J 2011. A programming language interface to describe transformations and code generation. *Lang Comp Parall Comp* **6548** 136-150

Scarpino M 2012 *Opencl in Action : How to Accelerate Graphics and Computation*. NY, USA: Manning

Sharma R, Gupta N, Narang V & Mittal A 2011 Parallel implementation of DNA sequences matching algorithms using pwm on Gpu architecture. *Int J Bioinform Res Appl* **7** 202-215

Stone JE, Gohara D & Shi G 2010 Opencl: a parallel programming standard for heterogeneous computing systems. *Comput Sci Eng* **12** 66-72

Stone JE, Hardy DJ, Ufimtsev IS & Schulten K 2010 Gpu-accelerated molecular modeling coming of age. *J Mol Graph Model* **29** 116-125

Stromme A, Carlson R & Newhall T 2012 Chestnut: A Gpu programming language for non-experts. *PMAM*

Ueng S, Lathara M, Baghsorkhi SS & Hwu WW 2008 Cuda-Lite: reducing Gpu programming complexity. *Lang Comp Parall Comp* **5335** 1-15

Vlachakis D, Bencurova E, Papangelopoulos N & Kossida S 2014 Current state-of-the-art molecular dynamics methods and applications. *Adv Protein Chem Struct Biol* **94** 269-313

Vlachakis D & Kossida S 2013 Molecular modeling and pharmacophore elucidation study of the classical swine fever virus helicase as a promising pharmacological target. *PeerJ* **1** e85

Vlachakis D, Tsagrasoulis D, Megalooikonomou V & Kossida S 2013 Introducing drugster: a comprehensive and fully integrated drug design, lead and structure optimization toolkit. *Bioinformatics* **29** 126-128

7

Cloning and functional characterization of a vertebrate low-density lipoprotein receptor homolog from eri silkmoth, *Samia ricini*

Rajni Bala[*], Ulfath Saba[*], Meenakshi Varma, Dyna Susan Thomas, Deepak Kumar Sinha[#], Guruprasad Rao, Kanika Trivedy, Vijayan Kunjupillai and Ravikumar Gopalapillai

Seri-biotech Research Laboratory, Central Silk Board, Kodathi, Carmelaram Post, Kodathi, Bengaluru 560 035, India.

Correspondence should be addressed to Ravikumar Gopalapillai; E-mail: ravikumarpillai@gmail.com

*These authors contributed equally

Present address: Plant-insect interaction group, ICGEB, New Delhi, India

Abstract

The lipophorin receptor (LpR) is the insect lipoprotein receptor and belongs to the low-density lipoprotein receptor (LDLR) superfamily. It has a vital role in the uptake of lipophorin (Lp) into various tissues. Here we report the full length cloning and functional characterization of an LpR from eri silkmoth, *Samia ricini*. The full length cDNA of *Sr*LpR7-1 is 4132 bp including an open reading frame (ORF) of 2595 bp. The deduced amino acid sequence revealed well structured ligand binding, epidermal growth factor, glycosylation, transmembrane and cytoplasmic domains. The ligand binding domain consisted of seven cysteine repeats instead of the common eight cysteine repeats indicating it as a homolog of human LDLR. We identified another splice variant, *Sr*LpR7-2 with a deletion of 27 amino acids in the *O*-glycosylation domain. Apart from the fat body, both isoforms are expressed in ovary, brain and other tissues at different developmental stages of the silkworm. RNAi experiments did not show any marked effects except that the adult emergence was delayed compared to controls. In addition, the *Sr*LpR7 cDNA was recombinantly expressed and ligand binding experiments confirmed that the receptor protein binds not only to *Sr*Lp but also to *Bombyx mori* Lp.

Introduction

Lipophorin (Lp) is critical in insects for lipid transport between tissues. The loading and unloading of lipids into and from fat body is accomplished by Lp and another protein called lipid transfer particle (LTP) (Arrese *et al.* 2001, Ryan & Van der Horst 2000). Lp functions as a reusable lipid shuttle and the delivery of lipids to various organs occurs without its internalization, accumulation or degradation (Van der Horst *et al.* 2002). Based on its density, Lp is classified as low, high and very high-density. Lp incorporates into the cells through receptor-mediated endocytosis or without endocytosis (Parra-Peralbo & Culi 2011, Raikhel & Dhadialla 1992, Rodenburg & Van der Horst 2005, Van Hoof *et al.* 2003). During lepidopteran vitellogenesis, Lp transports lipids and other yolk precursors from the fat body to the ovaries and, in some species, Lp itself becomes one of the major constituents of the protein yolk bodies (Sun *et al.* 2000, Swevers *et al.* 2005, Telfer *et al.* 1991).

The Lipophorin receptor (LpR) belongs to the low-density lipoprotein receptor (LDLR) superfamily. The members of this family bind to various ligands and are involved in the lipid metabolism of both vertebrates and invertebrates (Schneider & Nimpf 2003). The defining structural features of the family consist of i) a cysteine-rich ligand binding domain (LBD), ii) an epidermal growth factor domain (EGFD), iii) an *O*-linked sugar domain (OSD), iv) a transmembrane domain (TMD), and v) a cytoplasmic domain (CD). The number of cysteine-rich repeats in LBD varies among LDLR family members. LDLR contains seven cysteine-rich repeats whereas VLDLR has eight such repeats and large receptors like megalin and LDLR-related proteins have more cysteine-repeats in their LBDs (Willnow *et al.* 1994). LpRs have been cloned and characterized from *Locusta migratoria* (Dantuma *et al.* 1999), *Aedes aegypti* (Cheon *et al.* 2001, Seo *et al.* 2003), *Galleria mellonella* (Lee *et al.* 2003), *Bombyx mori* (Gopalapillai *et al.* 2006), *Blattella germanica* (Ciudad *et al.* 2007), *Apis mellifera* (Guiugli-Lazzarini *et al.* 2008) and *Leucophaea maderae* (Tufail *et al.* 2009). Most of these LpRs are homologues to

VLDLR, having eight cysteine-rich repeats in their LBDs. However, *Aedes aegypti* (fat body) (Seo *et al.* 2003), *Blattella germanica* (Ciudad *et al.* 2007) and *Drosophila melanogaster* (Parra-Peralbo & Culi 2011) LpRs have seven cysteine rich-repeats in their LBDs and thus resemble human LDLR.

Samia ricini (Lepidoptera: Saturniidae), the Indian eri silkworm, contributes significantly to the production of commercial silk (known for its white or brick-red eri silk) and is the only domesticated non-mulberry silkworm. It is distributed in India, Japan and China. The eri silk is also used for studies in biomedical applications of silk proteins (Pal *et al.* 2013). The Lp of the *S. ricini* is a high density lipoprotein composed of two apolipophorin subunits with a molecular mass of approximately 260 (Apo-I) and 80 kDa (Apo-II), respectively (manuscript submitted). As noted above, several LpR genes have been cloned and characterized in recent years including our work on *B. mori*. However, knowledge on the functional aspects of LpR and its interactions remain limited. The present report describes the primary structure of the *Sr*LpR gene, tissue and stage-specific expression patterns, recombinant protein expression and functional characterization. Additionally, apart from showing variations in RNAi efficiency, the cross reactivity of *Sr*LpR with *Bm*Lp is also demonstrated.

Materials and Methods

Cloning of *Sr*LpR cDNA

The eri silkworms, *S. ricini* were maintained in our laboratory and fed on castor leaves. Total RNA was purified from the fat body of Day 10 pupae using RNeasy mini kit (Qiagen). Reverse transcription was performed using oligo (dT) with Primescript RT-PCR system (Takara). Two degenerative primers (Forward: 5'GTITAYTGGACIGAYTGGGAYAA-3'; Reverse: 5'- TAIACIGGRTTRTCRAARTTCAT-3'; I-inosine) designed from the consensus sequences of LpRs were used to amplify a partial sequence of *Sr*LpR. The PCR product was cloned and sequenced. The 5' and 3' RACE kit (Roche) was used to obtain the full length *Sr*LpR cDNA using gene specific primers (SP1-5: 5'-CAGCATCGCGGCCACTGTTTGAGA-3'; SP2-5: 5'-CGCACATCTGGTTGTCGGGGAGT-3'; SP3-5: 5'-TCCTGGGTGCAGGTAAGCAGAGGT-3'; SP5-3: 5'-ACTCCCCGACAACCAGATGTGCG-3'; SP6-3: 5'-TCT CAAACAGTGGCCGCGATGCTG-3') and adapter primers from the kit according to the manufacturer's instructions. The amplified products were cloned and sequenced. Several independent clones were used to eliminate possible PCR errors.

Full-length cDNA was obtained by PCR using high-fidelity Accuprime DNA polymerase (Invitrogen).

RT-PCR

Tissues from fifth instar larvae, pupae (Day 12) and adult moths (Day 1) were dissected and RNA was extracted as above. Equal amount of RNA was used for reverse transcription. First-strand cDNA was synthesized using Primescript RT-PCR system (Takara) and was used as template for PCR. The primers employed for the *Sr*LpR-1 and *Sr*LpR-2 were: Forward: 5'-ATACCAGAGGAGGTGCATATAACG-3' and Reverse: 5'-CGATGTAACATTATGGTGAACGT-3'; Forward: 5'-ATTCGCCCCGCGTCTCCTGCGCTTG-3' and Reverse: 5'-CGATGTAACATTATGGTGA-ACGT-3', respectively. The forward primer for the *Sr*LpR7-1 expression was designed from the glycosylation domain having an addition of 81 nucleotides, the only divergent region specific to *Sr*LpR7-1. However, for the detection of *Sr*LpR7-2, primers were designed from a region common to both isoforms. The 18S rRNA was used as an internal control (Forward: 5'-TTGACGGAAGGGCACCACCAG-3' and Reverse: 5'- AGACGCCGGTCCCTCTAAGAAG-3'). Random hexamers were used for reverse transcription for 18S rRNA amplification. Care was taken to avoid genomic DNA contamination. Each RT-PCR product was verified by sequencing.

RNA interference (RNAi)

To obtain dsRNA targeted to both LpR mRNAs, a fragment was amplified by PCR using T7 promoter and gene-specific primers (Forward: 5'-TAATACGACTCACTATAGGGAGAACGTGCCA-GCCGCAGTCACATTTCG-3'; Reverse: 5'-TCTCCCTATAGTGAGTCGTATTAACAACGTGT-ACGGTCGCGTGGATCAC-3') and GFP as control (Forward: 5'-TAATACGACTCACTATAGGGAGA GCTACCTGTTCCATGGCCAACACTTG- 3'; Reverse: 5'- TCTCCCTATAGTGAGTCGTAT-TATGTCTGGTAAAAGGACAGGGCCATCGC-3'). The GFP plasmid was kindly given by Professor K. P. Gopinathan, Indian Institute of Science, Bangalore, India. The strategy employed was a single PCR with T7 promoter appended to 5' ends of both forward and reverse PCR primers as indicted above, to generate transcription template for both strands of the dsRNA. One µg of each PCR product was subjected to *in vitro* transcription. The *in vitro* transcriptions were carried out using MEGAscript RNAi Kit (Ambion) following the manufacturer's instructions except that the reaction mixes were incubated for 6 h at 37°C to increase dsRNA yield. This was followed by nuclease digestion to remove DNA and ssRNA, and purification of

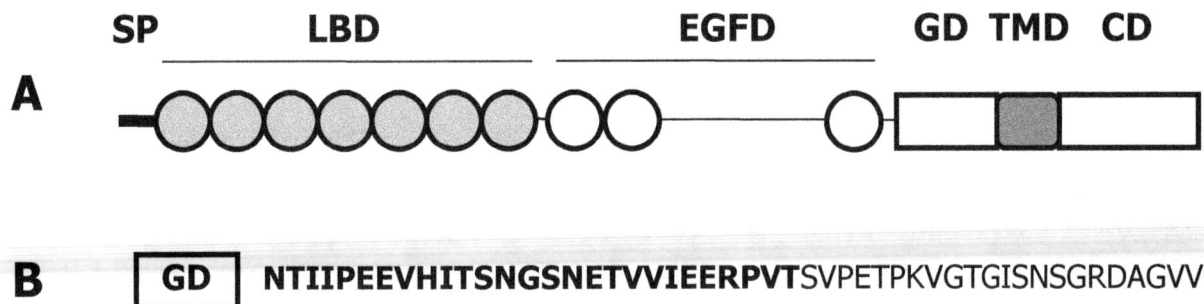

Figure 1. A: *Sr*LpR with seven cysteine repeats. SP: Signal peptide; LBD: Ligand binding domain; EGFD: Epidermal growth factor domain; GD: Glycosylation domain; TMD: Transmembrane domain; CD: Cytoplasmic domain. B: Addition of 27 amino acids (in bold) in GD.

dsRNA was carried out according to manufacturer's instructions. The quantity and integrity of dsRNA were analysed using Nanodrop 2000 (Thermo Fisher Scientific) and agarose gel electrophoresis. Female pupae (Day 1) of *S. ricini* were injected with 50 µg of dsRNA, into intersegmental membrane between 8th and 9th abdominal segments. Pupae were again injected with the same dose on Day 10. Water injected pupae and normal females without any injection were also used as negative controls. RT-PCR was used to investigate the expression of LpR mRNA in ovaries of Day 12 pupae using the above gene-specific primers. The female moths were mated with normal males in all groups. Observations on ovarian development, adult emergence, fecundity and hatching were recorded.

Isolation of lipophorin (Lp) and preparation of antibodies

Hemolymph from mid fifth instar female larvae (non-vitellogenic) of *S. ricini and B. mori* were collected, hemocytes were removed, and Lp was collected by KBr density gradient ultracentrifugation (Shapiro *et al.* 1984, Ravikumar *et al.* 2011). HDLp which formed a clear yellow band was collected, desalted, and used immediately for the binding assays. The density was 1.103 g/ml for *Sr*Lp and 1.064 g/ml for *Bm*Lp. The purified *Sr*Lp was confirmed by 5 % SDS-PAGE and polyclonal antibodies (Apo I and II) were produced in rabbits by Bhat Biotech India Pvt Limited, Bangalore, India. The *B. mori* HDLp antibody was kindly given by Dr. Kozo Tsuchida, National Institute of Infectious Diseases, Tokyo, Japan.

Protein expression and ligand binding

The cDNA of *Sr*LpR7-1 (without signal sequence) and with polyhistidine tag was cloned into pF25A ICE T7 Flexi Vector (Promega) between *Sgf*I and *Pme*I sites. After sequence verification, the constructed plasmid was used for *in vitro* transcription coupled translation using TNT-T7 Insect Cell Extract Protein Expression System (Promega). The protein was purified using ProBond Purification System (Invitrogen), separated on a 7.5 % SDS-PAGE under non-reducing conditions, and then transferred to PVDF membrane. After blocking, the membranes were incubated with 10 mg/ml of *S. ricini* or *B. mori* HDLp in binding buffer (20 mM HEPES, 150 mM NaCl, and 2 mM CaCl2, pH 7.5) containing 0.5% BSA and then incubated with rabbit anti-*Sr*HDLp antibody. *Sr*LpR7-1 bound antibodies were detected with alkaline phosphatase conjugated goat anti-rabbit IgG (Invitrogen) and NBT/BCIP (Roche). The control reactions consisted of protein analyzed under reducing conditions and also with Lp antibody omitted.

General methods

Unless indicated otherwise, all molecular biology techniques were performed essentially as described (Sambrook *et al.* 1989). Protein estimation was performed with the BCA Protein Assay Kit (Invitrogen) using BSA as the standard. Tools from NCBI/ExPASy and SignalP 3.0 were used for nucleotide and amino acid analyses.

Results

Sequence analyses

The full-length cDNA of *Sr*LpR7-1 contained a 201 bp 5'-UTR followed by a 2598 bp ORF and a 1333 bp 3'-UTR (Accession number-KU936050). The putative start codon is preceded by in-frame stop codons, indicating that the sequence represents a full-length open reading frame. It encoded a protein of 865 aa with a predicted molecular mass of 94 kDa without a signal peptide and an isoelectric point (pI) of 4.24. The putative signal peptide is 22 aa residues with a predicted cleavage site between residues 22 and 23 followed by five domains, characteristics of the LDLR superfamily

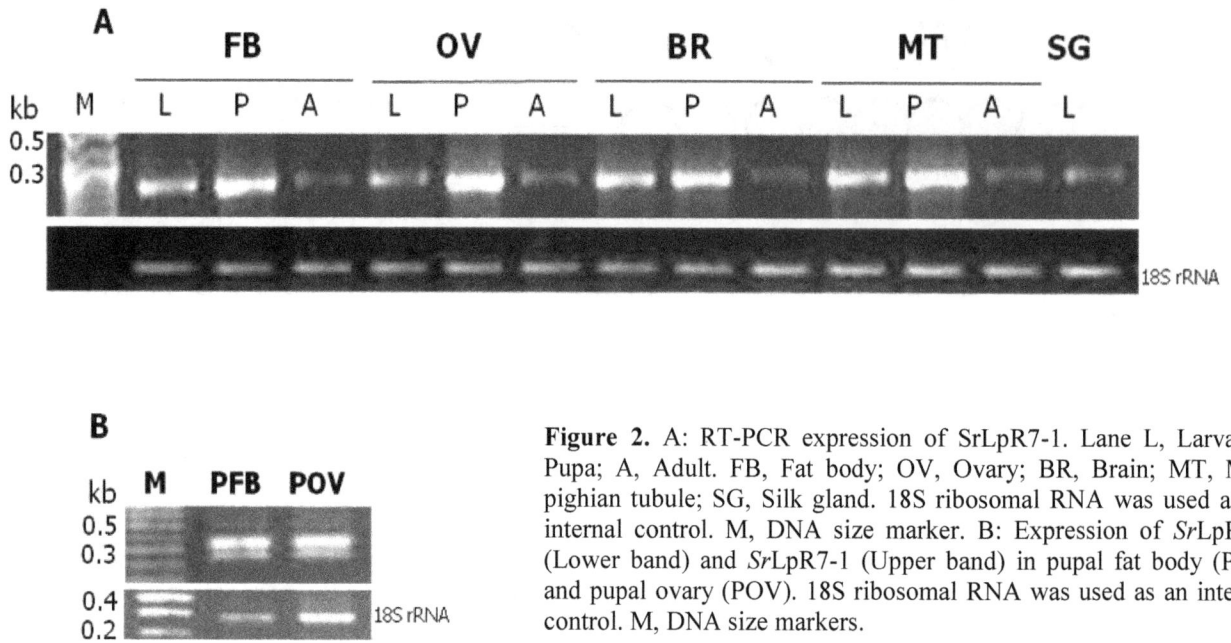

Figure 2. A: RT-PCR expression of SrLpR7-1. Lane L, Larva; P, Pupa; A, Adult. FB, Fat body; OV, Ovary; BR, Brain; MT, Malpighian tubule; SG, Silk gland. 18S ribosomal RNA was used as an internal control. M, DNA size marker. B: Expression of SrLpR7-2 (Lower band) and SrLpR7-1 (Upper band) in pupal fat body (PFB) and pupal ovary (POV). 18S ribosomal RNA was used as an internal control. M, DNA size markers.

(Figure 1A). Remarkably the first domain, the LBD consisted of seven cysteine-rich repeats in contrast to eight in most LpR and hence designated as SrLpR7. The sequence analysis indicated that the first cysteine repeat is absent. The LBD was characterised by conserved SDE residues which are involved in disulphide bond formation and folding. The LBD was followed by an EGF-like domain with A, B, and C type cysteine -rich repeats containing five copies of F/YWXD sequence. The role of this domain is the acid dependent dissociation of ligand (s) in the endosomes. The third was the O-linked glycosylation domain with 49 amino acids containing O-linked sugar chains and phosphorylation sites, followed by a short transmembrane domain. The exact role of the O-linked sugar domain is not known but it is often characterized by deletions and insertions in LDLR family members. The transmembrane domain plays an important role in receptor recycling, as it passes between plasma membrane and various endocytic vesicles. The fifth cytoplasmic domain contained the conserved peptide motif FDNPVY required for receptor internalization via clathrin-coated pits. We have also obtained another splice variant, SrLpR7-2, having a deletion of 27 amino acids in the O-linked sugar domain (Accession number-

KU936051; Figure 1B). The two cDNAs encoded identical proteins except for an insertion/deletion of 27 amino acids. The SrLpR7-1 showed a high degree of sequence similarity at the amino acid level with LpRs of other insects, such as B. mori (89.10 and 85.00 %; accession nos.- BAE71406 and XP_012551061, respectively), G. mellonella (87.40%, accession no.- ABF20542), B. germanica (68%, accession No.- CAL47126), L. maderae (67.70 %, accession No.- BAE00010) and A. aegypti (Ov- 66.53%, Fb- 61.50%; accession nos.- AAK72954 and AAQ16410, respectively).

Tissue distribution

Tissue and stage-specific mRNA expression levels were carried out by PCR using different primer pairs that amplify the two isoforms. As shown in Figure 2A, the SrLpR7-1 transcripts were detected in all selected tissues at various developmental stages, with highest transcript levels seen in larval and pupal fat bodies, brain, Malpighian tubules and pupal ovary. Moderate levels of expression were observed in the silk gland and all adult tissues. The transcript level was low in non-vitellogenic ovary (Larval ovary) which was increased during the vitellogenic phase (Pupal ovary),

Table 1. Effect of RNAi on adult emergence, fecundity and hatching. The values shown are mean ± S.D. (n = 3).

Treatment	Day of Adult Emergence	Fecundity (No.)	Hatching %
dsLpR- injected	18-19	210 ± 1.4	81± 2.3
dsGFP-injected	15-16	208± 2.0	80± 1.9
Water-injected	15-16	211± 1.9	79±2.3
Normal	15-16	208± 2.1	81±2.6

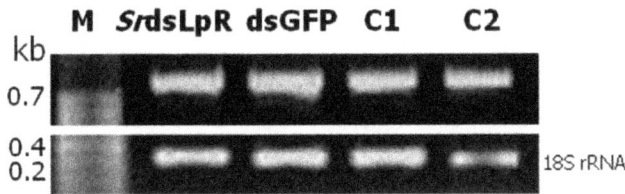

Figure 3. *Sr*LpR7 mRNA expression levels after RNAi. dsLpR, dsGFP, water injected (C1) and normal females (C2); 18S ribosomal RNA used as an internal control; M: DNA size markers.

coincident with the peak of yolk protein uptake. The mRNA expression of the *Sr*LpR7-2 showed low levels in most tissues (data not shown) except in pupal fat body and pupal ovary (Figure 2B).

RNAi
To determine the effects of RNAi of *Sr*LpR7, female pupae were injected with LpR dsRNA, GFP dsRNA and water as control groups. The results are presented in Table 1. We did not find considerable decrease in mRNA transcript levels in the experimental compared with the GFP and water-injected and normal females (Figure 3). However, the adult emergence was delayed for three days when compared to controls. Examination of ovaries revealed no abnormalities in any group which was followed by normal egg laying and hatching.

Ligand binding
The *Sr*LpR7-1 cDNA without its signal sequence was expressed and a protein of approximately 95 kDa was obtained. The protein was first incubated with Lps from *S. ricini* or *B. mori* and was subsequently detected using *S. ricini* anti-Lp antibody. The expressed receptor protein showed binding to Lp (Figure 4). Interestingly, the *Sr*LpR7-1 was also bound to *Bm*Lp and the *Sr*Lp antiserum reacted with *Bm*Lp (Figure 4). The cross reactivity of the *Sr*LpR to *Bm*Lp and the recognition of *Bm*Lp by *Sr*Lp antibody suggest the homology of structure at the amino acid level of these proteins between the two species. No binding was detected in the control reaction which was performed with SDS-PAGE under reducing conditions (Figure 4), indicating intact disulphide bonds are necessary for receptor binding. No binding was also detected in another control reaction in which Lp antibody was omitted (data not shown).

Discussion

We have cloned and characterized an LpR from eri silkworm that shares the typical modular domains, the hallmark of the members of LDLR superfamily. The

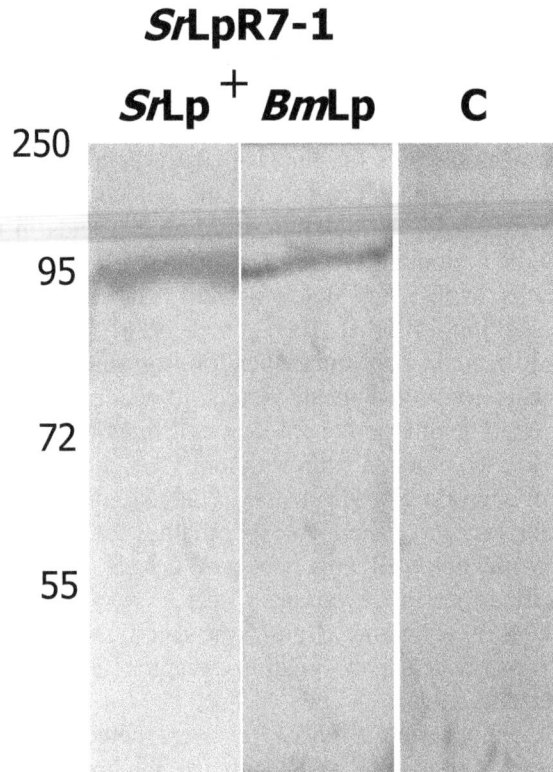

Figure 4. Functional expression of *Sr*LpR7-1 and its cross reactivity with *Bm*Lp. Lane 1: *Sr*LpR7-1 binds to *Sr*Lp; Lane 2: *Sr*LpR7-1 binds to *Bm*Lp; lane 3: Control showing no receptor binding. Numbers on the left indicate protein size markers in kDa.

LBD mediates the interaction between the receptor and the ligand, and usually consists of eight cysteine-rich repeats. While most LpRs contain eight cysteine-repeats in their LBDs and therefore identical to VLDLRs, the *Sr*LpR7-1 and 2 contain only seven such repeats and structurally resemble LDLRs. The LpRs of *A. aegypti* (fat body-specific) and *B. germanica* have also been reported to have seven cysteine-rich repeats (Cheon *et al*. 2001, Ciudad *et al*. 2007). A predicted *Bm*LpR isoform with seven cysteine repeats in LBD was found in the database (Accession Number: XP_012551061). As in *B. germanica*, the occurrence of seven cysteine repeats does not determine tissue or stage-specificity; *Sr*LpR7-1 was expressed in fat body, ovary and as well as other tissues and in all developmental stages. In *A. aegypti*, another splice variant of LpR specific to the ovary with eight cysteine repeats has been characterized (Seo *et al*. 2003). We could not isolate or amplify a full-length cDNA of *Sr*LpR with eight cysteine repeats from any tissues including ovary and fat body, whereas *Sr*LpR7-1 was frequently amplified indicating its abundance. The significance of the number of cysteine repeats in LBD is not known. In *D.*

melanogaster, lpr1 and *lpr2* are transcribed as multiple isoforms (having seven and eight cysteine repeats) and are essential for the cellular acquisition of neutral lipids. In addition, it is the presence of cysteine repeats in LBD and not the number that defines the ability of the receptor to mediate lipid uptake (Parra-Peralbo & Culi 2011). The variant form of *Sr*LpR is identical to the other except by the deletion of 27 amino acids in the OSD. This domain is the most divergent among five domains, with many deletions and insertions having occurred in the LDLR (Davis *et al*. 1986). The CD of *Sr*LpR contains an internalization signal, FDNPVY which is conserved in all vertebrate and invertebrate LDLRs. The alternative splicing and functional diversity are common features among the members of LDLR super family (Tveten *et al*. 2006). Compared to vertebrates, the functional specificities of LpR isoforms are not well understood. We have previously identified four splice variants of the *B. mori* LpR gene products, in which one, LpR4, was specific to the brain and possibly acts as a signalling receptor (Gopalapillai *et al*. 2006, 2014).

Transcripts of *Sr*LpR-1 were detected in all tissues tested by RT-PCR with the fat body, ovary, brain and Malpighian tubule being the highest sites of expression. The high level of transcripts in the pupal ovary suggests that the receptor is involved in the uptake of Lp into the ovary as vitellogenesis takes place in the pupal stage. Similar findings were reported from *B. mori* and *B. germanica* (Ciudad *et al*. 2007, Gopalapillai *et al*. 2006). In *A. aegypti*, the *Aa*LpROv is present during previtellogenic and vitellogenic stages whereas *Aa*LpRfb expression is restricted to the post-vitellogenic period (Cheon *et al*. 2001, Seo *et al*. 2003). Apart from most tissues, honey bee LpR mRNA has been specifically detected in the hypopharyngeal gland of adult females (Guidugli- Lazzarini *et al*. 2008). Expression of the LpR genes has been observed in a wide range of tissues including ovary, fat body, midgut, brain, Malpighian tubules, testis, embryo, muscle and silk glands (Tufail & Takeda 2009). Because of the functional diversity of Lps, the presence of LpR transcripts in a wide range of tissues is not surprising.

Lp plays an important role in lepidopteran vitellogenesis. Since *Sr*LpR7-1 was highly expressed in the vitellogenic ovary, we carried out RNAi experiments to determine its effect on ovarian development and reproductive success. In spite of having a delay in the transformation from pupa to adult moth, the dsLpR treated eri silkmoths showed normal fecundity and hatching. High or no silencing can occur at very different concentrations of dsRNA and lepidopteran insects often require relatively high dosage of RNAi mole-

cules (Terenius *et al*. 2011). The concentration of injection was 50 µg dsRNA as lower concentrations were ineffective and higher concentrations (>50 µg, injected at a time) induced pupal mortality (data not shown). Hence, each dose of 50 µg dsRNA was administered at intervals. In a similar experiment, the same doses 50 µg dsRNA of vitellogenin receptor of *S. ricini* (*Sr*VgR) severally interfered with the ovarian development and resulted in poorly developed ovarioles (Manuscript under preparation). There is a mixed response to RNAi in lepidopteran insects and an accumulating body of literature has revealed that the efficiency of RNAi varies between different species and genes being targeted (Terenius *et al*. 2011). In *B. germanica* the treatment with 10 µg of ds*Bg*LpR reduced the lipophorin levels in the ovary but had no significant effect on ovarian development and fertility (Ciudad *et al*. 2007). The extension of pupal period by the dsLpR treatment in eri silkworm may indicate its involvement in the pupal-adult transformation. While no other reports are available on RNAi using LpR genes from other insects, further studies are required to establish the sensitivity to RNAi with LpR genes.

Unlike vertebrate counterparts, functional studies on LpRs are scarce. The recombinant receptor was successfully expressed and the results of the binding experiments indicated that *Sr*LpR7-1 bound to Lp of eri silkworm. Furthermore, the disulphide bonds in the receptor are required for binding, as no binding was observed under reducing conditions. Our results are consistent with the bindings of mosquito *Aa*LpRfb and *Aa*LpRov (seven and eight cysteine repeats respectively); both receptors bind to mosquito Lp, despite the differences in their LBDs (Cheon *et al*. 2001, Seo *et al*. 2003). We further demonstrate that *Sr*LpR not only binds to *Sr*Lp but also to *Bm*Lp and the polyclonal antibodies of *Sr*Lp react with *Bm*Lp. The cross-reactivity of *Sr*Lp/LpR indicates they are biochemically and immunologically similar to *Bm*Lp/LpR, a closely related species. The amino acid sequence identity between the two LpRs is 89.10 %. There is only one report available on the cross-reactivity of LpR in which *Manduca sexta* LpR did not bind to human low-density lipoprotein, in spite of its structural similarities with vertebrate lipoprotein receptors (Tsuchida & Wells 1990). However, cross-reactivity of LDL receptors among vertebrates has been reported (Beisiegel *et al*. 1981).

Conclusions

The present study provides the first evidence of the presence of an LDLR homolog in a lepidopteran insect. We have further shown that the *Sr*LpR7 is able to cross-interact with Lp of another lepidopteron species.

Future studies on the protein expressions of LpR variants and their ligand specificities will add to a better understanding of the functionality of this ancient receptor.

Competing interests

The authors state no conflict of interest.

Authors' contributions

R. Bala, U. Saba, M. Varma, D. S. Thomas, D. K. Sinha were involved with experiments, G. Rao, K. Trivedy and V. Kunjupillai were assisted in data analysis, R. Gopalapillai was involved in project planning, evaluation of the data and writing of the paper.

Acknowledgements

We are grateful to Professor K.P. Gopinathan, Dr. Kazuei Mita, and Dr. Kozo Tsuchida for providing the plasmid and antibodies. Thanks are also due to Professor L.S. Shashidhara, IISER, Pune, India for help in the work. This work was supported by grants from Department of Biotechnology (DBT) (No. BT/PR12957/PBD/19/204/2009), New Delhi, Govt. of India. RB, US, and MV are thankful to DBT for support in the form of Junior Research Fellowships.

References

Arrese EL, Canavoso LE, Jouni ZE, Pennington JE, Tsuchida K & Wells MA 2001 Lipid storage and mobilization in insects: current status and future directions. *Insect Biochem Mol Biol* **31** 7-17

Beisiegel U, Toru K, Anderson RGW, Schneider WJ, Brown MS & Goldstein JL 1981 Immunologic cross-reactivity of the low density lipoprotein receptor from bovine adrenal cortex, human fibroblasts, adrenal gland, canine liver and adrenal gland, and rat liver. *J Biol Chem* **256** 4071- 4078

Cheon HM, Seo SJ, Sun J, Sappington TW & Raikhel AS 2001 Molecular characterization of the VLDL receptor homolog mediating binding of lipophorin in oocyte of the mosquito, *Aedes aegypti. Insect Biochem Mol Biol* **31** 753-760

Ciudad L, Bellés X & Piulachs MD 2007 Structural and RNAi characterization of the German cockroach lipophorin receptor, and the evolutionary relationships of lipoprotein receptors. *BMC Molecular Biology* **8** 53

Dantuma N P, Potters M, De Winther MP, Tensen CP, Kooiman FP, Bogerd J & Van der Horst DJ 1999 An insect homolog of the vertebrate very low-density lipoprotein receptor mediates endocytosis of lipophorins. *J Lipid Res* **40** 973-978

Davis, CG, Elhammer A, Russell DW, Schneider WJ, Kornfeld S, Brown MS & Goldstein J L 1986 Deletion of clustered O-linked carbohydrates does not impair function of low density lipoprotein receptor in transfected fibroblasts. *J Biol Chem* **261** 2828 -2838

Gopalapillai R, Kadono-Okuda K, Tsuchida K, Yamamoto K, Nohata J, Ajimura M & Mita K 2006 Lipophorin receptor of *Bombyx mori*: cDNA cloning, genomic structure, alternative splicing, and isolation of a new isoform. *J Lipid Res* **47**1005-1013

Gopalapillai R, Vardhana KV, Bala R, Modala V, Rao G, & Kumar V 2014 Yeast two-hybrid screen reveals novel protein interactions of the cytoplasmic tail of lipophorin receptor in silkworm brain. *J Mol Recognit* **27** 190-196

Guidugli-Lazzarini KR, do Nascimento AM, Tanaka ED, Piulachs MD, Hartfelder K, Bitondi MG & Simões ZL 2008 Expression analysis of putative vitellogenin and lipophorin receptors in honey bee (*Apis mellifera*) queens and workers. *J Insect Physiol* **54** 1138-1147

Lee CS, Han JH, Kim BS, Lee SM, Hwang JS, Kang SW, Lee BH & Kim HR 2003 Wax moth, *Galleria mellonella*, high density lipophorin receptor: alternative splicing, tissue-specific expression, and developmental regulation. *Insect Biochem Mol Biol* **33** 761-771

Pal S, Kundu J, Talukdar S, Thomas T & Kundu SC 2013 An emerging functional natural silk biomaterial from the only domesticated non-mulberry silkworm *Samia ricini. Macromol Biosci* **13** 1020-1035

Parra-Peralbo E & Culi J 2011 Drosophila lipophorin receptors mediate the uptake of neutral lipids in oocytes and imaginal disc cells by an endocytosis-independent mechanism. *PLoS Genet* **7** e1001297

Raikhel AS & Dhadialla TS 1992 Accumulation of yolk proteins in insect oocytes. *Annu Rev Entomol* **42** 527-548

Ravikumar G, Vardhana KV & Basavaraja HK 2011 Characterization of lipophorin receptor (LpR) mediating the binding of high density lipophorin (HDLp) in the silkworm, *Bombyx mori. Journal of Insect Science* **11** 150-158

Rodenburg KW & Van der Horst DJ 2005 Lipoprotein-mediated lipid transport in insects: analogy to the mammalian lipid carrier system and novel concepts for the functioning of LDL receptor family members. *Biochim Biophys Acta* **1736** 10-29

Ryan RO & Van der Horst DJ 2000 Lipid transport biochemistry and its role in energy production. *Annu Rev Entomol* **45** 233-260

Sambrook J, Fritsch EF & Maniatis T 1989. Molecular Cloning: A Laboratory Manual. 2nd edition. Cold

Spring Harbor Laboratory Press, Cold Spring Harbor, NY

Schneider WJ & Nimpf J 2003 LDL receptor relatives at the crossroad of endocytosis and signalling. *Cell Mol Life Sci* **60** 892-903

Seo S J, Cheon, HM, Sun J, Sappington TW & Raikhel AS 2003 Tissue- and stage-specific expression of two lipophorin receptor variants with seven and eight ligand-binding repeats in the adult mosquito. *J Biol Chem* **278** 41954-41962

Shapiro, JP, Keim PS & Law JH 1984 Structural studies on lipophorin, an insect lipoprotein. *J Biol Chem* **259** 3680-3685

Sun J, Hiraoka T, Dittmer NT, Cho KH & Raikhel AS 2000 Lipophorin as a yolk protein precursor in the mosquito, Aedes aegypti. *Insect Biochem Mol Biol* **30** 1161-1171

Swevers L. Raikhel AS, Sappington TW, Shirk P & Iatrou K 2005 Vitellogenesis and post-vitellogenic maturation of the insect ovarian follicle. In *Comprehensive molecular insect science Volume 1*. pp. 87-156. Edited by: Gilbert, L.I., Iatrou, K. and Gill, S.S.: Elsevier, Amsterdam.

Telfer WH, Pan ML & Law JH 1991 Lipophorin in developing adults of *Hyalophora cecropia*: support of yolk formation and preparation for flight. *Insect Biochem* **21** 653-663

Terenius O, Papanicolaou A, Garbutt JS, Eleftherianos I, Huvenne H, Kanginakudru S, Albrechtsen M, An C, Aymeric JL, Barthel A, Bebas P, Bitra K, Bravo A, Chevalier F, Collinge DP, Crava CM, de Maagd RA, Duvic B, Erlandson M, Faye I, Felföldi G, Fujiwara H, Futahashi R, Gandhe AS, Gatehouse HS, Gatehouse LN, Giebultowicz JM, Gómez I, Grimmelikhuijzen CJ, Groot AT, Hauser F, Heckel DG, Hegedus DD, Hrycaj S, Huang L, Hull JJ, Iatrou K, Iga M, Kanost MR, Kotwica J, Li C, Li J, Liu J, Lundmark M, Matsumoto S, Meyering-Vos M, Millichap PJ, Monteiro A, Mrinal N, Niimi T, Nowara D, Ohnishi A, Oostra V, Ozaki K, Papakonstantinou M, Popadic A, Rajam MV, Saenko S, Simpson RM, Soberón M, Strand MR, Tomita S, Toprak U, Wang P, Wee CW, Whyard S, Zhang W, Nagaraju J, Ffrench-Constant RH, Herrero S, Gordon K, Swevers L & Smagghe G 2011 RNA interference in Lepidoptera: An overview of successful and unsuccessful studies and implications for experimental design. *J Insect Physiol* **57** 231-245

Tsuchida K & Wells MA 1990 Isolation and characterization of a lipophorin receptor from the fat body of an insect, *Manduca sexta*. *J Biol Chem* **265** 5761-5767

Tufail M, Elmogy M, Fouda MM, Elgendy AM, Bembenek J, Trang LT, Shao QM & Takeda M 2009 Molecular cloning, characterization, expression pattern and cellular distribution of an ovarian lipophorin receptor in the cockroach, *Leucophaea maderae*. *Insect Mol Biol* **18** 281-294

Tufail M & Takeda M 2009 Insect vitellogenin/ lipophorin receptors: Molecular structures, role in oogenesis, and regulatory mechanisms. *J Insect Physiol* **55** 87-103

Tveten K, Ranheim T, Berge KE, Leren TP & Kulseth M A 2006 Analysis of alternatively spliced isoforms of human LDL receptor mRNA. *Clin Chim Acta* **373** 151-157

Van der Horst DJ, Van Hoof D, Van Marrewijk WJ & Rodenburg KW 2002 Alternative lipid mobilization: the insect shuttle system. *Mol Cell Biochem* **239** 113-119

Van Hoof D, Rodenburg KW & Van der Horst DJ 2003 Lipophorin receptor- mediated lipoprotein endocytosis in insect fat body cells. *J Lipid Res* **44** 1431-1440

Willnow TE, Orth K & Herz J 1994 Molecular dissection of ligand binding sites on the low density lipoprotein receptor-related protein. *J Biol Chem* **269** 15827-15832

8

Biochemical and cytogenetic changes in postovulatory and in vitro aged mammalian oocytes

John B. Mailhes

Department of Obstetrics and Gynecology, Louisiana State University Health Sciences Center, Shreveport, Louisiana, USA 71130

Correspondence should be addressed to John B. Mailhes; E-mail: jmailh@lsuhsc.edu

Abstract

Aneuploidy represents the most prevalent genetic disorder of man. Its association with spontaneous abortions, mental and physical retardation, and numerous malignant cells is well-known. Unfortunately, little is known about the causes and even less about the underlying molecular mechanisms of aneuploidy, especially in mammalian germ cells. Although several etiologies have been proposed for describing human aneuploidy, the only consistent finding remains its positive correlation with maternal age. At the outset, it is essential to point out that there exist numerous potential causes and mechanisms for the etiology of aneuploidy. Nevertheless, information about the molecular mechanisms of chromosome segregation in various species is providing a foundation for research designed to investigate the causes and mechanisms of aneuploidy. The intent of this review is to propose that the biochemical reactions and cellular organelles responsible for accurate chromosome segregation become compromised during postovulatory and in vitro oocyte aging; thus, increasing the probability of faulty chromosome segregation. Recent data have shown that the efficacies of the spindle assembly checkpoint and the chromosome cohesion proteins diminish as oocytes age postovulation and during in vitro culture. Such changes represent potential models for studying aneuploidy. Prior to describing the biochemical and cellular organelle changes found in aged oocytes and their effect on chromosome segregation, an overview of the molecular details surrounding chromosome segregation is presented.

Introduction

This review is based on the premise that postovulatory and in vitro oocyte aging are accompanied by a progressive and functional deterioration of the biochemical pathways and cellular organelles responsible for chromosome segregation. This review does not delve into the extensive literature dealing with the relationship between maternal age and aneuploidy.

The identification, chronology, and interaction of the biochemical events associated with chromosome segregation are undergoing extensive investigation - mainly in models other than mammalian oocytes. Such data provide a foundation for designing experiments to study aneuploidy in mammalian germ cells. To preserve genomic integrity, the interactions between unique biochemical pathways and cellular organelles must be choreographed precisely during mitosis and meiosis. Disarray among these events can result in aneuploidy. At the least, accurate chromosome segregation requires the temporally-coordinated interaction among: protein kinases and phosphatases, topoisomerases, DNA decantination, chiasmata resolution, chromatin condensation, microtubule kinetics, centrosomes, kinetochore-microtubule attachment and tension, kinetochores and their associated proteins, motor and passenger proteins, chromosome biorientation, spindle checkpoint proteins, anaphase promoting complex, proteasomes, securin, cohesin, and separin proteins. Furthermore, the complexity of studying aneuploidy is illustrated by the approximate 5,000 yeast genes that have been directly or indirectly associated with chromosome segregation. Many of these genes are also found in humans, and it appears that the basic mechanisms of chromosome segregation are similar between budding yeast (*Saccharomyces cervisiae*), fission yeast (*Schizosaccharomyces pombe*), and other eukaryotes (Yanagida 2005).

Numerous reports have described biochemical and cellular organelle changes in aged oocytes. However, relatively few studies were designed to concomitantly study such changes and aneuploidy under the same experimental design. Those that have done so, found a positive correlation between biochemical and cellular organelle alterations and aneuploidy (Emery *et al.* 2005, Mailhes *et al.* 1998, Plachot *et al.* 1988, Rodman 1971, Sakurada *et al.* 1996, Yamamoto & Ingalls 1972). It is suggested that experimental manipulation of the reported changes in aged oocytes can be used as models for investigating some of the numerous potential mechanisms of aneuploidy. Recent data from mammalian oocytes showed that degradation of spindle-assembly checkpoint (SAC) proteins (Homer *et al.* 2005a, Steuerwald *et al.* 2005) and chromosome cohesion proteins (Hodges *et al.* 2001, 2005, Prieto *et al.* 2004) increased the probability of premature centromere separation (PCS) and aneuploidy. The reader is referred to other reviews involving aneuploidy in male germ cells (Adler *et al.* 2002, Handel *et al.* 1999) and neoplastic cells (Bharadwaj & Yu 2004, Rajagopalan & Lengauer 2004, Yuen *et al.* 2005).

Overview of Mammalian Oocyte Aneuploidy Research

Although aneuploidy represents the greatest genetic affliction of man, little is known about its causes and even less about its underlying molecular mechanisms, especially in mammalian germ cells. Human aneuploidy is linked with embryonic loss, mental and physical anomalies in newborns, and cancer. Approximately 10-30% of human zygotes (Ford 1981, Hansmann 1983, Hassold & Hunt 2001), 50% of spontaneous abortuses (Bond & Chandley 1983, Hook 1985a), and 0.31% (204/64887) of human newborns (Hecht & Hecht 1987) have an abnormal chromosome number. Furthermore, data based on cytogenetic analyses of human oocytes and preimplantation embryos indicate that over 50% are aneuploid (Kuliev *et al.* 2003, Magli *et al.* 2001, Munne 2002). When considering preimplantation genetic diagnosis for aneuploidy, it seems relevant to note that within embryonic variation was reported among blastomeres when FISH technology was employed (Coulam *et al.* 2007).

For decades, numerous hypotheses have been proposed for the etiology of human germ cell aneuploidy. However, the only constant finding remains its positive correlation with maternal age (Bond & Chandley 1983, Chandley 1987, Hook 1985b). Even for this relationship, definitive data about the underlying mechanisms are lacking (Pellestor *et al.* 2005, Warburton 2005). Besides maternal age, other findings also appear relevant for understanding the genesis of germ cell aneuploidy. Earlier reports showed that the

incidence of aneuploidy for specific chromosomes occurred more frequently during female meiosis I than either meiosis II or male meiosis (Bond & Chandley 1983, Hassold & Sherman 1993, Hook 1985b). However, this finding has recently been questioned (Rosenbusch 2004). Also pertinent are the reports showing that certain human chromosomes are more susceptible to missegregation than others (Hassold 1985, Hassold & Hunt 2001, Hassold *et al.* 1984, Lamson & Hook 1980, Nicolaidis & Petersen 1998). Such variability among chromosomes requires prudence when only specific chromosomes (instead of the entire complement) are used to estimate the incidence of aneuploidy.

Another significant factor for consideration is the sexual dimorphism that exists for both spontaneous and induced germ cell aneuploidy (Eichenlaub-Ritter *et al.* 1996, Pacchierotti *et al.* 2007). Currently, data are unavailable from a study that was specifically designed to evaluate gender differences for mammalian germ cell aneuploidy. Other distinctive features of meiosis include the influence of neighboring somatic cells on germ cell differentiation and entry into meiosis (Geijsen *et al.* 2004, Toyooka *et al.* 2003) and the intrinsic sexual dimorphism between oogenesis and spermatogenesis (Handel & Sun 2005, Hodges *et al.* 2001). Thus, the fundamental distinctions between oogenesis and spermatogenesis, the differential susceptibility among chromosomes, the differences between mitosis and meiosis, and the numerous potential mechanisms demonstrate the complexity of studying aneuploidy.

Although it is not surprising that a unique precept explaining the etiology of aneuploidy has been adopted, it is now generally accepted that both nondisjunction and PCS represent primary events that lead to human aneuploidy (Anahory *et al.* 2003, Angell 1991, Angell *et al.* 1994, Pellestor *et al.* 2006, Wolstenholme & Angell 2000). Moreover, it appears that aneuploidy more often results from PCS of sister chromatids than from nondisjunction of whole chromosomes (Plachot 2003, Rosenbusch 2004). Additionally, variation has been shown to exist among specific chromosomes for the probabilities of both PCS and nondisjunction (Eichenlaub-Ritter 2003, Pellestor *et al.* 2003, Sun *et al.* 2000).

Early germ cell studies concentrated on the ability of various chemicals (mainly those that damaged microtubules) to induce aneuploidy (Adler 1990, 1993, Allen *et al.* 1986, Eichenlaub-Ritter 1996, Mailhes 1995, Mailhes *et al.* 1986, Miller & Adler 1992). Other investigations involved assay development and validation (Pacchierotti 1988, Mailhes & Marchetti, 1994, Eastmond *et al.* 1995, Parry *et al.* 1995) and gender differences (Eichenlaub-Ritter 1996, Eichenlaub-Ritter et al. 1996, Pacchierotti et al. 2007, Wyrobek et al. 1996). Based on the results from some of these studies, a broad-working hypothesis emerged. Several investigators proposed that endogenous-and exogenous-induced perturbations during the temporal sequence of oocyte maturation (OM) predispose oocytes to aneuploidy (Eichenlaub-Ritter 1993, Hansmann & Pabst, 1992, Mailhes & Marchetti 1994a). This proposal suggested that an induced temporal disarray (usually detected as a transient delay during metaphase I) among the cellular organelles and biochemical reactions controlling OM increased the probability of aneuploidy.

Although considerable data have shown that a chemically-induced delay during OM is often associated with chromosome missegregation, exceptions can be found. Phorbol 12,13-dibutyrate (Eichenlaub-Ritter 1993), colchicine (Mailhes & Yuan 1987), vinblastine sulfate (Mailhes & Marchetti 1994a, Russo & Pacchierotti 1988), griseofulvin (Marchetti & Mailhes 1995, Mailhes et al. 1993, Tiveron et al. 1992) induced both meiotic delay and aneuploidy. Conversely, isobutyl-1-methylxanthine and forskolin caused meiotic delay, but not aneuploidy (Eichenlaub-Ritter 1993), while etoposide treatment resulted in aneuploidy without meiotic delay (Mailhes et al. 1994, Tateno & Kamiguchi 2001). Consistency among the results from different studies regarding chemically-induced meiotic delay and oocyte aneuploidy cannot necessarily be expected due to: different experimental protocols, exposure of cells to compounds with diverse and often multiple modes of action, and the numerous potential mechanisms of aneuploidy (Mailhes 1995, Pacchierotti & Ranaldi 2006).

More recent studies have combined immunocytochemical techniques with cytogenetic analyses to affirm a positive correlation between oocyte meiotic spindle abnormalities and aneuploidy (Eichenlaub Ritter et al. 1996, Mailhes et al. 1999). Additional studies employing small molecule, cell-permeable inhibitors of specific biochemical reactions during cell division showed that the proteasome and calpain inhibitor MG-132 (Mailhes et al. 2002), the protein phosphatase 1 and 2A inhibitor okadaic acid (Mailhes et al. 2003a), and the Eg5 kinesin inhibitor monastrol (Mailhes et al. 2004) induced aneuploidy in mouse oocytes, while the tyrosine inhibitor vanadate resulted in spontaneous oocyte activation (Mailhes et al. 2003b).

Considerable research is being devoted to unraveling the molecular events underlying chromosome segregation, mainly in non-mammalian somatic cells (Lee & Orr-Weaver 2001, Nasmyth 2001, Uhlmann 2003a). The complexity of understanding the multifaceted events comprising chromosome segregation is illustrated by the approximately 5,000 yeast genes involved with chromosome segregation; many of these yeast genes have also been found in humans (Yanagida 2005). Chromosome segregation requires the temporally-coordinated interaction among: topoisomerases, chiasmata resolution, chromatin condensation, protein kinase and phosphatase reactions, microtubule kinetics, centrosomes, kinetochore-microtubule attachment and bipolar tension, kinetochores and their associated proteins, anaphase promoting complex, proteasomes, and cohesion, securin, and separin proteins. Although these events generally transcend among species and cell types (Yanagida 2005), little is actually known about the molecular mechanisms of chromosome segregation in mammalian oocytes (Collins & Crosignani 2005).

Thus, it emerges that the current state of mammalian germ cell aneuploidy research is mainly descriptive with little information about the underlying molecular mechanisms. It seems that the status of aneuploidy research can be summarized by an earlier statement, "The fact is that we are really not very much nearer today to pinning down the responsible mechanisms than we were twenty years ago when the human aneuploid conditions were first identified" (Bond & Chandley 1983).

Oocyte maturation

Disarray among the numerous events that occur during OM may lead to faulty chromosome segregation. Mammalian oogenesis is controlled by FSH, LH, autocrine and paracrine signaling, and unique growth factors (Anderiesz et al. 2000, Hiller 2001). Meiosis begins in the fetal ovary and is later arrested postpartum at the diplotene/ dictyate stage during meiosis I. Unless human oocytes undergo atresia, they remain in diplotene for decades until meiosis resumes prior to ovulation. Following proper hormonal stimulation, oocytes undergo the transition from diplotene to metaphase II (MII). This transition represents OM and involves nuclear and cytoplasmic remodeling and reduction to the haploid state (Dekel 1988, Racowsky 1993, Schultz 1986, Schultz 1988, Schultz et al. 1983). Upon completing OM, mammalian oocytes remain in MII for a limited time period until fertilization, spontaneous activation, or atresia. In most mammals, MII oocytes are ovulated and primed for fertilization, which initiates anaphase II. Among marine invertebrates, amphibians, fish, and mammals, species-dependent protein modifications by kinases and phosphatases account for differences in the initiation and the orderly temporal sequence of events during OM (Yamashita *et al.* 2000).

The intraoocyte titer of cyclic adenosine monophosphate (cAMP) influences the initiation of mammalian OM. Elevated levels of cAMP favor cAMP-dependent kinase activity and the retention of oocytes in the diplotene/dictyate stage of meiotic prophase. Conversely, low cAMP levels shift the equilibrium toward cAMP-dependent phosphatase activity, which is needed for activating maturation promoting factor (MPF) and the progression of OM (Boernslaeger *et al.* 1986, Dekel 1988, Dekel 2005, Downs *et al.* 1989, Racowsky 1993, Schultz 1988, Schultz *et al.* 1983). MPF is composed of a 34 kDa catalytic subunit (p34^{cdc2}) that exhibits serine-threonine kinase activity and a 45 kDa cyclin B regulatory subunit. In addition to low cAMP levels, MPF activation also requires that p34^{cdc2} be dephosphorylated at the tyrosine 15 residue and coupled with cyclin B. Conversely, tyrosine phosphorylation deactivates MPF (Dunphy & Kumagai 1991, Gautier *et al.* 1991, Strausfeld *et al.* 1991). MPF activity oscillates; it is highest during metaphase and lowest during anaphase (Arion *et al.* 1988, Draetta & Beach 1988), fertilization (Choi *et al.* 1991, Collas *et al.* 1993, Fulka *et al.* 1992), and partheneogenesis (Barnes *et al.* 1993, Collas *et al.* 1993, Kikuchi *et al.* 1995).

Besides MPF, other kinases and phosphatases also play significant roles during OM (Swain & Smith 2007). Mitogen-activated protein kinases (MAPKs) represent serine-threonine protein kinases that phosphorylate many of the same sites as active MPF (Fan & Sun 2004, Lee *et al.* 2000, Murray 1998, Takenaka *et al.* 1998). MAPKs mediate intracellular signal transmission in response to external stimuli, participate in assembling the first meiotic spindle, and prevent rodent oocytes from entering interphase during the interval between meiosis I and II (Gordo *et al.* 2001, Sobajima *et al.* 1993, Verlhac *et al.* 1994). Unlike MPF, MAPK activity remains high throughout OM.

Mos, the c-mos protooncogene product, represents another serine-threonine kinase that is active during OM (Paules *et al.* 1989, Sagata 1997, Singh & Arlinhgaus 1997). It helps activate the MAPK pathway (Dekel 1996) and functions as a cytostatic factor by preventing oocytes from prematurely exiting MII (Hashimoto 1996, Sagata 1996). Oocytes from c-mos deficient mice fail to arrest at MII and subsequently undergo spontaneous partheneogenic activation (Colledge *et al.* 1994, Hashimoto 1996, Hashimoto *et al.* 1994). In addition to their roles during OM, the kinases MPF, MAPKs, and Mos also have essential roles during the SAC, the anaphase promoting complex/cyclosome (APC), and the metaphase-anaphase

transition (MAT) (Dekel 1996, Dorée *et al.* 1995, Hyman & Mitchison 1991, Karsenti 1991, Murray 1998).

Correct temporal and synchronous interactions between specific enzymes and their target compounds are required for OM and the MAT, and faulty kinase and phosphatase activities have been shown to lead to downstream errors resulting in chromosome missegregation. Based on their antagonistic effects, relative to the degree of tyrosine p34^{cdc2} phosphorylation, unique kinase and phosphatase inhibitors have the potential for altering the rate of OM and for inducing spindle defects and aneuploidy in rodent oocytes (Jesus *et al.* 1991). Okadaic acid (OA) specifically inhibits the protein phosphatases 1 (PP1) and 2A (PP2A) that dephosphorylate serine and threonine residues (Cohen *et al.* 1990, Schönthal 1992). Following OA treatment of mouse oocytes and one-cell zygotes, hyperphosphorylation was noted in conjunction with abnormalities involving spindle fibers, multipolar spindles, kinetochores, and chromosome alignment (De Pennart *et al.* 1993, Schwartz & Schultz 1991, Vandre & Willis 1992, Zernicka-Goetz *et al.* 1993). Also, elevated frequencies of PCS and aneuploidy were found in mouse oocytes exposed to OA (Mailhes *et al.* 2003a). These effects may have been influenced by OA-induced hyperphosphorylation of microtubule organizing centers and microtubule-associated proteins (MAPs) (Schwartz & Schultz 1991, Vandre and Willis 1992) and that hyperphosphorylated MAPs have a reduced affinity for microtubules (Zernicka-Goetz *et al.* 1993).

Furthermore, the kinase inhibitor 6-dimethylaminopurine (6-DMAP) disrupts p34^{cdc2} kinase and MAPK activities and prevents meiotic progression of mouse oocytes (Rime *et al.* 1989, Szollosi *et al.* 1991, 1993). Protein phosphorylation and germinal vesicle breakdown (GVBD) were repressed when dictyate mouse oocytes were exposed to 6-DMAP prior to (GVBD); conversely, expulsion of the first polar body was inhibited when oocytes were exposed after GVBD (Rime *et al.* 1989). Other data showed that 6-DMAP inhibited protein phosphorylation in activated mouse MII oocytes and resulted in premature disappearance of phosphorylated proteins coupled with abnormalities involving polar body extrusion and pronuclei formation (Szollosi *et al.* 1993). Additionally, the pattern of protein dephosphorylation events noted in postovulatory and in vitro aged oocytes was correlated with increased frequencies of spontaneous oocyte activation and PCS (Angell 1994, Dailey *et al.* 1996).

A selected list of compounds associated with the metaphase-anaphase transition (MAT) during mitosis and meiosis and their general function is presented in Tables 1A and B. Such a listing is non-comprehensive and will certainly be modified and expanded as additional data become available.

The metaphase-anaphase transition (MAT) during mitosis and meiosis

Prior to the MAT and chromosome segregation, numerous events require coordination. These include: chromatin condensation, microtubule polymerization and their capture by kinetochores, correction of erroneous microtubule-kinetochore interactions, generation of microtubule-kinetochore tension, formation of a stable bipolar spindle, satisfaction of the spindle assembly checkpoint, removal of linkages between sister chromatid arms, and temporally-coordinated removal of centromeric cohesion proteins.

Although chromosome segregation during meiosis appears to largely depend on mechanisms analogous to those of mitosis, both general cell-cycle regulators and unique proteins have been identified during meiosis (Nasmyth 2001). Three major modifications of the mitotic machinery occur during meiosis. First, synapsis and recombination (chiasmata formation) occur between homologues prior to anaphase I. Second, the two sister chromatids of each chromosome must segregate syntelically while the homologues segregate amphitelically at anaphase I. Third, the cohesion between sister chromatid centromeres must remain intact until anaphase II onset in order for sister chromatids to segregate amphitelically.

Before discussing the cytologic and biochemical changes reported in aged oocytes and their effect on chromosome segregation, an overview of the physical and chemical linkages between chromosomes, kinetochore-microtubule interactions, the spindle checkpoint assembly complex, and the metaphase-anaphase transition is presented.

Resolution of DNA catenations, chromatin condensation, and removal of cohesion arm proteins

Following DNA replication, sister chromatids are linked by DNA double-strand catenations and cohesion proteins. These physical and chemical linkages help prevent precocious separation prior to anaphase onset, which can result in aneuploidy. However, these linkages must be timely removed so that sister chromatids orient syntelically at meiotic anaphase I and undergo amphitelic orientation during meiotic anaphase II and mitotic anaphase. Most of the DNA catenations on chromosome arms are lost prior to prophase; whereas, the majority of chromosome arm cohesin proteins are removed during prophase. However, it is essential that centromeric catenations and cohesions remain intact until correct kinetochore-microtubule attachment and tension have been attained. Otherwise, premature loss of centromeric cohesion inevitably predisposes cells to abnormal chromosome segregation.

Abnormal function of proteins required for establishing and maintaining the physical linkages between sister chromatids may result in aneuploidy and apoptosis. The Spo11 protein helps initiate meiotic recombination by generating DNA double-strand breaks, and disruption of Spo11 activity in mouse spermatocytes and oocytes resulted in synaptic-deficient germ cells and apoptosis (Baudat et al. 2000, Romanienko et al. 2000). Also, the synaptonemal complex protein 3 (Sycp3) helps maintain the structural integrity of meiotic chromosome axes. Mutant Sycp3 mammalian oocytes were ineffective in repairing DNA double-strand breaks and exhibited higher frequencies of aneuploidy (Wang & Hoog 2006).

Table 1A. Selected regulators of mitosis and meiosis.

Cohesion complex subunit proteins identified during mitosis	Cohesion complex subunit proteins identified during meiosis	Spindle assembly checkpoint (SAC) proteins
Smc1 and Smc3 (structural maintenance of chromosomes) – core cohesion complex subunit proteins.	**Smc1α** – replaces mitotic Smc1.	**Mad1** (mitotic-arrest deficient) – helps recruit Mad2 to kinetochores that lack tension and attachment. Forms a complex with Cdc20, Mad2, and Mad3.
Scc1/Rad21/Mcd1 (sister chromatid cohesion) – cleaved by separase at mitotic anaphase onset.	**Smc1β** – replaces mitotic Smc3.	**Mad2** – forms a complex with Cdc20, Mad1, and Mad3 and inhibits APC^{Cdc20} activity.
Scc3 (SA1/STAG1, SA2/STAG2) – phosphorylated by Aurora B and Plk1 kinases.	**STAG3** – replaces mitotic Scc3.	**Mad3/BubR1** – helps recruit Mad1 and Mad2 to kinetochores that lack attachment and tension; forms a complex with Cdc20, Mad1, Mad2, and Bub3.
Scc2 and Scc4 – enhance the binding of Scc1 and Scc3 to kinetochores that lack attachment and tension.	**Rec8** – replaces mitotic Scc1.	**Bub1** (budding inhibited by benzimidazole) – a serine-threonine protein kinase that binds with Bub3, Mad1, Mad2, Mad3, and CenP-E and helps recruit Shugoshin proteins to kinetochores.
		Bub2/Mps1 – helps regulate APC^{Cdh1}, mitotic exit, chromosome replication, and cytokinesis.
		Bub3 – binds with Bub1 and Mad3 and helps regulate APC activity.

Table 1B (continued from **1A**). Selected regulators of mitosis and meiosis.

Other compounds	Function
APC/C (anaphase promoting complex/cyclosome)	A 20S multi-subunit ligase that ubiquinates specific proteins targeted for proteolysis by proteasomes. APC^{cdc20} targets securin for proteolysis at the MAT; whereas, APC^{cdh1} targets mitotic cyclins and other substrates for degradation at mitotic exit. The cdh1 protein activates the APC from late anaphase through G1.
Astrin	A microtubule and kinetochore protein that has roles involving sister chromatid adhesion, centrosome integrity, and separase activity.
Aurora B kinase-Survivin-Inner Centromeric Protein-Borelin	A chromosome passenger protein complex with multiple roles: recruits SAC proteins and CenP-E to kinetochores lacking tension, reduces the affinity of Scc1 and Scc3 to chromatin via phosphorylation, helps coordinate correct kinetochore-microtubule attachments, and cytokinesis.
Cdc20 (cell division cycle 20)	Helps activate the APC when not bound by SAC proteins, recruits substrates to the APC, and forms a complex with Mad2, Mad3, and Bub3.
Cdks (Cyclin-dependent kinases)	Enzymes composed of a kinase subunit and an activating cyclin subunit. Cdks are needed for kinase activity.
CenP-E (centromeric protein E)	A motor protein that facilitates kinetochore-microtubule stabilization, binding of SAC proteins to kinetochores, and enhanced Mad3 activity.
Dynein/Dynactin	A microtubule motor protein required for the removal of the Rod-Zw10-Zwilch complex, Mad1, Mad2, and Mad3 from properly aligned kinetochores.
Kinesin	A microtubule motor protein.
MAPK/Mps1 (mitogen-activated protein kinase)	A serine-threonine kinase that helps recruit CenP-E to kinetochores. It also interacts with Mos protein for MPF activation.
MCAK/Kip 2-3 (microtubule centromere-associated kinesin)	Depolarizes microtubules and helps correct aberrant kinetochore-microtubule attachments.
Monopolin/Mam1/CdcPlk	Facilitates amphitelic orientation of homologues and syntelic orientation of sister chromatids during meiosis I.
Mos	The protein product of the *c-mos* proto-oncogene. Mos is an active component of a cytostatic factor. In conjunction with cyclin-dependent kinase 2, Mos is required for the metaphase II arrest of mature mouse oocytes and for activating MAPK.
MPF (maturation promoting factor)	A protein kinase comprising $p34^{cdc2}$/Cdk1 and cyclin B. MPF phosphorylates and helps regulate chromosome condensation, nuclear envelope breakdown, and spindle formation.
Op18/Stathmin (oncoprotein 18)	A protein that destabilizes microtubules; it is inhibited by phosphorylation.
P31/Cmt2	A protein involved with changing the stereo-configuration of Mad2.
Plk1 (Polo-like kinase 1)	A serine-threonine kinase that phosphorylates Scc1, Scc3, and Rec8 and reduces their affinity to chromosome arms.
PP2A (protein phosphatase 2A)	Dephosphorylates Sgo1 and supports Rec8 maintenance.
Proteasomes	Proteinase complexes that degrade intracellular ubiquinated compounds.
Rod-Zw10-Zwilch	A protein complex that helps recruit dynein, Mad1, and Mad2 to unaligned kinetochores.
Securin/Pds1/Cut2p	An APC substrate that binds to and inhibits separase activity.
Separase/Esp1	A protease that is inactive when bound by securin. However, upon securin proteolysis, separase is free to cleave centromeric Scc1 cohesions at mitotic anaphase onset, Rec8 at chromosome arms at meiotic anaphase I onset, and centromeric Rec8 at meiotic anaphase II onset.
Shugoshins (Sgo1 & Sgo2)	Sgo1 is a conserved eukaryotic kinetochore protein that protects centromeric Rec8 from separase activity during meiosis I, but not during meiosis II. Sgo1 enhances dephosphorylation and cohesion removal by recruiting PP2A to kinetochores. Shugoshins also have roles in chromosome congression, kinetochore-microtubule attachment, and syntelic orientation of sister chromatids during meiotic anaphase I
Slk19p	The *Saccharomyces cervisiae* Slk19p gene product is needed for proper chromosome segregation during meiosis I.
Sororin	An APC protein substrate that interacts with Shugoshins to facilitate cohesion binding to chromatin.
Spindly	A protein that helps inactivate the APC and participates with dynactin in recruiting dynein to kinetochores.
Spo11	Helps initiate meiotic recombination.
Sycp3	Helps maintain the structural integrity of meiotic chromosome axes.
Topoisomerase II (Topo II)	An enzyme that disrupts intercalated loops of DNA and then reanneals the DNA broken ends.
UbcH10	An enzyme that ubiquinates Cdc20. This facilitates the release of Mad2 and BubR1 from Cdc20, inactivates the SAC, and helps activate the APC.
Usp44	An enzyme that deubiquinates Cdc20. This enhances the retention of Mad2 by Cdc20, promotes SAC activity, and inhibits APC activation.

Besides Spo11 and Sycp3, other proteins also participate in resolving chiasmata, condensing chromatin, and facilitating chromatid cohesion and separation. Topoisomerase II (topo II) disrupts the intercalated loops on adjacent chromatids by catalyzing a DNA double-strand break in one of the sister chromatids. This enables the other sister chromatid to pass through the broken ends followed by topo II re-annealing the broken ends (Champoux 2001, Downes *et al.* 1991, Holm *et al.* 1989, Rose *et al.* 1990, Wang 2002). Sister chromatids remained physically linked and fail to separate during anaphase in cells lacking topo II activity (Dinardo *et al.* 1984). Thus, topo II activity is required for the transition from prophase to metaphase I (MI) in mouse spermatocytes (Cobb *et al.* 1997) and for proper chromosome segregation in mammalian somatic cells (Gorbsky 1994), mouse oocytes (Mailhes *et al.* 1994), and mouse spermatocytes (Marchetti *et al.* 2001). Besides topo II, other proteins also help disentangle and condense chromatin. The structural maintenance of chromosome proteins (Smc 2 and Smc4) bind to chromatid axes and help disentangle and condense sister chromatids and homologues during prophase and prometaphase (Hagstrom *et al.* 2002, Lavoie *et al.* 2002, Ono *et al.* 2004).

In addition to the physical linkages between sister chromatids, highly-conserved, multi-subunit protein cohesin complexes adhere to eukaryotic chromosomes and help conjoin sister chromatids and homologues (Marston & Amon 2004, Nasmyth & Schleiffer 2004). Some of the cohesin subunits differ between mitotic and meiotic cells (van Heemst & Heyting 2000). Such distinctions may reflect the need for maintaining cohesion during meiotic recombination and the requirement for sister chromatids to undergo syntelic segregation during meiotic AI and amphitelic segregation during AII (Revenkova & Jessberger 2005). Eukaryotic mitotic cells encode homologs of the Scc1/Rad21, Scc3 (SA1/STAG1, SA2/STAG2), Smc1, and Smc3 cohesion protein subunits (Haering & Nasmyth 2003, Parra *et al.* 2004, Prieto *et al.* 2002). Both Scc1 and Scc3 enhance cohesion by binding to numerous sites on chromosomes, while the core subunit proteins Smc1 and Smc3 are needed for both sister chromosome cohesion and DNA recombination (Eijpe *et al.* 2000, Haering *et al.* 2002, Lavoie *et al.* 2002, Petronczki *et al.* 2003). Scc2 and Scc4 represent a separate protein complex in yeast that facilitates the binding of cohesion proteins to centromeres and chromosome arms (Ciosk *et al.* 2000). The following differences in cohesion subunits have been found in meiotic cells: Rec8 replaces Scc1 in both budding yeast (Klein *et al.* 1999, Watanabe & Kitajima 2005) and mammals (Eijpe *et al.* 2003, Parisi *et al.* 1999); STAG3 replaces the Scc3 subunits SA1 and SA2 in mammals (Prieto *et al.* 2001, 2004); Smc1α and Smc1β replace Smc1 (Revenkova *et al.* 2001); and homologs for Smc3 have not been identified.

Cohesin proteins must remain located on centromeres until anaphase onset. Otherwise, early or non-removal can result in PCS or nondisjunction, respectively. The retention and removal of cohesion proteins require the activities of unique kinase, phosphatase, separase, and Shugoshin proteins. During mitotic prophase-prometaphase and meiotic MII, most of the arm cohesins are lost following phosphorylation of the Scc1 and Scc3 cohesin subunits by Aurora B kinase and Polo-like kinases (PLKs) (Alexandru *et al.* 2001, Clyne *et al.* 2003, Hauf *et al.* 2005, Lee & Amon 2003, Losada *et al.* 2002, Sumara *et al.* 2002, Yu & Koshland 2005). On the other hand, loss of centromeric cohesin is mediated by separase cleavage of Scc1 during mitotic anaphase onset (Uhlmann *et al.* 2000a, Uhlmann 2001, Waizenegger *et al.* 2000). Additionally, phosphorylation by PLKs also enhances the removal of centromeric cohesins (Clarke *et al.* 2005, Dai *et al.* 2003, Goldstein 1980, Lee *et al.* 2005). Although PLK phosphorylation has been detected during meiosis in female mice and the first zygotic division, its multi-faceted role requires additional investigation (Pahlavan *et al.* 2000). As will be mentioned later, both Aurora B kinase and PLKs have additional functions during cell division.

During meiosis I, DNA catenations and chromosome arm cohesins must be removed prior to anaphase so that homologues segregate amphitelically and sister chromatids segregate syntelically. Such removal of the meiosis-specific Rec8 cohesin protein on chromosome arms during meiosis I is facilitated by separase. However, it is essential that centromeric Rec8 remain intact between sister chromatids during anaphase I so that they can undergo syntelic orientation (Pasierbek *et al.* 2001, Siomos *et al.* 2001). Rec8 displays a similar pattern of localization in mammalian oocytes and spermatocytes and yeast; it is lost from chromosome arms during the MI-AI transition and from sister centromeres at the onset of AII (Lee *et al.* 2003, 2006).

Mammalian and yeast cells that lack cohesin proteins exhibited elevated frequencies of PCS and chromosome missegregation (Hoque & Ishikawa 2002, Michaelis *et al.* 1997, Sonoda *et al.* 2001, Tanaka *et al.* 2000). The *Saccharomyces cerevisiae* Slk19p gene is required for proper chromosome segregation during meiosis I. Slk19p mutants failed to maintain Rec8 at centromeres during anaphase I and displayed elevated levels of PCS and improper amphitelic segregation of sister chromatids (Kamieniecki *et al.* 2000). Phosphorylation of Rec8 facilitates its cleavage; whereas, dephosphorylation of Rec 8 by PP2A maintains centromeric cohesion during meiosis I (Kitajima *et al.* 2006, Riedel *et al.* 2006). Okadaic acid (OA) functions as a phosphatase 1 and 2A inhibitor, and exposure of mouse oocytes to OA resulted in elevated frequencies of PCS in both MI and MII oocytes and in aneuploid MII oocytes. The higher frequencies of PCS noted in oocytes was proposed to result from an OA-induced shift in the kinase-phosphatase equilibrium that favored enhanced kinase activity (Mailhes *et al.* 2003a).

Centromeric Rec8 must be protected from separase activity during meiosis I in order to facilitate syntelic orientation of sister chromatids during AI. This is enhanced by a group of evolutionarily-conserved eukaryotic Shugoshin (Sgo) proteins and

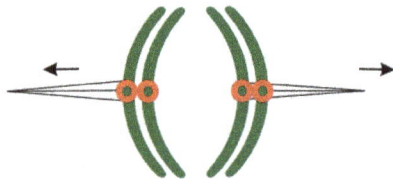

Amphitelic - Proper attachment of homologous chromosomes to a bipolar spindle and their orientation to opposite poles. Each daughter cell is expected to receive one chromosome (composed of two chromatids) resulting in a haploid state.

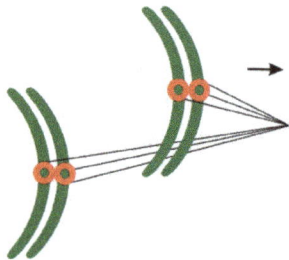

Syntelic - Improper attachment of both chromosomes to a monoastral spindle and their orientation to the same pole. One daughter cell is expected to receive both chromosomes (hyperhaploid), while the other cell will be minus a chromosome (hypohaploid).

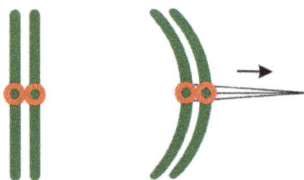

Monotelic - Improper attachment of one chromosome to a monoastral spindle and its orientation to one pole. The other chromosome is neither attached nor oriented. One daughter cell is expected to receive one chromosome (haploid), while the other daughter cell will be minus a chromosome (hypohaploid).

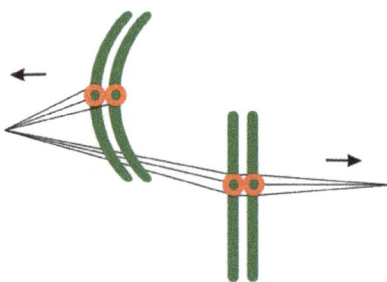

Merotelic - Improper attchment of one chromosome to a bipolar spindle and its non-orientation. The other chromosome is attached to a monoastral spindle and oriented to one pole. Onne daughter cell is expected to receive one chromosome (haploid) while the fate of the other chromosome is uncertain. Merotelic attachments are believed not to activate the SAC and may be corrected prior to anaphase onset (Cimini et al., 2004; Cimini, 2007). Also, anaphase can still occur in the presence of unattached kinetochores, microtubule disruption, and abnormal chromosome orientation (Rieder and Palazzo, 1992; Rieder et al., 1994).

Figure 1. Kinetochore-microtubule attachments and probable outcomes during meiosis I.

their orthologs (Katis *et al.* 2004a, Kitajima *et al.* 2004, Salic *et al.* 2004). Sgo1 in budding yeast (Kitajima *et al.* 2004) and its paralogue Sgo2 in fission yeast (Rabitsch *et al.* 2004) were initially identified and require Bub1 for proper centromeric localization (Kitajima *et al.* 2004). Subsequently, human and mouse Sgo1 and Sgo2 proteins were recognized (McGuinness *et al.* 2005, Tang *et al.* 2004, Watanabe & Kitajima 2005). During mitosis and meiosis in higher eukaryotes, Sgo1 helps maintain sister centromere cohesion by protecting centromeric Rec 8 from separase until sister chromatids undergo amphitelic segregation at anaphase II onset (Goulding & Earnshaw 2005, Kitajima *et al.* 2004, Marston *et al.* 2004, Tang *et al.* 2004, Watanabe & Kitajima 2005). In budding yeast, Sgo1disappears during anaphase I (Kitajima *et al.* 2004, Rabitsch *et al.* 2004); whereas, fission yeast Sgo2 persists until meiosis II (Katis *et al.* 2004a, Kitajima *et al.* 2004). Sgo2 in fission yeast represents a paralogue of Sgo1 and is required for chromosome congression at metaphase, proper kinetochore-microtubule attachment, and syntelic orientation of sister chromatids during AI (Kitajima *et al.* 2004, Rabitsch *et al.* 2004). Depletion of either Sgo1 (Wang & Dai 2005) or Sgo2 (Kitajima *et al.* 2004, Rabitsch *et al.* 2004) during meiosis I led to PCS and chromosome missegregation, and knock-out of Sgo1 in fission yeast resulted in chromosome missegregation (Gregan *et al.* 2005).

PP2A colocalizes with centromeric Sgo1 in human mitotic and meiotic cells. This enhances efficient PP2A dephosphorylation of Rec8, which renders it resistant to subsequent phosphorylation and cleavage. Furthermore, reduced PP2A activity resulted in loss of centromeric cohesion during mitosis and meiotic anaphase I accompanied by random sister chromatid segregation during meiotic anaphase II (Kitajima *et al.* 2006, Riedel *et al.* 2006, Tang *et al.* 2006). A human shugoshin-like protein (possibly orthologous to yeast Sgo1) localized to HeLa cell centromeres during prophase prevented phosphorylation of the Scc3 cohesin subunit. This protein normally disappears at anaphase onset, and its depletion by RNAi resulted in PCS (McGuinness *et al.* 2005).

Kinetochore-microtubule interaction, correction of faulty attachments, generation of tension and stabilization, and biorientation

Kinetochores help regulate chromosome segregation during mitosis and meiosis by mediating three main functions: attaching chromosomes to microtubules, facilitating microtubule dynamics essential for chromosome movement, and providing the site for spindle checkpoint activity. Kinetochores may initially capture

microtubules by four different modes (Biggins & Walczak 2003, Cinini *et al.* 2001): (1) *Amphitelic* – sister kinetochores orientated to opposite poles of a bipolar spindle, (2) *Syntelic* – both kinetochores of sister chromatids attached to a monastral spindle, (3) *Monotelic* – only one kinetochore is orientated to a pole while the other is unattached, and (4) *Merotelic* – one kinetochore is attached to both poles (Figures 1 & 2). During metaphase of mitosis and meiosis II, amphitelic orientation of sister chromatids is needed; whereas during meiosis I, syntelic attachment of sister chromatids and amphitelic attachment of homologues are required. Persistent monotelic and merotelic attachments, if not corrected, can lead to chromosome missegregation; whereas, merotelic attachments are not detected by the spindle checkpoint (Cimini 2007, 2008, Cimini *et al.* 2001, 2004, Rieder & Maiato 2004, Salmon *et al.* 2005).

Kinetochores contain both constitutive (structural) proteins (Amor *et al.* 2004) and transient (passenger) proteins that help coordinate various events during mitosis and meiosis (Duesbery *et al.* 1997, Vagnarelli & Earnshaw 2004). The constitutive centromeric proteins (CENP-A, B, C, D) are involved with: microtubule capture, correcting aberrant interactions, binding of spindle checkpoint proteins, and chromosome congression to the metaphase plate (Craig *et al.* 1999, Rieder & Salmon 1998, Simerly *et al.* 1990, Vagnarelli & Earnshaw 2004). Whereas, the transient proteins reside in the nucleus during G2, associate with chromosomes during prophase, localize to centromeres during metaphase, and transfer to the spindle at anaphase onset (Earnshaw & Cooke 1991).

The Aurora A and Aurora B serine-threonine protein kinases help support mitotic spindle assembly by phosphorylating the structural and motor proteins that are essential for spindle assembly and anaphase onset (Giet *et al.* 2005, Meraldi *et al.* 2004). The biorientation of homologues during meiotic MI and that of sister chromatids during mitosis and meiotic MII resembles a state of equilibrium between sister chromatid cohesion and microtubule-kinetochore tension (Miyazaki & Orr-Weaver 1994, Tanaka *et al.* 2000, Toth *et al.* 1999). Plk1 and Aurora B kinases are also involved with a fundamental function that decreases the incidence of chromosome missegregation. These kinases help correct aberrant microtubule-kinetochore attachments by generating kinetochore-microtubule tension (Ahonen *et al.* 2005, Stern 2002, Tanaka *et al.* 2002). After proper correct microtubule-kinetochore attachment and tension have been attained, microtubule polymerization-depolymerization is minimized while centromeres bi-orient and align on the metaphase plate. In addition to Aurora B kinase, the depolymerase activity of mitotic centromere associated kinesin (MCAK) helps coordinate the release of merotelic kinetochore-microtubule attachments (Kallio *et al.* 2002, Knowlton *et al.* 2006). Also, the SPO13 protein and the monopolin protein complex found during meiosis I in fission yeast facilitate syntelic orientation of sister chromatids (Katis *et al.* 2004b, Lee *et al.* 2004).

The Aurora B-inner-centromeric protein (INCENP)-Survivin-Plk1-Borealin transient protein kinase complex is involved with several functions involving chromosome segregation and cytokinesis; these include: (1) chromatin decondensation, (2) reducing the affinity of Scc1 for chromatin at chromosome arms, (3) generating tension at kinetochores, (4) organizing a bipolar spindle, (5) targeting SAC proteins to kinetochores, (6) initiating cytokinesis, (7) inhibiting the APC, (8) sensing and correcting abnormal microtubule-kinetochore attachments, and (9) influencing spindle geometry by phosphorylating MCAK (Adams *et al.* 2001a, b, Dewar *et al.* 2004, Shang *et al.* 2003, Tanaka *et al.* 2002, Vagnarelli & Earnshaw 2004). INCENP, Survivin, and Plk1 are needed for the proper kinetochore localization of Aurora B and for correcting merotelic microtubule-kinetochore attachments (Bolton *et al.* 2002, Ditchfield *et al.* 2003, Goto *et al.* 2006, Tong *et al.* 2002). Survivin also has important roles during spindle checkpoint signaling and in correcting abnormal kinetochore-spindle fiber attachments (Carvalho *et al.* 2003, Hwang *et al.* 1998, Johnson *et al.* 2004, Lampson *et al.* 2004, Lens & Medema 2003, Taylor *et al.* 2001, 2004). Aurora B kinase activity helps to destabilize syntelic attachments of sister chromatids during meiosis II and mitosis; this enhances the re-formation of correct amphitelic orientation (Hauf *et al.* 2003, Tanaka *et al.* 2002). Furthermore, overexpression of a stable form of Aurora B in mammalian somatic cells led to aneuploidy (Nguyen *et al.* 2005).

Spindle assembly checkpoint (SAC) protein complex and correction of faulty kinetochore-microtubule attachments
Chromosome segregation represents an irreversible event; orientation errors cannot be rectified after anaphase onset. In order to reduce the risk of missegregation, it is essential that a bipolar spindle be formed following correct microtubule-kinetochore attachment and tension. This is not left to chance. A transient mechanico-chemo surveillance mechanism or spindle-assembly checkpoint (SAC) protein complex helps insure that proper chromosome alignment and kinetochore-microtubule tension are attained prior to anaphase onset. However, the SAC is not foolproof; it can be overridden. Anaphase can still occur following exposure of cells to microtubule disrupting drugs, in the presence of abnormal spindle bipolarity, and in the presence of unattached kinetochores and abnormal chromosome orientation (Andreassen *et al.* 1996, Rieder & Palazzo 1992, Rieder *et al.* 1994).

Most SAC data have been derived from non-mammalian somatic cells, and although differences between mitotic and meiotic SAC proteins have been found, it appears that the basic molecular pathways are similar between mitosis and meiosis and among species (Dai *et al.* 2003a, Lee & Orr-Weaver 2001, Nasmyth 2001, Uhlmann 2001, 2003a). Three broad groups of interacting proteins comprise the SAC: (1) transport/motor proteins [dynein, Zw10, Rod] that convey unique SAC proteins from the cytoplasm to kinetochores, microtubules, and spindle poles; (2) binding proteins [Aurora B, MAPK, Mps1, Bub1, CENP-E] that bind certain SAC proteins to kinetochores; and (3) SAC proteins [Mad1, Mad2, Mad3/BubR1,Bub1, Bub3,] that transiently localize to kinetochores and temporally inhibit the MAT.

If defects in the integrity of kinetochore-spindle tension and attachment are detected, Mad1, Mad2, Mad3/ BubR1, Bub1, and Bub3 transiently associate with kinetochores by binding to Cdc20 (Fang 2002, Vigneron *et al.* 2004). Such binding inhibits APC activity and delays anaphase by blocking the ubiquination and subsequent proteolysis of securin and cyclin B by proteasomes (Bharadwaj & Yu 2004, Howell *et al.* 2004, Li & Benezra 1996, Luo *et al.* 2000, Musacchio & Hardwick 2002, Nasmyth 2005, Nicklas 1997, Rieder *et al.* 1994, Shah *et al.* 2004, Sluder & McCollum 2000, Taylor *et al.* 1998, 2004, Weiss & Winey 1996, Zhou *et al.* 2002).

Although less information is available about SAC proteins in mammalian germ cells relative to other cell types, several SAC proteins have been identified in mammalian oocytes. A functional Mad2-dependent spindle checkpoint was identified during meiosis in both mouse (Homer *et al.* 2005a, Tsurumi *et al.* 2004, Wassmann *et al.* 2003) and rat (Zhang *et al.* 2004) oocytes. Mad2 binds to unattached kinetochores and is released following proper microtubule-kinetochore tension and attachment (Homer *et al.* 2005a, Kallio *et al.* 2000, Ma *et al.* 2005, Steuerwald *et al.* 2005, Wassmann *et al.* 2003, Zhang *et al.* 2004, 2005). Mad1 helps recruit Mad2 to unattached kinetochores and was detected in mouse oocytes from the GV stage to MII (Chen *et al.* 1998, Chung & Chen 2002, Zhang *et al.* 2005). In addition to Mad1 and Mad2, Mad3/BubR1 activity was also detected in mouse oocytes (Tsurumi *et al.* 2004). Lastly, Bub1 was found on kinetochores from GVBD until early AI;

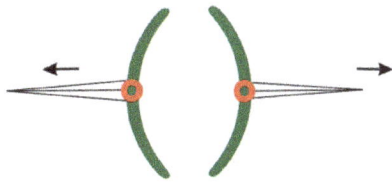

Amphitelic - Proper attachment of chromatids to a bipolar spindle and their orientation to opposite poles. Each daughter cell receives one chromatid.

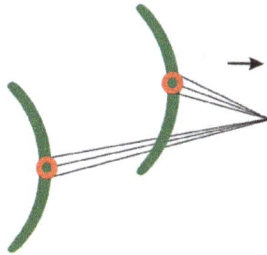

Syntelic - Improper attachment of both chromatids to a monoastral spindle and their orientation to the same pole. One daughter cell is expected to receive both chromatids (hyperhaploid), while the other cell will be minus a chromatid (hypohaploid).

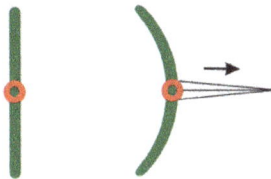

Monotelic - Improper attachment of one chromatid to a monoastral spindle and its orientation to one pole. The other chromatid is neither attached nor oriented. One daughter cell is expected to receive one chromatid (haploid), while the other daughter cell will be minus a chromatid (hypohaploid).

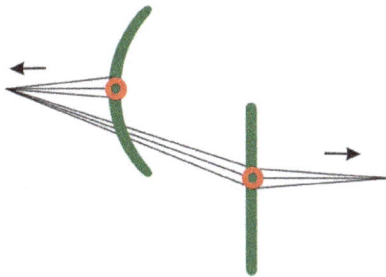

Merotelic - Improper attachment of one chromatid to a bipolar spindle and not oriented to either pole. The other chromatid is properly oriented to one pole. One daughter is expected to receive one chromatid (haploid), while the fate of the merotelically-oriented chromatid is uncertain.

Figure 2. Kinetochore-microtubule attachments and probable outcomes during meiosis II and mitosis.

then, it disappeared at late AI and re-appeared at MII (Brunet *et al.* 2003). Although SAC proteins are required for checkpoint functions during meiosis I and II in mouse oocytes, they appear non-essential for maintaining the cytostatic factor arrest during MII (Tsurumi *et al.* 2004).

Defective SAC function can lead to aneuploidy and abnormal cell cycle progression in mitotic and meiotic cells. Diminished Mad2 and Mad3/BubR1 activities resulted in PCS and aneuploidy in mammal-

ian oocytes (Dai *et al.* 2004) and somatic cells (Michel *et al.* 2001), as well as malignant transformation in human cells (Hanks *et al.* 2004). Chromosome missegregation followed ablation of Mad2 activity during budding yeast meiosis I (Shonn *et al.* 2000), and deletion of one MAD2 allele led to faulty SAC activity, PCS, and chromosome missegregation in human cancer cells and mouse fibroblasts (Michel *et al.* 2001). Also, RNA-interference reduction of Mad2 protein levels in human somatic cells induced premature cyclin B degradation, abnormal spindles, and cell death (Michel *et al.* 2004). Knockout of Mad2 in mouse embryonic cells resulted in aneuploidy and apopotosis (Dobles *et al.* 2000). Microinjection of anti-Mad1 or anti-Mad2 into GV-stage rodent and pig oocytes induced abnormalities in spindle morphology, chromosome alignment, and chromosome segregation (Ma *et al.* 2005, Zhang *et al.* 2004, 2005). Other data from mouse oocytes showed that depletion of Mad2 protein during meiosis I resulted in premature loss of securin proteins and cyclin B and elevated levels of aneuploidy; whereas, microinjection of hMad2-GFP mRNA during meiosis I inhibited homolog segregation (Homer *et al.* 2005a). Finally, an excess of Mad2 in *Xenopus* oocytes caused in a delay of chromatid segregation during anaphase II (Peter *et al.* 2001).

Apart from alterations to Mad2, anomalies in other SAC proteins resulted in cell cycle perturbations and chromosome missegregation. Partial down regulation of Mad1 in human somatic cells led to spindle checkpoint inactivation and aneuploidy (Kienitz *et al.* 2005). Deletion of the *Bub1* gene in fission yeast led to loss of centromeric Rec8 and amphitelic segregation of sister chromatids during meiosis I (Bernard *et al.* 2001); whereas, biallelic mutations of human *BUB1B* were associated with aneuploidy and cancer (Hanks *et al.* 2004). Knockout of BubR1 alleles in mice resulted in reduced BubR1 protein expression that was correlated with elevated levels of aneuploidy in fibroblasts, spermatocytes, and oocytes (Baker *et al.* 2004).

Other data from mice showed that disruption of Bub3 led to cytogenetic anomalies and embryonic lethality (Kalitsis *et al.* 2000). Exposure of HeLa cells to 5-10 nM taxol was followed by disassociation of Mad2 and BubR1 complexes, cell-cycle delay and chromosome missegregation (Ikui *et al.* 2005). Earlier work also showed that the antineoplastic agent taxol can induce dose-response effects of maturation delay, spindle defects, and aneuploidy in mouse oocytes and one-cell zygotes (Mailhes *et al.* 1999).

Recent data have shown that oocyte aging is correlated with altered Mad2 titers and cytogenetic abnormalities. Postovulatory aging of mouse oocytes resulted in a time-dependent reduction in the number of Mad2 transcripts and a concomitant elevation in the frequencies of PCS and premature anaphase (Steuerwald *et al.* 2005). Also, in vitro aging of pig oocytes led to a reduction of Mad2 expression in conjunction with abnormal chromosome segregation (Ma *et al.* 2005). In human oocytes, hMAD2 was detected during meiosis I (Homer *et al.* 2005b), and hMAD2 mRNA titers were shown to decrease with advancing maternal age (Steuerwald *et al.* 2001). These findings suggest that altered SAC activity, as detected in oo-

cytes aged in vivo and in vitro, represents one of many potential molecular mechanisms responsible for the genesis of aneuploidy.

Removal of centromeric cohesions and the metaphase-anaphase transition (MAT)

After proper microtubule-kinetochore tension and attachment has been attained or the SAC over-ridden, SAC proteins detach from Cdc20. This enables APC activation - a large protein complex that ubiquinates specific proteins (cyclin B, Securin, and possibly Sgo1) that are subsequently proteolyzed by proteasomes (Craig & Choo 2005, Glickman & Ciechanover 2002, Kotani *et al.* 1999, Salic *et al.* 2004). Proteasomes consist of multicatalytic 26S proteases and a 20S central core catalytic subunit bordered by two 19S components that hydrolyze C-terminal peptide bonds to acidic, basic, and hydrophobic amino-acid residues (Coux *et al.* 1996, Glickman & Ciechanover 2002, Goldberg 1995). This ubiquination and degradation of cellular proteins represent a tightly-regulated, temporally-controlled process that oversees numerous cellular processes including cell division (Glickman & Ciechanover 2002).

APC-mediated proteolysis during the somatic cell cycle depends upon both APC^{Cdc20} and APC^{Cdh1}. APC^{Cdc20} is active from prometaphase until the MAT; whereas, APC^{Cdh1} becomes active during anaphase and persists until the S phase. Various regulatory pathways control APC^{Cdc20} and APC^{Cdh1} activities. Phosphorylation of APC subunits by Cdk1 and Plk1 facilitate Cdc20 binding and APC activation (Glover *et al.* 1998, Sumara *et al.* 2004). Conversely, Emi1 inhibits Cdc20 binding to APC. Prior to mitosis, phosphorylation by Cdk1 and Cdk2 kinases inactivates Cdh1. However, as cells exit mitosis following cyclin B proteolysis, Cdh1 is dephosphorylated and APC^{Cdh1} mediates the proteolysis of Cdc20 and Plk1 (Peters 2002, Zachariae & Nasmyth 1999). APC^{Cdc20} targets cyclin B for degradation, which leads to Cdk1 inactivation. Also, APC^{Cdc20} activity leads to securin inactivation, which liberates separase upon satisfaction of the SAC.

Prior to normal chromatid segregation, the securin proteins, which inhibit separase activity, are ubiquinated by the APC and subsequently proteolyzed by proteasomes (Cohen-Fix *et al.* 1996, Uhlmann *et al.* 1999). Securin (Pds1p in budding yeast) activity is abrogated after each meiotic anaphase onset (Salah & Nasmyth 2000). In human somatic cells, D-box mutants of securin that were not degraded during metaphase resulted in chromosome missegregation (Hagting *et al.* 2002). This proteolysis of securin liberates the cysteine protease separase, which cleaves centromeric Scc1 during mitotic anaphase onset (Nasmyth

2002, Uhlmann *et al.* 1999, Waizenegger *et al.* 2002), Rec8 from chromosome arms during anaphase I (Agarwal & Cohen-Fix 2002, Buonomo *et al.* 2000, Jallepalli *et al.* 2001, Uhlmann 2003b), and Rec8 from centromeres during anaphase II (Waizenegger *et al.* 2000). Similar to mitosis, both APC and separase activities have been shown essential for proteolyzing securin and cyclin B prior to homolog segregation in mouse oocytes (Herbert *et al.* 2003, Terret *et al.* 2003).

Following inactivation or overriding of the SAC, the MAT represents a point-of-no-return. The temporal coordination of the MAT is directed by the interaction of unique biochemical events (kinases, phosphatases, proteolysis, topoisomerases, and motor proteins) with cellular organelles (kinetochores, centromeres, centrosomes, spindle fibers) (Dorée *et al.* 1995, Kirsch-Volders *et al.* 1998). During mitosis, the positive ends of microtubules are embedded in kinetochores and the negative ends are lodged in centrosomes. In conjunction with motor proteins, chromosome movement towards centrosomes arises from depolymerization of both the minus and positive ends of microtubules. Even though the MAT appears straightforward from a cytogenetic viewpoint, it is actually a complex series of events involving the coordination of independent processes that depend on prior checkpoint release and APC activation.

Separation of sister chromatids occurs by two independent processes: removal of cohesins from chromosomes and microtubule-dependent movement of chromatids to opposite poles. Chromatid arm separation and centromere separation (anaphase A) are independent events with different mechanisms (Rieder & Salmon 1998, Sluder & Rieder 1993), and chromatid separation does not initiate poleward movement of chromatids (anaphase B) (Zhang & Nicklas 1996). Also, sister chromatid separation does not directly depend on spindle formation because chromatids can separate in the absence of spindle attachment (Nasmyth *et al.* 2000, Rieder & Palazzo 1992) and even when MPF activity is elevated (Sluder & Rieder 1993).

Mishaps can occur during the MAT. If sister chromatids separate too early, they may both segregate to the same pole resulting in aneuploidy. Conversely, if sister chromatids fail to segregate, the outcome can range from aneuploidy to diploid gametes. In order to reduce the occurrence of such cytogenetic abnormalities, the MAT is not left to chance alone; it normally depends on satisfaction of the SAC and APC activation.

Premature Centromere Separation (PCS)

PCS and nondisjunction represent the major cytogenetic errors that lead to aneuploidy (Angell 1994, Dailey *et al.* 1996, Fragouli *et al.* 2006, Lim *et al.* 1995, Pellestor *et al.* 2005, Plachot 2003, Vialard *et al.* 2006, Wolstenholme & Angell 2000). PCS denotes the separation of sister chromatids or homologues prior to anaphase; whereas, nondisjunction results from the failure of chromatids or homologues to properly separate during anaphase. The link between PCS and aneuploidy during meiosis is that if homologues or sister chromatids separate prior to anaphase I, each of the homologues or sister chromatids may undergo random segregation (Figure 3). Also, PCS of sister chromatids prior to anaphase II onset can result in random segregation instead of amphitelic segregation. Experimental data have demonstrated a positive correlation between time postovulation and elevated frequencies of PCS in MII oocytes and aneuploidy in one-cell mouse zygotes (Mailhes *et al.* 1998).

The degree of PCS should be noted when considering the possible chromosome segregation patterns of a primary oocyte with PCS and the probability of aneuploidy. This can range from only the sister chromatids of one dyad to complete separation of all chromatids (Mailhes *et al.* 2003a). Considering the most elementary situation whereby the sister chromatids of one homologous chromosome separate prematurely and the other homologues segregate normally, three potential events may occur during anaphase I: (1) both of the disjoined sisters may segregate to the secondary oocyte, while the homologue segregates to the first polar body or *vice versa;* (2) both sisters may segregate along with its homologue to the secondary oocyte; and (3) both sisters may segregate to the first polar body along with its homologue. Thus, following anaphase I, the latter two outcomes would result in aneuploid secondary oocytes. Now, considering the case of a MII oocyte with two single chromatids (PCS of one dyad), three possible outcomes may occur during anaphase II: (1) one sister may segregate to the oocyte pronucleus, while the other goes to the second polar body; (2) both sisters may segregate to the oocyte pronucleus; or (3) both sisters may segregate to the second polar body. Again, the latter two segregation possibilities would result in aneuploidy. A noteworthy finding is that a bipolar spindle is not required for PCS because both chromatid arm and centromere separation can occur in the absence of a spindle (Rieder & Palazzo 1992, Sluder 1979).

The occurrence of PCS is not new. Rodman (1971) noted that postovulatory-aged mouse oocytes had higher frequencies of PCS than freshly ovulated

Figure 3. Possible segregation patterns during meiosis I and II when one homologue undergoes PCS prior to anaphase I.

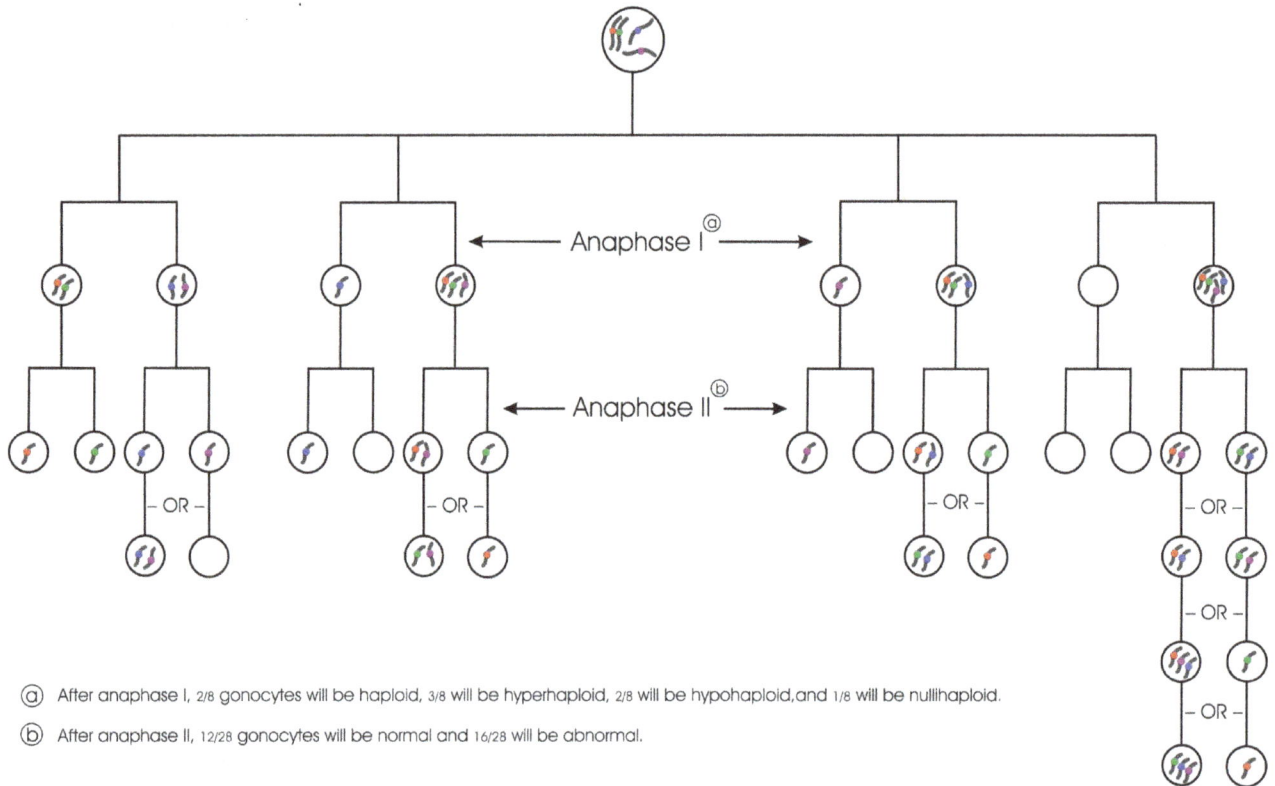

ⓐ After anaphase I, 2/8 gonocytes will be haploid, 3/8 will be hyperhaploid, 2/8 will be hypohaploid, and 1/8 will be nullihaploid.

ⓑ After anaphase II, 12/28 gonocytes will be normal and 16/28 will be abnormal.

oocytes. Subsequently, several groups found a positive correlation between postovulatory and in vitro oocyte aging with elevated levels of PCS in human (Angell 1991, 1994, Cupisti *et al.* 2003, Dailey *et al.* 1996, Pellestor *et al.*, 2002, 2003, Rosenbusch 2004), rodent (Mailhes *et al.* 1997b, 1998, Yin *et al.* 1998), and Drosophila oocytes (Jeffreys *et al.* 2003). Also, experimental data have supported a correlation between chemically-induced PCS in MII oocytes and aneuploidy in one-cell mouse zygotes (Mailhes *et al.* 1997b).

The molecular events underlying PCS are receiving considerable attention. The precocious loss of cohesin proteins from sister chromatids and homologues during mitosis and meiosis has been shown to result in PCS (Hoque & Ishikawa 2002, Sonada *et al.* 2001, Uhlmann 2003a). Mutants of the *ord* and *Mei-S322* Drosophila proteins, which help hold sister chromatids together prior to anaphase, exhibited higher frequencies of PCS and aneuploidy (Kerrebrock *et al.* 1992, Miyazaki & Orr-Weaver 1992). Also, abnormalities in other proteins involved with chromosome cohesion, such as SMC1 beta in mice (Hodges *et al.* 2005) and the yeast Pds5 protein (Hartman *et al.* 2000, Panizza *et al.* 2000) can alter normal segregation patterns. Recent results with HeLa cells showed that depletion of the microtubule and kinetochore protein astrin resulted in checkpoint arrested cells with PCS

(Thein *et al.* 2007). Although specific proteins have central roles in sister chromatid and homologue cohesion, other compounds also appear to be involved. Both culture media and the follicular fluid-meiosis-activating sterol were reported to affect the incidence of PCS in mouse oocytes in vitro (Cukurcam *et al.* 2003).

In addition to defects in cohesion proteins, PCS and aneuploidy can also result from abnormal SAC protein activity. Deficient Mad2 activity resulted in MPF degradation, APC activation, loss of sister chromatid cohesion, and PCS in both *Xenopus* oocytes (Peter *et al.* 2001) and aged mammalian oocytes (O'Neill & Kaufman 1988). Other data indicated that as time postovulation increased in mouse oocytes, the frequencies of PCS and premature anaphase (PA) increased, while the intraoocyte titer of MAD2 transcripts decreased (Steuerwald *et al.* 2005). Elevated PCS levels was also reported following the exposure of mouse oocytes to propylene glycol (Mailhes *et al.* 1997b) and tamoxifen (London & Mailhes 2001). Furthermore, when mouse oocytes were exposed to the phosphatase 1 and 2A inhibitor OA prior to metaphase I, complete separation of homologues into 80 chromatids and elevated levels of aneuploidy in MII oocytes were found (Mailhes *et al.* 2003a). A possible explanation for the elevated levels of PCS found in OA-exposed oocytes may involve protein hyperphosphorylation, as noted in hepatocytes (Cohen *et al.* 1990) and rat oocytes (Zernicka-Goetz & Maro 1993) following OA treatment. During mitosis and meiosis, phosphorylation of cohesins facilitates their removal prior to anaphase onset (Alexandru *et al.* 2001, Hoque & Ishikawa 2001, Lee & Amon 2003, Losada *et al.* 2000, Tomonaga *et al.* 2000, Yu & Koshland 2005). Finally, PP2A is found at yeast centromeres during mitosis and meiosis, and decreased PP2A activity led to loss of centromeric cohesion at anaphase I and random segregation of chromatids during anaphase II (Kitajima *et al.* 2006, Riedel *et al.* 2006, Tang *et al.* 2006).

Postovulatory and In Vitro Oocyte Aging

The broad focus of this review is that postovulatory or in vitro oocyte aging leads to a progressive and functional deterioration of the biochemical and cellular organelles required for accurate chromosome segregation, normal fertilization, and embryonic development (Austin 1967, 1970, Wilcox *et al.* 1998). Some of these age-related changes may serve as models for studying the numerous potential mechanisms of aneuploidy.

Mature mammalian oocytes remain capable of fertilization for a longer period of time than their time for expressing optimal gamete physiology. The fertilizable lifespan of mammalian oocytes ranges from 12 to 24 h (Hafez 1993). Although the fertilizable average lifespan for both induced-and naturally-ovulated mouse oocytes is approximately 15 h postovulation, their optimal time for fertilization lies between 4 to 6 h postovulation (Edwards & Gates 1959, Lewis & Wright 1935, Marston & Chang 1964). After ovulation, time-dependent intraoocyte changes occur that can lead to apoptosis (Exley *et al.* 1999, Gordo *et al.* 2002, Morita & Tilly 1999, Perez *et al.* 1999) and nuclear fragmentation (Gordo *et al.* 2002). Also, the time from insemination to fertilization, the rate of pronuclear formation, and the first cleavage division were shorter in postovulatory aged mouse oocytes than in freshly ovulated oocytes (Fraser 1979, Boerjan & de Boer 1990).

Most mammals, excluding humans and induced-ovulators, ovulate during or shortly after the estrus period of their estrous cycle; this facilitates fertilization of freshly ovulated oocytes (Hafez 1993). Since this situation does not occur in humans, a probability exists that postovulatory aged oocytes will be fertilized. Indeed, several groups have proposed that fertilization of postovulatory aged oocytes (delayed fertilization) represents a predisposition to aneuploidy (Blazak 1987, Hecht & Hecht 1987, Juberg 1983, Mailhes 1987, Pellestor 1991, Zenzes & Casper 1992).

Two human epidemiologic studies offered support for an association between delayed fertilization and early embryonic failure (Wilcox *et al.* 1998) and trisomic offspring (Juberg 1983).

Chemical Alterations in Aged Oocytes

Although freshly-ovulated and postovulatory aged oocytes appear morphologically similar, differences exist among certain cellular organelles and biochemical activities. Some of these dissimilarities resemble those found following fertilization or partheneogenic activation (Tarin *et al.* 1996, Xu *et al.* 1997), while others involve alterations to cellular organelles and biochemical events that can affect chromosome segregation.

Mammalian oocytes possess a time- and species-dependent predisposition to spontaneous activation if fertilization does not occur within a limited time following ovulation or in vitro culture. Numerous studies have shown that the incidence of spontaneous oocyte activation in mice begins to increase four hours postovulation (Homa *et al.* 1993, Kaufman 1983, Kubiak 1989, Moses & Masui 1994, Nagai 1987, Whittingham & Siracusa 1978, Winston *et al.* 1991, Yanagimachi & Chang 1961). Aged oocytes also had lower ATP levels at fertilization (Igrashi *et al.* 2005), higher sensitivities to: oxidative stress (Boerjan & de-Boer 1990, Takahashi *et al.* 2003, Tarin *et al.* 1996), calcium ionophores (Fulton & Whittingham 1978, McConnell *et al.* 1995, Vincent *et al.* 1992), partheneogenetic activation following chemical or mechanical stimuli (Cutherbertson & Cubbold 1985, Kaufman 1983, Kline & Kline 1992, Kubiak 1989, Nagai 1987), and spontaneous calcium release (Beatrice *et al.* 1984, Orrenius *et al.* 1992, Tombes *et al.* 1992). The higher titers of calcium found in aged oocytes were proposed to inhibit both tubulin polymerization and the depolymerization of existing microtubules (Kosower & Kosower 1978).

Relative to fresh oocytes, aged oocytes displayed higher calmodulin-dependent protein kinase II activities, but lower activities of MPF and MAPKs (Lorca *et al.* 1993, Moos *et al.* 1995, Verlhac *et al.* 1994). The diminished MPF activity (resulting from phosphorylation and conversion to pre-MPF) in aged porcine (Kikuchi *et al.* 1995, 2000) and bovine oocytes (Liu *et al.* 1998) and of MAPKs in both aged mouse (Xu *et al.* 1997) and porcine oocytes in vitro (Ma *et al.* 2005) were proposed to lead to spontaneous activation, abnormal chromosome segregation, and apoptosis. Furthermore, it was shown that the levels of active and inactive MPF could be regulated by exposing porcine oocytes to certain phosphatase and kinase inhibitors (Kikuchi *et al.* 2000). Such exogenous manipulation of phosphorylation-dephosphorylation events appear to offer another venue for investigating the events associated with oocyte aging and chromosome segregation. Additionally, both MPF and MAPK titers were reported to decrease more rapidly in oocytes cultured from biologically aged mice than those from young mice (Tatone *et al.* 2006).

Differences in kinase and phosphatase activities, protein synthesis, and maternal mRNA recruitment were also noted between fresh and aged bovine oocytes (Liu *et al.* 1998). Mos kinase (the product of the c-mos protooncogene) is needed for stabilizing MPF during the MII arrest of mouse oocytes (Gabrielli *et al.* 1993, Sagata 1996, 1997) and for microtubule spindle assembly (Sagata 1996, Wang *et al.* 1994, Zhao *et al.* 1991), and in vitro aging of bovine oocytes was shown to reduce the activity of Mos kinase (Wu *et al.* 1997).

When immature porcine oocytes were cultured for 40 to 72h in vitro, the levels of tubulin and the centromere protein B (CENP-B) remain unchanged as oocytes aged; whereas, the expressions of the Mad2 spindle checkpoint protein, the BCL2 antiapoptotic protein, and the mitogen-activated protein kinase (MAPK) decreased as culture time increased. Also, the proportions of oocytes with abnormal spindles and chromosomes increased with oocyte aging (Ma *et al.* 2005). Other data have shown that postovulatory aging of mouse oocytes resulted in a time-dependent reduction in the number of Mad2 transcripts and a concomitant elevation in the frequencies of PCS and PA (Steuerwald *et al.* 2005). A recent report utilized bisulfite sequencing and COBRA methods to evaluate the DNA methylation status of differentially methylated regions (DMRs) of two maternally imprinted genes – *Snrpn* and *Peg1/Mest*. Mouse oocytes aged in vivo for 29 h post-hCG exhibited demethylation of *Snrpn* DMRs. However, no change in the methylation status of *Peg1/Mest* was found at 29 h (Liang *et al.* 2008).

Histone deacetylase inhibitors are powerful anti-proliferative compounds undergoing clinical studies as antitumor drugs. Enhanced acetylation of lysines on histone H3 and H4 occurs during postovulatory oocyte aging, and the histone deacetylase inhibitor trichostatin A (TSA) can accelerate the rate of in vivo aging in mouse oocytes (Huang *et al.* 2007). Also, mouse oocytes cultured in the presence of TSA exhibited elevated levels of aneuploidy and early embryonic death (Akiyama *et al.* 2006). Another study found that exposure of HeLa cells to TSA led to loss of the Mad2 SAC protein from kinetochores and elevated levels of PCS (Magnaghi-Jaulin *et al.* 2007).

Cytologic and Cytogenetic Alterations in Aged Oocytes

Numerous cytologic and cytogenetic alterations have been described in aged mammalian oocytes. Relative to freshly ovulated oocytes, aged oocytes displayed alterations in cortical granule exocytosis and the zona pellucida (Cascio & Wassarman 1982, Diaz & Esponda 2004, Gianfortoni & Gulyas 1985, Howlett 1986, Longo 1981, Szollosi 1975, Xu *et al.* 1997, Yanagimachi & Chang 1961) and elevated levels of cytoplasmic asters and spindle anomalies (Eichenlaub-Ritter *et al.* 1986, 1988, George *et al.* 1996, Kim *et al.* 1996, Pickering *et al.* 1988, Segers *et al.* 2008). Furthermore, aged oocytes displayed higher frequencies of premature extrusion of the second polar body and apoptosis (Fissore *et al.* 2002, Gordo *et al.* 2000).

When exogenous calcium was added to *Xenopus* egg extracts, elevated frequencies of PCS and PA were detected (Shamu & Murray 1992). Others proposed that an excess of intracellular calcium, as found in aged oocytes, triggers a cascade of events resulting in PCS, PA, and chromosome missegregation (Fissore *et al.* 2002, Gordo *et al.* 2000, Tarin *et al.* 1996). Both PCS and PA have been proposed to represent cytogenetic manifestations of spontaneous activation in aged oocytes (Mailhes *et al.* 1997a, 1998). Aged oocytes displayed higher frequencies of chromosome displacement from the metaphase plate (Saito *et al.* 1993, Webb *et al.* 1986), and the levels of PCS and PA were higher in postovulatory and in vitro aged mammalian oocytes (Angell 1991, Cupisti *et al.* 2003, Dailey *et al.* 1996, Mailhes *et al.* 1997b, 1998, Pellestor *et al.* 2002, 2003, Rosenbusch 2004, Yin *et al.* 1998). Fertilization of aged oocytes was correlated with higher frequencies of fragmented female pronuclei (Fissore *et al.* 2002, Kikuchi *et al.* 2000, Szollosi 1971), decreased fertilization rates (Smith & Lodge 1987, Wolf *et al.* 1996), and embryonic viability (Ekins & Shaver 1975, Sakai & Endo 1988, Wilcox *et al.* 1998). Also, the frequencies of polyploidy (Austin 1967, Ishikawa & Endo 1995, Juetten & Bavister 1983, Shaver & Carr 1967, Vickers 1969) and aneuploidy (Mailhes *et al.* 1998, Plachot *et al.* 1988, Rodman 1971, Sakurada *et al.* 1996, Yamamoto & Ingalls 1972) were higher following delayed fertilization of mammalian oocytes.

Although most studies found a positive correlation between postovulatory aged oocytes and cytological and cytogenetic abnormalities, two studies reported that aneuploidy was not elevated in aged oocytes. Although an increase in aneuploidy was not detected when mouse oocytes were aged in vivo for 0 -14 hrs prior to in vitro fertilization, only 1 and 2 zygotes were analyzed from the 14 and 10 hr aged groups, respectively (Zackowski & Martin-Deleon 1988). Another study involving in vivo aging of mouse oocytes and cytogenetic analysis of single pronuclear haploid partheneogenones reported no association between oocyte ageing and aneuploidy (O'Neill & Kaufman 1988). However, these findings may be compromised by the difficulty of distinguishing between MII oocyte chromosomes and partheneogenome chromosomes as well as that between a first and a second polar body. It is noted that analysis of MII chromosomes cannot detect aneuploidy in postovulatory aged oocytes because an intervening cell division is needed between the induction and expression of aneuploidy.

When the developmental potential of a limited number of aged, failed-to-fertilize human oocytes were compared with fresh, ovulation-induced oocytes, higher levels of aneuploidy, aberrant spindles, and cleavage failure were noted in the aged oocytes (Hall *et al.* 2007). Also, human embryos resulting from in vitro maturation and delayed intracytoplasmic sperm injection exhibited higher levels of aneuploidy when compared with control embryos (Emery *et al.* 2005).

Conclusion

At each stage of mitosis and meiosis, the correct order and temporal interaction among various chemical reactions and cellular organelles are needed to preserve genomic integrity. Considerable experimental data and human epidemiological studies have shown that the probability of successful chromosome segregation and zygotic development are compromised when oocytes undergo in vivo or in vitro aging prior to fertilization. These biochemical and cytological changes reported in aged oocytes offer unique models for studying some of the numerous molecular aspects of aneuploidy.

Several innovative technologies have been used to study the molecular aspects of mitosis and meiosis. High-density oligonucleotide microarrays and PCR microarrays (Schlecht & Primig 2003) can be used to identified loci that regulate the cell cycle in eukaryotes, including mice and humans. Also, double-strand RNA-mediated post-transcriptional gene silencing (RNA interference) offers promise for investigating the pathways controlling cell cycle progression and chromosome segregation (Bettencourt-Dias *et al.* 2004, Prawitt *et al.* 2004). RNA silencing/knockdown has been used to alter the expression of Mos mRNA (Stein *et al.* 2003) and Mad2 (Homer *et al.* 2005a) in order to study the role of genes involved with oocyte maturation and chromosome segregation in mouse oocytes. When employing RNA interference technologies, the possibility of off-target effects and the effi-

ciency of gene silencing should be considered. Gene knockout strategies for genes upregulated during yeast meiosis showed that deletion of specific genes required for maintaining centromeric cohesion during anaphase I resulted in chromosome missegregation (Gregan *et al.* 2005, Marston *et al.* 2004). Furthermore, genomic and proteomic analyses have the ability to expand our knowledge about gene expression. Analyses of cancer cells showed that a subset of genes are universally activated in most cancers (Rhodes *et al.* 2004), and that overexpression of cell division regulatory genes were linked with chromosome aberrations and neoplastic progression (Rajagopalan & Lengauer 2004). Finally, the use of unique chemical inhibitors that block a specific pathway during chromosome segregation are helping to advance our knowledge about aneuploidy (Dorer *et al.* 2005, Mailhes *et al.* 2003a, 2004).

The present and future challenge will be to understand the complex molecular mechanisms of aneuploidy and genomic instability and to apply such knowledge to reducing the incidence of human genetic disease and cancer.

References

Adams RR, Maiato H, Earnshaw WC & Carmena M. 2001a. Essential roles of Drosophila inner centromere protein (INCENP) and aurora B in histone H3 phosphorylation, metaphase chromosome alignment, kinetochore disjunction, and chromosome segregation. *J Cell Biol* **153** 865-880

Adams RR, Eckley DM, Vagnarelli P, Wheatley SP, Gerloff DL, Mackay AM, Svingen PA, Kaufman SH & Earnshaw WC 2001b Human INCENP colocalizes with the Aurora-B/AIRK kinase on chromosomes and is overexpressed in tumor cells. *Chromosoma* **110** 65-74

Adler I-D 1990 Aneuploidy studies in mammals. In Mendelsohn ML, Albertini RJ, editors. Mutation and the Environment, Part B: Metabolism, Testing Methods, and Chromosomes, New York: Wiley-Liss, p 285-293

Adler I-D 1993 Synopsis of the in vivo results obtained with the 10 known or suspected aneugens tested in the CEC collaborative study. *Mutat Res* **287** 131-137

Adler I-D, Schmid TE & Baumgartner A 2002 Induction of aneuploidy in male mouse germ cells detected by the sperm–FISH assay: a review of the present data base. *Mutat Res* **504** 173-182

Agarwal R & Cohen-Fix O 2002 Mitotic regulation: the fine tuning of separase activity. *Cell Cycle* **1** 255-257

Ahonen LJ, Kallio MJ, Daum JR, Bolton M, Manke IA, Yaffe MB, Stukenberg PT & Gorbsky GJ 2005

Polo-like kinase 1 creates the tension-sensing 3F3/2 phosphoepitope and modulates the association of spindle-checkpoint proteins at kinetochores. *Curr Biol* **15** 1078-1089

Akiyama T, Nagata M & Aoki F 2006 Inadequate histone deactylation during oocyte meiosis causes aneuploidy and embryo death in mice. *Proc Natl Acad Sci USA* **103** 7339-7344

Allen JW, Liang JC, Carrano AV & Preston RJ 1986 Review of literature on chemical-induced aneuploidy in mammalian male germ cells. *Mutat Res* **167** 123-137

Alexandru G, Uhlmann F, Mechtler K, Poupart MA & Nasmyth K 2001 Phosphorylation of the cohesion subunit Scc1 by Polo/Cdc5 kinase regulates sister chromatid separation in yeast. *Cell* **105** 459-472

Amor DJ, Kalitsis P, Sumer H & Choo KH 2004 Building the centromere: from foundation proteins to 3D organization. *Trends Cell Biol* **14** 359-368

Anahory T, Andreo B, Regnier-Vigourouox G, Souile JP, Baudouin M, Demaille J & Pellestor F 2003 Sequential multiple probe fluorescence in-situ hybridization analysis of human oocytes and polar bodies by combining centromeric labeling and whole chromosome painting. *Mol Hum Report* **9** 577-585

Anderiesz C, Ferraretti A, Magli C, Fiorentino A, Fortini D, Gianaroli L, Jones GM & Trounson AO 2000 Effect of recombinant human gonadotrophins on human, bovine and marine oocyte meiosis. *Hum Report* **15** 1140-1148

Andreassen PR, Martina SN & Margolis RL 1996 Chemical induction of mitotic checkpoint override in mammalian cells results in aneuploidy following a transient metalloid state. *Mutat Res* **372** 181-194

Angell RR 1991 Predivision in human oocytes at meiosis I: a mechanism for trisomy formation in man. *Hum Genet* **86** 383-387

Angell RR 1994 Aneuploidy in older women; higher rates of aneuploidy in oocytes from older women. *Hum Report* **9** 1199-1201

Angell RR, Xian J, Keith J, Ledger W & Baird DT 1994 First meiotic division abnormalities in human oocytes: mechanism of trisomy formation. *Cytogenet Cell Genet* **65** 194-202

Arion D, Meier L, Brizuela L & Beach D 1988 Cdc2 is a component of the M phase-specific histone H1 kinase: evidence for identity with MPF. *Cell* **55** 371-378

Austin CR 1967 Chromosome deterioration in aging eggs of the rabbit. *Nature* **213** 1018-1019

Austin CR 1970 Ageing and reproduction: post-ovulatory deterioration of the egg. *J Reprod Fert Suppl* **12** 39-53

Baker DJ, Jeganathan KB, Cameron JD, Thompson M,

Juneja S, Kopecka A, Kumar R, Jenkins RB, de Groen PC, Roche P & van Deursen JM 2004 BubR1 insufficiency causes early onset of aging-associated phenotypes and infertility in mice. *Nat Genet* **36** 744-749

Barnes FL, Collas P, Powell R, King WA, Westhusin M & Shepherd D 1993 Influence of recipient oocyte cell cycle stage on DNA synthesis, nuclear envelope breakdown, chromosome constitution, and development in nuclear transplant bovine embryos. *Mol Rep Dev* **36** 33-41

Baudat F, Manova K, Yuen JP, Jasin M & Keeney S 2000 Chromosome synapsis defects and sexually dimorphic meiotic progression in mice lacking Spo11. *Mol Cell* **6** 989-998

Beatrice MC, Stiers DL & Pfeiffer DR 1984 The role of glutathione in the retention of Ca^{2+} by liver mitochondria. *J Biol Chem* **259** 1279-1287

Bernard P, Maure JF & Javerzat JP 2001 Fission yeast Bub1 is essential in setting up the meiotic pattern of chromosome segregation. *Nat Cell Biol* **3** 522-526

Bettencourt-Dias M, Giet R, Sinka R, Mazumdar A, Lock WG, Balloux PJ, Safiropoulos PJ, Yamaguchi S, Winter S, Carthew RW, Cooper M, Jones D, Frenz L & Glover DM 2004 Genome-wide survey of protein kinases required for cell cycle progression. *Nature* **432** 980-987

Bharadwal R & Yu H 2004 The spindle checkpoint, aneuploidy, and cancer. *Oncogene* **23** 2016-2027

Biggins S & Walczak CE 2003 Captivating capture: how microtubules attach to kinetochores. *Curr Biol* **13** R449-R460

Blazak WF 1987 Incidence of aneuploidy in farm animals. In Vig BK, Sandberg AA, editors. Aneuploidy, Part A: Incidence and Etiology. New York: Alan R. Liss. p.103-116

Boerjan ML & deBoer P 1990 First cell cycle of zygotes of the mouse derived from oocytes aged postovulation in vivo and fertilized in vivo. *Mol Rep Dev* **25** 155-163

Boernslaeger EA, Mattei P & Schultz RM 1986 Involvement of cAMP-dependent protein kinase and protein phosphorylation in regulation of mouse oocyte maturation. *Dev Biol* **114** 453-462

Bolton MA, Lan W, Powers SE, McCleland ML, Kuang J & Stukenberg PT 2002 Aurora B kinase exists in a complex with survivin and INCENP and its kinase activity is stimulated by survivin binding and phosphorylation. *Mol Biol Cell* **13** 3064-3077

Bond DJ & Chandley AC 1983 Aneuploidy, Oxford Monographs on Medical Genetics, No. 11, Oxford: Oxford University Press, p 1-198

Brunet S, Pahlavan G, Taylor S & Maro B 2003 Functionality of the spindle checkpoint during the first meiotic division of mammalian oocytes. *Reproduction* **126** 443-450

Buonomo SB, Clyne RK, Fuchs J, Loidl J, Uhlmann F & Nasmyth K 2000 Disjunction of homologous chromosomes in meiosis I depends on proteolytic cleavage of the meiotic cohesion Rec8 by separin. *Cell* **103** 387-398

Carvalho A, Carmena M, Sambade C, Earnshaw WC & Wheatley SP 2003 Survivin is required for stable checkpoint activation in taxol-treated HeLa cells. *J Cell Sci* **116** 2987-2998

Cascio SM & Wassarman PM 1982 Program of early development in the mammal: post-transcriptional control of a class of proteins synthesized by mouse oocytes and early embryos. *Dev Biol* **89** 397-408

Champoux JJ 2001 DNA topoisomerases: structure, function, and mechanism. *Annu Rev Biochem* **70** 369-413

Chandley AC 1987 Aneuploidy: an overview, In Vig BK, Sandberg AA, editors. Aneuploidy: Part A: Incidence and Etiology. New York: Alan R. Liss. p 1-8

Chen RH, Shevehenko A, Mann M & Murray AW 1998 Spindle checkpoint protein Xmad1 recruits Xmad2 to unattached kinetochores. *J Cell Biol* **143** 283-295

Choi T, Aoki F, Mori M, Yamashita M, Nagahama Y & Kohmoto K 1991 Activation of p34^{cdc2} protein kinase activity in meiotic and mitotic cell cycles in mouse oocytes and embryos. *Development* **113** 789-795

Chung E & Chen RH 2002 Spindle checkpoint requires Mad1-bound and Mad1-free Mad2. *Mol Biol Cell* **13** 1501-1511

Cimini D 2007 Detection and correction of merotelic kinetochore orientation by Aurora B, its partners. *Cell Cycle* **6** 1558-1564

Cimini D 2008 Merotelic kinetochore orientation, aneuploidy, and cancer. *Biochem Biophys Acta* doi:10.1016/j.bbcan.2008.05.003

Ciosk R, Shirayama M, Shevchenko A, Tanaka T, Toth A, Shevchenko A & Nasmyth K 2000 Cohesin's binding to chromosomes depends on a separate complex consisting of Scc2 and Scc4 proteins. *Mol Cell* **4** 243-254

Clarke AS, Tang TT, Ooi DL & Orr-Weaver TL 2005 POLO kinase regulates the Drosophila centromere cohesion protein MEI-S332. *Dev Cell* **8** 53-64

Clyne RK, Katis VL, Jessop L, Benjamin KR, Herskowitz I, Lichten M & Nasmyth K 2003 Polo-like kinase Cdc5 promotes chiasmata formation and cosegregation of sister chromatids at meiosis I. *Nat Cell Biol* **5** 379-382

Cobb J, Reddy RK, Park C & Handel MA 1997 Analysis of expression and function of topoisomerase I and II during meiosis in male mice. *Mol Rep Dev* **46** 489-498

Cohen P, Holmes CF & Tsukitani Y 1990 Okadaic acid: a new probe for the study of cellular regulation. *Trends Biochem Sci* **15** 98-102

Cohen-Fix O, Peters JM, Kirschner MW & Koshland D 1996 Anaphase initiation in Saccharomyces cerevisiae is controlled by the APC-dependent degradation of the anaphase inhibitor Pds1p. *Genes Dev* **10** 3081-3093

Collas P, Sullivan EJ & Barnes FL 1993 Histone H1 kinase activity in bovine oocytes following calcium stimulation. *Mol Rep Dev* **34** 224-231

Colledge WH, Carlton MBL, Udy GB & Evans MJ 1994. Disruption of *c-mos* causes parthenogenetic development of unfertilized mouse eggs. *Nature* **370** 65-68

Collins J & Crosignani PG 2005 Fertility and ageing. *Hum Reprod Update* **11** 261-276

Coulam CB, Jeyendran RS, Fiddler M & Pergament E 2007. Discordance among blastomeres renders preimplantation genetic diagnosis for aneuploidy ineffective. *J Assist Reprod Genet* **24** 37-41

Coux O, Tanaka K & Goldberg AL 1996 Structure and function of the 20S and 26S proteasomes. *Annu Rev Biochem* **65** 801-847

Craig JM, Earnshaw WC & Vagnarelli P 1999 Mammalian centromeres: DNA sequence, protein composition, and role in cell cycle progression. *Exptl Cell Res* **246** 249-262

Craig JM & Choo KHA 2005 Kiss and break up – a safe passage to anaphase in mitosis and meiosis. *Chromosoma* **114** 252-262

Cukurcam S, Hegele-Hartung C & Eichenlaub-Ritter U 2003 Meiosis-activating sterol protects oocytes from precocious chromosome segregation. *Hum Reprod* **18** 1908-1917

Cupisti S, Conn CM, Fragouli E, Whalley K, Mills JA, Faed MJ & Delhanty JD 2003 Sequential FISH analysis of oocytes and polar bodies reveals aneuploidy mechanisms. *Prenat Diagn* **23** 663-668

Cuthbertson KSR & Cubbbold PH 1985 Phorbol ester and sperm activate mouse oocytes by inducing sustained oscillations in cell Ca^{2+}. *Nature* **316** 541-542

Dai Y, Fulka Jr. J & Moor R 2003a Checkpoint controls in mammalian oocytes. In Trounson AO, Gosden RG, editors. Biology and Pathology of the Oocyte. Cambridge UK: Cambridge University Press. p 120-132

Dai W, Huang X & Ruan Q 2003b Polo-like kinases in cell cycle checkpoint control. *Frontiers Biosci* **8** 1128-1133

Dai W, Wang Q, Liu TY, Swamy M, Fang YQ, Xie SQ, Mahmood R, Yang YM, Xu M & Rao CV 2004 Slippage of mitotic arrest and enhanced tumor development in mice with BUBR1 haploinsufficiency. *Can-*

cer Res **15** 440-445

Dailey T, Dale B, Cohen J & Nunné S 1996 Association between nondisjunction and maternal age in meiosis II oocytes. *Am J Hum Genet* **59** 176-184

Dekel N 1988 Regulation of oocyte maturation: the role of cAMP. *Ann NY Acad Sci* **541** 211-216

Dekel N 1996 Protein phosphorylation-dephosphorylation in the meiotic cell cycle of mammalian oocytes. *Rev Reprod 1* 82-88

Dekel N 2005 Cellular, biochemical and molecular mechanisms regulating oocyte maturation. *Mol Cell Endocrinol* **234** 19-25

De Pennart H, Helene-Verlhac M, Cibert C, Santa Maria A & Maro B 1993 Okadaic acid induces spindle lengthening and disrupts the interaction of microtubules with the kinetochores in metaphase II-arrested mouse oocytes. *Dev Biol* **157** 170-181

Dewar H, Tanaka K, Nasmyth K & Tanaka TU 2004 Tension between two kinetochores suffices for their biorientation on the mitotic spindle. *Nature 428* 93-97

Diaz H & Esponda P 2004 Postovulatory ageing induces structural changes in the mouse zona pellucida. *J Submicrosc Cytol Pathol* **36** 211-217

Dinardo S, Voelkel KA & Sternglanz RL 1984 DNA topoisomerase mutant of *S. cerevisiae*: topoisomerase II is required for segregation of daughter molecules at the termination of DNA replication. *Proc Natl Acad Sci USA* **81** 2616-2620

Ditchfield C, Johnson VL, Tighe A, Ellston R, Haworth C, Johnson T, Mortlock A, Keen N & Taylor SS 2003 Aurora B couples chromosome alignment with anaphase by targeting BubR1, Mad2, and Cenp-E to kinetochores. *J Cell Biol* **161** 267-280

Dobles M, Liberal V, Scott ML, Benezra R & Sorger PK 2000 Chromosome missegregation and apoptosis in mice lacking the mitotic checkpoint protein Mad2. *Cell* **101** 635-645

Dorée M, Le Peuch C & Morin N 1995 Onset of chromosome segregation at the metaphase to anaphase transition of the cell cycle. *Prog Cell Cycle Res* **1** 309-318

Dorer RK, Zhong S, Tallarico JA, Wong WH, Mitchison TJ & Murray AW 2005 A small-molecule inhibitor of Mps1 blocks the spindle-checkpoint response to a lack of tension on mitotic chromosomes. *Curr Biol* **15** 1070-1076

Downes CS, Mullinger AM & Johnson RT 1991 Inhibitors of DNA topoisomerase II prevent chromatid separation in mammalian cells but do not prevent exit from mitosis. *Proc Natl Acad Sci USA* **88** 8895-8899

Downs SM, Daniel SAJ, Bornslaeger EA, Hoppe PC & Eppig JJ 1989 Maintenance of meiotic arrest in mouse oocytes by purines: modulation of cAMP levels and cAMP phosphodiesterase activity. *Gamete Res* **23** 323-334

Draetta G & Beach D 1988 Activation of cdc2 protein kinase during mitosis in human cells: Cell cycle dependent phosphorylation and subunit rearrangement. *Cell* **54** 17-26

Duesbery NS, Choi T, Brown KD, Wood KW, Resau J, Fukasawa K, Cleveland DW & Vande-Woude GF 1997 CENP-E is an essential kinetochore motor in maturing oocytes and is masked during mos-dependent, cell cycle arrest at metaphase II. *Proc Natl Acad Sci USA* **94** 9165-9170

Dunphy WG & Kumagai A 1991 The cdc25 protein contains an intrinsic phosphatase activity. *Cell 67* 189-196

Earnshaw WC & Cooke CA 1991 Analysis of the distribution of the INCENPs throughout mitosis reveals the existence of a pathway of structural changes in the chromosomes during metaphase and early events in cleavage furrow formation. *J Cell Sci* **98** 443-461

Eastmond DA, Schuler M & Rupa DS 1995 Advantages and limitations of using fluorescence in situ hybridization for the detection of aneuploidy in interphase human cells. *Mutat Res* **348** 153-162

Edwards RG & Gates AH 1959 Timing of the stages of the maturation divisions, ovulation, fertilization and the first cleavage of eggs of adult mice treated with gonadotrophins. *J Endocrin* **18** 292-304

Eichenlaub-Ritter U 1993 Studies on maternal age-related aneuploidy in mammalian oocytes and cell cycle control. In Sumner AT, Chandley AC, editors. Chromosomes Today, Vol. 11. London: Chapman & Hall. p 323-336

Eichenlaub-Ritter U 1996 Parental age-related aneuploidy in human germ cells and offspring: a story of past and present. *Environ Mol Mutagen* **28** 211-236

Eichenlaub-Ritter U, Chandley AC & Gosden RG 1986 Alterations to the microtubular cytoskeleton and increased disorder of chromosome alignment in spontaneously ovulated mouse oocytes in vivo: an immunofluorescence study. *Chromosoma* **94** 337-345

Eichenlaub-Ritter U, Stahl A & Luciani JM 1988 The microtubular cytoskeleton and chromosomes of unfertilized human oocyte aged in vitro. *Hum Genet* **80** 259-264

Eichenlaub-Ritter U, Baart E, Yin H & Betzendahl I 1996 Mechanisms of spontaneous and chemically-induced aneuploidy in mammalian oogenesis: basis of sex-specific differences in response to aneugens and the necessity for further tests. *Mutat Res 372* 279-294

Eichenlaub-Ritter U 2003 Aneuploidy in aging oocytes and after toxic insult. In Trounson A, Gosden RG, editors. Biology and pathology of the oocyte; its role in fertility and reproductive medicine. Cambridge, UK: Cambridge University Press. p 220-257

Eijpe M, Heyting C, Gross B & Jessberger R 2000 Association of mammalian SMC1 and SMC3 proteins with meiotic chromosomes and synaptonemal complexes. *J Cell Sci* **113** 673-682

Eijpe M, Offenberg H, Jessberger R, Revenkova E & Heyting C 2003 Meiotic cohesion REC8 marks the axial elements of rat synaptonemal complexes before cohesions SMC1β and SMC3. *J Cell Biol* **160** 657-670

Ekins JG & Shaver EL 1975 Cytogenetics of postimplantation rabbit conceptuses following delayed fertilization. *Teratology* **13** 57-64

Emery BR, Wilcox AL, Aoki VW, Peterson CM & Carrell DT 2005 In vitro oocyte maturation and subsequent delayed fertilization is associated with increased embryo aneuploidy. *Fertil Steril* **84** 1027-1029

Exley GE, Tang C, McElhinny AS & Warner CM 1999 Expression of caspase and BCL-2 apoptotic family members in mouse preimplantation embryos. *Biol Rep* **61** 231-239

Fan HY & Sun QY 2004 Involvement of mitogen-activated protein kinase cascade during oocyte maturation and fertilization in mammals. *Biol Rep* **70** 535-547

Fang G 2002 Checkpoint protein BubR1 acts synergistically with Mad2 to inhibit anaphase-promoting complex. *Mol Biol Cell* **13** 755-766

Fissore RA, Kurokawa M, Knott J, Zhang M & Smyth J 2002 Mechanisms underlying oocyte activation and postovulatory aging. *Reproduction* **124** 745-754

Ford JH 1981 Nondisjunction. In Burgio GR, Fraccaro M, Tiepolo L, Wolf U, editor. Trisomy 21. Berlin: Springer. p 103-143

Fragouli E, Wells D, Doshi A, Gotts S, Harper JC & Delhanty JD 2006 Complete cytogenetic investigation of oocytes from a young cancer patient with the use of comparative genomic hybridization reveals meiotic errors. *Prenat Diagn* **26** 71-76

Fraser LR 1979 Rate of fertilization in vitro and subsequent nuclear development as a function of the postovulatory age of the mouse egg. *J Rep Fertil* **55** 153-160

Fulka Jr. J, Jung T & Moor RM 1992 The fall of biological maturation promoting factor (MPF) and histone H1 kinase activity during anaphase and telophase in mouse oocytes. *Mol Rep Dev* **32** 378-382

Fulton BP & Whittingham DG 1978 Activation of mammalian oocytes by intracellular injection of calcium. *Nature* **273** 149-151

Gabrielli BG, Roy LM & Maller JL 1993 Requirement for *cdk2* in cytostatic factor-mediated metaphase II arrest. *Science* **259** 1766-1769

Gautier J, Solomon MJ, Booher RN, Bazan JF & Kirschner ME 1991 cdc25 is a specific tyrosine phosphatase that directly activates $p34^{cdc2}$. *Cell* **67** 197-211

Geijsen N, Horoschak M, Kim K, Gribnau J, Eggan K & Daley GQ 2004 Derivation of embryonic germ cells and male gametes from embryonic stem cells. *Nature* **427** 148-154

George MA, Pickering SJ, Braude PR & Johnson MH 1996 The distribution of α- and λ tubulin in fresh and aged human and mouse oocytes exposed to cryoprotectant. *Mol Hum Rep* **6** 445-456

Gianfortoni JG & Gulyas BJ 1985 The effects of short-term incubation (aging) of mouse oocytes on in vitro fertilization, zona solubility, and embryonic development. *Gamete Res* **11** 59-68

Giet R, Petretti C & Prignet C 2005 Aurora kinases, aneuploidy and cancer, a coincidence or a real link? *Trends Cell Biol* **15** 241-250

Glickman MH & Ciechanover A 2002 The ubiquitin-proteasome proteolytic pathway: destruction for the sake of construction. *Physiol Rev* **82** 373-428

Glover DM, Ohkura H & Tavares A 1998 Polo-like kinases: a team that plays throughout mitosis. *Genes Dev* **12** 3777-3787

Goldberg AL 1995 Functions of the proteasome: the lysis at the end of the tunnel. *Science* **268** 522- 523

Goldstein LS 1980 Mechanisms of chromosome orientation revealed by two meiotic mutants in Drosophila melanogaster. *Chromosoma* **78** 79-111

Gorbsky GJ 1994 Cell cycle progression and chromosome segregation in mammalian cells cultured in the presence of the topoisomerase II inhibitors ICRF-187 [(+)-1,2-bis(3,5-dioxopiperazinyl-1-yl) propane; ADR-529] and ICRF-159 (Razoxane). *Cancer Res* **54** 1042-1048

Gordo AC, Wu H, He CL & Fissore RA 2000 Injection of sperm cytosolic factor into mouse metaphase II oocyte induces different development fates according to the frequency of $[Ca^{2+}]$ oscillations and oocyte age. *Biol Reprod* **62** 1370-1379

Gordo A, He CL, Smith S & Fissore RA 2001 Mitogen-activated protein kinase plays a significant role in metaphase II arrest, spindle morphology and maintenance of maturation promoting factor activity in bovine oocytes. *Mol Rep Dev* **59** 106-114

Gordo AC, Rodrigues P, Kurokawa M, Jellerette T, Exley GE, Warner C & Fissore R 2002 Intracellular calcium oscillations signal apoptosis rather than activation in in vitro aged mouse eggs. *Biol Reprod* **66** 1828-1837

Goto H, Kiyono T, Tomono Y, Kawajiri A, Urano T, Furukawa K, Nigg EA & Inagaki M 2006 Complex formation of Plk1 and INCENP required for metaphase-anaphase transition. *Nat Cell Biol* **8** 180-187

Goulding SE & Earnshaw WC 2005 Shugoshin: a centromeric guardian senses tension. Bioessays 27 588-591

Gregan J, Rabitsch PK, Sakem B, Csutak O, Latypov V, Lehmann E, Kohli J & Nasmyth K 2005 Novel genes required for meiotic chromosome segregation are identified by a high-throughput knockout screen in fission yeast. *Curr Biol* **15** 1663-1669

Haering CH, Lowe J, Hochwagen A & Nasmyth K 2002

Molecular architecture of SMC proteins and the yeast cohesion complex. *Mol Cell* **9** 773-788

Haering CH & Nasmyth K 2003 Building and breaking bridges between sister chromatids. *BioEssays* **25** 1178-1191

Hafez ESE 1993 Reproduction in Farm Animals, 6ᵗʰ edn. Philadelphia: Lea & Febiger. p 144-164

Hagstrom KA, Holmes VF, Cozzarelli NR & Meyer BJ 2002 C. elegans condensing promotes mitotic chromosome architecture, centromere organization, and sister chromatid segregation during mitosis. *Genes Dev* **16** 729-742

Hagting A, den Elzen N, Vodermaier HC, Waizenegger IC, Peters J-M & Pines J 2002 Human securin proteolysis is controlled by the spindle checkpoint and reveals when the APC/C switches from activation by Cdc20 to Cdh1. *J Cell Biol* **157** 1125-1137

Hall VJ, Compton D, Stojkovic P, Nesbitt M, Herbert M, Murdoch A & Stojkovic M 2007 Developmental competence of human *in vitro* aged oocytes as host cells for nuclear transfer. *Hum Reprod* **22** 52-62

Handel MA, Cobb J & Eaker S 1999 What are the spermatocyte's requirements for successful meiotic division? *J Exp Zool* **285** 243-250

Handel MA & Sun F 2005 Regulation of meiotic cell divisions and determination of gamete quality: impact of reproductive toxins. *Semin Reprod Med* **23** 213-221

Hanks S, Coleman K, Reid S, Plaja A, Firth H, Fitzpatrick D, Kidd A, Mehes K, Nash R, Robin N, Shannon N, Tolmie J, Swansbury J, Itthum A, Douglas J & Rahman N 2004 Constitutional aneuploidy and cancer predisposition caused by biallelic mutations in BUB1B. *Nat Gen* **36** 1159-1161

Hansmann I 1983 Factors and mechanisms involved in nondisjunction and X-chromosome loss. In Sandberg AA, editor. Cytogenetics of the Mammalian X Chromosome, Part A: Progress and Topics in Cytogenetics, Vol. 3A. New York: Alan R. Liss. p 131-170

Hansmann I & Pabst B 1992 Nondisjunction by failures in the molecular control of oocyte maturation. *Ann Anat* **174** 485-490

Hartman T, Stead K, Koshland D & Guacci V 2000 Pds5 is an essential chromosomal protein required for both sister chromatid cohesion and condensation in Saccharomyces cerevisiae. *J Cell Biol* **151** 613-626

Hashimoto N, Watanabe N, Furuta Y, Tamemoto H, Sagata N, Yokoyama M, Okazaki K, Nagayoshi M, Takeda N, Ikawa Y & Aizawa S 1994 Parthenogenetic activation of oocytes in c-mos deficient mice. *Nature* **370** 68-71

Hashimoto N 1996 Role of c-mos proto-oncogene product in the regulation of mouse oocyte maturation. *Horm Res* **46** 11-14

Hassold T, Chiu D & Yamane JA 1984 Parental origin of autosomal trisomies. *Ann Human Genet* **48** 129-144

Hassold TJ 1985 The origin of aneuploidy in humans. In Dellarco VL, Voytek PE, Hollaender A, editors. Aneuploidy: Etiology and Mechanisms. New York: Plenum Press. p 103-115

Hassold T & Sherman S 1993 The origin of nondisjunction in humans. In Sumner AT, Chandley AC, editors. Chromosomes Today. London, UK: Chapman and Hall. p 313-322

Hassold T & Hunt PA 2001 To err (meiotically) is human: the genesis of human aneuploidy. *Nat Rev Genet* **2** 280-291

Hauf S, Cole RW, LaTerra S, Zimmer C, Schnapp G, Walter R, Heckel A, van Meel J, Rieder CL & Peters JM 2003 The small molecule Hesperadin reveals a role for Aurora B in correcting kinetochore-microtubule attachment and in maintaining the spindle assembly checkpoint. *J Cell Biol* **161** 281-294

Hauf S, Roitinger E, Koch B, Dittrich CM, Mechtler K & Peters JM 2005 Dissociation of cohesion during early mitosis depends on phosphorylation of SA2. *PloS Biol* **3** 0419-0432

Hecht F & Hecht BK 1987 Aneuploidy in humans: dimensions, demography, and dangers of abnormal numbers of chromosomes. In Vig BK, Sandberg AA, editors. Aneuploidy, Part A: Incidence and Etiology. New York: Alan R. Liss. p 9-49

Herbert M, Levasseur M, Homer H, Yallop K, Murdoch A & McDougall A 2003 Homologue disjunction in mouse oocytes requires proteolysis of securin and cyclin B. *Nat Cell Biol* **5** 1023-1025

Hiller SG 2001 Gonadotropic control of ovarian follicular growth and development. *Mol Cell Endocrinol* **179** 39-46

Hodges CA, LeMaire-Adkins R & Hunt PA 2001 Coordinating the segregation of sister chromatids during the first meiotic division: evidence for sexual dimorphism. *J Cell Sci* **114** 2417-2426

Hodges CA, Revenkova E, Jessberger R, Hassold T & Hunt PA 2005 SMC1β-deficient female mice provide evidence that cohesions are a missing link in age-related nondisjunction. *Nat Genet* **37** 1351-1355

Holm C, Stearns T & Botstein D 1989 DNA topoisomerase II must act at mitosis to prevent nondisjunction and chromosome breakage. *Mol Cell Biol* **9** 159-168

Homa ST, Carroll J & Swann K 1993 The role of calcium in mammalian oocyte maturation and egg activation. *Hum Reprod* **8** 1274-1281

Homer HA, McDougall A, Levasseur M, Yallop K, Murdoch AP & Herbert M 2005a. Mad2 prevents aneuploidy and premature proteolysis of cyclin B and securing during meiosis I in mouse oocytes. *Genes Dev* **19** 202-207

Homer HA, McDougall A, Levasseur M, Murdoch AP

& Herbert M 2005b RNA interference in meiosis I human oocytes: towards and understanding of human aneuploidy. *Mol Hum Rep* **11** 397-404

Hook EB 1985a The impact of aneuploidy upon public health: mortality and morbidity associated with human chromosome abnormalities. In Dellarco VL, Voytek PE, Hollaender A, editors. Aneuploidy: Etiology and Mechanisms. New York: Plenum Press. p 7-33

Hook EB 1985b Maternal age, paternal age, and human chromosome abnormality: nature, magnitude, etiology, and mechanisms of effects, In Dellarco VL, Voytek PE, Hollaender A, editors. Aneuploidy: Etiology and Mechanisms, New York: Plenum Press. p 117-132

Hoque MT & Ishikawa F 2001 Human chromatid cohesion component hRad21 is phosphorylated in M phase and associated with metaphase chromosomes. *J Biol Chem* **276** 5059-5067

Hoque MT & Ishikawa F 2002 Cohesin defects lead to premature sister chromatid separation, kinetochore dysfunction, and spindle-assembly checkpoint activation. *J Biol Chem* **277** 42306-42314

Howell BJ, Moree B, Farrar EM, Stewart S, Fang G & Salmon ED 2004 Spindle checkpoint protein dynamics at kinetochores in living cells. *Curr Biol* **14** 953-964

Howlett SK 1986 A set of proteins showing cell cycle dependent modification in the early mouse embryo. *Cell* **45** 387-396

Huang JC, Yan LY, Lei ZL, Miao YL, Shi LH, Yang JW, Wang Q, Ouyang YC, Sun QY & Chen DY 2007 Changes in histone acetylation during postovulatory aging of mouse oocytes. *Biol Reprod* **77** 666-670

Hwang LH, Lau LF, Smith DL, Mistrot CA, Hardwick KG, Hwang ES, Amon A & Murray AW 1998. Budding yeast Cdc20: a target of the spindle checkpoint. *Science* **279** 1041-1044

Hyman AA & Mitchison TJ 1991 Two different microtubule-based motor activities with opposite polarities in kinetochores. *Nature* **351** 206-211

Igarashi H, Takahashi T, Takahashi E, Tezuka N, Nakahara K, Takahashi K & Kurachi H 2005 Aged mouse oocytes fail to readjust intracellular adenosine triphosphates at fertilization. *Biol Reprod* **72** 1256-1261

Ikui AE, Yang CP, Matsumoto T & Horwitz S 2005 Low concentrations of taxol cause mitotic delay followed by premature dissociation of p55CDC from Mad2 and BubR1 and abrogation of the spindle checkpoint, leading to aneuploidy. *Cell Cycle* **4** 1385-1388

Ishikawa H & Endo A 1995 Combined effects of maternal age and delayed fertilization on the frequency of chromosome anomalies in mice. *Hum Reprod* **10** 883-886

Jallepalli PV, Waizenegger I, Bunz F, Langer S, Speicher MR, Peters JM, Kinzler KW, Vogelstein B & Lengauer C 2001 Securin is required for chromosomal stability in human cells. *Cell* **105** 445-457

Jeffreys, CA, Burrage PS & Bickel SE 2003 A model system for increased meiotic nondisjunction in older oocytes. *Curr Biol* **13** 498-503

Jesus C, Rime H, Haccard O, Van Lint J, Goris J, Merlevede W & Ozon R 1991 Tyrosine phosphorylation of p34^{cdc2} and p42 during meiotic maturation of *Xenopus* oocyte: antagonistic action of okadaic acid and 6-DMAP. *Development* **111** 813-820

Johnson VL, Scott MI, Holt SV, Hussein D & Taylor SS 2004 Bub1 is required for kinetochore localization of BubR1, Cenp-E, Cenp-F and Mad2, and chromosome congression. *J Cell Sci* **117** 1577-1589

Juberg RC 1983 Origin of chromosomal abnormalities: evidence for delayed fertilization in meiotic nondisjunction. *Hum Genet* **64** 122-127

Juetten J & Bavister B 1983 Effects of egg aging on in vitro fertilization and first cleavage division in the hamster. *Gamete Res* **8** 219-230

Kalitsis P, Earle E, Fowler KJ & Choo KH 2000 Bub3 gene disruption in mice reveals essential mitotic spindle checkpoint function during early embryogenesis. *Genes Dev* **14** 2277-2282

Kallio M, Eriksson JE & Gorbsky GJ 2000 Differences in spindle association of the mitotic checkpoint protein Mad2 in mammalian spermatogenesis and oogenesis. *Dev Biol* **225** 112-123

Kallio MJ, McCleland ML, Stukenberg PT & Gorbsky GJ 2002 Inhibition of aurora B kinase blocks chromosome segregation, overrides the spindle checkpoint, and perturbs microtubule dynamics in mitosis. *Curr Biol* **12** 900-905

Kamieniecki RJ, Shanks RM & Dawson DS 2000 Slk19p is necessary to prevent separation of sister chromatids in meiosis I. *Curr Biol* **10** 1182-1190

Karsenti E 1991 Mitotic spindle morphogenesis in animal cells. *Semin Cell Biol* **4** 251-260

Katis VL, Galova M, Rabitsch KP & Nasmyth K 2004a Maintenance of cohesion at centromeres after meiosis I in budding yeast requires a kinetochore-associated protein related to MEI-S332. *Curr Biol* **14** 560-572

Katis VL, Matos J, Mori S, Shirahige K, Zachariae W & Nasymth K 2004b Spo13 facilitates monopolin recruitment to kinetochores and regulates maintenance of centromeric cohesion during yeast meiosis. *Curr Biol* **14** 2183-2196

Kaufman MH 1983 Early Mammalian Development: Parthenogenetic Studies. Cambridge: Cambridge Univ. Press. p 1-276

Kerrebrock AW, Miyazaki WY, Birnby D & Orr-Weaver TL 1992 The Drosophila mei-S332 gene promotes sister-chromatid cohesion in meiosis following kinetochore differentiation. *Genetics* **130** 827-841

Kienitz A, Vogel C, Morales I, Muller R & Bastians H

2005 Partial downregulation of MAD1 causes spindle checkpoint inactivation and aneuploidy, but does not confer resistance towards taxol. *Oncogene* **24** 4301-4310

Kikuchi K, Izaike Y, Noguchi J, Furukawa T, Daen FP, Naito K & Toyoda Y 1995 Decrease of histone H1 kinase activity in relation to parthenogenetic activation of pig follicular oocytes matured and aged in vitro. *J Reprod Fert* **105** 325-330

Kikuchi K, Naito K, Noguchi J, Shimada A, Kaneko H, Yamashita M, Aoki F, Tojo H & Toyoda Y 2000 Maturation/M-phase promoting factor: a regulator of aging in porcine oocytes. *Biol Rep* **63** 715-722

Kim NH, Moon SJ, Prather RS & Day BN 1996 Cytoskeletal alteration in aged porcine oocytes and parthenogenesis. *Mol Rep Dev* **43** 513-518

Kirsch-Volders M, Cundari E & Verdoodt B 1998 Towards a unifying model for the metaphase/anaphase transition. *Mutagenesis* **13** 321-335

Kitajima TS, Kawashima SA & Watanabe Y 2004 The conserved kinetochore protein shugoshin protects centromeric cohesion during meiosis. *Nature* **427** 510-517

Kitajima TS, Sakuno T, Ishiguro K, Iemura S, Natsume t, Kawashima SA & Watanabe Y 2006. Shugoshin collaborates with protein phosphatase 2A to protect cohesin. *Nature* **441** 46-52

Klein F, Mahr P, Galova M, Buonomo SB, Michaelis C, Nairz K & Nasmyth K 1999 A central role for cohesions in sister chromatid cohesion, formation of axial elements, and recombination during yeast meiosis. *Cell* **98** 91-103

Kline D & Kline JT 1992 Repetitive calcium transients and the role of calcium exocytosis and cell cycle activation in the mouse egg. *Dev Biol* **149** 80-89

Knowlton AL, Lan W & Stukenberg PT 2006 Aurora B is enriched at merotelic attachment sites, where it regulates MCAK. *Curr Biol* **16** 1705-1710

Kosower NS & Kosower EM 1978 The glutathione status of cells. *Int Rev Cytol* **54** 109-160

Kotani S, Tanaka H, Yasuda H & Todokoro K 1999 Regulation of APC by phosphorylation and regulatory factors. *J Cell Biol* **146** 791-800

Kubiak JZ 1989 Mouse oocytes gradually develop the capacity for activation during the metaphase II arrest. *Dev Biol* **136** 537-545

Kuliev A, Cieslak J, Ilkevitch Y & Verlinsky Y 2003 Chromosomal abnormalities in a series of 6733 human oocytes in preimplantation diagnosis for age-related aneuploidies. *Rep Biomed Online* **6** 54-59

Lampson MA, Renduchitala K, Khodjakov A & Kapoor TM 2004 Correcting improper chromosome-spindle attachments during cell division. *Nat Cell Biol* **6** 232-237

Lamson SH & Hook EB 1980 A simple function for maternal age-specific rates of Down syndrome in the 20- to-48 year age range and its biological implications. *Am J Hum Genet* **32** 743-753

Lavoie BD, Hogan E & Koshland D 2002 In vivo dissection of the chromosome condensation machinery: reversibility of condensation distinguishes contributions of condensin and cohesion. *J Cell Biol* **156** 805-815

Lee BH & Amon A 2003 Role of Polo-like kinase CDC5 in programming meiosis I chromosome segregation. *Science* **300** 482-486

Lee BH, Kiburz BM & Amon A 2004 Spo13 maintains centromeric cohesion and kinetochore coorientation during meiosis I. *Curr Biol* **14** 2168-2182

Lee J, Miyano T & Moor RM 2000 Localization of phosphorylated MAP kinase during transition from meiosis I to Meiosis II in pig oocytes. *Zygote* **8** 119-125

Lee J, Iwai T, Yokota T & Yamashita M 2003 Temporally and spatially selective loss of Rec8 protein from meiotic chromosomes during mammalian meiosis. *J Cell Sci* **116** 2781-2790

Lee J, Okada K, Ogushi S, Miyano T, Miyake M & Yamashita M 2006 Loss of Rec8 from chromosome arm and centromere region is required for homologous chromosome separation and sister chromatid separation, respectively, in mammalian oocytes. *Cell Cycle* **5** 1448-1455

Lee JY & Orr-Weaver TL 2001 The molecular basis of sister-chromatid cohesion. *Annu Rev Cell Dev Biol* **17** 753-777

Lee JY, Hayashi-Hagihara A & Orr-Weaver TL 2005 Roles and regulation of the Drosophila centromere cohesion protein MEI-332 family. *Phil Trans R Soc B* **360** 543-552

Lens SMA & Medema RH 2003 The survivin/Aurora B complex: its role in coordinating tension and attachment. *Cell Cycle* **2** 507- 510

Lewis WH & Wright ES 1935 On the early development of the mouse egg. *Carnegie Inst Contrib Embryol* **25** 113-143

Li Y & Benezra R 1996 Identification of a human mitotic checkpoint gene hsMAD2. *Science* **274** 246-248

Liang XW, Zhu JQ, Miao YL, Liu JH, Wei L, Lu SS, Hou Y, Schatten H, Lu KH & Sun QY 2008 Loss of methylation imprint of Snrpn in postovulatory aging mouse oocytes. *Biochem Biophys Res Comm* **371** 16-21

Lim AS, Ho AT & Tsakod MF 1995 Chromosomes of oocytes failing in-vitro fertilization. *Hum Rep* **10** 2570-2575

Liu L, Ju JC & Yang X 1998 Differential inactivation of maturation-promoting factor and mitogen-activated protein kinase following parthenogenetic activation of bovine oocytes. *Biol Rep* **59** 537-545

London SN & Mailhes JB 2001 Tamoxifen-induced alterations in meiotic maturation and cytogenetic abnormalities in mouse oocytes and 1-cell zygotes. *Zygote* **9** 97-104

Longo FJ 1981 Changes in the zona pellucida and plasmalemmae of aging mouse oocytes. *Biol Rep* **25** 399-411

Lorca T, Cruzalegul FH, Fesquet D, Cavadore D, Mery JC, Means A & Doree M 1993 Calmodulin-dependent protein kinase II mediates inactivation of MPF and CSF upon fertilization of Xenopus eggs. *Nature* **366** 270-273

Losada A, Yokochi T, Kobayashi R & Hirano T 2000 Identification and characterization of SA/Scc3p subunits in the Xenopus and human cohesion complexes. *J Cell Biol* **150** 405-416

Losada A, Hirano M & Hirano T 2002. Cohesin release is required for sister chromatid resolution, but not for condensing-mediated compaction at the onset of mitosis. *Genes Dev* **16** 3004-3016

Luo X, Fang G, Coldiron M, Lin Y, Yu H, Kirschner MW & Wagner G 2000 Structure of the Mad2 spindle assembly checkpoint protein and its interaction with Cdc20. *Nat Struct Biol* **7** 224-229

Ma W, Zhang D, Hou Y, Li Y-H, Sun Q-Y, Sun, X-F & Wang W-H 2005 Reduced expression of MAD2, BCL2, MAP kinase activity in pig oocytes after in vitro aging are associated with defects in sister chromatid segregation during meiosis II and embryo fragmentation after activation. *Biol Rep* **72** 373-383

Magli C, Gianaroli L & Ferraretti AP 2001 Chromosomal abnormalities in embryos. *Mol Cell Endocrinol* **183** 29-34

Magnaghi-Jaulin, Eot-Houllier G, Fulcrand G & Jaulin C 2007 Histone deacetylase inhibitors induce premature sister chromatid separation and override of the mitotic spindle assembly checkpoint. *Cancer Res* **67** 6360-6367

Mailhes JB, Preston RJ & Lavappa KS 1986 Mammalian in vivo assays for aneuploidy in female germ cells *Mutat Res* **320** 139-148

Mailhes JB 1987 Incidence of aneuploidy in rodents. In Vig BK, Sandberg AA, editors. Aneuploidy, Part A: Incidence and Etiology. New York: Alan R. Liss. p 67-101

Mailhes JB & Yuan ZP 1987 Differential sensitivity of mouse oocytes to colchicine-induced aneuploidy. *Environ Mol Mutagen* **10** 183-188

Mailhes JB, Marchetti F & Aardema MJ 1993 Griseofulvin-induced aneuploidy and meiotic delay in mouse oocytes: effect of dose and harvest time. *Mutat Res* **300** 155-163

Mailhes JB & Marchetti F 1994 Chemically-induced aneuploidy in mammalian oocytes. *Mutat Res* **320** 87-111

Mailhes JB & Marchetti F 1994a. The influence of postovulatory ageing on the retardation of mouse oocyte maturation and chromosome segregation induced by vinblastine. *Mutagenesis* **9** 541-545

Mailhes JB, Marchetti F, Phillips Jr. GL & Barnhill DR 1994 Preferential pericentric lesions and aneuploidy induced in mouse oocytes by the topoisomerase II inhibitor etoposide. *Terat Carcinogen Mutagen* **14** 39-51

Mailhes JB 1995 Important biological variables that can influence the degree of chemical-induced aneuploidy in mammalian oocytes and zygotes. *Mutat Res* **339** 155-176

Mailhes JB, Young D, Aardema MJ & London SN 1997a Thiabendazole-induced cytogenetic abnormalities in mouse oocytes. *Environ Mol Mutagen* **29** 367-371

Mailhes JB, Young D & London SN 1997b 1,2-Propanediol-induced premature centromere separation in mouse oocytes and aneuploidy in one-cell zygotes. *Biol Rep* **57** 92-98

Mailhes JB, Young D & London SN 1998 Postovulatory ageing of mouse oocytes in vivo and premature centromere separation and aneuploidy. *Biol Rep* **58** 1206-1210

Mailhes JB, Carabatsos MJ, Young D, London SN, Bell M & Albertini DF 1999 Taxol-induced meiotic maturation delay, spindle defects, and aneuploidy in mouse oocytes and zygotes. *Mutat Res* **423** 79-80

Mailhes JB, Hilliard C, Lowery M & London SN 2002 MG-132, an inhibitor of proteasomes and calpains, induced inhibition of oocyte maturation and aneuploidy in mouse oocytes. *Cell & Chromosome* **1** 1-7

Mailhes JB, Hilliard, Fuseler JW & London SN 2003a Okadaic acid, an inhibitor of protein phosphatase 1 and 2A, induces premature separation of sister chromatids during meiosis I and aneuploidy in mouse oocytes in vitro. *Chromosome Res* **11** 619-631

Mailhes JB, Hilliard C, Fuseler JW & London SN 2003b Vanadate, an inhibitor of tyrosine phosphatases, induced premature anaphase in oocytes and aneuploidy and polyploidy in mouse bone marrow cells. *Mutat Res* **538** 101-107

Mailhes JB, Mastromatteo C & Fuseler JW 2004 Transient exposure to the Eg5 kinesin inhibitor monastrol leads to syntelic orientation of chromosomes and aneuploidy in mouse oocytes. *Mutat Res* **559** 153-167

Marchetti F & Mailhes JB 1995 Variation of mouse oocyte sensitivity to griseofulvin-induced aneuploidy during the second meiotic division. *Mutagenesis* **10** 113-121

Marchetti F, Bishop JB, Lowe X, Generoso WM, Hozier J & Wyrobek AJ 2001 Etoposide induces heritable chromosomal aberrations and aneuploidy during male meiosis in the mouse. *Proc Natl Acad Sci USA* **98** 3952-3957

Marston JH & Chang MC 1964 The fertilizable life of ova and their morphology following delayed insemination in mature and immature mice. *J Exp Zool* **155** 237-252

Marston AL & Amon A 2004 Meiosis: cell-cycle controls shuffle and deal. *Nat Rev Mol Cell Biol* **5** 983-987

Marston AL, Tham WH, Shah H & Amon A 2004 A genome-wide screen identifies genes required for centromeric cohesion. *Science* **303** 1367-1370

McConnell JM, Campbell L & Vincent C 1995 Capacity of mouse oocytes to become activated depends on completion of cytoplasmic but not nuclear meiotic maturation. *Zygote* **3** 45-55

McGuinness BE, Hirota T, Kudo NR, Peters JM & Nasmyth K 2005 Shugoshin prevents dissociation of cohesin from centromeres during mitosis in vertebrate cells. *PLoS Biol* **3** e86

Meraldi P, Honda R & Nigg EA 2004. Aurora kinases link chromosome segregation and cell division to cancer susceptibility. *Curr Opin Genet Dev* **14** 29-36

Michaelis C, Ciosk R & Nasmyth K 1997 Cohesins: chromosomal proteins that prevent premature separation of sister chromatids. *Cell* **91** 35-45

Michel LS, Liberal V, Chatterjee A, Kirchwegger R, Pasche B, Gerald W, Dobles M, Lorger PK, Murty VVVS & Benerza R 2001 MAD2 haplo-insufficiency causes premature anaphase and chromosome instability in mammalian cells. *Nature* **409** 355-359

Michel L, Diaz-Rodriguez E, Narayan G, Hernando E, Vundavalli M & Benezra R 2004 Complete loss of the tumor suppressor MAD2 causes premature cyclin B degradation and mitotic failure in human somatic cells. *Proc Natl Acad Sci USA* **101** 4459-4464

Miller BM & Adler I-D 1992 Aneuploidy induction in mouse spermatocytes. *Mutagenesis* **7** 69-76

Miyazaki WY & Orr-Weaver TL 1992 Sister-chromatid misbehavior in Drosophila ord mutants. *Genetics* **132** 1047-1061

Miyazaki WY & Orr-Weaver TL 1994 Sister-chromatid cohesion in mitosis and meiosis. *Ann Rev Genet* **28** 167-187

Moos J, Visconti PW, Moore GD, Schultz RM & Kopf GS 1995 Potential role of mitogen-activated protein kinase (MAP) in pronuclear envelope assembly and disassembly following fertilization of mouse eggs. *Biol Reprod* **53** 692-699

Moses RM & Masui Y 1994 Enhancement of mouse egg activation by the kinase inhibitor, 6-dimethylainopurine (6-DMAP). *J Exp Zool* **270** 211-218

Morita Y & Tilly JL 1999 Oocyte apoptosis: like sand through an hourglass. *Dev Biol* **213** 1-17

Munne S 2002 Preimplantation genetic diagnosis of numerical and structural chromosome abnormalities. *Reprod Biomed Online* **4** 183-196

Murray AW 1998 MAP kinases in meiosis. *Cell* **92** 157-159

Musacchio A & Hardwick KG 2002 The spindle checkpoint: structural insights into dynamic signaling. *Nat Rev Mol Cell Biol* **3** 731-741

Nagai T 1987 Parthenogenetic activation of cattle follicular oocytes in vitro with ethanol. *Gamete Res* **16** 243-249

Nasmyth K, Peters JM & Uhlmann F 2000 Splitting the chromosome: Cutting the ties that bind sister chromatids. *Science* **288** 1379-1384

Nasmyth K 2001 Disseminating the genome: joining, resolving, and separating sister chromatids during mitosis and meiosis. *Annu Rev Genet* **35** 673-745

Nasmyth K 2002 Segregating sister genomes: the molecular biology of chromosome separation. *Science* **297** 559-565

Nasmyth K & Schleiffer A 2004 From a single double helix to paired double helices and back. *Phil Trans R Soc Lond B* **359** 99-108

Nasmyth K 2005 How do so few control so many? *Cell* **120** 739-746

Nguyen HG, Chinnappan D, Urano T & Ravid K 2005 Mechanism of Aurora-B degradation and its dependency on intact KEN and A-boxes: identification of an aneuploidy-promoting property. *Mol Cell Biol* **25** 4977-4992.

Nicolaidis P & Petersen MB 1998 Origin and mechanisms of non-disjunction in human autosomal trisomies. *Human Reprod* **13** 313-319

Nicklas RB 1997 How cells get the right chromosomes. *Science* **275** 632-637

O'Neill GT & Kaufman MH 1988 Influence of post-ovulatory aging on chromosome segregation during the second meiotic division in mouse oocytes: a parthenogenetic analysis. *J Exptl Zool* **248** 125-131

Ono T, Fang Y, Spector DL & Hirano T 2004 Spatial and temporal regulation of condensins I and II in mitotic chromosome assembly in human cells. *Mol Biol Cell* **15** 3296-3308

Orrenius S, Burkitt MJ, Kass GN, Dypbukt JM & Nicotera P 1992 Calcium ions and oxidative cell injury. *Ann Neurol* **32** S33-S42.

Pacchierotti F 1988 Chemically induced aneuploidy in germ cells of mouse. In Vig BK, Sandberg AA, editors. Aneuploidy, Part B: Induction and Test Systems, Progress and Topics in Cytogenetics, Vol. 7b. New York: Alan R. Liss. p 123-139

Pacchierotti F & Ranaldi R 2006 Mechanisms and risk of chemically induced aneuploidy in mammalian germ cells. *Curr Pharm Des* **12** 1489-1504

Pacchierotti F, Adler I-D, Eichenlaub-Ritter U & Mailhes JB 2007 Gender effects on the incidence of aneuploidy in mammalian germ cells. *Environ Res* **104** 46-69

Pahlavan G, Polanski Z, Kalab P, Golsteyn R, Nigg EA & Maro B 2000 Characterization of polo-like kinase 1 during meiotic maturation of the mouse oocyte. *Dev Biol* **220** 392-400

Panizza S, Tanaka T, Hochwagen A, Eisenhaber F & Nasmyth K 2000 Pds5 cooperates with cohesion in maintaining sister chromatid cohesion. *Curr Biol* **10** 1557-1564

Parisi S, McKay MJ, Molnar M, Thompson MA, van der Spek PJ, van Drunen-Schoenmaker E, Kanaar R, Lehmann E, Hoeijmakers JH & Kohli J 1999 Rec8p, a meiotic recombination and sister chromatid cohesion phosphoprotein of the Rad21p family conserved from fission yeast to humans. *Mol Cell Biol* **19** 3515-3528

Parra M, Viera A, Gomez R, Page J, Benavente R, Santos JL, Rufas JS & Suja JA 2004 Involvement of the cohesion Rad21 and SCP3 in monopolar attachment of sister kinetochores during mouse meiosis I. *J Cell Sci* **117** 1221-1234

Parry JM, Henderson L & Mckay JM 1995 Procedures for the detection of chemically induced aneuploidy: recommendations of a UK Environmental Mutagen Society working group. *Mutagenesis* **10** 1-14

Pasierbek P, Jantsch M, Melcher M, Schleiffer A, Schweizer D & Loidl J 2001 A *Caenorhaboditis elegans* cohesion protein with functions in meiotic chromosome pairing and disjunction. *Genes Dev* **15** 1349-1360.

Paules RS, Buccione R, Moschel RC, Vande Woude GF & Eppig JJ 1989 Mouse mos protooncogene product is present and functions during oogenesis. *Proc Natl Acad Sci USA* **86** 5395-5399

Pellestor F 1991 Frequency and distribution of aneuploidy in human female gametes. *Hum Genet* **86** 283-288

Pellestor F, Andreo B, Arnal F, Humeau C & Demaille J 2002 Mechanisms of non-disjunction in human female meiosis: the co-existence of two modes of malsegregation evidenced by the karyotyping of 1397 invitro unfertilized oocytes. *Hum Rep* **17** 2134-2145

Pellestor F, Andreo, Arnal F, Humeau C & Demaille J 2003 Maternal aging and chromosome abnormalities: new data drawn from in vitro unfertilized aged oocytes. *Hum Genet* **112** 195-203

Pellestor F, Anahory T & Hamamah S 2005 The chromosomal analysis of human oocytes. An overview of established procedures. Hum Rep Update 11 15-32

Pellestor F, Andreo B, Anahory T & Hammamah S 2006 The occurrence of aneuploidy in human: lessons from the cytogenetic studies of human oocytes. *Eur J Med Genet* **49** 103-116

Perez GI, Tao XJ & Tilly JL 1999. Fragmentation and death (a.k.a. apoptosis) of ovulated oocytes. *Mol Human Rep* **5** 414-420

Peter M, Castro A, Lorca T, Le Peuch C, Magnaghi-Jaulin Doree M & Labbe JC 2001 The APC is dispensable for first meiotic division in *Xenopus* oocytes. *Nature Cell Biol* **3** 83-87

Peters JM 2002 The anaphase-promoting complex: proteolysis in mitosis and beyond. *Mol Cell* **9** 931-943

Petronczki M, Siomos MF & Nasmyth K 2003 Un Ménage à Quatre: The molecular biology of chromosome segregation in meiosis. *Cell* **112** 423-440

Pickering SJ, Johnson MH, Braude Pr & Houliston E 1988 Cytoskeletal organization in fresh, aged and spontaneously activated human oocytes. *Hum Rep* **3** 978-989

Plachot M 2003 Genetic analysis of the oocyte - A review. *Placenta* **24** S66-S69

Plachot M, deGrouchy J, Junca AM, Mandelbaum J, Salat-Baroux J & Cohen J 1988. Chromosome analysis of human oocytes and embryos: does delayed fertilization increase chromosome imbalance? *Hum Rep* **3** 125-127

Prawitt D, Brixel L, Spangenberg C. Eshkind L, Heck R, Oesch F, Zabel B & Bockamp E 2004 RNAi knockdown mice: an emerging technology for post-genomic functional genetics. *Cytogenet Genome Res* **105** 412-421

Prieto I, Suja JA, Pezzi N, Kremer L, Martinez AC, Rufas JS & Barbero JL 2001 Mammalian STAG3 is a cohesion specific to sister chromatid arms in meiosis I. *Nat Cell Biol* **3** 761-766

Prieto I, Tease C, Pezzi N, Buesa JM, Ortega S, Kremer L, Martinez A, Martinez AC, Hulten MA & Barbero JL 2004 Cohesion component dynamics during meiotic prophase I in mammalian oocytes. *Chromosome Res* **12** 197-213

Prieto I, Pezzi N, Buesa JM, Kremer L, Barthelemy I, Carreiro C, Roncal F, Martinez A, Gomez L, Fernandez R, Martinez AC & Barbero JL 2002 STAG2 and Rad21 mammalian mitotic cohesions are implicated in meiosis. *EMBO Rep* **3** 543-550

Rabitsch KP, Gregan J, Schleiffer A, Javerzat JP, Eisenhaber F & Nasmyth K 2004 Two fission yeast homologs of *Drosophila* Mei-S332 are required for chromosome segregation during meiosis I and II. *Curr Biol* **14** 287-301

Racowsky C 1993 Somatic control of meiotic status in mammalian oocytes, In Haseltine FP, Heyner S, editors. Meiosis II, Contemporary Approaches to the Study of Meiosis. Wash. D.C.: Amer Assoc Adv Science Press. p 107-116

Rajagopalan H & Lengauer C 2004 Aneuploidy and cancer. *Nature* **432** 338-341

Revenkova E, Eijpe M, Heyting C, Gross B & Jessberger R 2001 Novel meiosis-specific isoform of mammalian SMC1. *Mol Cell Biol* **21** 6984-6998

Revenkova E & Jessberger R 2005 Keeping sister chromatids together: cohesins in meiosis. *Reproduction* **130** 783-790

Rhodes DR, Yu J, Shanker K, Deshpande N, Varambally R, Ghosh D, Barrette T, Pandey A & Chinnaiyan AM 2004 Large-scale meta-analysis of cancer microarray data identifies common transcriptional profiles of neoplastic transformation and progression. *Proc Natl Acad Sci USA* **101** 9309-9314

Riedel CG, Katis VL, Katou Y, Itoh T, Helmhart W, Galova M, Petronczki M, Gregan J, Cetin B, Mudrak I, Ogris E, Mechtler K, Pelletier L, Bucholz F, Shirahige K & Nasmyth K 2006 Protein phosphatase 2A protects centromeric sister chromatid cohesion during meiosis I. *Nature* **441** 53-61

Rieder CL & Palazzo RE 1992 Colcemid and the mitotic cycle. *J Cell Sci* **102** 387-392

Rieder CL & Salmon ED 1998 The vertebrate cell kinetochore and its role during mitosis. *Trends Cell Biol* **8** 310-318

Rieder CL, Schultz A, Cole R & Sluder G 1994 Anaphase onset in vertebrate somatic cells is controlled by a checkpoint that monitors sister kinetochore attachment at the spindle. *J Cell Biol* **127** 1301-1310

Rieder CL & Maiato H 2004 Stuck in division or passing through: what happens when cells cannot satisfy the spindle assembly checkpoint. *Dev Cell* **7** 637-651

Rime H, Neant I, Guerrier P & Ozon R 1989 6-dimethylaminopurine (6-DMAP), a reversible inhibitor of the transition to metaphase during the first meiotic cell division of the mouse oocyte. *Dev Biol* **133** 169-179

Rodman TC 1971 Chromatid disjunction in unfertilized ageing oocytes. *Nature* **233** 191-193

Romanienko PJ & Camerini-Otero RD 2000 The mouse Spo11 gene is required for meiotic chromosome synapsis. *Mol Cell* **6** 975-987

Rose D, Thomas W & Holm C 1990 Segregation of recombined chromosomes in meiosis requires DNA topoisomerase II. *Cell* **60** 1009-1017

Rosenbusch B 2004 The incidence of aneuploidy in human oocytes assessed by conventional cytogenetic analysis. *Hereditas* **141** 97-105

Russo A & Pacchierotti F 1988 Meiotic arrest and aneuploidy induced by vinblastine in mouse oocytes. *Mutat Res* **202** 215-221

Sagata N 1996 Meiotic metaphase arrest in animal oocytes: its mechanisms and biological significance. *Trends Cell Biol* **6** 22-28

Sagata N 1997 What does Mos do in oocytes and somatic cells? *BioEssays* **19** 13-21

Saito H, Koike K, Saito T, Nohara M, Kawagoe S & Hiroi M 1993 Aging changes in the alignment of chromosomes after human chorionic gonadotropin stimulation may be a possible cause of decreased fertility in mice. *Horm Res* **39** 28-31

Sakai N & Endo A 1988 Effects of delayed mating on preimplantation embryos in spontaneously ovulated mice. *Gamete Res* **19** 381-385

Sakurada K, Ishikawa H & Endo A 1996 Cytogenetic

effects of advanced maternal age and delayed fertilization on first-cleavage mouse embryos. *Cytogen Cell Genet* **72** 46-49

Salah SN & Nasmyth K 2000 Destruction of the securing Pds1p occurs at the onset of anaphase during both meiotic divisions in yeast. *Chromosoma* **109** 27-34

Salic A, Waters JC & Mitchison TJ 2004 Vertebrate shugoshin links sister centromere cohesion and kinetochore microtubule stabilization in mitosis. *Cell* **118** 567-578

Salmon ED, Cimini D, Cameron LA & DeLuca JG 2005 Merotelic kinetochores in mammalian tissue cells. *Phil Trans R Soc B* **360** 553-568

Schlecht U & Primig M 2003 Mining meiosis and gametogenesis with DNA microarrays. *Reproduction* **125** 447-456

Schönthal S 1992 Okadaic acid-a valuable new tool for the study of signal transduction and cell cycle regulation? *New Biol* **4** 16-21

Schultz RM 1988 Regulatory functions of protein phosphorylation in meiotic maturation of mouse oocytes in vitro. In Haseltine FP, First NL, editors. Meiotic Inhibition: Molecular Control of Meiosis. New York: Alan R Liss. p 137-151

Schultz RM, Montgomery RR & Belanoff JR 1983 Regulation of mouse oocyte maturation: implication of a decrease in oocyte cAMP and protein dephosphorylation in commitment to resume meiosis. *Dev Biol* **97** 264-273

Schultz RM 1986 Molecular aspects of mammalian oocyte growth and maturation, in Rossant J, Pederson RA, editors. Experimental Approach to Mammalian Embryonic Development. Cambridge UK: Cambridge University Press. p 195-237

Schwartz DA & Schultz RM 1991 Stimulatory effect of okadaic acid, an inhibitor of protein phosphatases, on nuclear envelope breakdown and protein phosphorylation in mouse oocytes and one-cell zygotes. *Dev Biol* **145** 119-127

Segers I, Adriaenssens T, Coucke W, Cortvrindt R & Smitz J 2008 Timing of nuclear maturation and post-ovulatory aging in oocytes of in vitro-grown mouse follicles with of without oil overlay. *Biol Rep* **78** 859-868

Shah JV, Botvinick E, Bonday Z, Fumari F, Berns M & Cleveland DW 2004 Dynamics of centromere and kinetochore proteins; implications for checkpoint signaling and silencing. *Curr Biol* **14** 942-952

Shamu CE & Murray AW 1992 Sister chromatid separation in frog extracts requires DNA topoisomerase II activity during anaphase. *J Cell Biol* **117** 921-934

Shang C, Hazbun TR, Cheesman IM, Arand J, fields S, Drubin DG & Barnes G 2003 Kinetochore protein interactions and their regulation by Aurora kinase Ipl1p.

Mol Biol Cell **14** 3342-3355

Shaver EL & Carr DH 1967 Chromosome abnormalities in rabbit blastocysts following delayed fertilization. *J Reprod Fert* **14** 415-420

Shonn MA, McCarroll R & Murray AW 2000 Requirement of the spindle checkpoint for proper chromosome segregation in budding yeast meiosis. *Science* **289** 300-303

Simerly C, Balczon R, Brinkley BR & Schatten G 1990 Microinjected centromere [corrected] kinetochore antibodies interfere with chromosome movement in meiotic and mitotic mouse oocytes. *J Cell Biol* **111** 1491-1504

Singh B & Arlinghaus RB 1997 Mos and the cell cycle. *Prog Cell Cycle Res* **3** 251-259

Siomos MF, Badrinath A, Pasierbek P, Livingstone D, White J, Glotzer M & Nasmyth K 2001 Separase is required for chromosome segregation during meiosis I in Caenorhabditis elegans. *Curr Biol* **11** 1825-1835

Sluder G 1979 Role of spindle microtubules in the control of cell cycle timing. *J Cell Biol* **80** 674-691

Sluder G & Rieder CL 1993 The events and regulation of anaphase onset. In Vig BK, editor. Chromosome Segregation and Aneuploidy, NATO ASI Series, Vol. H72. Berlin: Springer-Verlag. p 211-224

Sluder G & McCollum D 2000 The mad ways of meiosis. *Science* **289** 254-255

Smith AL & Lodge JR 1987 Interactions of aged gametes: in vitro fertilization using in vitro-aged sperm and in vivo-aged ova in the mouse. *Gamete Res* **16** 47-56

Sobajima T, Aoki F & Kohomoto K 1993 Activation of mitogen-activated protein kinase during meiotic maturation in mouse oocytes. *J Rep Fert* **97** 389-394

Sonada E, Matsusaka T, Morrison C, Vagnarelli P, Hoshi O, Ushiki T, Nojima K, Fukagawa T, Waizenegger IC, Peters JM, Earnshaw WC & Takeda S 2001 Scc1/Rad21/Mcd1 is required for sister chromatid cohesion and kinetochore function in vertebrate cells. *Dev Cell* **1** 759-770

Stein P, Svoboda P & Schultz RM 2003 Transgenic RNAi in mouse oocytes: a simple and fast approach to study gene function. *Dev Biol* **256** 187-193

Stern BM 2002 Mitosis: aurora gives chromosomes a healthy stretch. *Curr Biol* **12** R316-318

Steuerwald N, Cohen J, Herrera RJ, Sandalinas M & Brenner CA 2001 Association between spindle assembly checkpoint expression and maternal age in human oocytes. *Mol Hum Reprod* **7** 49-55

Steuerwald N, Steuerwald MD & Mailhes JB 2005 Postovulatory aging of mouse oocytes leads to decreased MAD2 transcripts and increased frequencies of premature centromere separation and anaphase. *Mol Hum Rep* **11** 623-630

Strausfeld U, Labbe JC, Fesquet D, Cavadore JC,

Picard A, Sadhu K, Russell P & Doree M 1991 Dephosphorylation and activation of a p34^{cdc2} / cyclin B complex in vitro by human CDC25 protein. *Nature* **351** 242-245

Sumara I, Vorlaufer E, Stukenberg PT, Kelm O, Redemann N, Nigg EA & Peters JM 2002 The dissociation of cohesion from chromosomes in prophase is regulated by Polo-like kinase. *Mol Cell* **9** 515-525

Sumara I, Gimenez-Abian JF, Gerlich D, Hirota T, Kraft C, de la torre C, Ellenberg J & Peters JM 2004 Roles of polo-like kinase 1 in the assembly of functional mitotic spindles. *Curr Biol* **14** 1712-1722

Sun FY, Schmid TE, Schmid E, Baumgartner A & Adler ID 2000 Trichlorfon induces spindle disturbances in V79 cells and aneuploidy in male mouse germ cells. *Mutagenesis* **15** 17-24

Swain JE & Smith GD 2007 Reversible phosphorylation and regulation of mammalian oocyte meiotic chromatin remodeling and segregation. *Soc Reprod Fert Suppl* **63** 343-358

Szollosi D 1971 Morphological changes in mouse eggs due to aging in the fallopian tube. *Am J Anat* **130** 209-226

Szollosi D 1975 Mammalian eggs ageing in the fallopian tubes. In Blandeau RJ, editor. Ageing Gametes. Basel: Karger. p 98-121

Szollosi MS, Debey P, Szollosi D, Rime H & Vautier D 1991 Chromatin behavior under influence of puromycin and 6-DMAP at different stages of mouse oocyte maturation. *Chromosoma* **100** 339-354

Szollosi MS, Kubiak JZ, Debey P, de Pennart H, Szollosi D & Maro B 1993 Inhibition of protein kinases by 6-dimethylaminopurine accelerates the transition to interphase in activated mouse oocytes. *J Cell Sci* **104** 861-872

Takahashi T, Takahashi E, Igarashi H, Tezuka N & Kurachi H 2003 Impact of oxidative stress in aged mouse oocytes on calcium oscillations at fertilization. *Mol Rep Dev* **66** 143-152

Takenaka K, Moriguchi T & Nishida E 1998 Activation of the protein kinase p38 in the spindle assembly checkpoint and mitotic arrest. *Science* **280** 599-602

Tanaka T, Fuchs J, Loidl J & Nasmyth K 2000 Cohesin ensures bipolar attachment of microtubules to sister centromeres and resists their precocious separation. *Nat Cell Biol* **2** 492-499

Tanaka TU, Rachidi N, Janke C, Pereira G, Galova M, Schiebel E, Stark MJ & Nasmyth K 2002 Evidence that the IpI1-Sli15 (Aurora kinase-INCENP) complex promotes chromosome bi-orientation by altering kinetochore-spindle pole connections. *Cell* **108** 317-329

Tang Z, Sun Y, Harley SE, Zou H & Yu H 2004 Human Bub1 protects centromeric sister-chromatid cohesion through Shugoshin during mitosis. *Proc Natl Acad Sci USA* **101** 18012-18017

Tang Z, Shu H, Qi W, Mahmood NA, Mumby MC & Yu H 2006 PP2A is required for centromeric location of Sgo1 and proper chromosome segregation. *Dev Cell* **10** 575-585

Tateno H & Kamiguchi Y 2001 Meiotic stage-dependent induction of chromosome aberrations in Chinese hamster primary oocytes exposed to topoisomerase II inhibitor etoposide. *Mutat Res* **476** 139-148

Tarin JJ, Vendrell FJ, Ten J, Blanes R, Van Blerkom J & Cano A 1996 The oxidizing agent tertiary butyl hydroperoxide induces disturbances in spindle organization, c-meiosis, and aneuploidy in mouse oocytes. *Mol Hum Rep* **2** 895-901

Tatone C, Carbone MC, Gallo R, Monache SD, Di Cola M, Alesse E & Amicarelli F 2006 Age-associated changes in mouse oocytes during postovulatory in vitro culture: possible role for meiotic kinases and survival factor BCL2. *Biol Rep* **74** 395-402

Taylor SS, Ha E & McKeon F 1998 The human homolog of Bub3 is required for kinetochore localization of Bub1 and a Mad3/Bub1-related protein kinase. *J Cell Biol* **142** 1-11

Taylor SS, Hussein D, Wang Y, Elderkin S & Morrow CJ 2001 Kinetochore localization and phosphorylation of the mitotic checkpoint components Bub1 and BubR1 are differentially regulated by spindle events in human cells. *J Cell Sci* **114** 4385-4395

Taylor SS, Maria I, Scott F & Holland AJ 2004 The spindle checkpoint: a quality control mechanism which ensures accurate chromosome segregation. *Chromosome Res* **12** 599-616

Terret ME, Wassmann K, Waizenegger I, Maro B, Peters JM & Verlhac MH 2003 The meiosis I-to-meiosis II transition in mouse oocytes requires separase activity. *Curr Biol* **13** 1797-1802

Thein KH, Kleylein-Sohn J, Nigg EA & Gruneberg U 2007 Astrin is required for the maintenance of sister chromatid cohesion and centrosome integrity. *J Cell Biol* **178** 345-354

Tiveron C, Marchetti F, Bassani B & Pacchierotti F 1992 Griseofulvin-induced aneuploidy and meiotic delay in female mouse germ cells. I. Cytogenetic analysis of metaphase II oocytes. *Mutat Res* **266** 143-150

Tombes RM, Semerly C, Borisy GG & Schatten G 1992. Meiosis, egg activation, and nuclear envelope breakdown are differentially reliant on Ca^{2+} independent in the mouse oocyte. *J Cell Biol* **117** 799-811

Tomonaga T, Nagao K, Kawasaki Y, Furuya K, Murakami A, Morishita J, Yuasa Y, Sutani T, Kearsey SE, Uhlmann F, Nasmyth K & Yanagida M 2000 *Genes*

Dev **14** 2757-2770

Tong C, Fan HY, Lian l, Li SW, Chen DY, Schatten H & Sun QY 2002 Polo-like kinase 1 is a pivotal regulator of microtubule assembly during mouse oocyte meiotic maturation, fertilization, and early embryonic mitosis. *Biol Rep* **67** 546-554

Toyooka Y, Tsunekawa N, Akasu R & Noce T 2003 Embryonic stem cells can form germ cells in vitro. *Proc Natl Acad Sci USA* **100** 11457-11462

Toth A, Ciosk R, Uhlmann F, Galova M, Schleiffer A & Nasmyth K 1999 Yeast cohesin complex requires a conserved protein, Eco1p(Ctf7), to establish cohesion between sister chromatids during DNA replication. *Genes Dev* **13** 320-333

Tsurumi C, Hoffmann S, Geley S, Graeser R & Polanski Z 2004 The spindle assembly checkpoint is not essential for CSF arrest of mouse oocytes. *J Cell Biol* **167** 1037-1050

Uhlmann F, Lottspeich F & Nasmyth K 1999 Sister-chromatid separation at anaphase onset is promoted by cleavage of the cohesion subunit Scc1. *Nature* **400** 37-42

Uhlmann F 2001 Chromosome cohesion and segregation in mitosis and meiosis. *Curr Opin Cell Biol* **13** 754-761

Uhlmann F 2003a Chromosome cohesion and separation: from men to molecules. *Curr Biol* **13** R104- R114

Uhlmann F 2003b Separase regulation during mitosis. *Biochem Soc Symp* **70** 243-251

Vagnarelli P & Earnshaw WC 2004 Chromosomal passengers: the four-dimensional regulation of mitotic events. *Chromosoma* **113** 211-222

Vandre DD & Willis VL 1992 Inhibition of mitosis by okadaic acid: possible involvement of a protein phosphatase 2A in the transition from metaphase to anaphase. *J Cell Sci* **101** 79-91

Van Heemst D & Heyting C 2000 Sister chromatid cohesion and recombination in meiosis. *Chromosoma* **109** 10-26

Verlhac M.H., Kubiak JZ, Clarke HJ & Maro BH 1994 Microtubule and chromatin behavior follow MAP kinase activity but not MPF activity during meiosis in mouse oocytes. *Development* **120** 1017-1025

Vialard F, Petit C, Bergere M, Gomes DM, Martel-Petit V, Lombroso R, Ville Y, Gerard H & Selva J 2006 Evidence of a high proportion of premature unbalanced separation of sister chromatids in the first polar bodies of women of advanced age. *Hum Rep* **21** 1172-1178

Vickers AD 1969 Delayed fertilization and chromosomal anomalies in mouse embryos. J Rep Fert 20 69-76

Vigneron S, Prieto S, Bernis C, Labbe JC, Castro A & Lorca T 2004 Kinetochore localization of spindle checkpoint proteins: Who controls whom. *Mol Biol*

Cell **15** 4584-4596

Vincent C, Cheek TR & Johnson MH 1992 Cell cycle progression of parthenogenetically activated mouse oocytes to interphase is dependent on the level of internal calcium. *J Cell Sci* **103** 389-396

Waizenegger I, Hauf S, Meinke A & Peters JM 2000 Two distinct pathways remove mammalian cohesion from chromosome arms in prophase and from centromeres in anaphase. *Cell* **103** 399-410

Waizenegger I, Gimenez-Abian JF, Wernic D & Peters JM 2002 Regulation of human separase by securin binding and autocleavage. *Curr Biol* **12** 1368-1378

Wang JC 2002 Cellular roles of DNA topoisomerase: a molecular perspective. *Nat Rev Mol Cell Biol* **3** 430-440

Wang H & Hoog C 2006 Structural damage to meiotic chromosomes impairs DNA recombination and checkpoint control in mammalian oocytes. *J Cell Biol* **173** 485-495

Wang XM, Yew N, Peloquin JG, Vande Woude GF & Borisy GG 1994 *Mos* oncogene product associates with kinetochores in mammalian somatic cells and disrupts mitotic progression. *Proc Natl Acad Sci USA* **91** 8329-8333

Wang X & Dai W 2005 Shugoshin, a guardian for sister chromatid segregation. *Exptl Cell Res* **310** 1-9

Warburton D 2005 Biological aging and the etiology of aneuploidy. *Cytogenet Genome Res* **111** 266-272

Wassmann K, Niault T & Maro B 2003 Metaphase I arrest upon activation of the Mad2-dependent spindle checkpoint in mouse oocytes. *Cur Biol* **13** 1596-1608

Watanabe Y & Kitajima TS 2005 Shugoshin protects cohesion complexes at centromeres. *Philos Trans R Soc Lond B Biol Sci* **360** 515-521

Webb M, Howlett SK & Maro B 1986 Parthenogenesis and cytoskeleton organization in aging mouse eggs. *J Embryol Exp Morphol* **95** 131-145

Weiss E & Winey M 1996 The S. cerevisiae SPB duplication gene MPS1 is part of a mitotic checkpoint. *J Cell Biol* **132** 111-123

Whittingham DG & Siracusa G 1978 The involvement of calcium in the activation of mammalian oocytes. *Exp Cell Res* **113** 311-317

Wilcox AJ, Weinberg CR & Baird DD 1998 Postovulatory ageing of the human oocyte and embryo failure. *Hum Rep* **13** 394-397

Winston N, Johnson M, Pickering S & Braude P 1991 Parthenogenetic activation and development of fresh and aged human oocytes. *Fert Ster* **56** 904-912

Wolf DP, Alexander M, Zelinski-Wooten M & Stouffer RL 1996 Maturity and fertility of rhesus monkey oocytes collected at different intervals after an ovulatory stimulus (human chorionic gonadotropin) in in vitro fertilization cycles. *Mol Rep Dev* **43** 76-81

Wolstenholme J & Angell RR 2000 Maternal age and trisomy – a unifying mechanism of formation. *Chromosoma* **109** 435-438

Wu B, Ignotz G, Currie WB & Yang X 1997 Expression of Mos proto-oncoprotein in bovine oocytes during maturation in vitro. *Biol Rep* **56** 260-265

Wyrobek AJ, Aardema M, Eichenlaub-Ritter U, Ferguson L & Marchetti F 1996 Mechanisms and targets involved in maternal and paternal age effects on numerical aneuploidy. *Environ Mol Mutagen* **28** 254-264

Xu Z, Abbott A, Kopf GS, Schultz RM & Ducibella T 1997 Spontaneous activation of ovulated mouse eggs: time-dependent effects on M-phase exit, cortical granule exocytosis, maternal messenger ribonucleic acid recruitment, and inositol 1,4,5-triphosphate sensitivity. *Biol Rep* **57** 743-750

Yamamoto M & Ingalls TH 1972 Delayed fertilization and chromosome anomalies in the hamster embryo. *Science* **176** 518-521

Yamasita M, Mita K, Yoshida N & Kondo T 2000 Molecular mechanisms of the initiation of oocyte maturation: general and species-specific aspects. *Prog Cell Cycle Res* **4** 115-129

Yanagimachi R & Chang MC 1961 Fertilizable life of golden hamster ova and their morphological changes at the time of losing fertilizability. *J Exp Zool* **148** 185-197

Yanagida M 2005 Basic mechanisms of eukaryotic chromosome segregation. *Phil Trans R Soc B* **360** 609-621

Yin H, Baart E, Betzendahl & Eichenlaub-Ritter U 1998 Diazepam induces meiotic delay, aneuploidy and predivision of homologues and chromatids in mammalian oocytes. *Mutagenesis* **13** 567-580

Yu HG & Koshland D 2005 Chromosome morphogenesis: condensin-dependent cohesion removal during meiosis. *Cell* **123** 397-407

Yuen KW, Montpetit B & Hieter P 2005 The kinetochore and cancer: what's the connection? *Curr Opin Cell Biol* **17** 576-582

Zachariae W & Nasmyth K 1999 Whose end is destruction: cell division and the anaphase-promoting complex. *Genes Dev* **13** 2039-2058

Zackowski J & Martin-Deleon PA 1988 Second meiotic nondisjunction is not increased in postovulatory aged murine oocytes fertilized in vitro. *In vitro Cell Dev Biol* **24** 133-137

Zenzes MT & Casper RF 1992 Cytogenetics of human oocytes, zygotes, and embryos after in vitro fertilization. *Hum Genet* **88** 367-375

Zernicka-Goetz M & Maro B 1993 Okadaic acid affects spindle organization in metaphase II-arrested rat oocytes. *Exptl Cell Res* **207** 189-193

Zernicka-Goetz M, Kubiak JZ, Antony C & Maro B 1993 Cytoskeletal organization of rat oocytes during metaphase II arrest and following abortive activation: a study by confocal laser scanning microscopy. *Mol Rep Dev* **35** 165-175

Zhang D & Nicklas R 1996 Anaphase and cytokinesis in the absence of chromosomes. *Nature* **382** 466-468

Zhang D, Ma W, LI Y-H, Hou Y, Li S-W, Meng X-Q, Sun X-F. Sun, Q-Y & Wang W-H 2004 Intra-oocyte localization of MAD2 and its relationship with kinetochores, microtubules, and chromosomes in rat oocytes during meiosis. *Biol Rep* **71** 740-748

Zhang D, Li M, Ma W, Yi H, Li Y-H, Li S-W, Sun Q-Y & Wang W-H 2005 Localization of mitotic arrest deficient 1 (MAD1) in mouse oocytes during the first meiosis and its functions as a spindle checkpoint protein. *Biol Rep* **72** 58-68

Zhao X, Singh B & Batten BE 1991 The role of *c-mos* proto-oncoprotein in mammalian meiotic maturation. *Oncogene* **6** 43-49

Zhou J, Yao J & Joshi HC 2002 Attachment and tension in the spindle checkpoint. *J Cell Sci* **115** 3547-3555

9

The Role of S-Palmitoylation of the Human Glucocorticoid Receptor (hGR) in Mediating the Nongenomic Glucocorticoid Actions

Nicolas C. Nicolaides[1,2] Tomoshige Kino[3], Michael L. Roberts[1], Eleni Katsantoni[4], Amalia Sertedaki[2], Paraskevi Moutsatsou[5], Anna-Maria G. Psarra[6], George P. Chrousos[1,2,7] and Evangelia Charmandari[1,2]

[1]Division of Endocrinology and Metabolism, Center of Clinical, Experimental Surgery and Translational Research, Biomedical Research Foundation of the Academy of Athens, Athens, Greece
[2]Division of Endocrinology, Metabolism and Diabetes, First Department of Pediatrics, National and Kapodistrian University of Athens Medical School, "Aghia Sophia" Children's Hospital, Athens, Greece
[3]Program in Reproductive and Adult Endocrinology, *Eunice Kennedy Shriver* National Institute of Child Health and Human Development, National Institutes of Health, Bethesda, Maryland
[4]Division of Hematology-Oncology, Basic Research Center, Biomedical Research Foundation of the Academy of Athens, Athens, Greece
[5]Department of Clinical Biochemistry, University of Athens Medical School, "Attiko" Hospital, Athens, 12462, Greece
[6]Department of Biochemistry and Biotechnology, University of Thessaly, Larissa, 41221, Greece
[7]Saudi Diabetes Study Research Group, King Fahd Center for Medical Research, King Abdulaziz University, Jeddah, Saudi Arabia

Correspondence should be addressed to Nicolas C. Nicolaides; E-mail: nnicolaides@bioacademy.gr

Abstract

Background: Many rapid nongenomic glucocorticoid actions are mediated by membrane-bound glucocorticoid receptors (GRs). S-palmitoylation is a lipid post-translational modification that mediates the membrane localization of some steroid receptors. A highly homologous amino acid sequence (663YLCMKTLLL671) is present in the ligand-binding domain of hGRα, suggesting that hGRα might also undergo S-palmitoylation.
Aim: To investigate the role of the motif 663YLCMKTLLL671 in membrane localization of the hGRα and in mediating rapid nongenomic glucocorticoid signaling.

Methods and Results: We showed that the mutant receptors hGRαY663A, hGRαC665A and hGRαLL670/671AA, and the addition of the palmitoylation inhibitor 2-bromopalmitate did not prevent membrane localization of hGRα and co-localization with caveolin-1, and did not influence the biphasic activation of mitogen-activated protein kinase (MAPK) signaling pathway in the early time points. Finally, the hGRα was not shown to undergo S-palmitoylation.
Conclusions: The motif 663YLCMKTLLL671 does not play a role in membrane localization of hGRα and does not mediate the nongenomic glucocorticoid actions.

Introduction

Glucocorticoids are steroid hormones that regulate a broad spectrum of physiologic functions essential for life, such as growth, reproduction, behavior, cognition, as well as many immune, metabolic and cardiovascular functions, through their ubiquitously expressed glucocorticoid receptor (hGR) (Charmandari *et al.* 2005, Chrousos *et al.* 2004, Chrousos & Kino 2005, 2007, Nicolaides *et al.* 2010, Rhen & Cidlowski 2005). The hGR belongs to the superfamily of steroid/thyroid/ retinoic acid receptor proteins that function as ligand-dependent transcription factors. The hGR regulates the transcription rate of many genes either by direct binding to their promoter regions or through interactions with other transcription factors, such as nuclear factor (NF)-κB, signal transducers and activators of transcription (STATs) and activator protein (AP)-1 (Nicolaides *et al.* 2010, Rhen & Cidlowski 2005).

In addition to the genomic actions, glucocorticoids exert rapid, 'non-genomic actions',

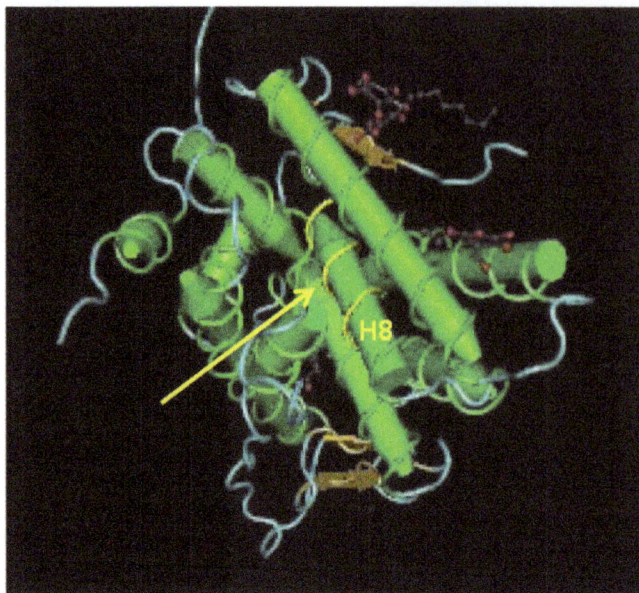

Figure 1. Stereotactic conformation of the agonist form of the LBD of hGRα. The yellow arrow indicates the 9-amino acid sequence, 663YLCMKTLL670, which is located on helix 8. H: Helix

which occur within seconds or minutes and are not inhibited by transcriptional and translational inhibitors (Ayroldi *et al.* 2012, Bellavance & Rivest 2012, Lee *et al.* 2012, Losel & Wehling 2003, Song & Buttgereit 2006, Stellato 2004). Examples of non-genomic actions of glucocorticoids include i) the immediate suppression of ACTH secretion from the anterior pituitary gland following a rise in glucocorticoid concentrations (Hinz & Hirschelmann 2000); ii) the increased frequency of excitatory post-synaptic potentials in the hippocampus (Karst *et al.* 2005); the rapid and transient decrease in blood pressure associated with a concomitant increase in coronary and cerebral blood flow in patients with myocardial infarction or stroke (Hafezi-Moghadam *et al.* 2002); and some T cell-related immune functions (Löwenberg *et al.* 2006). These rapid non-genomic actions of glucocorticoids are likely to be mediated by membrane glucocorticoid receptors that transduce the glucocorticoid signal via activation of downstream kinases (Ayroldi *et al.* 2012, Bellavance & Rivest 2012, Lee *et al.* 2012, Losel & Wehling 2003, Song & Buttgereit 2006, Stellato 2004).

Covalent lipid modifications anchor numerous signaling proteins to the cytoplasmic face of the plasma membrane (Linder & Deschenes 2007). S-palmitoylation refers to a reversible thioster linkage of palmitate (a C16 saturated fatty acid) to cysteine (Cys) residues of soluble proteins with hydrophobic moieties. This post-translational modification is catalyzed by membrane-bound palmitoyl-transferases (PATs) and increases protein hydrophobicity, thereby

enabling protein binding to membranes. Due to its reversible nature, this reaction provides a powerful on/off switch mechanism through anchoring/de-anchoring of proteins on the plasma membrane (Linder & Deschenes 2007).

Many heptahelical G protein-coupled receptors (GPCR) contain a conserved F(X6)LL sequence, where X is any amino acid and L is leucine or isoleucine. F, LL, and the precise 6-amino acid spacing between F and LL are required for protein export from the endoplasmic reticulum (Duvernay *et al.* 2004). A similar to the F(X6)LL sequence, highly conserved, 9-amino acid motif has been identified in the ligand-binding domains (LBDs) of most steroid receptors, and in particular of the human estrogen receptors (ER) α and β, progesterone receptors A (PR-A) and B (PR-B), and the androgen receptor (AR). In contrast to typical GPCRs, the third amino acid of this motif (YLCMKTLL) in all steroid receptors except the mineralocorticoid receptor (MR) is cysteine. Recent studies have demonstrated that this motif plays an important role in S-palmitoylation, membrane localization and the steroid signaling through activation of the mitogen-activated protein kinase (MAPK) and phosphatidylinositol 3-kinase (PI3K) (Marino *et al.* 2006, Pedram *et al.* 2007). Mutation of the phenylalanine or tyrosine at position -2, cysteine at position 0, and hydrophobic isoleucine/leucine or leucine/leucine combinations at positions +5/6, relative to cysteine, significantly reduced membrane localization, MAPK and PI3K activation, thymidine incorporation into DNA and cell viability stimulated by specific steroid receptor ligands (Pedram *et al.* 2007). The localization sequence mediated palmitoylation of some steroid receptors, which facilitated their caveolin-1 association, subsequent membrane localization, and nongenomic signaling by respective steroids. Therefore, S-palmitoylation within the LBD may be a crucial modification for membrane translocation and function of these steroid receptors (Pedram *et al.* 2007).

Intriguingly, in the LBD domain of the hGRα protein, a sequence highly homologous to the above-discussed motif is present (663YLCMKTLL670), suggesting that hGRα might also undergo S-palmitoylation (Marino & Ascenzi 2006). This post-translational modification might account for membrane-initiated rapid nongenomic glucocorticoid signaling. The position of this motif in the crystal structure of the LBD of hGRα is shown in Figure 1. The hGRα sequence 663YLCMKTLL670 has been further confirmed to be palmitoylable by analysis with CSS-Palm (available from: http://bioinformatics.lcd-ustc.org/css_palm/) (Marino & Ascenzi 2006).

The aim of our study was to determine the role of the motif 663YLCMKTLLL671 in mediating rapid nongenomic glucocorticoid signaling and to investigate whether the hGRα undergoes S-palmitoylation. We performed *in vitro* studies to determine the specific residues within the 9-amino acid motif of the LBD of hGRα that are crucial for rapid glucocorticoid signaling. Specifically, we investigated whether mutation of the amino acids Y, C and LL at positions -2, 0 and +5/6, respectively, relative to

cysteine (C) in the 9-amino acid motif, significantly reduce localization of the receptor to the plasma membrane, palmitoylation, association with caveolin-1, and MAPK and AKT activation.

Materials and Methods

Plasmids
The pRShGRα plasmid expresses hGRα under the control of the Rous sarcoma virus (RSV) promoter.

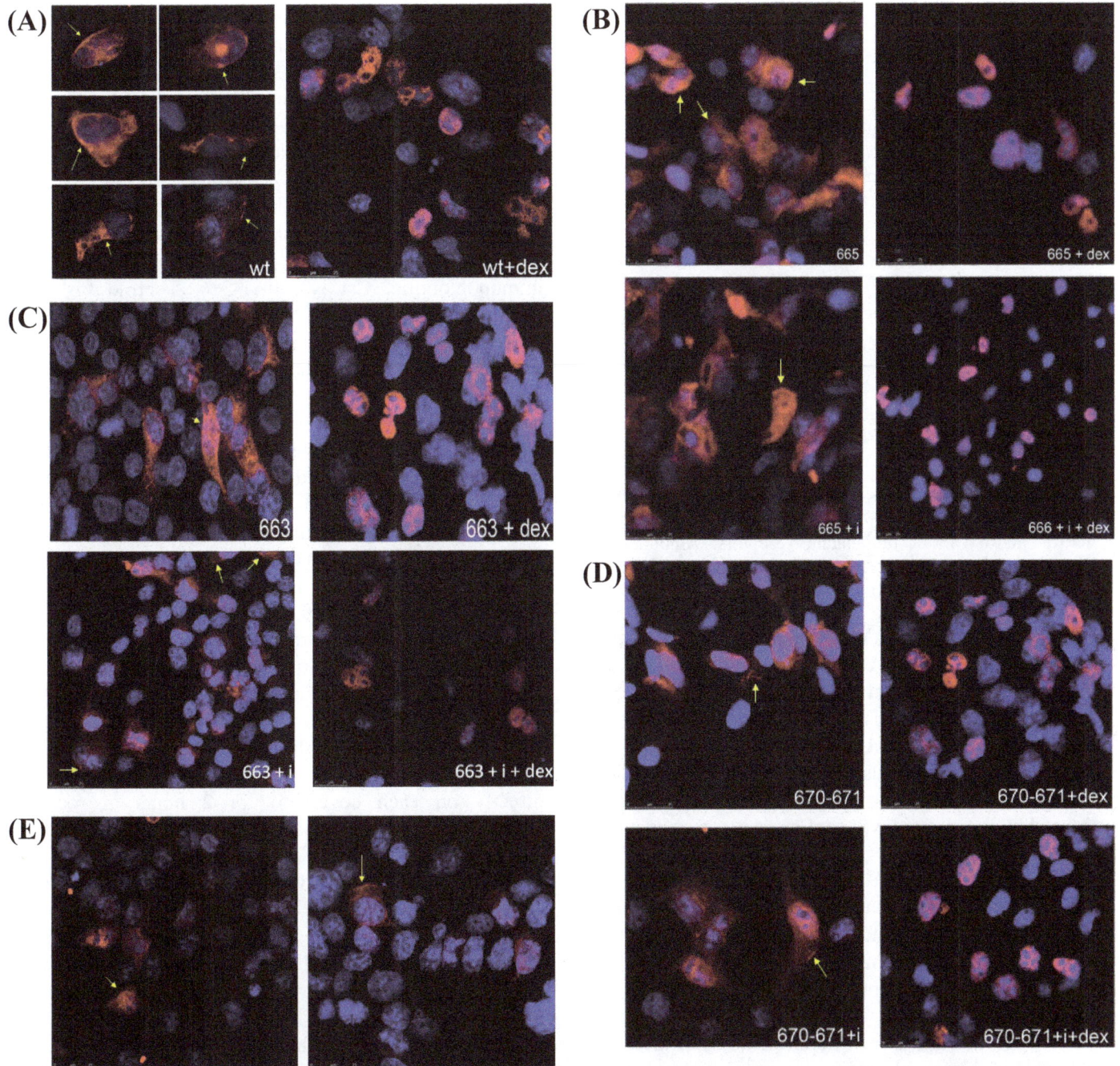

Figure 2. (A-D) Localization of the wild-type hGRα (hGRαWT) or its mutant receptors expressed in COS-7 cells in the absence or presence of dexamethasone. (E) Localization of hGRαWT in the presence of the palmitoylation inhibitor. The localization of hGRα-related proteins was examined with immunofluorescent staining. The yellow arrows point to membrane localization of hGRα. Dex: dexamethasone; i: palmitoylation inhibitor 2-Br.

Cytoplasm

C   WT   665   663   670/671   WT+Inh

hGRα ▶

Plasma Membrane

C   WT   665   663   670/671   WT+Inh

hGRα ▶

Nucleus

C   WT   665   663   670/671   WT+Inh

hGRα ▶

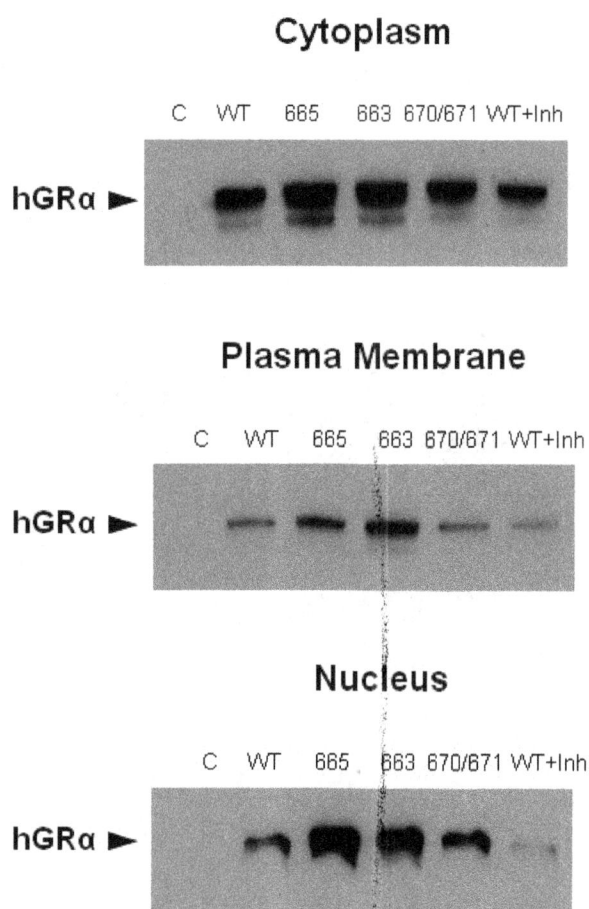

Figure 3. Cytoplasmic, membrane and nuclear localization of the hGRα in the COS-7 cells expressing the wild-type hGRα (hGRαWT) or the mutant receptors. Amounts of hGRα-related proteins were examined in subcellular fractionated samples with Western blotting.

The plasmids pRShGRαY663A, pRShGRαC665A and pRShGRαLL670/671AA were constructed by introducing the indicated mutations into the pRShGRα plasmid using PCR-assisted site-directed mutagenesis (Stratagene, La Jolla, CA, USA). The primers used for site-directed mutagenesis are shown in Table 1. The successful introduction of the indicated mutations into the pRShGRα plasmid was confirmed by sequencing.

Cell cultures

COS-7 embryonic African green monkey kidney cells, which do not express endogenously the glucocorticoid receptor, were grown in Dulbecco's Modified Eagle Medium (DMEM) supplemented with 10% Fetal Bovine Serum (FBS) and penicillin-streptomycin. Cells were incubated in a humidified atmosphere of 5% CO_2 at 37°C and passaged every 3-4 days using trypsin.

Immunofluorescence

COS-7 cells (3×10^5) were plated on glass coverslips in 6-well plates. Twenty-four hours later (confluency 90-95%), cells were transiently transfected with pRShGRα, pRShGRαC665A, pRShGRαY663A or pRShGRαLL670/671AA using lipofectamine 2000 (Invitrogen, Carlsbad, CA, USA). Forty-eight hours after transfection, dexamethasone 10^{-6} M was added for 2 hours in specific wells. When indicated, the palmitoyl-acetyl-transferase inhibitor bromohexa-decanoic acid (2-Br) (Sigma Aldrich, Gillingham, Dorset, UK) (100 μM, diluted in 100% DMSO) was added 30 min before dexamethasone administration. Subsequently, cells were washed with PBS, fixed with 4% fresh paraformaldehyde in PBS for 10 min at room temperature, and permeabilized with 0.2% Triton-X100 in PBS for 1 min. Prior to antibody incubation, cells were blocked with 2% BSA in PBS at room temperature. Following washing with PBS, cells were incubated with primary anti-hGRα antibody (dilution 1:100 in 0.5% BSA-PBS) (polyclonal anti-rabbit immunoglobulin, Cat No. PA1-516, Thermo Fisher Scientific, Boston, MA, USA) for 2 hours and subsequently with secondary antibody (dilution 1:500 in 0.5% BSA-PBS) (Alexa Fluor® 594 donkey anti-rabbit IgG, Cat No A21207, Invitrogen) for 1 hour at room temperature. Cells were washed with PBS and samples were mounted in 4',6-diamidino-2-phenylindole (DAPI) and stored at 4°C overnight. Samples were analyzed by confocal microscopy (DM IRB, Leica, Wetzlar, Germany). The confluency of the cells was 60-70% on each coverslip.

For double immunofluorescence staining, cells were co-incubated with 0.5% BSA-PBS containing a combination of primary anti-hGRα antibody (dilution 1:100) (polyclonal anti-rabbit immunoglobulin, Cat No. PA1-516, Thermo Fisher Scientific) and primary anti-caveolin-1 antibody (dilution 1:200) (monoclonal anti-mouse immunoglobulin, Cat No sc-53564, Santa Cruz Biotechnology, Santa Cruz, CA, USA) (placed directly on each coverslip) for 2 hours at room temperature. Following washing with PBS, cells were co-incubated with 0.5% BSA-PBS containing a combination of secondary antibodies (dilution 1:500 each) (Alexa Fluor® 594 donkey anti-rabbit IgG, Cat No A21207, Invitrogen, and Fluorescein goat anti-mouse IgG, Cat No F2761, Invitrogen) (placed directly on each coverslip) for 1 hour at room temperature. Samples were washed with PBS, mounted in DAPI and stored at 4°C overnight. Random fields of samples were selected, the cells with membrane localization of hGRα or co-localization of hGRα with caveolin-1 were counted manually, and the number of these cells was divided by the number of total cells within these fields.

(A)

(B)

(C)

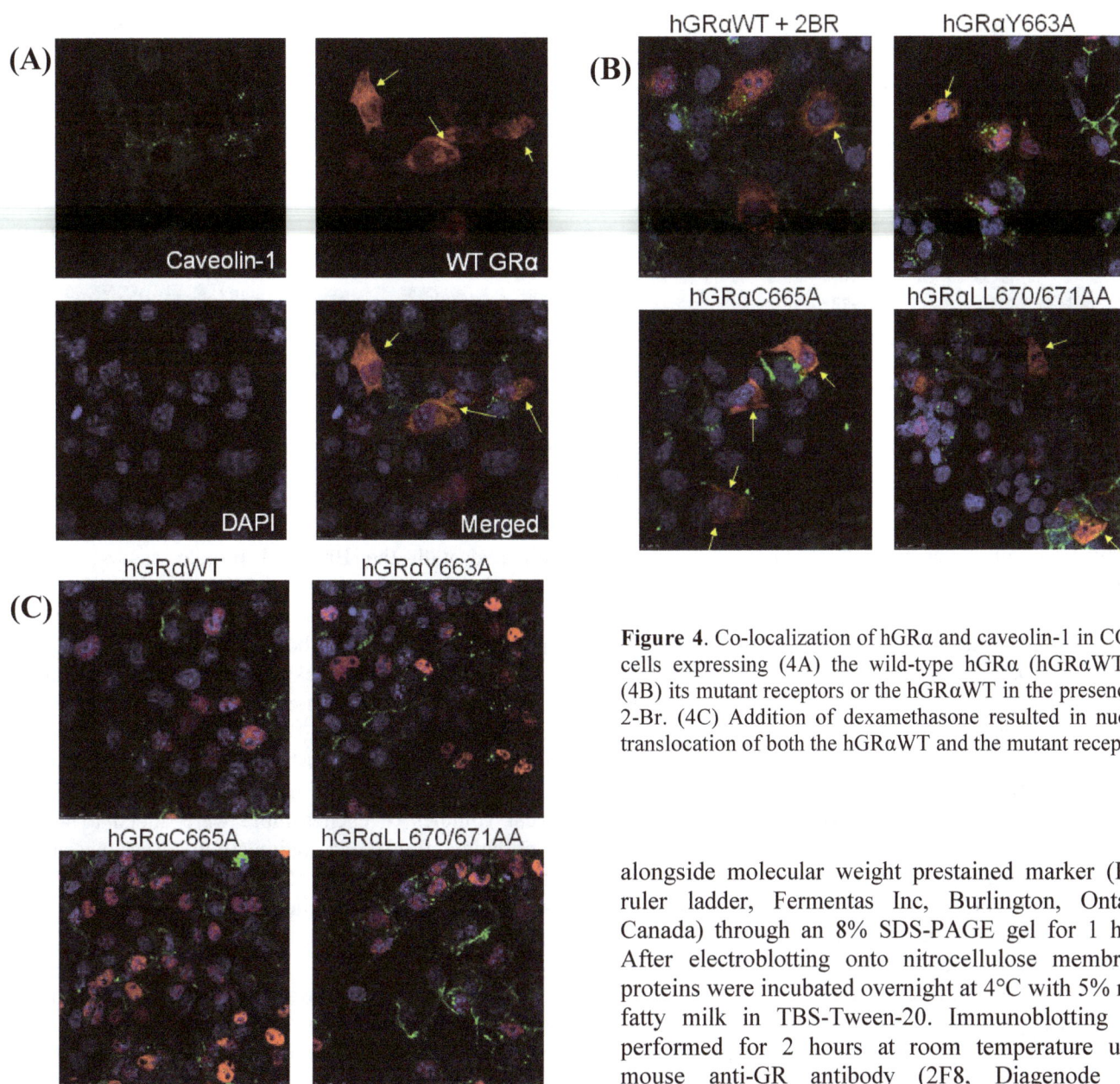

Figure 4. Co-localization of hGRα and caveolin-1 in COS-7 cells expressing (4A) the wild-type hGRα (hGRαWT) or (4B) its mutant receptors or the hGRαWT in the presence of 2-Br. (4C) Addition of dexamethasone resulted in nuclear translocation of both the hGRαWT and the mutant receptors.

Subcellular fractionation

COS-7 cells (2×10^6) were plated in 75-cm^2 flasks. Twenty-four hours later, cells were transiently transfected with pRShGRα, pRShGRαC665A, pRShGRαY663A or pRShGRαLL670/671AA using lipofectamine 2000 (Invitrogen). Forty-eight hours after transfection, cells were trypsinized and counted. From the total number of cells, 5×10^6 cells were washed once in PBS and were centrifuged (500 x g) for 3 min. PBS was carefully removed and cell pellets were mixed gently. The subcellular fractionation was performed using the Subcellular Protein Fractionation Kit (Pierce, Rockford, IL, USA) according to the instructions of the manufacturer. Equal amounts of protein (20 μg) were mixed with Laemli buffer (5X), heated at 95°C for 5 min and electrophoresed alongside molecular weight prestained marker (Page ruler ladder, Fermentas Inc, Burlington, Ontario, Canada) through an 8% SDS-PAGE gel for 1 hour. After electroblotting onto nitrocellulose membrane, proteins were incubated overnight at 4°C with 5% non-fatty milk in TBS-Tween-20. Immunoblotting was performed for 2 hours at room temperature using mouse anti-GR antibody (2F8, Diagenode Inc, Denville, USA) (1:500, diluted in 5% non-fatty milk in TBS-Tween-20). After washing with TBS-Tween-20, the membrane was incubated with horseradish peroxidase-conjugated goat anti-mouse secondary antibody (Dako, Glostrup, Denmark) (1:5000, diluted in 5% non-fatty milk in TBS-Tween-20) for 1 hour at room temperature. Proteins were visualized using the ECL Plus Western Blotting Detection System (Amersham Biosciences, Little Chalfont Buckinghamshire, UK) and exposed to high performance chemiluminescence film (Kodak, Rochester, New York, USA).

MAPK and AKT kinase assays

COS-7 cells (3×10^5) were plated in 6-well plates. Twenty-four hours later, cells were transiently transfected with pRShGRα, pRShGRαC665A,

(A)

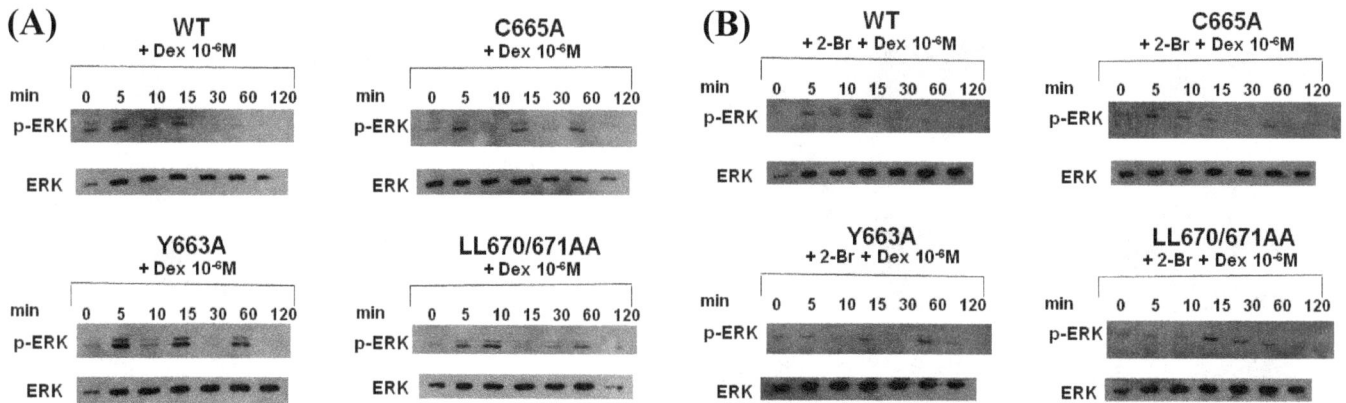

Figure 5. (A) Time-course effects of dexamethasone on the MAPK activity in COS-7 cells expressing the wild-type hGRα (hGRαWT) or its mutant receptors. (B) The effect of 2-Br on the MAPK activity in COS7 cells expressing the wild-type hGRα (hGRαWT) or its mutant receptors. Expression levels of total ERK and p-ERK were examined by Western blotting.

pRShGRαY663A or pRShGRαLL670/671AA using lipofectamine 2000 (Invitrogen), as described above. Twenty-four hours after transfection, the inhibitor 2-Br (100 μM diluted in 100% DMSO) was added in half of the wells. Sixteen hours later, cells were incubated with dexamethasone 10^{-6} M for different periods of time (0, 5, 10, 15, 30, 60 and 120 min). Cells were washed twice with PBS and lysed in 150 μl Complete Lysis-M (Roche, Basel, Switzerland) supplemented with Na_2VO_4 (0.2mM) and NaF (1mM). The homogenates were centrifuged (13,000 rpm at 4°C) for 15 min to obtain whole cell extracts. Western blotting was performed using mouse anti-p-ERK antibody (Cell Signaling, Beverly, MA, USA) (1:1000, diluted in 5% non-fatty milk in TBS-Tween-20) or rabbit anti-p-Akt antibody (Cell Signaling) (1:1000, diluted in 5% non-fatty milk in TBS-Tween-20) for 2 hours at room temperature. After stripping, the membrane was re-blotted using rabbit anti-ERK antibody (Cell Signaling) (1:1000, diluted in 5% non-fatty milk in TBS-Tween-20) or rabbit anti-Akt antibody (Cell Signaling) (1:250, diluted in 5% non-fatty milk in TBS -Tween-20) for 2 hours at room temperature. Following washing, the membrane was incubated with horseradish peroxidase-conjugated goat anti-rabbit secondary antibody (1:5000, diluted in 5% non-fatty milk in TBS-Tween-20, Dako) for 1 hour at room temperature. Finally, proteins were visualized as described above.

Palmitoylation assays

COS-7 cells (5×10^6) were plated in 100-mm cell dishes. Twenty-four hours later, cells were transiently transfected with pRShGRα or erbA^{-1} using lipofectamine 2000 (Invitrogen). Forty-eight hours after transfection, supplemented DMEM was replaced by plain DMEM for 1 hour. Cells were then exposed to 9,10-^{3}H(N) palmitic acid (0,1 mCi/mL) and incubated

for 2 hours at 37°C. Protein extracts were prepared as previously described, and protein concentrations were estimated with the Bradford protein assays. Equal amounts of protein extracts (500 μg) were used for immunoprecipitation assays. The anti-hGRα or anti-caveolin-1 antibody was bound to magnetic beads for 1 hour at 4°C. Protein extracts were then incubated with anti-hGRα- or anti-caveolin-1-bound magnetic beads overnight at 4°C. The next day the protein immunocomplex was washed six times, mixed with Laemli buffer (5X) and heated for 5 min at 95°C. Equal volumes of each sample were added to 4 ml scintillation fluid, and radioactivity was measured using a β-counter (Beckman LS6000IC counter, Beckman Coulter Inc., Fullerton, CA, USA).

Statistical analysis

All experiments were performed at least three times. Statistical analyses were carried out using the unpaired Student *t* test with a two-tailed P value.

Results

Neither the mutant receptors nor the addition of 2-Br prevented membrane localization of hGRα

To examine the membrane localization of the hGRα, COS-7 cells were transiently transfected with pRShGRα, pRShGRαC665A, pRShGRαY663A or pRShGRαLL670/671AA, and the immunofluorescence microscopy analysis was performed. Membrane localization of the hGRα was observed in 15% of the cells transfected with pRShGRα (Figure 2A) and in 13% of the cells transfected with pRShGRαC665A (Figure 2B), whereas the percentage of membrane localization of the hGRα in cells transfected with pRShGRαY663A or pRShGRαLL670/671AA was 11% in each case (Figures 2C & 2D). Addition of the inhibitor 2-Br did not prevent the membrane

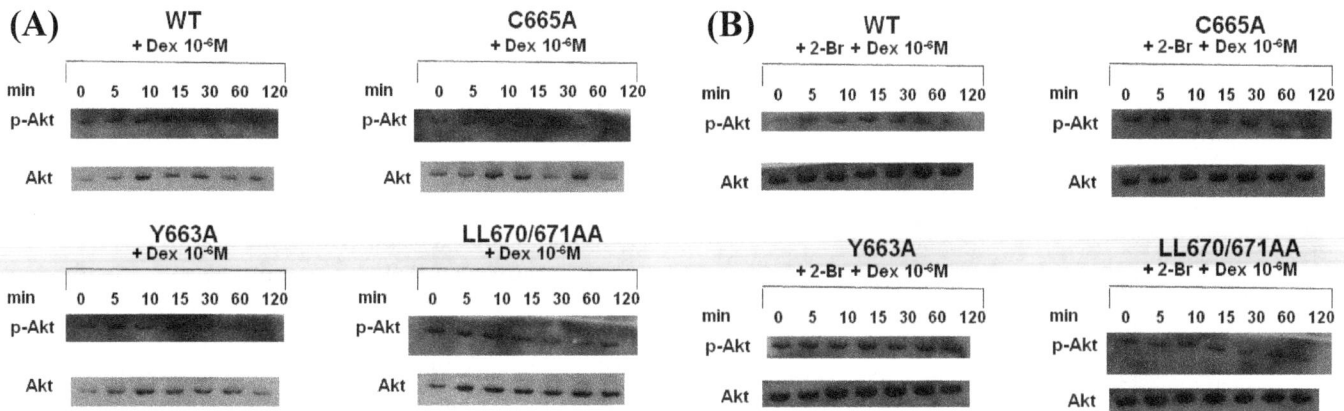

Figure 6. (A) Time-course effect of dexamethasone on the AKT activity in COS-7 cells expressing the wild-type hGRα (hGRαWT) or its mutant receptors. (B) The effect of 2-Br on the AKT activity in COS-7 cells expressing the wild-type hGRα (hGRαWT) or its mutant receptors. Expression levels of the total AKT and p-Akt were examined with Western blotting.

localization in cells transfected either with the wild-type or the mutant receptors (percentage of membrane localization 13%) (Figure 2E). Addition of dexamethasone resulted in nuclear translocation of the hGRα (wild-type or mutant hGRs) (percentage of membrane localization 0%) (Figures 2A-2D). Our immunofluorescence results were further confirmed by subcellular fractionation of COS-7 cells transfected with the pRShGRα or the mutant receptors (Figure 3).

The wild-type hGRα co-localized with caveolin-1 at the plasma membrane

We investigated whether the hGRα co-localizes with caveolin-1 using the double immunostaining method. Caveolin-1 was detected strongly as "dots" on the plasma membrane, as well as weakly in the cytoplasm in the COS-7 untransfected cells, suggesting a role of caveolin-1 in endocytosis/exocytosis processes. Co-localization of the hGRα with caveolin-1 was observed at the plasma membrane of the COS-7 cells transiently transfected either with the hGRα or its mutants at the same percentages as above (Figures 4A & 4B). Addition of the inhibitor 2-Br did not prevent the co-localization of hGRα with caveolin-1 (Figure 4B), whereas addition of dexamethasone resulted in nuclear translocation of hGRα without any co-migration with caveolin-1 (Figure 4C).

The mutant receptors preserved their ability to induce rapid nongenomic glucocorticoid actions through MAPK activation

To examine the role of the 663YLCMKTLLL671 motif in the induction of rapid nongenomic glucocorticoid actions, kinase signaling assays were performed in COS-7 cells transiently transfected either with the pRShGRα or the mutant receptors. Addition of dexamethasone for different periods of time resulted in a biphasic activation of the MAPK activity through

induction of ERK phosphorylation at 5 and 15 minutes, either in the absence or presence of 2-Br (Figures 5A, 5B). All mutant receptors tested did not prevent this dexamethasone-induced biphasic MAPK activation; however the double-mutant receptor hGRαLL670/671AA activated this kinase pathway at 5 and 10 minutes (Figures 5A, 5B). Interestingly, the activity of the AKT pathway was not induced by the dexamethasone-activated hGRα or the mutant receptors both in the absence or presence of the palmitoylation inhibitor (Figures 6A, 6B).

The hGRα did not undergo S-palmitoylation

To examine whether the hGRα was a palmitoylated protein, COS-7 cells transiently transfected with the pRShGRα or erbA^{-1} were incubated with 9,10-^{3}H(N) palmitic acid. The immunoprecipitated caveolin-1 was used as positive control, whereas the empty vector was used as a negative control. Caveolin-1 was shown to incorporate 9,10-^{3}H(N) palmitic acid, whereas the wild-type hGRα did not incorporate the tritiated substrate, indicating that the hGRα did not undergo S-palmitoylation (Figure 7).

Discussion

In the present study, we investigated the role of the highly conserved motif 663YLCMKTLLL671 of the LBD of hGRα in mediating the rapid, nongenomic glucocorticoid actions. We demonstrated that this 9-amino acid motif did not play a role in membrane localization of the hGRα and did not influence the co-localization of the receptor with caveolin-1 or the ligand-induced activation of MAPK pathway. We also showed that the hGRα did not undergo S-palmitoylation.

The 663YLCMKTLLL671 motif did not mediate the membrane localization of hGRα, granted

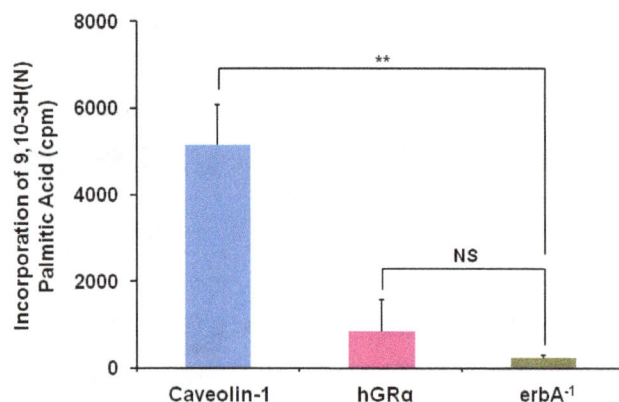

Figure 7. Palmitoylation of wild-type hGRα and caveolin-1. Caveolin-1 incorporated $9,10\text{-}^3\text{H(N)}$ palmitic acid, whereas the wild-type hGRα (hGRαWT) did not incorporate the tritiated substrate. Bars represent mean ± SEM of at least three independent experiments. **: P<0.01, n.s.: not significant, compared to control.

that the mutant receptors pRShGRαC665A, pRShGRαY663A and pRShGRαLL670/671AA were localized at the plasma membrane in the same percentage compared to that of the wild-type hGRα. Moreover, this motif was not responsible for co-localization of hGRα with caveolin-1. Indeed, in the absence of ligand, all the mutant receptors were still co-localized with caveolin-1 at the membrane, while addition of dexamethasone triggered their nuclear translocation. Furthermore, the activation of MAPK signaling pathway was not influenced by the mutant receptors, while neither the wild-type receptor nor the mutant receptors activated the PI3K/AKT pathway in COS-7 cells. These findings clearly showed that the 9-amino acid motif lying between amino acids 663 and 671 of the hGRα LBD did not influence the membrane localization of the receptor or the MAPK-mediated nongenomic glucocorticoid signaling.

To examine whether the hGRα undergoes S-palmitoylation by incorporating palmitic acid to the cysteine residue of the motif 663YLCMKTLLL671, we used the palmitoylation inhibitor 2-Br in some experiments and we performed palmitoylation assays using $9,10\text{-}^3\text{H(N)}$ palmitic acid. The membrane localization of the receptor and the nongenomic glucocorticoid signaling were not affected by the presence of the inhibitor, suggesting that this lipid post-translational modification did not play an important role for these actions. In addition, palmitoylation assays showed that compared with caveolin-1, which is a known palmitoylated protein, the hGRα did not incorporate palmitic acid.

We have chosen to investigate whether the 663YLCMKTLLL671 was a palmitoylated motif,

because this amino acid sequence was highly homologous to the conserved peptide located in the LBD of ERα and ERβ, PR-A and PR-B, and the AR. Pedram *et al.* (2007) demonstrated that these steroid receptors underwent S-palmitoylation through thioester linkage between palmitic acid and the cysteine residue located within the 9-amino acid motif. They showed that S-palmitoylation facilitated membrane localization, co-localization with caveolin-1, thymidine incorporation into DNA, cell viability, and MAPK and PI3K activation. In subsequent studies, they demonstrated that heat shock protein 27 promoted S-palmitoylation of the above steroid receptors and was required for their membrane localization (Razandi *et al.* 2010). They also showed that DHHC-7 and -21 were the specific PATs that catalyzed S-palmitoylation of ER, PR and AR (Pedram *et al.* 2012) and revealed the important role of estrogen membrane signaling in organ development and metabolism (Pedram *et al.* 2013, 2014). Based on these findings, we hypothesized that the hGRα might be a palmitoylated protein mediating the rapid nongenomic glucocorticoid signaling. However, our results showed that the motif 663YLCMKTLLL671 was not palmitoylated.

Our findings concur with two recent studies (Deng *et al.* 2015, Samarasinghe *et al.* 2011). In the latter study, Samarasinghe and colleagues (2011) investigated the role of nongenomic glucocorticoid actions in regulating the gap junction-mediated intercellular communication and neural progenitor cell proliferation. They demonstrated that a short time exposure of neural progenitor cells to dexamethasone reduced the gap junction-mediated intercellular communication through a signaling pathway initiated by membrane-bound glucocorticoid receptors, which co-localized with caveolin-1 and cellular Src (c-Src) kinase, and triggered a MAPK-mediated phosphorylation of connexin 43, an important gap junction protein (Samarasinghe *et al.* 2011). To examine whether this membrane localization of the receptor was palmitoylation-dependent, they expressed the hGRαC665A in CHO cells and found no difference in the number of cells with membrane localized hGRα, suggesting that the receptor might not undergo S-palmitoylation (Samarasinghe *et al.* 2011). The second study by Deng *et al.* showed that the endogenously expressed GRα in 4B cells did not incorporate tritiated palmitic acid, indicating that the GRα did not undergo S-palmitoylation (Deng *et al.* 2015). Moreover, they showed that substitution of the highly conserved cysteine 683 to alanine preserved membrane localization of the receptor, suggesting that this amino acid did not play a role in the localization of the GRα at the plasma membrane (Deng *et al.* 2015). However, they demonstrated that leucines located within this 9-amino acid motif were essential for interaction with HSP90 and for ligand-binding, thereby influencing the transcriptional activity of the GR (Deng *et al.* 2015). Although this study showed for the first time that GR was not a palmitoylated protein, the authors did not further investigate the role of this 9-amino acid motif in nongenomic glucocorticoid actions. Our study shows that neither point mutations in this motif nor the addition of 2-Br influenced the biphasic activation of MAPK pathway, which is involved in mediating the nongenomic glucocorticoid effects.

Delineating the molecular mechanisms of nongenomic glucocorticoid actions is extremely important, because membrane-initiated glucocorticoid signaling pathways are implicated in several pathophysiologic mechanisms. Given that several cardioprotective and immunosuppressive actions of glucocorticoids are non-genomic (Hafezi-Moghadam *et al.* 2002, Löwenberg *et al.* 2006), it is likely that the development of specific agonists and antagonists with well-characterized effects, which are able to modify discrete glucocorticoid-induced cellular functions, will help significantly towards the therapeutic management of myocardial infarction, stroke and several immunological conditions.

In summary, we demonstrated that the 663YLCMKTLLLL671 did not play a role in membrane localization of the hGRα and did not influence the co-localization of the receptor with caveolin-1 or the ligand-induced activation of the MAPK pathway. We also showed that the hGRα did not undergo S-palmitoylation. The membrane localization of the receptor, as well as the molecular mechanisms underlying the nongenomic glucocorticoid actions still remain an enigma. Further studies are necessary to investigate these research questions and to explore their clinical implications.

Acknowledgements

This work was supported by i) the European Union (European Social Fund - ESF) and Greek national funds of the National Strategic Reference Framework (NSRF) - Research Funding Program: Heracleitus II; and ii) the European Union (European Social Fund - ESF) and Greek national funds through the Operational Program "Education and Lifelong Learning" of the National Strategic Reference Framework (NSRF) - Research Funding Program: THALIS - University of Athens (UOA), Athens, Greece.

References

Ayroldi E, Cannarile L, Migliorati G, Nocentini G, Delfino DV & Riccardi C 2012 Mechanisms of the anti-inflammatory effects of glucocorticoids: genomic and nongenomic interference with MAPK signaling pathways. *FASEB J* **26** 4805-4820

Bellavance MA & Rivest S 2012 The neuroendocrine control of the innate immune system in health and brain diseases. *Immunol Rev* **248** 36-55

Charmandari E, Tsigos C & Chrousos GP 2005 Endocrinology of the stress response. *Annu Rev Physiol* **67** 259-284

Chrousos GP, Charmandari E & Kino T 2004 Glucocorticoid action networks-an introduction to systems biology. *J Clin Endocrinol Metab* **89** 563-564

Chrousos GP & Kino T 2005 Intracellular glucocorticoid signaling: a formerly simple system turns stochastic. *Sci STKE* **2005** pe48

Chrousos GP & Kino T 2007 Glucocorticoid action networks and complex psychiatric and /or somatic disorders. *Stress* **10** 213-219

Deng Q, Waxse B, Riquelme D, Zhang J & Aguilera G 2015 Helix 8 of the ligand binding domain of the glucocorticoid receptor (GR) is essential for ligand binding. *Mol Cell Endocrinol* **408** 23-32

Duvernay MT, Zhou F & Wu G 2004 A conserved motif for the transport of G protein-coupled receptors from the endoplasmic reticulum to the cell surface. *J Biol Chem* **279** 30741-30750

Hafezi-Moghadam A, Simoncini T, Yang Z, Limbourg FP, Plumier JC, Rebsamen MC, Hsieh CM, Chui DS, Thomas KL, Prorock AJ, Laubach VE, Moskowitz MA, French BA, Ley K& Liao JK 2002 Acute cardiovascular protective effects of corticosteroids are mediated by non-transcriptional activation of endothelial nitric oxide synthase. *Nat Med* **8** 473-479

Hinz B & Hirschelmann R 2000 Rapid non-genomic feedback effects of glucocorticoids on CRF-induced ACTH secretion in rats. *Pharm Res* **17** 1273-1277

Karst H, Berger S, Turiault M, Tronche F, Schütz G & Joëls M 2005 Mineralocorticoid receptors are indispensable for nongenomic modulation of hippocampal glutamate transmission by corticosterone. *Proc Natl Acad Sci U S A* **102** 19204-19207

Lee SR, Kim HK, Youm JB, Dizon LA, Song IS, Jeong SH, Seo DY, Ko KS, Rhee BD, Kim N & Han J 2012 Non-genomic effect of glucocorticoids on cardiovascular system. *Pflugers Arch* **464** 549-559

Linder ME & Deschenes RJ 2007 Palmitoylation: policing protein stability and traffic. *Nat Rev Mol Cell*

Biol **8** 74-84

Losel R & Wehling M 2003 Nongenomic actions of steroid hormones. *Nat Rev Mol Cell Biol* **4** 46-56

Löwenberg M, Verhaar AP, Bilderbeek J, Marle Jv, Buttgereit F, Peppelenbosch MP, van Deventer SJ & Hommes DW 2006 Glucocorticoids cause rapid dissociation of a T-cell-receptor-associated protein complex containing LCK and FYN. *EMBO Rep* **7** 1023-1029

Marino M & Ascenzi P 2006 Steroid hormone rapid signaling: the pivotal role of S-palmitoylation. *IUBMB Life* **58** 716-719

Marino M, Ascenzi P & Acconcia F 2006 S-palmitoylation modulates estrogen receptor alpha localization and functions. *Steroids* **71** 298-303

Nicolaides NC, Galata Z, Kino T, Chrousos GP & Charmandari E 2010 The human glucocorticoid receptor: Molecular basis of biologic function. *Steroids* **75** 1-12.

Pedram A, Razandi M, Deschenes RJ & Levin ER 2012 DHHC-7 and -21 are palmitoylacyltransferases for sex steroid receptors. *Mol Biol Cell* **23** 188-199

Pedram A, Razandi M, Lewis M, Hammes S & Levin ER 2014 Membrane-localized estrogen receptor α is required for normal organ development and function. *Dev Cell* **29** 482-490

Pedram A, Razandi M, O'Mahony F, Harvey H, Harvey BJ & Levin ER 2013 Estrogen reduces lipid content in the liver exclusively from membrane receptor signaling. *Sci Signal* **6** ra36

Pedram A, Razandi M, Sainson RC, Kim JK, Hughes CC & Levin ER 2007 A conserved mechanism for steroid receptor translocation to the plasma membrane. *J Biol Chem* **282** 22278-22288

Razandi M, Pedram A & Levin ER 2010 Heat shock protein 27 is required for sex steroid receptor trafficking to and functioning at the plasma membrane. *Mol Cell Biol* **30** 3249-3261

Rhen T & Cidlowski JA 2005 Anti-inflammatory action of glucocorticoids-new mechanisms for old drugs. *N Engl J Med* **353** 1711-1723

Samarasinghe RA, Di Maio R, Volonte D, Galbiati F, Lewis M, Romero G & DeFranco DB 2011 Nongenomic glucocorticoid receptor action regulates gap junction intercellular communication and neural progenitor cell proliferation. *Proc Natl Acad Sci U S A* **108** 16657-16662

Song IH & Buttgereit F 2006 Non-genomic glucocorticoid effects to provide the basis for new drug developments. *Mol Cell Endocrinol* **246** 142-146

Stellato C 2004 Post-transcriptional and nongenomic effects of glucocorticoids. *Proc Am Thorac Soc* **1** 255-263

10

Protein kinase CK1 interacts with and phosphorylates RanBPM *in vitro*

Sonja Wolff[§], Balbina García-Reyes[§], Doris Henne-Bruns, Joachim Bischof[#] and Uwe Knippschild[#]

Department of General and Visceral Surgery, Surgery Center, Ulm University Hospital, Ulm, Germany

§ These authors contributed equally
# These authors share senior authorship

Correspondence should be addressed to Uwe Knippschild; E-mail: uwe.knippschild@uniklinik-ulm.de

Abstract

Members of the casein kinase 1 (CK1) family of serine/threonine kinases are highly conserved from yeast to mammals and are involved in the regulation of various cellular processes. Specifically, the CK1 isoforms δ and ε have been shown to be involved in the regulation of proliferative processes, differentiation, circadian rhythm, as well as in the regulation of nuclear transport. In this report we show that CK1δ and ε interact with murine RanBPM in the yeast two-hybrid system (YTH) and that the putative CK1δ-interacting domains of RanBPM are located between aa 155-386 and aa 515-653. Furthermore, in mammalian cells CK1δ partially co-localizes with RanBPM and can be co-immunoprecipitated with RanBPM. In addition, CK1δ strongly phosphorylates RanBPM within aa 436-514 *in vitro*. The identification of the interacting and scaffolding protein RanBPM as a new substrate of CK1δ points towards a possible function for CK1δ in modulating RanBPM specific functions.

Introduction

Members of the casein kinase 1 (CK1) family form a group of highly related, ubiquitously expressed serine/threonine-specific kinases which can be found in all eukaryotic organisms (Knippschild *et al.* 2014, Venerando *et al.* 2014). CK1 isoforms can be detected in the nucleus and cytoplasm, and are associated with cellular structures like the cytoskeleton, the centrosomes, transport vesicles and the plasma membrane. Members of the CK1 family are able to phosphorylate a wide variety of substrates, suggesting an involvement of CK1 in the regulation of various cellular processes, including chromosome segregation, circadian rhythm, centrosome-specific functions, differentiation and apoptotic processes, and nuclear import processes (Cruciat 2014, Knippschild *et al.* 2014, Venerando *et al.* 2014). Furthermore, deregulation of CK1 expression and activity has been associated with several diseases, among them neurodegenerative diseases (Flajolet *et al.* 2007, Hanger *et al.* 2007, Perez *et al.* 2011, Rubio de la Torre *et al.* 2009) and cancer (Knippschild *et al.* 2014, Schittek & Sinnberg 2014). On protein level the activity of CK1 is modulated by several mechanisms, among them subcellular localization, (auto-) phosphorylation, dephosphorylation, site-specific cleavage by endoproteases, and interaction with cellular structures and proteins (Cruciat 2014, Knippschild *et al.* 2014).

The Ran Binding Protein in the Microtubule-organizing center (MTOC), RanBPM, has first been described as a centrosomal protein with a molecular mass of 55 kDa (RanBPM55) (Nakamura *et al.* 1998). Later it was determined that RanBPM55 was a truncated variant of a 90 kDa protein (RanBP9) with both nuclear and cytoplasmic distribution (Nishitani *et al.* 2001). The majority of Ran binding proteins (RanBPs) are related to the importin-β family of receptors, which regulate nuclear trafficking, and are regulated by the Ras-like nuclear small GTPase Ran (Murrin & Talbot 2007). However, RanBPM lacks the consensus Ran-binding domain (Beddow *et al.* 1995) and rather acts as an interaction partner for multiple receptors and as scaffolding protein for signal transduction components (Suresh *et al.* 2012). RanBPM contains several functional domains, including a SPRY domain as well as LisH and CTLH motifs. The SPRY domain is involved in mediating several of the known protein–protein interactions of RanBPM (Cheng *et al.* 2005, Hafizi *et al.* 2005, Mikolajczyk *et al.* 2003, Rao *et al.* 2002, Wang *et al.* 2002a, Yuan *et al.* 2006), whereas the LisH and CTLH motifs are important for the regulation of homo-dimerization and binding to microtubules (Emes & Ponting 2001, Gerlitz *et al.* 2005, Mateja *et al.*

Table 1. Sequences of RanBPM55-specific primers (RanBPM, a 55 kDa protein, the nucleotide position refers to the sequence of the murine RanBP9 NM_019930.2) (nt: nucleotide).

Primer	Sequence	RanBP9 Sequence
5'RanBPM-1	5´-TCTCGAGACCCGGGTATGGGAATTGGTCTTT-3´ (5´ RanBP9, XhoI/XmaI)	nt 463-478
5'RanBPM-2	5´-TCTCGAGACCCGGGTATGTTTGTGTTTGATAT-3´ (5´ RanBP9, XhoI/XmaI)	nt 775-788
5'RanBPM-3	5´-TCTCGAGACCCGGGTATGCGGTGTTTGGGA-3´ (5´ RanBP9, XhoI/XmaI)	nt 1156-1167
5'RanBPM-4	5´-TCTCGAGACCCGGGTATGGAAAGTTGTAGC-3´ (5´ RanBP9, XhoI/XmaI)	nt 1306-1317
5'RanBPM-5	5´-TCTCGAGACCCGGGTATGGAAGATTGTGAC-3´ (5´ RanBP9, XhoI/XmaI)	nt 1543-1557
3'RanBPM-6	5´-TCTCGAGGGTACCTAAGGATGTTGCCCAAA-3´ (3´ RanBP9, XhoI/KpnI)	nt 760-774
3'RanBPM-7	5´-TCTCGAGGGTACCTACACTTCACTGTCTGT-3´ (3´ RanBP9, XhoI/KpnI)	nt 1141-1155
3'RanBPM-8	5´-TCTCGAGGGTACCTATACACCATTGCTACAAC-3´ (3´ RanBP9, XhoI/KpnI)	nt 1310-1326
3'RanBPM-9	5´-TCTCGAGGGTACCTATTCGTGGTCATGTTTAG-3´ (3´ RanBP9, XhoI/KpnI)	nt 1526-1542
3'RanBPM-10	5´-TCTCGAGGGTACCTAATGTAGGTAGTCTTCC-3´ (3´ RanBP9, XhoI/KpnI)	nt 1944-1959

2006).

RanBPM interacts with a variety of receptors and signal transduction components, such as MET (Wang *et al.* 2002b), CDK11p46 (Mikolajczyk *et al.* 2003), Mirk/Dyrk1B (Zou *et al.* 2003), Axl (Hafizi *et al.* 2002, 2005), and TrkA/B (Yin *et al.* 2010; Yuan *et al.* 2006). Furthermore, RanBPM acts as a scaffolding protein modulating downstream signaling of neural cell adhesion molecule L1 (Cheng *et al.* 2005), β2-integrin LFA-1 (Denti *et al.* 2004), semaphorins (Togashi *et al.* 2006), and BM88/Cend1 (Tsioras *et al.* 2013). By interacting with p73α, RanBPM seems to enhance the pro-apoptotic activity of p73α, by preventing its degradation via the ubiquitin proteasome pathway (Kramer *et al.* 2005). In addition, RanBPM has been shown to exhibit enhancing effects on amyloid β peptide generation in Alzheimer's disease by scaffolding APP (amyloid precursor protein), BACE1 and LRP

(lipoprotein receptor-related protein) (Lakshmana *et al.* 2008, 2009, 2010, 2012). Moreover, RanBPM interacts with HDAC6, a component of the aggresome complex. There is increasing evidence that RanBPM promotes aggresome formation since silencing of RanBPM impairs aggresome formation (Salemi *et al.* 2014).

In the present study we identified RanBPM as an interaction partner of CK1 family members in the yeast two-hybrid (YTH) system, especially of CK1δ and ε. The putative CK1δ-interacting domains of RanBPM are located between aa 155-386 and aa 515-653. Furthermore, RanBPM co-immunoprecipitates with CK1δ, and is phosphorylated by CK1δ at several amino acids within aa 436-514 *in vitro*. These findings point towards a possible function for CK1δ in regulating RanBPM specific functions.

Materials and methods

Expression vectors

Construction of the bait plasmids pGBKT7-wt-CK1δ and pGBKT7-mt-CK1δ has been described previously (Wolff *et al.* 2005). pGBKT7-CK1α, pGBKT7-CK1γ3 and pGBKT7-CK1ε were kindly provided by Dr. David Meek, Dundee, Scotland (Sillibourne *et al.* 2002). Primers listed in Table 1 were used to amplify cDNAs of several fragments of murine RanBPM55 from a GAL4 Matchmaker mouse testis library in pACT2 (Clontech, Takara Bio Inc., USA). PCR products were ligated into pcDNA3.1/V5-His©-TOPO® (Invitrogen, USA) and subcloned into

Table 2. Generation of murine RanBPM55 expression vectors (bp: base pair; aa: amino acid).

Primer pair	bp RanBP9	aa RanBP9	Vector names	Fusion Protein (FP)
5'RanBPM-1 / 3'RanBPM-10	1497	155-653	pcDNA3.1-RanBPM pGADT7-RanBPM pGEX-RanBPM	FP 974 FP 980 FP 986
5'RanBPM-1 / 3'RanBPM-6	312	155-258	pGADT7-RanBPM pGEX-RanBPM	FP 969 FP 975
5'RanBPM-2 / 3'RanBPM-7	381	259-385	pGADT7-RanBPM pGEX-RanBPM	FP 970 FP 976
5'RanBPM-3 / 3'RanBPM-8	171	386-442	pGADT7-RanBPM pGEX-RanBPM	FP 971 FP 977
5'RanBPM-4 / 3'RanBPM-9	237	436-514	pGADT7-RanBPM pGEX-RanBPM	FP 972 FP 978
5'RanBPM-5 / 3'RanBPM-10	417	515-653	pGADT7-RanBPM pGEX-RanBPM	FP 973 FP 979

```
pVII16      ------------LKRLYPAVDEQETPLPRSWSPKDKFSYIGLSQNNLRVHYKGHGKTPKD 48
NP_064314.2 ALNEQEKELQRRLKRLYPAVDEQETPLPRSWSPKDKFSYIGLSQNNLRVHYKGHGKTPKD 180
                        ************************************************

pVII16      AASVRATHPIPAACGIYYFEVKIVSKGRDGYMGIGLSAQGVNMNRLPGWDKHSYGYHGDD 108
NP_064314.2 AASVRATHPIPAACGIYYFEVKIVSKGRDGYMGIGLSAQGVNMNRLPGWDKHSYGYHGDD 240
            ************************************************************

pVII16      GHSFCSSGTGQPYGPTFTTGDVIGCCVNLINNTCFYTKNGHSLGIAFTDLPPNLYPTVGL 168
NP_064314.2 GHSFCSSGTGQPYGPTFTTGDVIGCCVNLINNTCFYTKNGHSLGIAFTDLPPNLYPTVGL 300
            ************************************************************

pVII16      QTPGEVVDANFGQHPFVFDIEDYMREWRTKIQAQIDRFPIGDREGEWQTMIQKMVSSYLV 228
NP_064314.2 QTPGEVVDANFGQHPFVFDIEDYMREWRTKIQAQIDRFPIGDREGEWQTMIQKMVSSYLV 360
            ************************************************************

pVII16      HHGYCATAEAFARSTDQTVLEELASIKNRQRIQKLVLAGRMGEAIETTQQLYPSLLERNP 288
NP_064314.2 HHGYCATAEAFARSTDQTVLEELASIKNRQRIQKLVLAGRMGEAIETTQQLYPSLLERNP 420
            ************************************************************

pVII16      NLLFTLKVRQFIEMVNGTDSEVRCLGGRSPKSQDSYPVSPRPFSSPSMSPSHGMSIHSLA 348
NP_064314.2 NLLFTLKVRQFIEMVNGTDSEVRCLGGRSPKSQDSYPVSPRPFSSPSMSPSHGMSIHSLA 480
            ************************************************************

pVII16      PGKSSTAHFSGFESCSNGVISNKAHQSYCHSKHQLSSLTVPELNSLNVSRSQQVNNFTSN 408
NP_064314.2 PGKSSTAHFSGFESCSNGVISNKAHQSYCHSKHQLSSLTVPELNSLNVSRSQQVNNFTSN 540
            ************************************************************

pVII16      DVDMETDHYSNGVGETSSNGFLNGSSKHDHEMEDCDTEMEVDCSQLRRQLCGGSQAAIER 468
NP_064314.2 DVDMETDHYSNGVGETSSNGFLNGSSKHDHEMEDCDTEMEVDCSQLRRQLCGGSQAAIER 600
            ************************************************************

pVII16      MIHFGRELQAMSEQLRRECGKNTANKKMLKDAFSLLAYSDPWNSPVGNQLDPIQREPVCS 528
NP_064314.2 MIHFGRELQAMSEQLRRECGKNTANKKMLKDAFSLLAYSDPWNSPVGNQLDPIQREPVCS 660
            ************************************************************

pVII16      ALNSAILETHNLPKQPPLALAMGQATQCLGLMARSGVGSCAFATVEDYLH 578
NP_064314.2 ALNSAILETHNLPKQPPLALAMGQATQCLGLMARSGVGSCAFATVEDYLH 710
            **************************************************
```

Figure 1. RanBPM sequence aligments on protein level. Comparison of the amino acid sequence of the cDNA sequence cloned into pVII16 with the amino acid sequence of murine RanBPM (gi: 161353515, accession number NP_064314.2).

pGADT7, pEYFP-C1 (both Clontech, Takara Bio Inc., USA) and pGEX-4T-3 (GE Healthcare, USA) (Table 2).

Yeast two-hybrid assay

To screen for novel CK1δ interacting proteins, yeast two-hybrid analyses were carried out as described previously (Wolff *et al.* 2005). The yeast strain AH109 was co-transformed with pGBKT7-CK1δ and the mouse testis library in pACT2 (Clontech, Takara Bio Inc, USA). To eliminate auto-activating library plasmids or plasmids containing proteins, which interacted with the GAL4 DNA binding domain, library plasmids isolated from positive yeast clones were co-

transformed with pGBKT7 into AH109. False positive library plasmids were identified after co-transformation with pAS2-LaminC. The cDNA inserts of true positive plasmids were sequenced.

Cell culture and transfection of cells

MiaPaCa2 (Yunis *et al.* 1977), HeLa (Gey, 1952), and CV1 cells (Manteuil *et al.* 1973) were maintained in Dulbecco's modified Eagle's medium (DMEM) whereas BxPC3 cells (Loor *et al.* 1982) were maintained in a 1:1 mixture of DMEM and Roswell Park Memorial Institute medium (RPMI). All media contained 10% heat-inactivated fetal calf serum (FCS) (Gibco, USA), 100 units/mL penicillin, 100 μg/mL streptomycin (Gibco, Germany), and 2 mM glutamine. Cell cultures were incubated at 37°C in a humidified 5% (v/v) CO_2 atmosphere. Cell lines HeLa and CV1 were transfected with the plasmid pEYFP-C1-RanBPM55. Transient transfections were performed using subconfluent cultures in 10-cm dishes or six-well tissue culture plates using Effectene Transfection Reagent (Qiagen, Germany) according to the manufacturer's instructions.

Antibodies and immunocytochemistry

The peptide GFLNCSSKHDHEMED (aa 579 - 593 of human or aa 503 - 517 of murine RanBP9, respectively) was coupled to KLH, emulsified in TiterMax Gold (Sigma-Aldrich, Germany) and used for subsequent subcutaneous injections of rabbits. After three to four boosts the animals were bled out. Depletion experiments using the RanBPM-specific peptide for preincubation with the RanBPM-specific serum clearly demonstrated the specificity of the anti-RanBPM serum.

To determine the subcellular localization of CK1δ in HeLa and CV1 cells, the monoclonal mouse

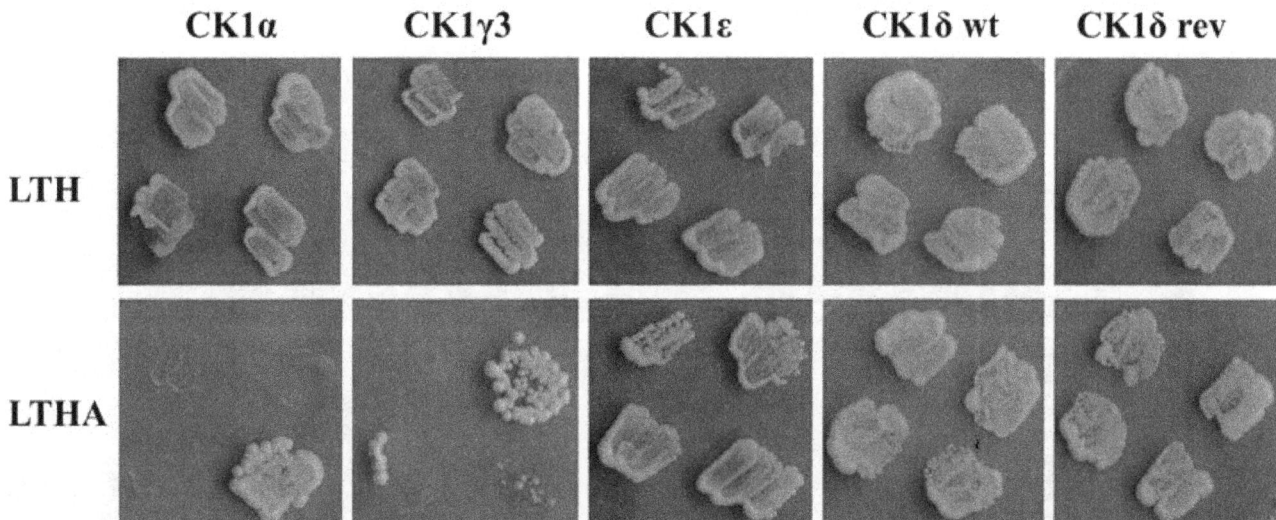

Figure 2. Interaction of CK1 isoforms and RanBPM. Vector pVII16 expressing RanBPM55 was cotransformed with pGBKT7-CK1α, pGBKT7-CK1γ3, pGBKT7-CK1ε pGBKT7-CK1δrev or pGBKT7-CK1δwt into the yeast strain AH109. Positive clones were selected by growing the transformed yeast first on SD/-Trp/-Leu/-His plates (LTH) followed by platting onto more stringent SD/-Trp/-Leu/-His/-Ade plates (LTHA) The ability of the transformants to grow was analyzed.

anti-CK1δ antibody 128A (ICOS Corporation, now Lilly, USA) was used. An Alexa 568-conjugated anti-mouse antibody (MoBiTec GmbH, Germany) was used as fluorophore-labeled secondary antibody. Cells were fixed and permeabilized with ice-cold methanol. Stained cells were analyzed using an epifluorescence microscope (IX70 Olympus, Germany).

Cell lysis, immunoprecipitation, SDS-PAGE, Western blotting, and IP-Western analysis

Cellular lysates were obtained as described previously (Wolff et al. 2005). Cleared extracts were subjected to immunoprecipitation using RanBPM-specific serum (CGV1) and protein A sepharose beads (PAS) (GE Healthcare, USA). Negative controls were done using mouse IgG (Jackson ImmunoResearch Europe Ltd., UK). Immunoprecipitates were analyzed as described before (Wolff et al. 2005). For IP-Western experiments, the endogenous RanBPM was detected in BxPC3 and MiaPaCa2 cells by using the RanBPM specific rabbit serum (CGV1). After immunoprecipitation of RanBPM, the precipitates were separated by SDS-PAGE and transfered to nitrocellulose membranes. The detection of CK1δ in the immune complex was performed using the antibody 128A (ICOS Corporation, now Lilly, USA).

Expression and purification of recombinant proteins

Expression and purification of GST-RanBPM55 fusion proteins was done as described previously (Knippschild et al. 1997). In vitro transcription/translation of His-RanBPM55 was done using TNT reticulocyte extract (Promega, USA) according to the manufacturer's instructions.

In vitro kinase assays

In vitro kinase assays were carried out as described previously (Knippschild et al. 1996, Wolff et al. 2005) using His-RanBPM55 or GST-RanBPM55 fusion proteins (expressed from pGEX-4T-3) encoded by the pGEX-vectors shown in Table 2. Reactions were started by adding CK1δ kinase domain (CK1δKD; NEB, USA) as enzyme and incubated at 30°C for 30 min or as indicated. Phosphorylated proteins were separated by SDS-PAGE and detected by autoradiography.

Results

CK1 isoforms interact with RanBPM55

In order to identify new protein interaction partners for CK1δ, a cDNA library from mouse testis was screened for interactions with the mutant CK1δ variant CK1δrev (Hirner et al. 2012) as bait in a yeast-two-hybrid (YTH) screen. Variant CK1δrev was chosen for the

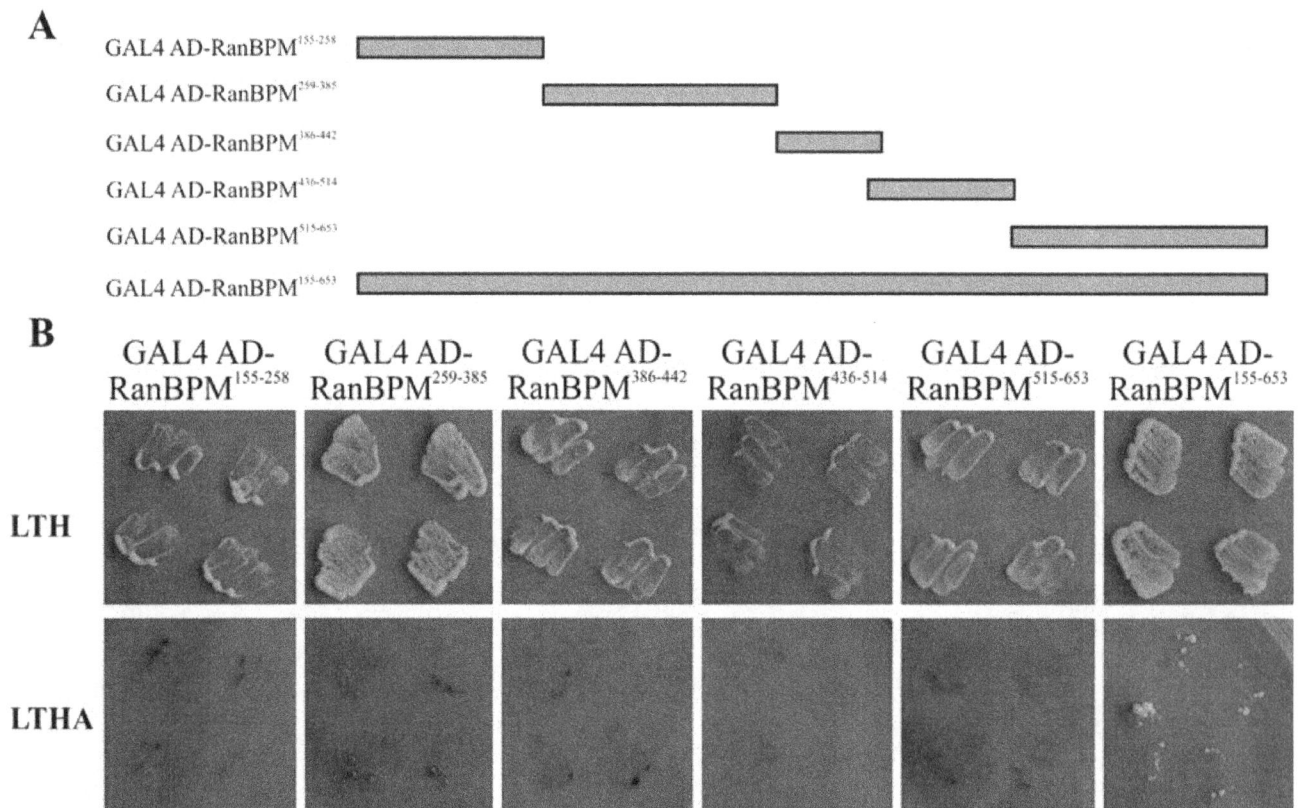

Figure 3. Interaction of CK1δ and RanBPM. (A) Schematic representation of the RanBPM55 fragments. (B) The vectors with RanBPM fragments were cotransformed with pGBKT7-CK1δ rev into the yeast strain AH109. Positive clones were selected by growing the transformed yeast first on SD/-Trp/-Leu/-His plates (LTH) followed by platting onto more stringent SD/-Trp/-Leu/-His/-Ade plates (LTHA). The ability of the transformants to grow was analyzed.

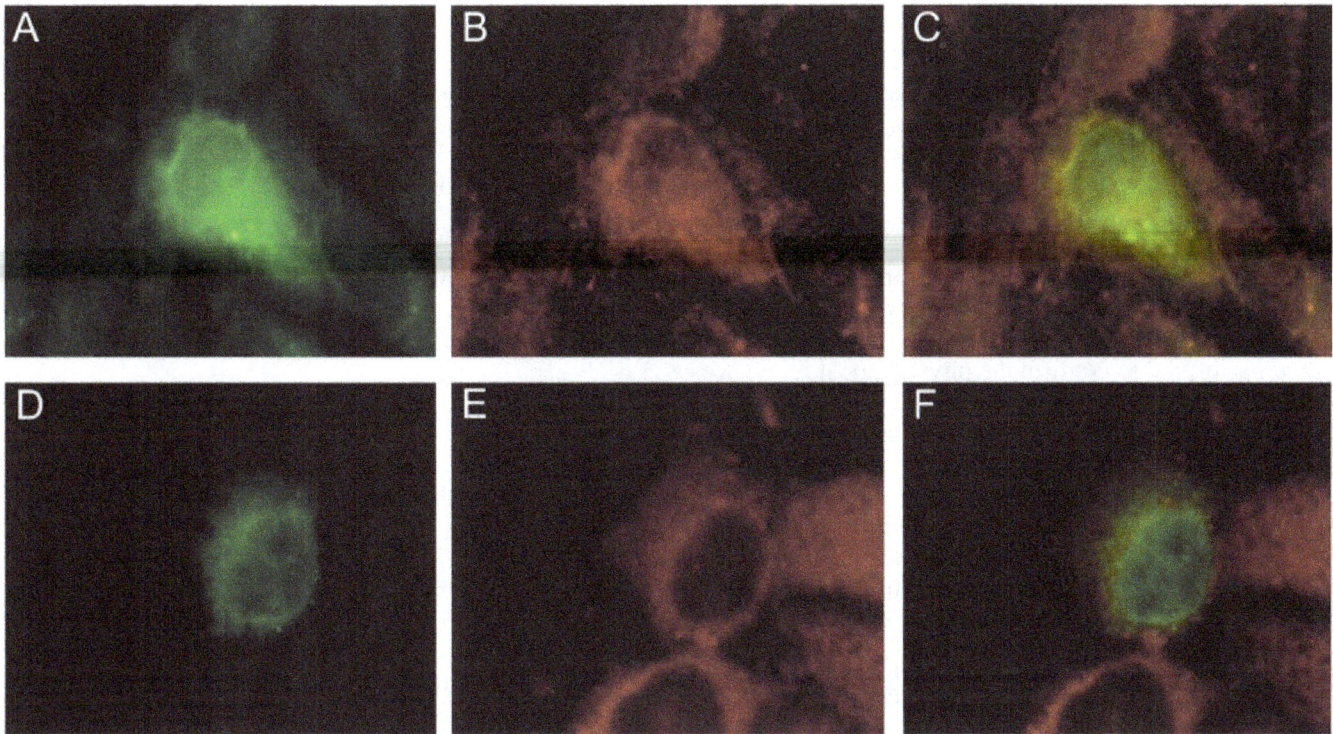

Figure 4. Subcellular localization of CK1δ and EYFP-RanBPM55 in CV1 and HeLa cell lines. Two days after transfection CV1 cells and HeLa cells were fixed with ice-cold methanol and permeabilized. (A) and (D) show EYFP-RanBPM55. In panels (B) and (E), CK1δ was labeled with the antibody 128A and an Alexa 568-conjugated secondary antibody. In the superposition of the fluorescence channels shown in panels (C) and (F), co-localization of RanBPM55 and CK1δ appears as a yellow color.

initial experiments since this mutant only exhibits one third of the kinase activity of wt CK1δ and therefore its overexpression does not cause the same cytotoxic effects which can be observed for overexpression of wt CK1δ.

Transformed yeast cells were selected on semi-stringent SD/-Trp/-Leu/-His (LTH) plates, followed by re-platting positive colonies onto SD/-Trp/-Leu/-His/-Ade (LTHA) plates for a more stringent selection. From one positive clone the plasmid termed pVII16 was isolated containing a N-terminally truncated sequence of murine RanBPM (Figure 1). pVII16 together with either pGBKT7-CK1α, pGBKT7-CK1γ3, pGBKT7-CK1ε, pGBKT7-rev-CK1δ, or pGBKT7-CK1δwt, were cotransformed into AH109 to analyze its interaction with the different CK1 isoforms in the YTH system. Cells cotransformed with RanBPM/CK1δrev, RanBPM/CK1δwt, and RanBPM/CK1ε exhibited growth on SC/-Leu/-Trp/-His/-Ade medium. RanBPM/CK1α, RanBPM/CK1γ3 showed only partial growth on more stringent medium (Figure 2). The specificity of the interaction of RanBPM with the indicated CK1 isoforms was confirmed by the inability of RanBPM to interact with Gal4 DBD or GAL4 DBD-lamin C (data not shown). For further studies the full-length sequence of murine RanBPM55, a N-terminally truncated form of RanBP9 (accession ID AAD01272.1), was amplified from the mouse testis cDNA library.

Identification of CK1δ and RanBPM55 interaction domains

Fragments of RanBPM55 were generated (Figure 3a) and used in the YTH system in order to identify the CK1δ domain(s) necessary for the interaction with RanBPM55. On LTH plates (Figure 3b), fragment GAL4 AD-RanBPM[259-385] could mediate an interaction with CK1δ as strongly as full length RanBPM55

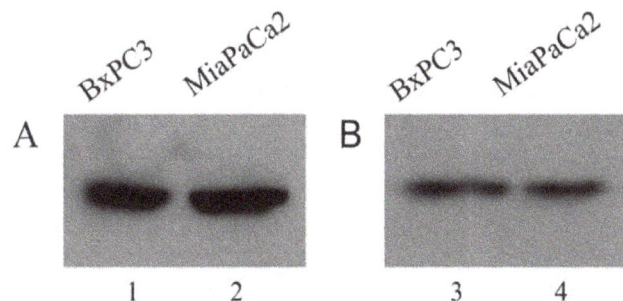

Figure 5. CK1δ co-immunoprecipitates with RanBPM. (A) Detection of RanBPM in cellular lysates of BxPC3 and MiaPaCa2 cells were separated by SDS-PAGE and transferred to a nitrocellulose membrane. The detection of RanBPM in Western blot analysis was performed using the polyclonal rabbit serum CGV1 (lane A1: BxPC3, anti-RanBPM; lane A2: MiaPaCa2, anti-RanBPM). (B) RanBPM was immunoprecipitated from BxPC3 cells and MiaPaCa-2 cells using an anti-RanBPM antibody. The detection of CK1δ in the immune complex was performed using the antibody 128A. (lane B1: BxPC3, IP against RanBPM; lane B2: MiaPaCa2, IP against RanBPM).

(GAL4 AD-RanBPM[155-653]). Fragments including aa 155-258, aa 386-442 and aa 515-653 of RanBPM55 also showed interaction whereas GAL4 AD-RanBPM[436-514] only showed a weak interaction with CK1δ. All described interactions were not observed using more stringent SD/-Trp/-Leu/-His/-Ade medium (LTHA).

Characterization of the interaction between RanBPM and CK1δ in mammalian cells

Additional cell biological and biochemical analyses were performed to confirm the physiological relevance of the interaction between RanBPM and CK1δ detected in the YTH system. First, analysis of the subcellular localization of CK1δ and RanBPM55 in HeLa and CV1 cells transfected with the plasmid pEYFP-C1-RanBPM55 by immunofluorescence microscopy, using the CK1δ-specific antibody 128A, revealed a partial co-localization of both proteins in the perinuclear area (Figure 4). Secondly, IP-Western-analysis revealed that CK1δ co-immunoprecipitates with RanBPM in cellular lysates of BxPC3 and MiaPaCa2 cells (Figure 5).

RanBPM55 is phosphorylated by CK1δ in vitro

Since sequence based analysis of RanBPM55 identified several serine and threonine residues as putative targets for CK1 mediated phosphorylation, in vitro kinase assays were performed to verify RanBPM55 phosphorylation by CK1. His-tagged RanBPM55, in vitro translated from pcDNA3.1-RanBPM (His-RanBPM[155-653]), was phosphorylated using C-terminally truncated CK1δ (CK1δKD). The appearance of a phosphorylated band at around 55 kDa indicated that CK1δ can phosphorylate RanBPM55 in vitro (Figure 6a). No phosphate incorporation into the substrate was detected for control reactions performed without kinase (data not shown).

In order to localize the phosphorylation sites targeted by CK1δ, different GST-RanBPM55 fragments were purified and same protein amounts of these were used as substrates for in vitro kinase assays. As shown in Figure 6b CK1δKD differently phosphorylated the different GST-RanBPM fusion proteins.

Highest phosphate incorporation was detected for the GST-RanBPM[436-514] protein pointing to the existence of one or more major phosphorylation sites in the respective region. GST-RanBPM[515-653] as well as GST-RanBPM259-385 showed significantly weaker phosphorylation intensity. Nearly all GST-RanBPM fusion proteins showed different degrees of degradation after the performed purification steps, especially GST-RanBPM[259-385] and GST-RanBPM[515-653] were highly degraded, and most phosphate incorporation could be detected in degradation products.

Discussion

CK1δ, a member of the CK1 family, plays an important role in the regulation of various cellular processes including circadian rhythm, vesicle transport, chromosome segregation, and centrosome specific functions. Therefore, CK1δ has to be tightly regulated by several mechanisms, including interaction with cellular structures and proteins (reviewed by Knippschild et al. 2014). However, little is known about protein-protein interactions which may influence activity or localization of CK1 isoforms. In the present study we identified RanBPM as a new interaction partner for CK1 isoforms α, γ3, δ, and ε in the yeast two-hybrid system. However, interaction of RanBPM with CK1δ and ε was significantly stronger compared to those with CK1α and γ3. Using different RanBPM fragments the domain responsible for mediating the interaction with CK1δ could be located between aa 259-385. This region is located directly N-terminal to the SPRY domain (aa 212–333) of RanBPM, which has been shown to mediate most of the so far reported protein-protein interactions (Ponting et al. 1997). In subsequent experimental approaches, we were able to confirm the interaction of RanBPM and CK1δ using mammalian cell lines. Whereas our immunofluorescence analyses only revealed a partial co-localization in the perinuclear region, our IP-Western analyses clearly showed that CK1δ co-immunoprecipitates with RanBPM. The partial co-localization of both proteins could be explained by the fact that RanBPM is predominantly localized in chromatin-free areas, the so-called nuclear speckles

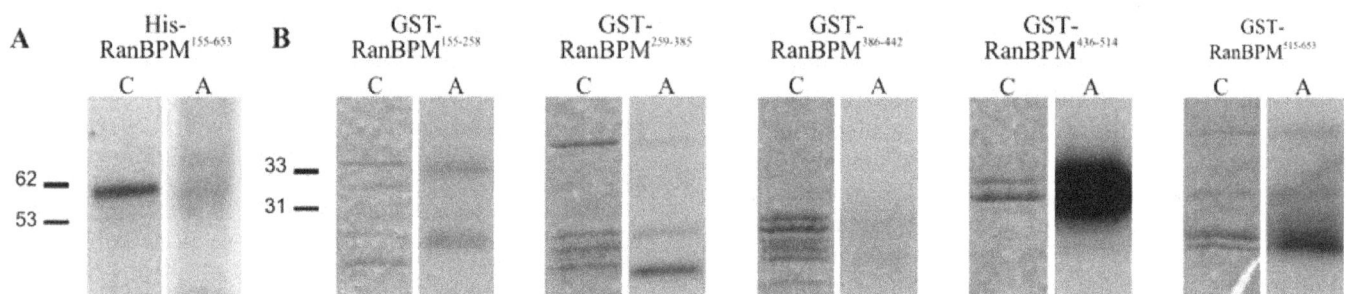

Figure 6. In vitro phosphorylation of RanBPM by CK1δ. (A) In vitro kinase assays were performed using His-RanBPM55 as substrate and C-terminally truncated CK1δ. Proteins were separated in SDS-PAGE, stained with Coomassie staining solution (C), and the phosphorylation was detected by autoradiography (A). (B) In vitro kinase assays were performed using GST-RanBPM55 fragments (described in Figure 1) as substrate and C-terminally truncated CK1δ. The proteins were separated in SDS-PAGE, stained with Coomassie staining solution (C), and the phosphorylation was detected by autoradiography (A).

(Lamond & Spector 2003) in association with various proteins (Ideguchi *et al.* 2002, Mikolajczyk *et al.* 2003, Rao *et al.* 2002, Wang *et al.* 2002b, Zou *et al.* 2003), whereas CK1δ can be mainly detected in the perinuclear region in association with membrane structures, transport vesicles, microtubules, and centrosomes (reviewed by Knippschild *et al.* 2014). However, in some cases it has been reported that RanBPM is located in the cytoplasm thereby interacting with various cellular structures and the plasma membrane (Lamond & Spector 2003). RanBPM often acts as scaffold protein containing a LisH/CTLH motif, which is present in proteins involved in microtubule dynamics, cell migration, nucleokinesis, and chromosome segregation (Kobayashi *et al.* 2007). By phosphorylation CK1δ might regulate these functions of RanBPM. CK1δ actually is able to phosphorylate RanBPM *in vitro* especially between aa 436-514. Interestingly, this strongly phosphorylated RanBPM fragment exhibits the weakest interaction with CK1δ as determined by YTH assays. This fact indicates, that binding of RanBPM55 to CK1δ is mediated by a certain domain (i.e. aa 259-385) which is not identical to the domain which appears to be the main target for CK1δ-mediated phosphorylation (aa 436-514). RanBPM has been reported to be phosphorylated by several kinases, including CDK11p46, Plk (Mikolajczyk *et al.* 2003), Dyrk1B (Tsioras *et al.* 2013), PKC gamma/delta (Rex *et al.* 2010), and p38 (Denti *et al.* 2004) pointing to an important role of site-specific phosphorylation in regulating RanBPM-specific functions as well as its subcellular localization (Suresh *et al.* 2012). However, additional experiments have to be set up to identify the amino acids of RanBPM targeted by CK1δ and to identify the physiological consequences of CK1δ-mediated site-specific phosphorylation on RanBPM cellular functions.

In summary, RanBPM has been identified as a new interaction partner of CK1δ. The specificity of the biological relevance is underlined by the following facts: (i) RanBPM specifically interacts with various CK1 isoforms, namely with CK1α, γ3, δ, and ε, (ii) RanBPM is phosphorylated by CK1δ *in vitro*, and (iii) RanBPM interacts with CK1δ in pancreatic tumor cell lines, as confirmed by IP-Western analysis. However, additional analysis is required to evaluate the physiological relevance of the interaction between CK1δ and RanBPM in detail.

Acknowledgements

This work was supported by a grant from the Deutsche Krebshilfe, Dr. Mildred Scheel Stiftung, awarded to Uwe Knippschild (108489) and by a DAAD fellowship awarded to Balbina García-Reyes (A1297377).

The expression vectors pGBKT7-CK1γ3, pGBKT7-KD-CK1δ, pGBKT7-CK1δ and pGBKT7-CK1ε were kindly provided by Dr. David Meek. We would also like to thank Matthias Piesche and Annette Blatz for technical assistance.

References

Beddow AL, Richards SA, Orem NR & Macara IG 1995 The Ran/TC4 GTPase-binding domain: identification by expression cloning and characterization of a conserved sequence motif. *Proc Natl Acad Sci U S A* **92** 3328-3332

Cheng L, Lemmon S & Lemmon V 2005 RanBPM is an L1-interacting protein that regulates L1-mediated mitogen-activated protein kinase activation. *J Neurochem* **94** 1102-1110

Cruciat CM 2014 Casein kinase 1 and Wnt/beta-catenin signaling. *Curr Opin Cell Biol* **31C** 46-55

Denti S, Sirri A, Cheli A, Rogge L, Innamorati G, Putignano S, Fabbri M, Pardi R & Bianchi E 2004 RanBPM is a phosphoprotein that associates with the plasma membrane and interacts with the integrin LFA-1. *J Biol Chem* **279** 13027-13034

Emes RD & Ponting CP 2001 A new sequence motif linking lissencephaly, Treacher Collins and oral-facial-digital type 1 syndromes, microtubule dynamics and cell migration. *Hum Mol Genet* **10** 2813-2820

Flajolet M, He G, Heiman M, Lin A, Nairn AC & Greengard P 2007 Regulation of Alzheimer's disease amyloid-beta formation by casein kinase I. *Proc Natl Acad Sci U S A* **104** 4159-4164

Gerlitz G, Darhin E, Giorgio G, Franco B & Reiner O 2005 Novel functional features of the Lis-H domain: role in protein dimerization, half-life and cellular localization. *Cell Cycle* **4** 1632-1640

Gey GO, Coffman WD & Kubicek MT 1952 Tissue culture studies of the proliferative capacity of cervical carcinoma and normal epithelium. *Cancer Res* **12** 264-265

Hafizi S, Alindri F, Karlsson & Dahlback B 2002 Interaction of Axl receptor tyrosine kinase with C1-TEN, a novel C1 domain-containing protein with homology to tensin. *Biochem Biophys Res Commun* **299** 793-800

Hafizi S, Gustafsson A, Stenhoff J & Dahlback B 2005 The Ran binding protein RanBPM interacts with Axl and Sky receptor tyrosine kinases. *Int J Biochem Cell Biol* **37** 2344-2356

Hanger DP, Byers HL, Wray S, Leung KY, Saxton MJ, Seereeram A, Reynolds CH, Ward MA & Anderton BH 2007 Novel phosphorylation sites in tau from Alzheimer brain support a role for casein kinase 1 in disease pathogenesis. *J Biol Chem* **282** 23645-23654

Hirner H1, Günes C, Bischof J, Wolff S, Grothey A, Kühl M, Oswald F, Wegwitz F, Bösl MR, Trauzold A, Henne-Bruns D, Peifer C, Leithäuser F, Deppert W & Knippschild U 2012 Impaired CK1 delta activity attenuates SV40-induced cellular transformation in vitro and mouse mammary carcinogenesis in vivo. *PLoS One* **7** e29709

Ideguchi H, Ueda A, Tanaka M, Yang J, Tsuji T, Ohno S, Hagiwara E, Aoki A & Ishigatsubo Y 2002 Structural and functional characterization of the USP11 deubiquitinating enzyme, which interacts with the RanGTP-associated protein RanBPM. *Biochem J* **367** 87-95

Knippschild U, Krüger M, Richter J, Xu P, García-Reyes B, Peifer C, Halekotte J, Bakulev V & Bischof J 2014 The CK1 Family: Contribution to Cellular Stress Response and Its Role in Carcinogenesis. *Front Oncol* **4** 96

Knippschild U, Milne D, Campbell L & Meek D 1996 p53 N-terminus-targeted protein kinase activity is stimulated in response to wild type p53 and DNA damage. *Oncogene* **13** 1387-1393

Knippschild U, Milne DM, Campbell LE, DeMaggio AJ, Christenson E, Hoekstra MF & Meek DW 1997 p53 is phosphorylated in vitro and in vivo by the delta and epsilon isoforms of casein kinase 1 and enhances the level of casein kinase 1 delta in response to topoisomerase-directed drugs. *Oncogene* **15** 1727-1736

Kobayashi N, Yang J, Ueda A, Suzuki T, Tomaru K, Takeno M, Okuda K & Ishigatsubo Y 2007 RanBPM, Muskelin, p48EMLP, p44CTLH, and the armadillo-repeat proteins ARMC8alpha and ARMC8beta are components of the CTLH complex. *Gene* **396** 236-247

Kramer S, Ozaki T, Miyazaki K, Kato C, Hanamoto T & Nakagawara A 2005 Protein stability and function of p73 are modulated by a physical interaction with RanBPM in mammalian cultured cells. *Oncogene* **24** 938-944

Lakshmana MK, Chen E, Yoon IS & Kang DE 2008 C-terminal 37 residues of LRP promote the amyloidogenic processing of APP independent of FE65. *J Cell Mol Med* **12** 2665-2674

Lakshmana MK, Chung JY, Wickramarachchi S, Tak E, Bianchi E, Koo EH & Kang DE 2010 A fragment of the scaffolding protein RanBP9 is increased in Alzheimer's disease brains and strongly potentiates amyloid-beta peptide generation. *FASEB J* **24** 119-127

Lakshmana MK, Hayes CD, Bennett SP, Bianchi E, Reddy KM, Koo EH & Kang DE 2012 Role of RanBP9 on amyloidogenic processing of APP and synaptic protein levels in the mouse brain. *FASEB J* **26** 2072-2083

Lakshmana MK, Yoon IS, Chen E, Bianchi E, Koo EH & Kang DE 2009 Novel role of RanBP9 in BACE1 processing of amyloid precursor protein and amyloid beta peptide generation. *J Biol Chem* **284** 11863-11872

Lamond AI & Spector DL 2003 Nuclear speckles: a model for nuclear organelles. *Nat Rev Mol Cell Biol* **4** 605-612

Loor R, Nowak NJ, Manzo ML, Douglass HO & Chu TM 1982 Use of pancreas-specific antigen in immuno-diagnosis of pancreatic cancer. *Clin Lab Med* **2** 567-578

Manteuil S, Pages J, Stehelin D & Girard M 1973 Replication of simian virus 40 deoxyribonucleic acid: analysis of the one-step growth cycle. *J Virol* **11** 98-106

Mateja A, Cierpicki T, Paduch M, Derewenda ZS & Otlewski J 2006 The dimerization mechanism of LIS1 and its implication for proteins containing the LisH motif. *J Mol Biol* **357** 621-631

Mikolajczyk M, Shi J, Vaillancourt RR, Sachs NA & Nelson M 2003 The cyclin-dependent kinase 11(p46) isoform interacts with RanBPM. *Biochem Biophys Res Commun* **310** 14-18

Murrin LC & Talbot JN 2007 RanBPM, a scaffolding protein in the immune and nervous systems. *J Neuroimmune Pharmacol* **2** 290-295

Nakamura M, Masuda H, Horii J, Kuma Ki, Yokoyama N, Ohba T, Nishitani H, Miyata T, Tanaka M & Nishimoto T 1998 When overexpressed, a novel centrosomal protein, RanBPM, causes ectopic microtubule nucleation similar to gamma-tubulin. *J Cell Biol* **143** 1041-1052

Nishitani H, Hirose E, Uchimura Y, Nakamura M, Umeda M, Nishii K, Mori N & Nishimoto T 2001 Full-sized RanBPM cDNA encodes a protein possessing a long stretch of proline and glutamine within the N-terminal region, comprising a large protein complex. *Gene* **272** 25-33

Perez DI, Gil C & Martinez A 2011 Protein kinases CK1 and CK2 as new targets for neurodegenerative diseases. *Med Res Rev* **31** 924-954

Ponting C, Schultz J & Bork P 1997 SPRY domains in ryanodine receptors (Ca(2+)-release channels). *Trends Biochem Sci* **22** 193-194

Rao MA, Cheng H, Quayle AN, Nishitani H, Nelson CC & Rennie PS 2002 RanBPM, a nuclear protein that interacts with and regulates transcriptional activity of androgen receptor and glucocorticoid receptor. *J Biol Chem* **277** 48020-48027

Rex EB, Rankin ML, Yang Y, Lu Q, Gerfen CR, Jose PA & Sibley DR 2010 Identification of RanBP 9/10 as interacting partners for protein kinase C (PKC) gamma/delta and the D1 dopamine receptor: regulation of PKC-mediated receptor phosphorylation. *Mol Pharmacol* **78** 69-80

Rubio de la Torre E, Luzon-Toro B, Forte-Lago I, Minguez-Castellanos A, Ferrer I & Hilfiker S 2009 Combined kinase inhibition modulates parkin inactivation. *Hum Mol Genet* **18** 809-823

Salemi LM, Almawi AW, Lefebvre KJ & Schild-Poulter C 2014 Aggresome formation is regulated by RanBPM through an interaction with HDAC6. *Biol Open* **3** 418-430

Schittek B & Sinnberg T 2014 Biological functions of casein kinase 1 isoforms and putative roles in tumorigenesis. *Mol Cancer* **13** 231

Sillibourne JE, Milne DM, Takahashi M, Ono Y & Meek DW 2002 Centrosomal anchoring of the protein kinase CK1delta mediated by attachment to the large, coiled-coil scaffolding protein CG-NAP/AKAP450. *J Mol Biol* **322** 785-797

Suresh B, Ramakrishna S & Baek KH 2012 Diverse

roles of the scaffolding protein RanBPM. *Drug Discov Today* **17** 379-387

Togashi H, Schmidt EF & Strittmatter SM 2006 RanBPM contributes to Semaphorin3A signaling through plexin-A receptors. *J Neurosci* **26** 4961-4969

Tsioras K, Papastefanaki F, Politis PK, Matsas R & Gaitanou M 2013 Functional Interactions between BM88/Cend1, Ran-binding protein M and Dyrk1B kinase affect cyclin D1 levels and cell cycle progression/exit in mouse neuroblastoma cells. *PLoS One* **8** e82172

Venerando A, Ruzzene M & Pinna LA 2014 Casein kinase: the triple meaning of a misnomer. *Biochem J* **460** 141-156

Wang D, Li Z, Messing EM & Wu G 2002a Activation of Ras/Erk pathway by a novel MET-interacting protein RanBPM. *J Biol Chem* **277** 36216-36222

Wang Y, Marion Schneider E, Li X, Duttenhofer I, Debatin K & Hug H 2002b HIPK2 associates with RanBPM. *Biochem Biophys Res Commun* **297** 148-153

Wolff S, Xiao Z, Wittau M, Sussner N, Stoter M & Knippschild U 2005 Interaction of casein kinase 1 delta (CK1 delta) with the light chain LC2 of microtubule associated protein 1A (MAP1A). *Biochim Biophys Acta* **1745** 196-206

Yin YX, Sun ZP, Huang SH, Zhao L, Geng Z & Chen ZY 2010 RanBPM contributes to TrkB signaling and regulates brain-derived neurotrophic factor-induced neuronal morphogenesis and survival. *J Neurochem* **114** 110-121

Yuan Y, Fu C, Chen H, Wang X, Deng W & Huang BR 2006 The Ran binding protein RanBPM interacts with TrkA receptor. *Neurosci Lett* **407** 26-31

Yunis AA, Arimura GK & Russin DJ 1977 Human pancreatic carcinoma (MIA PaCa-2) in continuous culture: sensitivity to asparaginase. *Int J Cancer* **19** 128-135

Zou Y, Lim S, Lee K, Deng X & Friedman E 2003 Serine/threonine kinase Mirk/Dyrk1B is an inhibitor of epithelial cell migration and is negatively regulated by the Met adaptor Ran-binding protein M. *J Biol Chem* **278** 49573-49581

11

The Niemann-Pick C1 and caveolin-1 proteins interact to modulate efflux of low density lipoprotein-derived cholesterol from late endocytic compartments

David Jelinek[1], Randy A. Heidenreich[2], Robert A. Orlando[1], and William S. Garver[1]

[1]Department of Biochemistry and Molecular Biology, [2]Department of Pediatrics, School of Medicine, The University of New Mexico Health Sciences Center, Albuquerque, New Mexico, US

Correspondence should be addressed to William S. Garver; Email: wgarver@salud.unm.edu

Abstract

The Niemann-Pick C1 (NPC1) protein has a central role in regulating the efflux of lipoprotein-derived cholesterol from late endosomes/lysosomes and transport to other cellular compartments. Since the NPC1 protein has been shown to regulate the transport of cholesterol to cellular compartments enriched with the ubiquitous cholesterol-binding and transport protein caveolin-1, the present study was performed to determine whether the NPC1 and caveolin-1 proteins interact and function to modulate efflux of low density lipoprotein (LDL)-derived cholesterol from endocytic compartments. To perform these studies, normal human fibroblasts were grown in media with lipoprotein-deficient serum (LPDS) or media with LPDS supplemented with purified human LDL. The results indicated reciprocal co-immunoprecipitation and partial co-localization of the NPC1 and caveolin-1 proteins that was decreased when fibroblasts were grown in media with LDL. Consistent with interaction of the NPC1 and caveolin-1 proteins, a highly conserved caveolin-binding motif was identified within a cytoplasmic loop located adjacent to the sterol-sensing domain (SSD) of the NPC1 protein. To examine the functional relevance of this interaction, fibroblasts were transfected with caveolin-1 siRNA and found to accumulate increased amounts of LDL-derived cholesterol within late endosomes/lysosomes. Together, this report presents novel results demonstrating that the NPC1 and caveolin-1 proteins interact to modulate efflux of LDL-derived cholesterol from late endocytic compartments.

Introduction

A number of studies have indicated that most tissues with the exception of liver obtain cholesterol primarily through endogenous biosynthesis and to a lesser extent through receptor-mediated endocytosis of remnant lipoproteins (Dietschy et al. 1993, Osono et al. 1995, Turley et al. 1981). The endocytosis of lipoproteins, particularly cholesterol-enriched low-density lipoprotein (LDL) and the subsequent transport of LDL-derived cholesterol through cellular compartments, serves an important role in maintaining intracellular cholesterol homeostasis. Early studies performed using human fibroblasts demonstrated that LDL particles bind to both the LDL receptor and LDL receptor-related protein (LRP) present on the cell surface to facilitate the internalization of these particles into endocytic compartments (Brown and Goldstein 1975, Brown et al. 1976). It is within early endosomes, and to a lesser extent late endosomes/lysosomes, that the bulk of cholesteryl ester contained within the hydrophobic core of these particles is hydrolyzed by an acidic lipase to generate LDL-derived cholesterol (Goldstein et al. 1975, Sugii et al. 2003).

The transport of LDL-derived cholesterol from endocytic compartments is primarily regulated by both the Niemann-Pick C2 (NPC2) and Niemann-Pick C1 (NPC1) proteins, which have been shown to function in a non-redundant and cooperative manner (Infante et al. 2008b, Sleat et al. 2004). The soluble NPC2 protein is contained within the lumen of endocytic compartments, where it binds and facilitates the transport of cholesterol to the limiting membrane (Cheruku et al. 2006, Friedland et al. 2003, Ko et al. 2003). In contrast, the membrane-bound NPC1 protein is associated with a unique late-endosome, commonly referred to as the NPC1 compartment, that transiently interacts with cholesterol-enriched endocytic compartments (Garver et al. 2000, Neufeld et al. 1999). Although the human NPC1 cDNA opening reading frame predicts a protein of 1,278 amino acids with an estimated molecular weight of 142 kDa, the NPC1 protein migrates as a doublet corresponding to 170 and 190 kDa due to variable asparagine-linked glycosylation (Watari *et al.* 1999a, Watari *et al.* 1999b). Studies indicate that both the 170 and 190 kDa NPC1 proteins are capable of interacting with endocytic compartments but that the 170 kDa NPC1 protein is 50% less efficient in facilitating the efflux of cholesterol from these compartments (Watari *et al.* 1999b). The NPC1 protein is also capable of binding cholesterol through two distinct domains, referred to as the amino or N-terminal domain (NTD) and the sterol-sensing domain (SSD) (Infante *et al.* 2008a, Liu *et al.* 2009, Ohgami *et al.* 2004). A series of eloquent studies have demonstrated that the NPC2 protein transfers cholesterol to the NPC1-NTD positioned within the endocytic lumen, followed by the NPC1 protein inserting the hydrophobic isooctyl side chain of cholesterol into the endocytic membrane (Infante *et al.* 2008a, Infante *et al.* 2008b, Kwon *et al.* 2009).

In general, studies performed using either NPC1 or NPC2 human fibroblasts have demonstrated that these proteins regulate the transport of LDL-derived cholesterol to different cellular compartments, including the Golgi appartus, plasma membrane, and endoplasmic reticulum (Blanchette-Mackie *et al.* 1988, Coxey *et al.* 1993, Underwood *et al.* 1998, Wojtanik & Liscum 2003). More recent studies have indicated that a distinct region of the Golgi apparatus, the *trans*-Golgi network (TGN), preferentially receives cholesterol derived from the NPC1 compartment and distributes this cholesterol to the endoplasmic reticulum and plasma membrane (Garver *et al.* 2002, Urano *et al.* 2008). It should be noted that the TGN and TGN-derived vesicles are enriched with caveolin-1, which is known to be a ubiquitous cholesterol-binding and transport protein (Dupree *et al.* 1993, Li & Papadopoulos 1998, Murata *et al.* 1995). Consistent with this result, a number of studies indicate that caveolin-1 is actively involved in facilitating the transport of cholesterol between cellular compartments (Fielding & Fielding 1996, Smart *et al.* 1996, Uittenbogaard *et al.* 2002, Uittenbogaard *et al.* 1998). Moreover, studies have shown that caveolin-1 partially co-localizes with the NPC1 compartment and that expression of caveolin-1 is significantly increased in both human fibroblasts and mouse tissues as a result of decreased NPC1 protein function, thereby suggesting that the NPC1 protein and caveolin-1 may interact and function in an undefined manner (Garver *et al.* 1997a, Garver *et al.* 2000, Garver *et al.* 1997b, Higgins *et al.* 1999).

The present study was performed to determine whether the NPC1 and caveolin-1 proteins interact and

function to modulate the efflux of LDL-derived cholesterol from endocytic compartments. To perform these studies, normal human fibroblasts were grown in media containing lipoprotein-deficient serum (LPDS), which represented the control or basal culture condition, in addition to media containing LPDS supplemented with purified human LDL. In brief, the results indicated i) an inverse association between relative amounts of the NPC1 and caveolin-1 proteins when fibroblasts were grown in media with LDL, ii) reciprocal co-immunoprecipitation of the NPC1 and caveolin-1 proteins was decreased when fibroblasts were grown in media with LDL, iii) partial co-localization of the NPC1 and caveolin-1 proteins was decreased when fibroblasts were grown in media with LDL, iv) identification of a conserved caveolin-binding motif located adjacent to the SSD of the NPC1 protein, and v) an increased amount of LDL-derived cholesterol within late endosomes/lysosomes resulting from caveolin-1 siRNA knockdown. Together, this report presents novel results demonstrating that the NPC1 and caveolin-1 proteins interact to modulate efflux of LDL-derived cholesterol from endocytic compartments.

Materials and Methods

Materials
DMEM, PBS, trypsin-EDTA, and 100 U/ml penicillin/streptomycin (P/S) were purchased from Invitrogen Corporation (Carlsbad, CA). Fetal bovine serum (FBS) and lipoprotein-deficient serum (LPDS) were purchased from Cocalico Laboratories (Reamstown, PA). Complete protease inhibitor cocktail tablets were purchased from Roche Diagnostic (Indianapolis, IN). Human low-density lipoprotein was purchased from RayBiotech (Norcross, GA). The RNeasy Mini Kit was purchased from Qiagen (Valencia, CA). The TaqMan Gene Expression Assay including PCR primers (NPC1, Hs00264835_m1 and caveolin-1, Hs00184697_ml) were purchased from Applied Biosystems (Foster City, CA). The Niemann-Pick C1 (NPC1) antibody was generated in rabbits against a m i n o   a c i d s   1 2 5 4 - 1 2 7 3 (NKAKSCATEERYGTERER) of the human NPC1 protein and purchased from Invitrogen Corporation (Carlsbad, CA). The caveolin-1 antibodies (clones C060 and 2297) were purchased from BD Biosciences (San Jose, CA). The lysosome-associated membrane protein-1 (LAMP-1) antibody (clone H4A3) was purchased from Santa Cruz Biotechnology (Santa Cruz, CA). Stealth caveolin-1 siRNA duplex oligomers, Stealth non-specific siRNA Negative Control duplex oligomers, Lipofectamine 2000, and Opti-MEM I Reduced Serum Medium were purchased from Invitrogen

Corporation (Carlsbad, CA). The peroxidase, Cy2, and Cy3-conjugated goat secondary antibodies were purchased from Jackson ImmunoResearch Laboratories (West Grove, PA). The West Pico SuperSignal Substrate, Bicinchoninic Acid (BCA) protein assay, and Protein A-Sepharose beads were purchased from Pierce Chemical Company (Rockford, IL).

Cell culture and harvest

Normal human fibroblasts (CRL-2097) were purchased from the American Type Culture Collection (Manassas, VA) and cultured using basic media (DMEM, 10% FBS, and 1% P/S) in a humidified incubator at 37°C with 5% CO_2. At ~ 30% confluence, the cells were rinsed with PBS and media was changed to DMEM, 5% LPDS, and 1% P/S to deplete cellular sterol pools and increase expression of the LDL receptor. When the cells reached ~ 60% confluence (24 h), media was changed to fresh DMEM, 5% LPDS, and 1% P/S or the same media supplemented with human LDL (50 µg/ml LDL). The cells were allowed to incubate (24 h) and were harvested for experimentation.

RNA preparation

Total RNA was extracted from fibroblasts using the RNeasy Mini Kit followed by treatment of the RNA with RNase-free DNase to remove residual contaminating DNA. The concentration of RNA was determined by absorbance at 260 nm, with the purity of RNA determined by the ratio of absorbance at 260 nm and 280 nm.

Quantitative RT-PCR

The relative amounts of target mRNA were determined using quantitative reverse transcription polymerase-chain reaction (qRT-PCR) analysis. Reverse transcription was performed using 2.5 µM random hexamers, 4.0 mM dNTP's, 15 mM $MgCl_2$, 50 U reverse transcriptase, and 100 U RNAase inhibitor to produce cDNA. Pre-developed commercially available primers and probes were used for detection of NPC1 and caveolin-1 mRNA. The qRT-PCR was performed with a TaqMan Gene Expression Assay containing PCR primers and TaqMan MGB probe (FAM dye-labeled) using an ABI-PRISM Sequence Detection System (Applied Biosystems, Foster City, CA). The quantification of PCR products was normalized to 18S rRNA (internal control).

Reciprocal co-immunoprecipitation

Fibroblasts were solubilized using Buffer A (PBS containing 1% v/v Triton X-100, 60 mM octylglucoside, and protease inhibitor cocktail), homogenized using needle aspiration, and centrifuged (100,000 x g at 4°C

for 30 min) to remove insoluble material. The protein concentration of this solubilized homogenate was determined and 250 µg aliquots were diluted to 1.0 ml with Buffer A. The primary antibodies (rabbit anti-human NPC1 or mouse anti-human caveolin-1) were added to the solubilized and diluted homogenate (1:250 dilution) and incubated during nutation (4°C for 4 h). The Protein A-Sepharose beads were rinsed and blocked with Buffer A containing 10 mg/ml FAFA (4°C for 4 h) and collected by centrifugation (1,000 x g at 4°C for 1 min). The solubilized and diluted homogenate containing primary antibodies was added onto the Protein A-Sepharose beads and incubated during nutation (4°C for 16 h). The beads were collected by centrifugation (1,000 x g at 4°C for 5 min) and rinsed in a sequential manner with Buffer A containing 10 mg/ml FAFA and decreasing concentrations of NaCl (0.5 M, 0.4 M, 0.3 M, 0.2 M, and 0.1 M NaCl). Finally, the rinsed beads were mixed with SDS-PAGE sample buffer (50°C for 5 min to detect NPC1 and 100°C for 5 min to detect caveolin-1) in preparation for SDS-PAGE.

Caveolin-1 siRNA knockdown

Fibroblasts were transfected with caveolin-1 siRNA duplex oligomers to knockdown caveolin-1 gene expression. In brief, cells were plated using basic media without antibiotics (DMEM and 10% FBS) and grown to ~ 30% confluence. The cells were rinsed with PBS and media was changed to DMEM and 5% LPDS, followed by transfection (5 h) with either 2.0 nM Stealth caveolin-1 siRNA duplex oligomers or 2.0 nM non-specific siRNA Negative Control duplex oligomers using Lipofectamine 2000. After transfection, the cells were rinsed with PBS and incubated in DMEM and 5% LPDS (48 h). The media was changed to DMEM and 5% LPDS or the same media supplemented with human LDL (50 µg/ml LDL) and incubated (24 h). Finally, the cells were rinsed with PBS and harvested for experimentation.

Immunoblot analysis

The relative amounts of NPC1 and caveolin-1 protein were determined using immunoblot analysis. Protein samples were separated using either 7% or 13% SDS-PAGE under reduced conditions and transferred to a nitrocellulose membrane. In brief, blocking buffer (10 mM sodium phosphate pH 7.4, 150 mM NaCl, 0.05% Tween 20, and 5% non-fat dry milk) was used to block non-specific sites on the nitrocellulose membrane (2 h). The membranes were incubated in blocking buffer containing rabbit anti-human NPC1 antibody (1:1,500 dilution) or mouse anti-human caveolin-1 antibody (1:750 dilution) (4°C for 16 h). The membranes were rinsed (3 x 5 min) and incubated with blocking buffer containing peroxidase-conjugated goat secondary antibody (1:5,000 dilution) (1.5 h). This particular combination of host specific primary and secondary antibodies prevented cross-reaction while performing immunoblot analysis to detect reciprocal co-immunoprecipiation of the NPC1 or caveolin-1 proteins. Finally, the membranes were rinsed (3 x 5 min) and bound secondary antibodies were identified using enhanced chemiluminescence.

Fluorescence labeling

The fibroblasts were fixed onto coverslips using PBS containing 4% paraformaldehyde (30 min). After fixation, the coverslips were rinsed with PBS (3 x 5 min) and placed in quenching buffer (PBS and 50 mM NH_4Cl) for 15 min. The coverslips were then rinsed with PBS (3 x 5 min) and placed in Buffer B (PBS, 10% goat serum, and 0.05% w/v saponin) or Buffer C (PBS, 10% goat serum, and 0.05% w/v filipin) for 90 min. Following this, they were placed in Buffer B or C containing rabbit anti-human NPC1 antibody (1:250 dilution), mouse anti-human caveolin-1 antibody (1:250 dilution), or mouse anti-human LAMP-1 antibody (1:250 dilution) for 90 min. They were then rinsed with PBS (3 x 5 min) and placed in Buffer B or C containing Cy2 and Cy3-conjugated goat secondary antibody (1:500 dilution) for 90 min. Finally, the coverslips were rinsed with PBS (3 x 5 min) and mounted onto slides using Aqua Poly/Mount.

Deconvolution fluorescence microscopy

Deconvolution fluorescence microscopy is a relatively new technique that decreases the amount of unfocused and distorted fluorescence through computational processing, thereby promoting restoration of multiple focal planes into a high-resolution three-dimensional fluorescent image. The images were obtained using an Olympus IX-70 inverted microscope equipped with a 60X and 100X (NA 1.4) oil immersion objective (Olympus America, Melville, NY), Photometrics cooled CCD camera (Roper Scientific Instruments, Tucson, AZ), and DeltaVision RT restoration microscopy system software (Applied Precision, Issaquah, WA). The emission wavelengths used for obtaining fluorescent images were 350 nm for the DAPI filter, 528 nm for the Cy2 filter, and 617 nm for the Cy3 filter. For the different fluorescent probes, 10 sections were obtained using a distance of 0.2 μm (recommended step size for the NA of the objectives) set between focal planes. The data, subjected to 5 deconvolution iterations, was projected using SoftWoRX software and processed using identical parameters with Adobe Photoshop software CS2 (Adobe Sys-

tems, Mountain View, CA). Finally, the percentage of co-localization between the NPC1 and caveolin-1 proteins, and the intensity of filipin-staining for cholesterol within LAMP-1 vesicles, was determined using the color range selection and area measurement/intensity features with Adobe Photoshop software CS2.

Identification and sequence comparison of caveolin-binding motifs

The three potential caveolin-binding motifs, denoted by the principal consensus sequence jXjXXXXj and the less common consensus sequences jXXXXjXXj and jXjXXXXjXXj, where j represents one of three aromatic amino acids (Trp W, Phe F, or Tyr Y), were used to screen the human NPC1 amino acid sequence at the National Center for Biotechnology Information (NCBI). Moreover, sequence comparison of caveolin-binding motifs among orthologous NPC1 proteins was performed according to the following species and reference sequences: Human (GenBank: AAH63302.1), Chimpanzee (NCBI: XP_001155163.1), Mouse (NCBI: NP_032746.2), Pig (NCBI: NP_00999487.1), Dog (NCBI: NP_001003107.1), Hamster (GenBank: AAF31692.1), Cattle (Genbank: AAI51277.1), Horse (NCBI: XP_001490228.2), Bird (NCBI: XP_419162.2), Rat (GenBank: EDL86695.1), and Cat (NCBI: NP_001009829.2).

Statistical analysis

For all experiments, quantitative data is represented as the mean ± standard deviation (SD) using five plates of

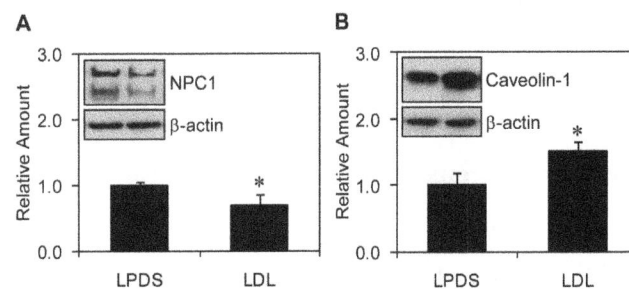

Figure 1. Relative amounts of the NPC1 and caveolin-1 proteins. The average amounts of NPC1 protein (A) and caveolin-1 protein (B) for fibroblasts grown in media with LPDS were normalized to β-actin (internal control) and assigned a value of 1.0, while the average amounts of NPC1 protein and caveolin-1 protein for fibroblasts grown in media with LDL were expressed as fold change. A representative immunoblot showing the relative amounts of NPC1 and caveolin-1 protein for fibroblasts grown in LPDS (left lane) and LDL (right lane) are included as insets within respective graphs. Quantitative data is represented as the mean ± S.D. using five plates of fibroblasts grown using both culture conditions. $*P \leq 0.05$ compared to fibroblasts grown in media with LPDS.

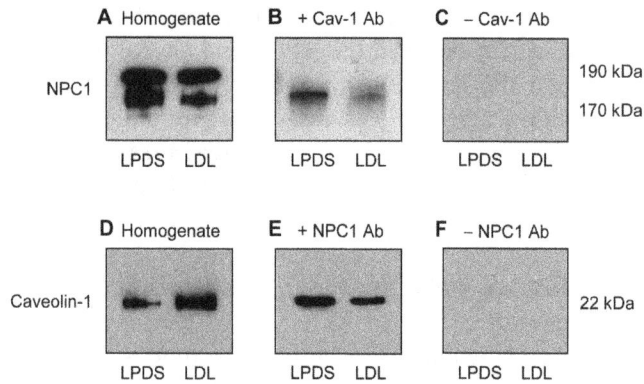

Figure 2. Reciprocal co-immunoprecipitation of the NPC1 and caveolin-1 proteins. An equivalent amount of solubilized and diluted homogenate protein from fibroblasts grown in media with LPDS or LDL was incubated in the presence or absence of rabbit anti-human NPC1 antibody or mouse anti-human caveolin-1 antibody and resulting antibody-antigen complexes were precipitated using Protein A-Sepharose beads. The beads were incubated with sample buffer and immunoblot analysis was performed using reciprocal antibodies (same caveolin-1 or NPC1 antibody, respectively). The solubilized and diluted homogenate protein (A, D) served as a positive control for identification of the co-immunoprecipitated proteins (B, E), while the solubilized and diluted homogenate protein incubated in the absence of NPC1 or caveolin-1 primary antibody (C, F) served as a negative control.

human fibroblasts (qRT-PCR and immunoblot analysis) or ten fluorescent images (quantification of co-localization) of human fibroblasts grown using both culture conditions (LPDS or LDL). Significant differences ($P \leq 0.05$) between two groups of data were determined using the two-tailed Student's t-test assuming equal variance.

Results

Relative amounts of the NPC1 and caveolin-1 proteins

Fibroblasts were grown using the two culture conditions (LPDS and LDL) and processed to determine relative amounts of the NPC1 and caveolin-1 proteins. The results indicated a significant average decrease (31%) in the relative amounts of NPC1 protein for fibroblasts grown in media with LDL compared to fibroblasts grown in media with LPDS (Figure 1A). In contrast, the results indicated a significant average increase (52%) in the relative amounts of caveolin-1 protein for fibroblasts grown in media with LDL compared to fibroblasts grown in media with LPDS (Figure 1B). Similarly, consistent with transcriptional regulation of the NPC1 and caveolin-1 genes, the results indicated a significant average decrease (33%) in the relative amounts of NPC1 mRNA and a significant

average increase (33%) in the relative amounts of caveolin-1 mRNA for fibroblasts grown in media with LDL compared to fibroblasts grown in media with LPDS (data not shown).

Reciprocal co-immunoprecipitation of the NPC1 and caveolin-1 proteins

Fibroblasts were grown using the two culture conditions (LPDS and LDL) and processed to perform reciprocal co-immunoprecipitation of the NPC1 and caveolin-1 proteins. The results indicated that immunoprecipitation of caveolin-1 promoted co-immunoprecipitation of the lower molecular weight NPC1 protein (170 kDa) when fibroblasts were grown in media with LPDS or LDL (Figure 2A, B, C). Similarly, immunoprecipitation of the NPC1 protein promoted co-immunoprecipitation of caveolin-1 when fibroblasts were grown in media with LPDS or LDL (Figure 2D, E, F). However, it must be noted that co-immunoprecipitation of either the NPC1 or caveolin-1 protein was decreased for fibroblasts grown in media with LDL compared to fibroblasts grown in media with LPDS (Figure 2B, E).

Cellular distribution of the NPC1 and caveolin-1 proteins

Fibroblasts were grown using the two culture conditions (LPDS and LDL) and processed to perform deconvolution fluorescence microscopy. The NPC1 protein, which is known to be primarily associated with a unique population of late endosomes, was represented by vesicles approximately 1.5 mm in diameter and distributed throughout the perinuclear region of the cytoplasm when fibroblasts were grown in media with LPDS or LDL (Figure 3A, D). In contrast, caveolin-1 was represented by a relatively increased number of similarly sized vesicles that were also distributed throughout the perinuclear region of the cytoplasm when fibroblasts were grown in media with LPDS or LDL (Figure 3B, E). The merged images indicated partial co-localization of these vesicles when fibroblasts were grown in media with LPDS or LDL (Figure 3C, F).

Percentage of co-localization between the NPC1 and caveolin-1 proteins

Fibroblasts were grown using the two culture conditions (LPDS and LDL) and processed to perform deconvolution fluorescence microscopy and determine the percentage of co-localization between the NPC1 and caveolin-1 proteins. The results indicated that, on average, caveolin-1 co-localized with 13% of the NPC1 protein when fibroblasts were grown in media with LPDS (Figure 4A). In contrast, the average

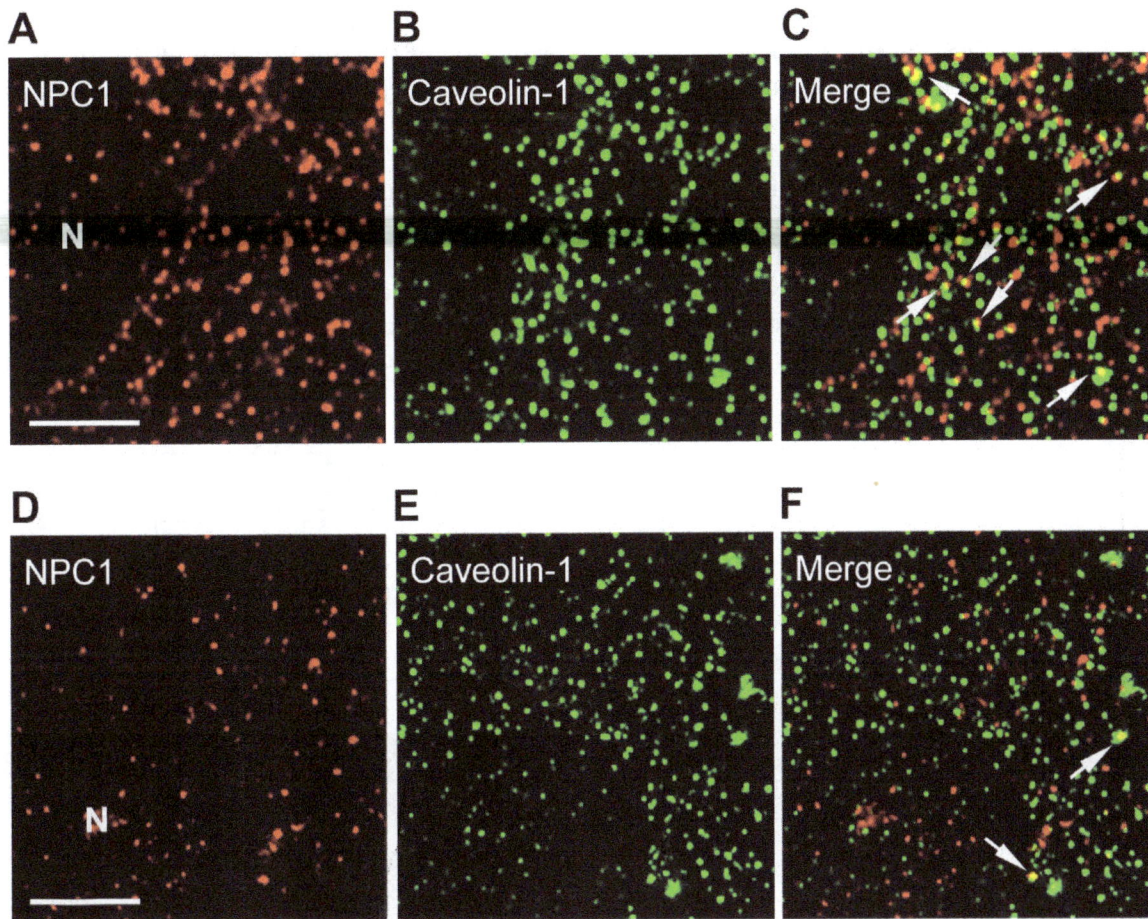

Figure 3. Cellular distribution of the NPC1 and caveolin-1 proteins. A representative fibroblast grown in LPDS is provided in the top set of images (A, B, C), while a representative fibroblast grown in LDL is provided in the bottom set of images (D, E, F). Partial co-localization of the NPC1 and caveolin-1 proteins are indicated using white arrows. Fluorescent images were obtained using a deconvolution fluorescence microscope equipped with an Olympus 100x (NA 1.4) oil-immersion objective. All images were acquired and processed using identical conditions. Bar represents 15 mm. N, nucleus.

amount of caveolin-1 that co-localized with the NPC1 protein was significantly decreased (62%) when fibroblasts were grown in media with LDL. Similarly, on average, the NPC1 protein co-localized with 10% of caveolin-1 when fibroblasts were grown in media with LPDS, but the average amount was significantly decreased (40%) when fibroblasts were grown in media with LDL (Figure 4B).

Identification and sequence comparison of caveolin-binding motifs associated with orthologous NPC1 proteins

The principal caveolin-binding motif, denoted by the consensus sequence jXjXXXXj, where j represents Trp (W), Phe (F), or Tyr (Y), but not the less common caveolin-binding motifs denoted by the consensus sequences jXXXXjXXj and jXjXXXXjXXj, was identified within the NPC1 protein amino acid sequence. The amino acid consensus sequence (818-FRFFKNSY -825) is encoded within cytoplasmic loop 4 adjacent to

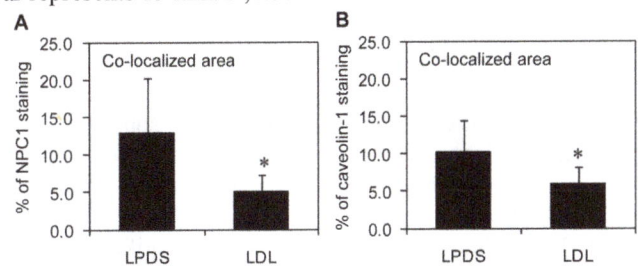

Figure 4. Percentage of co-localization between the NPC1 and caveolin-1 proteins. The total areas representing NPC1 or caveolin-1 staining were independently identified using the color range selection feature within Adobe Photoshop software, followed by determining the overlapping area and percentage of co-localization for caveolin-1 staining (A) and NPC1 staining (B) from fibroblasts grown in media with LPDS or LDL. Fluorescent images were obtained using a deconvolution fluorescence microscope equipped with an Olympus 100x (NA 1.4) oil-immersion objective. Quantitative data is represented as the mean ± S.D. using ten images acquired in a random manner and processed using identical conditions. *$P \leq 0.05$ compared to fibroblasts grown in media with LPDS.

the sterol-sensing domain (amino acids 616-797) of the human NPC1 protein. Since the NPC1 protein caveolin-binding motif is positioned within the cytoplasm, it would theoretically permit interaction with the caveolin-scaffolding domain (82-DGIWKASFTTFTVTKYWFYR-101) of caveolin-1 also positioned within the cytoplasm (Figure 5A). A comparison of the amino acid sequences from orthologous NPC1 proteins indicated that the NPC1 protein caveolin-binding motif is highly conserved among several different species (Figure 5B). The one species (cat) that does not encode the common caveolin-binding motif consensus sequence, instead encodes

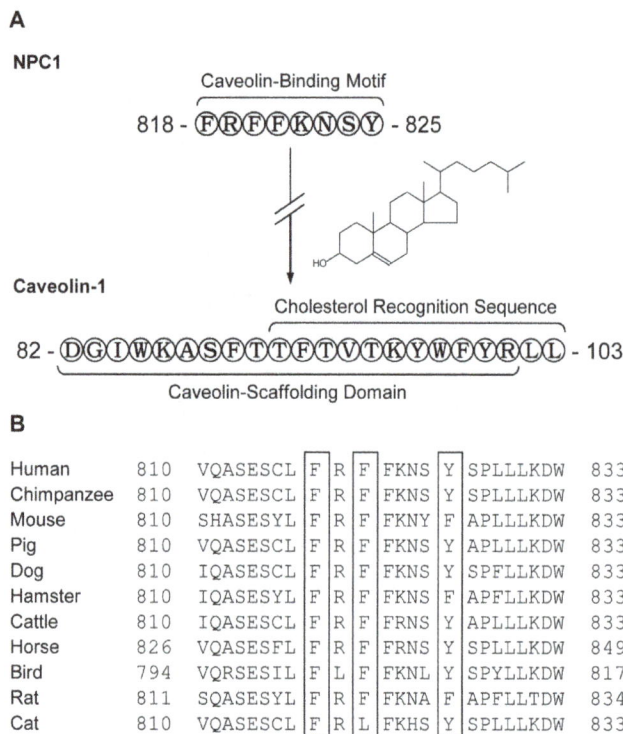

what has been reported as a permissible caveolin-like binding motif (I, L, or V instead of W, F or Y in place of the second j) also capable of interaction with the caveolin-scaffolding domain.

Relative amounts of the NPC1 and Caveolin-1 proteins resulting from caveolin-1 siRNA knockdown

Fibroblasts were transfected with either non-specific siRNA or caveolin-1 siRNA and grown using the two culture conditions (LPDS and LDL) to determine the relative average amounts of the NPC1 and caveolin-1

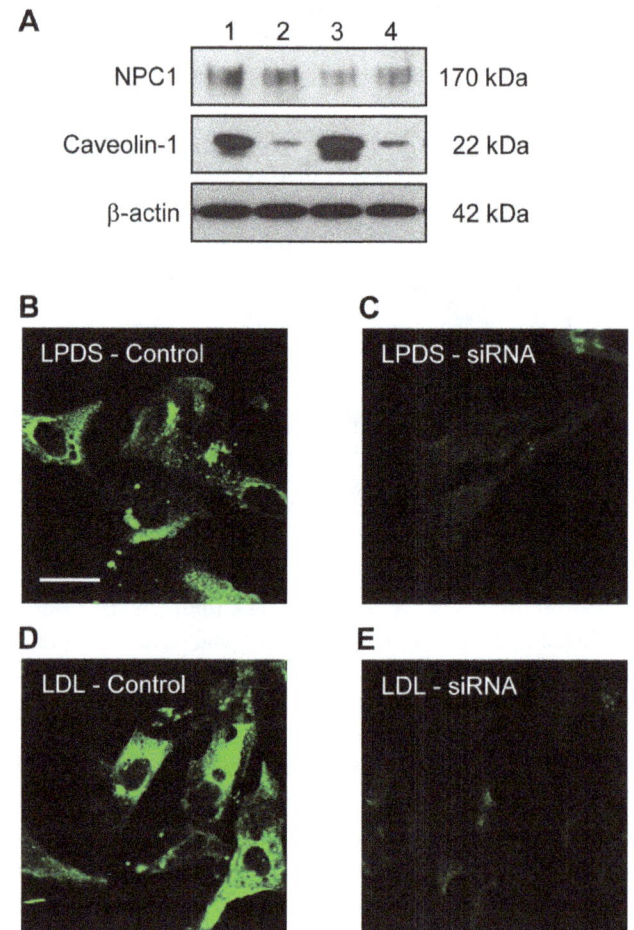

Figure 5. Identification and sequence comparison of caveolin-binding motifs associated with orthologous NPC1 proteins. A schematic representation of the topology for both the NPC1 and caveolin-1 proteins is provided (A). The human NPC1 protein contains a previously unidentified caveolin-binding motif (818-FRFFKNSY-825) located within cytoplasmic loop 4 adjacent to the sterol-sensing domain (amino acids 616-797). It is proposed that the NPC1 protein caveolin-binding motif interacts with the caveolin-1 protein through the caveolin-scaffolding domain (82-DGIWKASFTTFTVTKYWFYR-101). The caveolin-1 protein also possesses a cholesterol recognition sequence (91-TFTVTKYWFYRLL-103) located near the cytofacial membrane and cytoplasm boundary that partially overlaps with the caveolin-scaffolding domain. A comparison of the amino acid sequences between orthologous NPC1 proteins indicates that the NPC1 protein caveolin-binding motif (jXjXXXXj), where j mostly represents the amino acids Trp (W), Phe (F), or Tyr (Y), is conserved among several different species (B).

Figure 6. Relative amounts of the NPC1 and caveolin-1 proteins resulting from caveolin-1 siRNA knockdown. The relative amounts of the NPC1 and caveolin-1 proteins for fibroblasts transfected with non-specific siRNA (control) or caveolin-1 siRNA (siRNA) and grown in media with LPDS or LDL was determined using immunoblot analysis, where 1 = LPDS Control, 2 = LPDS siRNA, 3 = LDL Control, and 4 = LDL siRNA (A). A field of representative fibroblasts transfected with non-specific siRNA (control) or caveolin-1 siRNA (siRNA) and grown in media with LPDS (B and C) or LDL (D and E) is provided. Fluorescent images were obtained using a deconvolution fluorescence microscope equipped with an Olympus 60x (NA 1.4) oil-immersion objective. All images were acquired and processed using identical conditions.

proteins. The results indicated no significant difference in the relative amounts of the NPC1 protein for fibroblasts transfected with caveolin-1 siRNA when fibroblasts were grown in media with LPDS or LDL (Figure 6A). In contrast, the results indicated a significant decrease in the relative average amounts of caveolin-1 protein for fibroblasts transfected with caveolin-1 siRNA when fibroblasts were grown in media with LPDS or LDL (89% and 92%, respectively), compared to fibroblasts transfected with the non-specific (control) siRNA (Figure 6A). Consistent with the significantly decreased average amounts of caveolin-1 protein determined using immunoblot analysis, the relative amounts of caveolin-1 protein were noticeably decreased for fibroblasts transfected with caveolin-1 siRNA when fibroblasts were grown in media with LPDS or LDL compared to fibroblasts transfected with the non-specific (control) siRNA upon examination using fluorescence microscopy (Figure 6B, C, D, F).

Cellular distribution of cholesterol in relation to late endosomes/lysosome resulting from caveolin-1 siRNA knockdown

Fibroblasts were transfected with non-specific siRNA or caveolin-1 siRNA and grown in media with LDL to determine the relative cellular distribution of cholesterol in relation to late endosomes/lysosomes using deconvolution fluorescence microscopy. The images indicated one primary cellular distribution for cholesterol which was localized within perinuclear cytoplasmic vesicles (Figure 7A, D). With respect to fibroblasts transfected with the non-specific siRNA (control), the images indicated that only a certain population of the LAMP-1

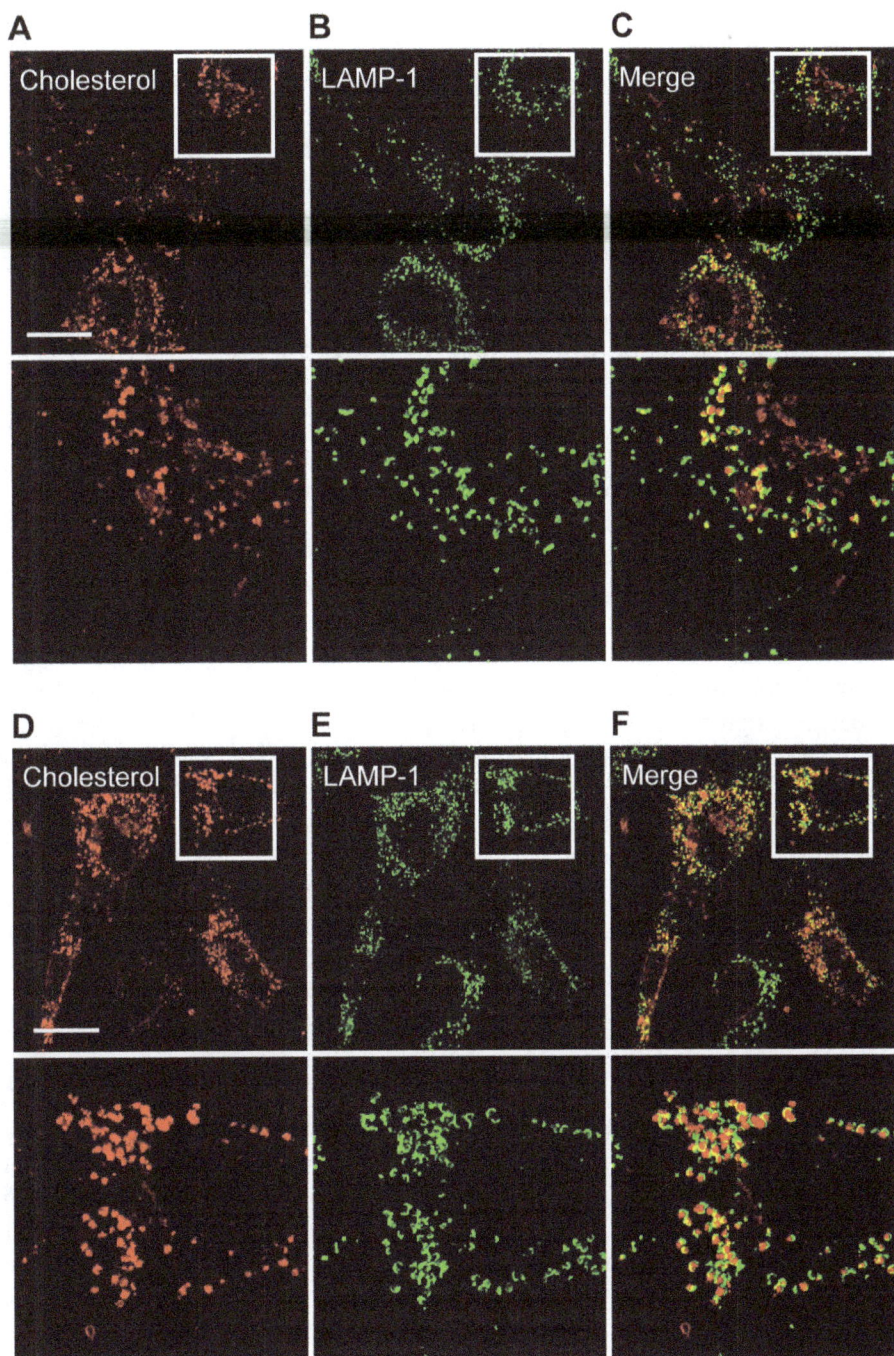

Figure 7. Cellular distribution of cholesterol in relation to late endosomes/lysosomes resulting from caveolin-1 siRNA knockdown. Fibroblasts were processed to label cholesterol using filipin and the LAMP-1 protein (a marker protein for late endosomes/lysosomes). Fibroblasts transfected with a non-specific siRNA (control) and grown in media with LDL are provided in the top set of images (A, B, C), while fibroblasts transfected with caveolin-1 siRNA and grown in media with LDL are provided in the bottom set of images (D, E, F). A defined area (square white outline) within fibroblasts are enlarged and placed directly below the corresponding image to visualize the cellular distribution of cholesterol (A, D), the LAMP-1 protein (B, E), and merged images (C, F). Fluorescent images were obtained using a deconvolution fluorescence microscope equipped with an Olympus 60x (NA 1.4) oil-immersion objective. All images were acquired and processed using identical conditions. Bar represents 15 mm.

containing vesicles, represented by semi-rounded and hollowed vesicles, co-localized with cholesterol (Figure 7A, B, C). In contrast, fibroblasts transfected with caveolin-1 siRNA had a noticeably increased population of LAMP-1 containing vesicles that co-localized with cholesterol (Figure 7D, E, F). There was no noticeable difference between fibroblasts transfected with the non-specific siRNA (control) and caveolin-1 siRNA when grown in media with LPDS (data not shown), thereby suggesting an accumulation of cholesterol within the LAMP-1 containing vesicles was derived from LDL.

Relative amounts of cholesterol associated with late endosomes/lysosomes resulting from caveolin-1 siRNA knockdown

Fibroblasts were transfected with non-specific siRNA or caveolin-1 siRNA and grown in media with LDL to determine the relative amount of LDL-derived cholesterol associated with late endosomes/lysosomes using deconvolution fluorescence microscopy. The results indicated a significant average increase (44%) in the relative amounts of cholesterol localized within LAMP-1 containing vesicles, which are known to represent late endosomes/lysosomes, for fibroblasts transfected with caveolin-1 siRNA compared to fibroblasts transfected with non-specific siRNA (control) (Figure 8). These results again suggest an accumulation of cholesterol within the LAMP-1 containing vesicles was derived from LDL.

Discussion

The present study was performed to determine whether the NPC1 and caveolin-1 proteins interact and function to modulate the efflux of LDL-derived cholesterol from endocytic compartments. To perform these studies, normal human fibroblasts were grown in media containing LPDS, which represented the control or basal culture condition, in addition to media containing LPDS supplemented with purified human LDL. In brief, the results indicated i) an inverse association between relative amounts of the NPC1 and caveolin-1 proteins when fibroblasts were grown in media with LDL; ii) that reciprocal co-immunoprecipitation of the NPC1 and caveolin-1 proteins was decreased when fibroblasts were grown in media with LDL; iii) that partial co-localization of the NPC1 and caveolin-1 proteins was decreased when fibroblasts were grown in media with LDL; iv) identification of a conserved caveolin-binding motif located adjacent to the SSD of the NPC1 protein, and v) an increased amount of LDL-derived cholesterol within late endosomes/lysosomes resulting from caveolin-1 siRNA knockdown. To-

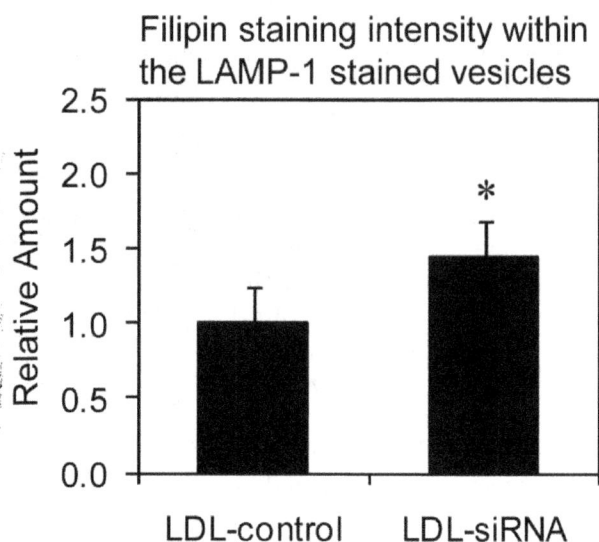

Figure 8. Relative amounts of cholesterol associated with late endosomes/lysosomes resulting from caveolin-1 siRNA knockdown. The total area representing LAMP-1 containing vesicles was identified using the color range selection feature, followed by determining the intensity of cholesterol staining within this area to calculate the relative amounts of cholesterol associated with late endosomes/lysosomes. Fluorescent images were obtained using a deconvolution fluorescence microscope equipped with an Olympus 60x (NA 1.4) oil-immersion objective. Quantitative data is represented as the mean ± S.D. using ten fields acquired in random manner and processed using identical conditions for fibroblasts grown in media with LDL. *$P \leq 0.05$ compared to fibroblasts transfected with non-specific siRNA (control) and grown in media with LDL.

gether, this report presents novel results demonstrating that the NPC1 and caveolin-1 proteins interact to modulate efflux of LDL-derived cholesterol from late endocytic compartments.

The results indicated an inverse association between relative amounts of the NPC1 and caveolin-1 proteins when fibroblasts were grown in media with LDL. Specifically, compared to fibroblasts grown in media with LPDS, supplementation of media with LDL significantly decreased the relative average amounts of NPC1 protein, but significantly increased the relative average amounts of caveolin-1 protein. A logical explanation for the inverse expression of the NPC1 and caveolin-1 genes is supported by studies indicating that LDL-dependent feedback inhibition of the SREBP pathway promotes downregulation of the NPC1 gene and upregulation of the caveolin-1 gene (Bist *et al*. 1997, Garver *et al*. 2008, Hailstones *et al*. 1998). However, it should be noted that feedback inhibition of the SREBP pathway and subsequent transcriptional regulation of the NPC1 and caveolin-1 genes may only serve as a partial explanation. It is well accepted that decreased NPC1 protein function alters the efflux of cholesterol from endocytic compartments and impairs feedback inhibition of the SREBP pathway (Garver *et al*. 2007, Kruth *et al*. 1986, Liscum *et al*. 1989, Pentchev *et al*. 1986). As a result, decreased NPC1 protein function should decrease expression of the caveolin-1 gene. This is contrary to previous results indicating that NPC1 heterozygous (NPC1+/-) and NPC1 homozygous affected (NPC1-/-) human fibroblasts and mouse livers have significantly increased amounts of caveolin-1 protein, suggesting that the caveolin-1 protein may compensate for decreased NPC1 protein function (Garver *et al*. 1997a, Garver *et al*. 1997b Garver *et al*. 2002). Consistent with this hypothesis, the relative average amounts of the NPC2 protein, which is well known to interact and function in a coordinate manner with the NPC1 protein, is likewise increased in NPC1-/- mouse livers (Klein *et al*. 2006).

To determine whether the NPC1 and caveolin-1 proteins interact, studies were performed using reciprocal co-immunoprecipitation. The results indicated that immunoprecipitation of either the NPC1 or caveolin-1 proteins resulted in co-immunoprecipitation of the caveolin-1 and NPC1 proteins, respectively. Interestingly, of the two NPC1 proteins (170 and 190 kDa) which differ due to asparagine-linked glycosylation within the NTD, only the lower molecular weight NPC1 protein co-immunoprecipitated with caveolin-1. Previous studies have indicated that this lower molecular weight NPC1 protein is ~50% less efficient in facilitating efflux of LDL-derived cholesterol from endocytic compartments (Watari *et al*. 1999a, Watari *et al*. 1999b). Moreover, the results indicated that reciprocal co-immunoprecipitation of the NPC1 or caveolin-1 proteins was decreased when fibroblasts were grown in media with LDL, suggesting that the increased efflux of cholesterol from endocytic compartments partially inhibits interaction of the NPC1 and caveolin-1 proteins.

Consistent with interaction of the NPC1 and caveolin-1 proteins, the results indicated partial co-localization of the NPC1 and caveolin-1 proteins when examined using deconvolution fluorescence microscopy. This particular result confirms earlier reports describing partial co-localization of the NPC1 and caveolin-1 proteins using both human and mouse fibroblasts (Garver *et al*. 2000, Higgins *et al*. 1999). Moreover, consistent with the decreased interaction of the NPC1 and caveolin-1 proteins when fibroblasts are grown in media with LDL, quantification of high-resolution fluorescent images revealed significantly less co-localization when fibroblasts were grown using this condition. Interestingly, a recent study has indi-

cated that caveolin-1 enriched vesicles that transiently interact with late endosomes/lysosomes are capable of forming caveolin-1 enriched subdomains that participate in the selective release of endosomal cargo (Pelkmans et al. 2004).

Studies have indicated that caveolin-1 is capable of specific and direct interaction with a number of proteins through a scaffolding domain (Human caveolin-1: 82-DGIWKASFTTFTVTKYWFYR-101) and that this domain partially overlaps with a cholesterol recognition sequence (Human caveolin-1: 91-TFTVTKYWFYRLL-103) that has stable and high-affinity for cholesterol (Li & Papadopoulos 1998, Murata *et al*. 1995). Both the caveolin-scaffolding domain and cholesterol recognition sequence are located at the cytofacial membrane and cytoplasm boundary near the amino-terminus of caveolin-1. This caveolin-scaffolding domain has been shown to interact with proteins that possess a caveolin-binding motif consensus sequence (jXjXXXXj), where j represents the amino acids Trp (W), Phe (F), or Tyr (Y) (Couet *et al*. 1997). Importantly, the human NPC1 protein was found to possess a previously unidentified caveolin-binding motif (818-FRFFKNSY-825) located within cytoplasmic loop 4 adjacent to the sterol-sensing domain (amino acids 616-797). A computer database search also revealed that the human NPC1 protein caveolin-binding motif is highly conserved among different species. This being the case, it is interesting to note that Patched, a membrane-bound protein with extensive sequence homology to the NPC1 protein, has also been shown to possess a caveolin-binding motif located within a cytoplasmic loop adjacent to the sterol-sensing domain and capable of interacting with caveolin-1 (Karpen *et al*. 2001).

To determine whether the caveolin-1 protein might have a role in modulating the efflux of LDL-derived cholesterol from endocytic compartments, fibroblasts were transfected with caveolin-1 siRNA to decrease expression of the caveolin-1 gene, and grown in media with LDL. The results clearly demonstrated a significant increase in the relative average amounts of LDL-derived cholesterol within LAMP-1 containing vesicles when compared to fibroblasts transfected with control siRNA and grown using similar conditions. However, compared to fibroblasts transfected with control siRNA, the total concentration of cellular cholesterol or cholesteryl ester was not significantly different for fibroblasts transfected with caveolin-1 siRNA, thereby suggesting that decreased amounts of caveolin-1 protein alters the distribution of cellular cholesterol (data not shown). This particular result is consistent with an earlier report indicating that a caveolin-1 dominant-negative mutant promotes altered intracellular cholesterol homeostasis characterized by an accumulation of cholesterol within late endosomes/lysosomes, although the concentration of cellular cholesterol remains unchanged (Pol *et al.* 2001, Roy *et al.* 1999). Moreover, a recent report using cells derived from the caveolin-1 deficient mouse indicated that caveolin-1 has a key role in maintaining intracellular cholesterol homeostasis, characterized by only a modest change in the concentration of cellular cholesterol and cholesteryl ester (Frank *et al.* 2006). This being the case, it is interesting to note that whole body cholesterol homeostasis, including plasma lipid levels, remain relatively normal for both the caveolin-1 deficient mouse and a patient diagnosed with a rare form of congenital lipodystrophy (Kim *et al.* 2008, Razani *et al.* 2002, Razani *et al.* 2001).

Conclusion

In summary, the present study determined that the NPC1 and caveolin-1 proteins interact and function to modulate efflux of LDL-derived cholesterol from endocytic compartments. A plausible model may be envisioned whereby the conserved NPC1 protein caveolin-binding motif allows interaction with the caveolin-1 scaffolding domain in the absence of LDL-derived cholesterol, but which is interrupted by the binding of cholesterol to the high-affinity caveolin-1 cholesterol recognition sequence which prompts dissociation of the NPC1 and caveolin-1 proteins and subsequent transport of cholesterol to other caveolin-1 cellular compartments, most likely the TGN. Clearly, additional studies will be necessary to further define how the NPC1 and caveolin-1 proteins modulate the efflux of LDL-derived cholesterol from endocytic compartments and maintain intracellular cholesterol homeostasis.

Acknowledgments

This work was supported in part by grants received from the National Institutes of Health (R21 DK071544), the Ara Parseghian Medical Research Foundation, and private donations for the investigation of childhood genetic and metabolic diseases. The authors would like to express their appreciation to Sarah Mount Patrick for expertise in performing deconvolution fluorescence microscopy and quantitation of fluorescence co-localization.

Author Contributions

DJ and WSG conceived and performed these studies, while DJ, RAH, RAO and WSG interpreted results and prepared the manuscript.

References

Bist A, Fielding PE & Fielding CJ 1997 Two sterol regulatory element-like sequences mediate up-regulation of caveolin gene transcription in response to low density lipoprotein free cholesterol. *Proc Natl Acad Sci USA* **94** 10693-10698

Blanchette-Mackie EJ, Dwyer NK, Amende LM, Kruth HS, Butler JD, Sokol J, Comly ME, Vanier MT, August JT, Brady RO & Pentchev PG 1988 Type C Niemann-Pick disease: Low density lipoprotein uptake is associated with premature cholesterol accumulation in the Golgi complex and excessive cholesterol storage in lysosomes. *Proc Natl Acad Sci USA* **85** 8022-8026

Brown MS & Goldstein JL 1975 Regulation of the activity of the low density lipoprotein receptor in human fibroblasts. *Cell* **6** 307-316

Brown MS, Ho YK & Goldstein JL 1976 The low-density lipoprotein pathway in human fibroblasts: Relation between cell surface receptor binding and endocytosis of low-density lipoprotein. *Ann NY Acad Sci* **275** 244-257

Cheruku SR, Xu Z, Dutia R, Lobel P & Storch J 2006 Mechanism of cholesterol transfer from the Niemann-Pick type C2 protein to model membranes supports a role in lysosomal cholesterol transport. *J Biol Chem* **281** 31594-31604

Couet J, Li S, Okamoto T, Ikezu T & Lisanti MP 1997 Identification of peptide and protein ligands for the caveolin-scaffolding domain. *J Biol Chem* **272** 6525-6533

Coxey RA, Pentchev PG, Campbell G & Blanchette-Mackie EJ 1993 Differential accumulation of cholesterol in Golgi compartments of normal and Niemann-Pick type C fibroblasts incubated with LDL: A cytochemical freeze-fracture study. *J Lipid Res* **34** 1165-1176

Dietschy JM, Turley SD, & Spady DK 1993 Role of liver in the maintenance of cholesterol and low density lipoprotein homeostasis in different animal species, including humans. *J Lipid Res* **34** 1637-1659

Dupree P, Parton RG, Raposo G, Kurzchalia TV & Simons K 1993 Caveolae and sorting in the trans-Golgi network of epithelial cells. *EMBO J* **12** 1597-1605

Fielding PE & Fielding CJ 1996 Intracellular transport of low density lipoprotein derived free cholesterol begins at clathrin-coated pits and terminates at cell surface caveolae. *Biochemistry* **35** 14932-14938

Friedland N, Liou H-L, Lobel P & Stock AM 2003 Structure of a cholesterol-binding protein deficient in Niemann-Pick type C2 disease. *Proc Nat Acad Sci USA* **100** 2512-2517

Garver WS, Erickson RP, Wilson JM, Colton TL, Hossain GS, Kozloski MA & Heidenreich RA 1997a Altered expression of caveolin-1 and increased cholesterol in detergent insoluble membrane fractions from liver in mice with Niemann-Pick disease type C. *Biochim Biophys Acta* **1361** 272-280

Garver WS, Heidenreich RA, Erickson RP, Thomas MA & Wilson JM 2000 Localization of the murine Niemann-Pick C1 protein to two distinct intracellular compartments. *J Lipid Res* **41** 673-687

Garver WS, Hsu SC, Erickson RP, Greer WL, Byers DM & Heidenreich RA 1997b Increased expression of caveolin-1 in heterozygous Niemann-Pick type II human fibroblasts. *Biochem Biophys Res Comm* **236** 189-193

Garver WS, Jelinek D, Oyarzo JN, Flynn J, Zuckerman M, Krishnan K, Chung BH & Heidenreich RA 2007 Characterization of liver disease and lipid metabolism in the Niemann-Pick C1 mouse. *J Cell Biochem* **101** 498-516

Garver WS, Krishnan K, Gallagos JR, Michikawa M, Francis GA & Heidenreich RA 2002 Niemann-Pick C1 protein regulates cholesterol transport to the trans-Golgi network and plasma membrane caveolae. *J Lipid Res* **43** 579-589

Garver WS, Jelinek D, Francis GA & Murphy BD 2008 The Niemann-Pick C1 gene is downregulated by feedback inhibition of the SREBP pathway in humans fibroblasts. *J Lipid Res* **49** 1090-1102

Goldstein JL, Dana SE, Faust JR, Beaudet AL & Brown MS 1975 Role of lysosomal acid lipase in the metabolism of plasma low density lipoprotein. Observations in cultured fibroblasts from a patient with cholesteryl ester storage disease. *J Biol Chem* **250** 8487-8495

Hailstone D, Sleer LS, Parton RG & Stanley KK 1998 Regulation of caveolin and caveolae by cholesterol in MDCK cells. *J Lipid Res* **39** 369-379

Higgins ME, Davies JP, Chen FW & Ioannou YA 1999 Niemann-Pick C1 is a late endosome-resident protein that transiently associates with lysosomes and the trans-Golgi network. *Mol Gene Metab* **68** 1-13

Infante RE, Radhakrishnan A, Abi-Mosleh L, Kinch LN, Wang ML, Grishin NV, Goldstein JL & Brown MS 2008a Purified NPC1 Protein II. Localization of sterol binding to a 240-amino acid soluble luminal loop. *J Biol Chem* **283** 1064-1075

Infante RE, Wang ML, Radhakrishnan A, Kwon HJ, Brown MS & Goldstein JL 2008b NPC2 facilitates bidirectional transfer of cholesterol between NPC1 and lipid bilayers, a step in cholesterol egress from lysosomes. *Proc Natl Acad Sci USA* **105** 15287-15292

Karpen HE, Bukowski JT, Hughes T, Gratton J, Sessa WC & Gailani MR 2001 The sonic hedgehog receptor patched associates with caveolin-1 in cholesterol-rich microdomains of the plasma membrane. *J Biol Chem* **276** 19503-19511

Kim CA, Delepin M, Boutet M, El Mourabit H, Le Lay S, Meier M, Nemani M, Bridel E, Leite CC, Bertola DR, Semple RK, O'Rahilly S, Dugail I, Capeau J, Lathrop M & Magre J 2008 Association of a homozygous nonsense caveolin mutation with Berardinelli-Seip congenital lipodystrophy. *J Clin Endo Metab* **93** 1129-1134

Klein A, Amigo L, Retamal MJ, Morales MG, Miquel JF, Rigotti A & Zanlungo S 2006 NPC2 is expressed in human and murine liver and secreted into bile: Potential implications for body cholesterol homeostasis. *Hepatology* **43** 126-133

Ko DC, Binkley J, Sidow A & Scott MP 2003 The integrity of a cholesterol-binding pocket in Niemann-Pick C2 protein is necessary to control lysosome cholesterol levels. *Proc Natl Acad Sci USA* **100** 2518-2525

Kruth HS, Comly ME, Butler JD, Vanier MT, Fink JK, Wenger DA, Patel S & Pentchev PG 1986 Type C Niemann-Pick disease. Abnormal metabolism of low density lipoprotein in homozygous and heterozygous fibroblasts. *J Biol Chem* **261** 16769-16774

Kwon HJ, Abi-Mosleh, Wang ML, Deisenhofer J, Goldstein JL, Brown MS & Infante RE 2009 Structure of N-terminal domain of NPC1 reveals distinct subdomains for binding and transfer of cholesterol *Cell* **137** 1213-1224

Li H & Papadopoulos V 1998 Peripheral-type benzodiazepine receptor function in cholesterol transport. Identification of a putative cholesterol recognition/interaction amino acid sequence and consensus pattern. *Endocrinology* **139** 4991-4997

Liscum L, Ruggiero RM & Faust JR 1989 The intracellular transport of low density lipoprotein-derived cholesterol is defective in Niemann-Pick type C fibroblasts. *J Cell Biol* **108** 1625-1636

Liu R, Lu P, Chu JWK & Sharom FJ 2009 Characterization of fluorescent sterol binding to purified human NPC1. *J Biol Chem* **284** 1840-1852

Murata M, Peranen J, Schreiner R, Wieland F, Kurzchalia TV & Simons K 1995 VIP21/caveolin is a cholesterol-binding protein. *Proc Natl Acad Sci USA* **92** 10339-10343

Neufeld EB, Wastney M, Patel S, Suresh S, Conney

AM, Dwyer NK, Roff CF, Ohno K, Morris JA, Carstea ED, Incardona JP, Strauss JF, Vanier MT, Patterson MC, Brady RO, Pentchev PG & Blanchette-Mackie EJ 1999 The Niemann-Pick C1 protein resides in a vesicular compartment linked to retrograde transport of multiple lysosomal cargo. *J Biol Chem* **274** 9627-9635

Ohgami N, Ko DC, Thomas M, Scott MP, Chang CCY & Chang TY 2004 Binding between the Niemann-Pick C1 protein and a photoactivatable cholesterol analog requires a functional sterol-sensing domain. *Proc Natl Acad Sci USA* **101** 12473-12478

Osono Y, Woollett LA, Herz J & Dietschy JM 1995 Role of the low density lipoprotein receptor in the flux of cholesterol through the plasma and across the tissues of the mouse. *J Clin Invest* **95** 1124-1132

Pelkmans L, Burli T, Zerial M & Helenius A 2004 Caveolin-stabilized membrane domains as multifunctional transport and sorting devices in endocytic membrane traffic. *Cell* **118** 767-780

Pentchev PG, Kruth HS, Comly ME, Butler JD, Vanier MT, Wenger DA & Patel S 1986 Type C Niemann-Pick disease. A parallel loss of regulatory responses in both the uptake and esterification of low density lipoprotein-derived cholesterol in cultured fibroblasts. *J Biol Chem* **261** 16775-16780

Pol A, Luetterforst R, Lindsay M, Heino S, Ikonen E & Parton RG 2001 A caveolin dominant negative mutant associates with lipid bodies and induces intracellular cholesterol imbalance. *J Cell Biol* **152** 1057-1070

Razani B, Engelman JA, Wang XB, Schubert W, Zhang XL, Marks CB, Macaluso F, Russell RG, Li M, Pestell RG, Di Vizio D, Hou H, Kneitz B, Lagaud G, Christ GJ, Edelmann W & Lisanti MP 2001 Caveolin-1 null mice are viable but show evidence of hyperproliferative and vascular abnormalities. *J Biol Chem* **276** 38121-38138

Razani B, Combs TP, Wang XB, Frank PG, Park DS, Russell RG, Li M, Tang B, Jelicks LA, Scherer PE & Lisanti MP 2002 Caveolin-1-deficient mice are lean, resistant to diet-induced obesity, and show hypertriglyceridemia with adipocyte abnormalities. *J Biol Chem* **277** 8635-8647

Roy S, Luetterforst R, Harding A, Apolloni A, Etheridge M, Stang E, Rolls B, Hancock JF & Parton RG 1999 Dominant-negative caveolin inhibits H-Ras function by disrupting cholesterol-rich plasma membrane domains. *Nat Cell Biol* **1** 98-105

Sleat DE, Wiseman JA, El-Banna M, Price SM, Verot L, Shen MM, Tint GS, Vanier MT, Walkley SU & Lobel P 2004 Genetic evidence for nonredundant functional cooperativity between NPC1 and NPC2 in lipid transport. *Proc Natl Acad Sci USA* **101** 5886-5891

Smart EJ, Ying Y, Donzell WC & Anderson RGW 1996 A role for caveolin in transport of cholesterol from endoplasmic reticulum to plasma membrane. *J Biol Chem* **271** 29427-29435

Sugii S, Reid PC, Ohgami N, Du H & Chang TY 2003 Distinct endosomal compartments in early trafficking of low density lipoprotein-derived cholesterol. *J Biol Chem* **278** 27180-27189

Turley SD, Spady DK & Dietschy JM 1995 Role of liver in the synthesis of cholesterol and the clearance of low density lipoproteins in the cynomolgus monkey. *J Lipid Res* **36** 67-79

Uittenbogaard A, Everson WV, Matveev SV & Smart EJ 2002 Cholesteryl ester is transported from caveolae to internal membranes as part of a caveolin-annexin II lipid-protein complex. *J Biol Chem* **277** 4925-4931

Uittenbogaard A, Ying Y & Smart EJ 1998 Characterization of a cytosolic heat-shock protein-caveolin chaperone complex. Involvement in cholesterol trafficking. *J Biol Chem* **273** 6525-6532

Underwood KW, Jacobs NL, Howley A & Liscum L 1998 Evidence for a cholesterol transport pathway from lysosomes to endoplasmic reticulum that is independent of the plasma membrane. *J Biol Chem* **273** 4266-4274

Urano Y, Watanabe H, Murphy SR, Shibuya Y, Geng Y, Peden AA, Chang CCY & Chang TY 2008 Transport of LDL-derived cholesterol from the NPC1 compartment to the ER involves the trans-Golgi network and the SNARE protein complex. *Proc Natl Acad Sci USA* **105** 16513-16518

Watari H, Blanchette-Mackie EJ, Dwyer NK, Glick JM, Patel S, Neufeld EB, Brady RO, Pentchev PG & Strauss III JF 1999a Niemann-Pick C1 protein: Obligatory roles for N-terminal domains and lysosomal targeting in cholesterol mobilization. *Proc Natl Acad Sci USA* **96** 805-810

Watari H, Blanchette-Mackie EJ, Dwyer NK, Watari M, Neufeld EB, Patel S, Pentchev PG & Strauss III JF 1999b Mutations in the leucine zipper motif and sterol-sensing domain inactivate the Niemann-Pick C1 glycoprotein. *J Biol Chem* **274** 21861-21866

Wojtanik KM & Liscum L 2003 The transport of low density lipoprotein-derived cholesterol to the plasma membrane is defective in NPC1 cells. *J Biol Chem* **278** 14850-14856

12

Zika virus infection: a review of available techniques towards early detection

Kathleen KM Glover[1,2] and Kevin M Coombs[1,2]

[1] Department of Medical Microbiology, Faculty of Health Sciences, University of Manitoba, Winnipeg, Manitoba R3E 0J6 Canada;
[2] Manitoba Centre for Proteomics and Systems Biology, Room 799, 715 McDermot Avenue, Winnipeg, Manitoba, Canada R3E 3P4 Canada

Correspondence should be addressed to Kathleen KM Glover; E-mail: gloverk@myumanitoba.ca

Abstract

Zika virus belongs to the family *Flaviviridae* as do other viruses like Dengue, West Nile and Yellow Fever. They are arboviruses transmitted by the *Aedes* species of mosquito. Zika virus was first isolated in rhesus monkeys in Uganda in 1947. Human infections of the virus were found between the 1960s and 1980s in Africa, the Americas, Asia and the Pacific. The similarity in clinical presentation in Zika-infected patients compared with Dengue caused infections to be previously misdiagnosed as Dengue infection. The Zika virus pandemic in 2015 created a lot of concern globally because of little information about available techniques, samples as well as no available antiviral and vaccines for treatment and vaccination against infection. In addition, the vectors identified for transmission, *Aedes aegypti* and *Aedes albopictus,* were of great concern due to their ability to survive both temperate and tropical climatic conditions, hence indicating the possible global spread of Zika virus infection. Almost two years after the report of infection in pregnant women in Brazil resulting in microcephalic babies, Zika virus was identified as a public health problem. Thus, a lot of research into early detection and prevention has been conducted to control the spread of the virus. This review paper highlights available information on techniques currently available for diagnosis of infection caused by Zika virus.

Introduction

Zika virus, a once neglected tropical microorganism, has been in existence since 1947 when it was first isolated from the blood of a sentinel rhesus monkey (Macacamulatta) in the Zika forest near Entebbe, Uganda (Dick *et al.* 1952). Human infections caused by this virus were found between the 1960s and 1980s in Africa, the Americas, Asia and the Pacific. The first major outbreak of Zika infection outside Africa was reported in the Yap Islands of Micronesia in 2007 (Lanciotti *et al.* 2008). Fagbami in 1979 published a report of detection of Zika virus infection in 30% of the patient sera in four communities in Oyo state, Nigeria, along with three other Flavivirus, namely Yellow Fever (50%), West Nile (46%), and Wesselsbron (59%) (Fagbami 1979).

In French Polynesia in 2013, a woman with Guillain-Barr syndrome who exhibited influenza-like symptoms was negative for all 4 serotypes of dengue by plaque reduction neutralization test but positive for Zika virus (Oehler *et al.* 2014).

In Yap state, patients were presenting dengue like symptoms with conjunctivitis. Thus, they were initially diagnosed serologically as being infected with dengue virus (Zika virus outbreak, 2007). However, molecular diagnoses by reverse transcription–PCR (RT-PCR) and sequencing confirmed Zika virus infection, based on approximately 90% nucleotide identity with the Zika virus genome.

In Brazil, the first confirmed case of Zika virus was reported in a patient who was initially exhibiting dengue-like symptoms (Zanluca *et al.* 2015). Serological and molecular analysis of patient samples were negative for Dengue and Chikungunya, pathogenic re-emerging arboviruses (Lanciotti *et al.* 2008), but were positive for Zika by RT-PCR after detecting a 364bp amplicon expected only for Zika (Zanluca *et al.* 2015). However, in 2015 the world's attention was drawn to Brazil, when there was an increase in the number of microcephaic babies being born by pregnant women (Victora 2016). The number was initially below 200 prior to 2015 (Brazil Ministry of Health 2016). However, this figure increased to 4783 towards the later part of 2015 and the earlier part of 2016 (Brazil Ministry of Health, 2016).

Classification

Zika virus belongs to the family *Flaviviridae* which also includes other viruses such as dengue (DENV), Yellow Fever and West Nile viruses. All of these are of public health importance (Pierson & Diamond 2013, Faye *et al.* 2014). This virus has a single-stranded RNA genome with a positive sense polarity (Faye *et al.* 2014) and falls in group four according to the Baltimore system of classifying viruses.

The Zika viral genome is composed of 10,794 nucleotides which code for 3,419 amino acids (Kuno *et al.* 2007). The Zika virus genome, which acts as mRNA upon reaching the cytoplasm, is immediately translated into a polyprotein which is cleaved into 3 structural proteins; capsid (C), precursor membrane (prM), envelope (E) and 7 non-structural proteins (Kuno *et al.* 2007).

Flavivirus are known to replicate near the endoplasmic reticulum of the infected cell. However, some studies have reported Zika viral antigens in the nucleus of an infected cell, which requires further investigation (Buckley *et al.* 1988). Zika virus was classified into African and Asian lineages by phylogenetic analysis (Gatherer 2016). The Asian lineage originated during the virus's migration from Africa to Southeast Asia, where it was first detected in Malaysia. From there, Zika virus spread to the Pacific Islands, separately to Yap and French Polynesia, and then to New Caledonia, Cook Islands, Easter Island, and the Americas (Gatherer 2016).

Transmission

The vector that has been globally identified to transmit Zika virus is the *Aedes* species of mosquito. They are able to survive both tropical and temperate climatic conditions (Ledermann *et al.* 2014, Kraemer *et al.* 2015, Gaffigan *et al.* 2016). *Aedes aegypti* and *Aedes albopictus* have a wider global coverage as they are able to survive both climatic conditions. However, some species, namely *Aedes luteocephalus* in Africa and *Aedes hensilli* in the Pacific Islands, have limited global distribution. The *Aedes aegypti* is the main species of mosquito for transmission of Zika virus as its feeding habit of biting multiple hosts to complete its blood meals tends to spread the virus in the process. The mosquito is infected during a blood meal from an infected host which could be humans or animals after which the virus replicates within the vector and is later transmitted to another host which could be man during its next blood meal (Ioos *et al.* 2014). This species of mosquito is also known to transmit Dengue virus (Thomas *et al.* 2016). *Aedes albopictus* has not been implicated in most cases of Zika infections. However, in 2007, it was implicated in an outbreak of Zika infection in Gabon. Also, *Aedes albopictus* is endemic in the U.S.A. and the ability of the virus to survive in any *Aedes* species of mosquito indicates the *Aedes*

albopictus vectoral role of Zika virus in the United States (Fauci *et al.* 2016).

Beside vector transmission, various research groups have shown evidence of other modes of transmission of the Zika virus. Transplacental transmission has been confirmed and is currently a global concern as infected babies are born with microcephaly (Schuler-Faccini *et al.* 2016). Sexual transmission of Zika virus has also been reported (Musso *et al.* 2015). Thus, some countries, like the U.S.A, have recommended couples who have visited endemic areas to delay child bearing, and advised couples to use condoms to reduce the spread of the virus. In an outbreak in French Polynesia, Zika virus was isolated from the semen of a patient diagnosed with hematospermia (Musso *et al.* 2015). In addition, a case of sexual transmission of Zika virus was reported in France in a woman who acquired Zika sexually from her partner who had recently visited Brazil (Elgot *et al.* 2016). CDC also is in the process of investigating 14 potential sexually acquired Zika virus cases (Duhaime-Ross 2016).

Zika virus has also been identified as one of the possible transfusion transmissible infections in addition to others like HIV and hepatitis B (Musso *et al.* 2014). The virus was detected by PCR in 42 (3%) of the 1505 blood samples donated from asymptomatic donors in French Polynesia (Musso *et al.* 2014). This finding suggested the Zika virus being a transfusion transmissible virus, which should be screened for before transfusing patients (Musso *et al.* 2014). Another report detected Zika virus serologically in blood donors in a Zika endemic area (Aubry *et al.* 2015).

Clinical manifestation

Concerning symptoms, Zika infection mimics that of Dengue (Ioos *et al.* 2014). Thus, Zika was previously misdiagnosed in patients as Dengue infection (Oehler *et al.* 2014). The period between infection and onset of clinical symptoms is approximately 3 to 12 days (Anna *et al.* 2016). However, infection is asymptomatic in approximately 80% of cases (Duffy *et al.* 2009, Ioos *et al.* 2014). The clinical symptoms manifest as a self-limited febrile syndrome associated with rash, conjunctivitis and arthralgias (Simpson 1964, Musso *et al.* 2016, Petersen *et al.* 2016). Rash, a prominent feature, is maculopapular and pruritic in most cases; it begins proximally and spreads to the extremities with spontaneous resolution within 1–4 days of onset (Simpson, 1964) while fever is typically low grade (37.4°C – 38.0°C) (Dupont-Rouzeyrol *et al.* 2014, Musso *et al.* 2015, Simpson 1964).

Pathogenesis

Currently, two diseases that have been reported to be associated with Zika virus infection are microcephaly and Guillain–Barré syndrome (Cao-Lormeau *et al.* 2016, Oliveira *et al.* 2016). Microcephaly is a condition that results from the virus attacking the neural progenitor cells, disrupting their development and thereby affecting the formation of the fetal brain which affects the baby's development (Brasil *et al.* 2016, Marrs *et al.* 2016, Dang *et al.* 2016, Garcez *et al.* 2016, Qian *et al.* 2016, Tang *et al.* 2016). Neural progenitor cells (NPC) are self-renewing neural stem cells in the brain which differentiate into neurons, astrocytes, and oligodendrocytes (Delvecchio *et al.* 2016). Several research findings have established the link between Zika virus and microcephaly such as detection of viral RNA in the amniotic fluid of Zika infected pregnant women as well as in brain tissues from microcephalic babies (Brasil *et al.* 2016, Calvet *et al.* 2016, Driggers *et al.* 2016, Mlakar *et al.* 2016, Melo *et al.* 2016). *In vivo* experiments where pregnant mice were infected with Zika led to neuronal death, cell cycle arrest and apoptosis of NPCs leading to embryonic microcephaly and growth restriction (Cugola *et al.* 2016, Li *et al.* 2016, Miner *et al.* 2016, Shao *et al.* 2016).

Guillain-Barre syndrome is a rare but serious autoimmune disorder in which the immune system attacks healthy nerve cells in the peripheral nervous system. This leads to weakness, numbness, and tingling. It can eventually cause paralysis. Aetologic agents that were previous associated with its incidence were *Campylobacter jejuni*, influenza, *Epstein-Barr virus, Cytomegalovirus* and *HIV*. Recently, clusters of the Guillain–Barré syndrome and microcephaly have been spatially and temporally related to the current outbreak of Zika infection in the Americas (Beatriz *et al.* 2016, World Health Organization 2016). The World Health Organization on Feb 1 2016, declared Zika associated microcephaly and Guillain-Barre syndrome as a Public Health Emergency of International Concern (PHEIC), not based on the current information of Zika associated cases but on what is not known of clusters of microcephaly, Guillain-Barré syndrome, and possibly other neurological defects reported by country representatives from Brazil and French Polynesia (Heymann *et al.* 2016, Oehler *et al.* 2014, Pan American Health Organization 2016).

Laboratory diagnosis

Clinical presentation of patients infected with Zika is not reliable for preliminary diagnoses due to its similarity with other flaviviruses like Dengue. Thus, laboratory diagnosis is used to confirm Zika infection. Detection is based on the type of patient sample being analyzed. Samples that are currently used for detection of Zika infection in patients are blood or serum. Viremia is detectable between 3 to 5 days post infection. Other samples with Zika diagnostic potential are cerebrospinal fluid (CSF), urine, saliva, amniotic fluid, semen and fetal brain tissue (Plourde 2016). Urine and saliva which are noninvasive samples offer a better alternative for diagnosis as compared to blood and cerebrospinal fluid. Also, in most cases, collection of urine and saliva does not need any technical expertise as compared to blood and CSF which is collected by trained health professional in a designated health facility. Urine samples are currently recommended for diagnosis of Zika virus infection as viral RNA persists longer in urine as compared to blood (Plourde 2016, Gourinat *et al.* 2014, Besnard *et al.* 2014). One study reported that Zika virus RNA was detected in urine up to 20 days post infection after viral RNA was undetectable in the blood (Plourde 2016). Thus a patient urine sample must also be tested to confirm diagnosis when a negative blood test is reported for a Zika suspected patient. Detection of Zika viral RNA in saliva is best during the acute phase of the infection but is not ideal during the late stage. Musso *et al.* (2015) screened 182 Zika suspected patient's saliva and urine samples. Zika viral RNA was positive for 35 (19.2%) of their saliva while negative in their blood. On the other hand, 16 (8.8%) of the patients tested positive for Zika viral RNA in their blood but were negative in their saliva. This indicates that blood samples should also be screened in cases where saliva is used for Zika diagnosis.

Techniques for diagnosis

Serology and molecular diagnoses are currently the techniques being used for confirming Zika infection in patients. Detection of Zika virus in the blood or serum by molecular techniques, namely RT-PCR, is best during the acute stage of the infection when the patient is viremic (Lanciotti *et al.* 2008). However, serology cannot be used as Zika virus IgM may be undetectable during the acute phase of the infection (Hayes 2009). Serologic testing has a major limitation. Cross-reactivity with other flaviviruses (particularly Dengue), limits specificity (Beatriz *et al.* 2016). Therefore, positive serologic test results should be confirmed with molecular testing. The plaque reduction neutralization assay, a seroneutralization assay, generally has improved specificity over serology, but may still yield cross-reactive results in secondary Flavivirus infections (Hayes 2009, Anna *et al.* 2016).

Currently according to the CDC, symptomatic Zika suspected infected patients are diagnosed by RNA NAT (nucleic acid testing). Serum samples are tested two weeks after the onset of dengue-like symptoms. Urine samples are also tested before the 14th day after the onset of symptoms. A positive result is confirmatory of Zika infection. However, a negative result does not exclude infection. Additional serum IgM testing must be done to rule out infection (Diagnostic Tests for Zika Virus, 2017).

In the case of pregnant women who fall under the high risk group, RNA nucleic acid testing must be done on both serum and urine two weeks after any visit to a Zika endemic region. In addition, RNA testing must be done for those who are IgM positive for Zika after exposure. Pregnant women are routinely screened for Zika as part of their antenatal care in areas where Zika is endemic, especially during the 1st and 2nd trimester of pregnancy as most of the developmental processes of the growing fetus takes place at this stage (Diagnostic Tests for Zika Virus 2017).

The Trioplex Real-time RT-PCR Assay (Trioplex rRT-PCR) is a laboratory test designed to detect Zika virus, Dengue virus, and Chikungunya virus RNA (Centers for Disease Control and Prevention, 2016). This method has not yet received approval for Zika diagnosis by the Food and Drug Administration (FDA) except on Emergency Use Authorization (EUA) according to the CDC. The Trioplex rRT-PCR qualitatively detects and differentiates Zika from other Flavivirus in human sera, whole blood (EDTA), cerebrospinal fluid, urine and amniotic fluid. According to the CDC, patient samples analyzed by this assay must meet certain criteria; namely clinical signs and symptoms associated with Zika virus infection, history of residence in or travel to a geographic region with active Zika transmission at the time of travel, or other epidemiologic criteria for which Zika virus testing may be indicated as part of a public health investigation. A negative result must be combined with clinical observations, patient history, and epidemiological information (Centers for Disease Control and Prevention, 2016).

However, this assay can give both false positive and false negative results. False positive results can arise from contamination from previous positive samples. However, inclusion of negative controls helps detect any possible contamination of patient samples, thereby reducing their incidence. False negative results occur as a result of the method of sample collection which affects the quality of the sample analyzed, degradation of the viral RNA during sample transportation, the time of sample collection and analysis when viral RNA might no longer be detectable in some samples after the onset of symptoms (approximately 14 days post-onset of symptoms for serum, whole blood, and/or urine), failure to follow the authorized assay procedures and failure to use authorized extraction kits and platforms (Centers for Disease Control and Prevention CDC 2016, Plourde 2016, Musso et al. 2015). Addition of controls helps to assess the quality of samples as well as adhering to CDC guidelines on the use of assays helps reduce false negative and false positive results.

Proteomics, a tool that utilizes genetic signatures, is currently also being explored by many research groups for early diagnosis of diseases, vaccine and antiviral production. This molecular based technique has led to early identification and quantification of functionally active biomarkers at specific time points during disease progression. The technique can be used to detect the disease process caused by the virus early after infection and to prevent the pathology observed after infection. Also, the high sensitivity and specificity of this technique reduces false positive and negative results. Garcez et al. (2016) used combined proteomics to analyze human neurospheres derived from neural stem cells exposed to Zika virus (ZIKV) isolated in Brazil and identified 500 genes and proteins altered after viral infection, which contribute to the development of microcephaly in infected babies. Amongst the proteins detected, 199 were downregulated and 259 were upregulated when mock and Zika infected neurosphere protein expression profiles were compared (Garcez et al. 2016). Although biomarkers can help for the early diagnosis of a disease, they must be adequately validated over a period of time before being used for confirmation of a disease process.

The Zika outbreak in Brazil raised a lot of concern globally as there was little information about methods for diagnosing Zika as well as vaccine and antivirals for prevention. Currently, however, samples for detection of Zika infection as well as methods for diagnosing infection have been identified, thereby improving management of infected patients. In addition, two DNA ZIKV vaccine candidates have entered phase 1 human safety testing (ClinicalTrials.gov numbers, NCT01099852 and NCT0284048) (Marston et al. 2016). Thus, getting an effective vaccine for immunological protection is closer than expected. An effective antiviral is also being researched currently at the National Institutes of Health (NIH); compounds having antiviral properties are being screened to determine their antiviral effect on Zika (Awasthi 2016).

Authors Contribution

Authors made equal contributions in writing up this review paper.

References

Aubry M, Finke J, Teissier A, Roche C, Broult J, Paulous S, Desprès P, Cao-Lormeau VM & Musso D 2015 Seroprevalence of arboviruses among blood donors in French Polynesia, 2011–2013. *Int J Infect Dis* **41** 11-12

Awasthi S 2016 Zika Virus: Prospects for the Development of Vaccine and Antiviral Agents. *Journal of Antivirals and Antiretrovirals* **8** 8–10

Parra B, Lizarazo J, Jiménez-Arango JA, Zea-Vera AF, González-Manrique G, Vargas J, Angarita JA,

Zuñiga G, Lopez-Gonzalez R, Beltran CL, Rizcala KH, Morales MT, Pacheco O, Ospina ML, Kumar A, Cornblath DR, Muñoz LS, Osorio L, Barreras P, Pardo CA 2016 Guillain-Barré Syndrome Associated with Zika Virus Infection in Colombia. *N Engl J Med* **375** 1513-1523

Besnard M., Lastère S, Teissier A, Cao-Lormeau VM, Musso D, 2014. Evidence of perinatal transmission of Zika virus, French Polynesia December 2013 and January 2014, Euro Surveill. 19, pii:20751

Brasil P, Pereira JP Jr, Moreira ME, Ribeiro Nogueira RM, Damasceno L, Wakimoto M, Rabello RS, Valderramos SG, Halai UA, Salles TS, Zin AA, Horovitz D, Daltro P, Boechat M, Raja Gabaglia C, Carvalho de Sequeira P, Pilotto JH, Medialdea-Carrera R, Cotrim da Cunha D, Abreu de Carvalho LM, Pone M, Machado Siqueira A, Calvet GA, Rodrigues Baião AE, Neves ES, Nassar de Carvalho PR, Hasue RH, Marschik PB, Einspieler C, Janzen C, Cherry JD, Bispo de Filippis AM, Nielsen-Saines K 2016 Zika Virus Infection in Pregnant Women in Rio de Janeiro. *N Engl J Med* **375** 2321-2334

Brazil Ministry of Health. Emergency Health Operations Center On Microcephaly. Monitoring of Microcephaly Cases in the Brazil. Epidemiological report 11-week epidemiological study 04/2016 (24 a1/30/2016). Brasília: Ministry of Health, 2016.

Brazil Ministry of Health. DATASUS. Anomaly or fault tab Congenital in live births. http://www2.aids.gov.br/cgi/tabcgi.exe?caumul/anoma.def (accessed Feb 4, 2016).

Buckley A, Gould EA 1988 Detection of virus-specific antigen in the nuclei or nucleoli of cells infected with Zika or Langat virus. *J Gen Virol* **69** 1913-1920

Cao-Lormeau VM, Blake A, Mons S, Lastere S, Roche C, Vanhomwegen J, Dub T, Baudouin L, Teissier A, Larre P, Vial AL, Decam C, Choumet V, Halstead SK, Willison HJ, Musset L, Manuguerra JC, Despres P, Fournier E, Mallet HP, Musso D, Fontanet A, Neil J, Ghawché F 2016 Guillain-Barr Syndrome outbreak associated with Zika virus infection in French Polynesia: a case control study. *Lancet* **387** 1531-1539

Centers for Disease Control and Prevention (CDC), 2016. Trioplex Real-time RT-PCR Assay. Retrieved from https://www.cdc.gov/zika/pdfs/trioplex-real-time-rt-pcr-assay-instructions-for-use.pdf

Cugola FR, Fernandes IR, Russo FB, Freitas BC, Dias JL, Guimarães KP, Benazzato C, Almeida N, Pignatari GC, Romero S, Polonio CM, Cunha I, Freitas CL, Brandão WN, Rossato C, Andrade DG, Faria Dde P, Garcez AT, Buchpigel CA, Braconi CT, Mendes E, Sall AA, Zanotto PM, Peron JP, Muotri AR, Beltrão-Braga PC 2016 The Brazilian Zika virus strain causes birth defects in experimental models. *Nature* **534** 267-271

Elgot J 2016. France records first sexually transmitted case of Zika in Europe. The guardian 2016. [Online] Available from: http://www.theguardian.com/world/2016/feb/27/zika-france-records-firstsexually-transmitted-case-europe [Accessed on 28 February 2016].

Fauci AS, Morens DM 2016. Zika virus in the Americas - yet another arbovirus threat. *N Engl J Med* **374** 601-604

Delvecchio R, Higa LM, Pezzuto P, Valadão AL, Garcez PP, Monteiro FL, Loiola EC, Dias AA, Silva FJ, Aliota MT, Caine EA, Osorio JE, Bellio M, O'Connor DH, Rehen S, de Aguiar RS, Savarino A, Campanati L, Tanuri A 2016 Chloroquine, an Endocytosis Blocking Agent, Inhibits Zika Virus Infection in Different Cell Models. *Viruses* **8** pii: E322

Diagnostic Tests for Zika Virus. Centre for Disease Control and Prevention. Accessed February 21, 2017.

Dick GW, Kitchen S, Haddow A 1952 Zika virus. Isolations and serological specificity. *Trans R Soc Trop Med Hyg* **46** 509–520

Driggers RW, Ho CY, Korhonen EM, Kuivanen S, Jääskeläinen AJ, Smura T, Rosenberg A, Hill DA, DeBiasi RL, Vezina G, Timofeev J, Rodriguez FJ, Levanov L, Razak J, Iyengar P, Hennenfent A, Kennedy R, Lanciotti R, du Plessis A, Vapalahti O 2016 Zika Virus Infection with Prolonged Maternal Viremia and Fetal Brain Abnormalities. *N Engl J Med* **374** 2142-2151

Duffy MR, Chen TH, Hancock WT, Powers AM, Kool JL, Lanciotti RS, Pretrick M, Marfel M, Holzbauer S, Dubray C, Guillaumot L, Griggs A, Bel M, Lambert AJ, Laven J, Kosoy O, Panella A, Biggerstaff BJ, Fischer M, Hayes EB 2009 Zika virus outbreak on Yap Island, Federated States of Micronesia. *N Engl J Med* **360** 2536-2543

Duhaime-Ross A 2016. Zika linked to more birth defects than just microcephaly. The verge, 2016. [Online] Available from: http://www.theverge.com/2016/3/8/11181088/zika-birth-defects-fetaldeath-growth-retardation-who [Accessed on 9 March 2016].

Dupont-Rouzeyrol M, O'Connor O, Calvez E, Daurès M, John M, Grangeon JP, Gourinat AC 2015 Co-infection with Zika and dengue viruses in 2 patients, New Caledonia, 2014. *Emerg Infect Dis* **21** 381-382

Fagbami AH 1979 Zika virus infections in Nigeria: virological and seroepidemiological investigations in Oyo State. *J Hyg (Lond)* **83** 213-219

Faye O, Freire CC, Iamarino A, Faye O, Oliveira JVC, Diallo M, Zanotto PMA, Sall AA 2014 Molecular evolution of Zika virus during its emergence in the 20th century. *PLOS Negl Trop Dis* **8** e2636

Gaffigan TVWR, Pecor JE, Stoffer JA, Anderson T 2016 Systematic catalog of culicidae. Suitland (MD): Smithsonian Institution; Silver Spring (MD): Walter Reed Army Institute of Research;. [cited 2016 Feb 10]. http://mosquitocatalog.org.

Garcez PP, Nascimento JM, de Vasconcelos JM, da Costa RM, Delvecchio R, Trindade P, Loiola EC, Higa LM, Cassoli JS, Vitória G, Sequeira P, Sochacki J, Aguiar RS, Fuzii HT, Bispo de Filippis AM, JLS Gonçalves Vianez Júnior, Martins-de-Souza D, Tanuri

A, Rehen SK 2016 Combined proteome and transcriptome analyses reveal that Zika virus circulating in Brazil alters cell cycle and neurogenic programmes in human neurospheres. *PeerJ Preprints* **4** e2033v1

Garcez PP, Loiola EC, Madeiro da Costa R, Higa LM, Trindade P, Delvecchio R, Nascimento JM, Brindeiro R, Tanuri A, Rehen SK 2016 Zika virus impairs growth in human neurospheres and brain organoids. *Science* **352** 816-818

Gatherer D, Kohl A 2016 Zika virus: a previously slow pandemic spreads rapidly through the Americas. *J Gen Virol* **97** 269–273

Gourinat AC, O'Connor O, Calvez E, Goarant C, Dupont-Rouzeyrol M 2015 Detection of Zika virus in urine. *Emerg Infect Dis* **21** 84-86

Hayes EB 2009 Zika virus outside Africa. *Emerg Infect Dis* **15** 1347-1350

Heymann DL, Hodgson A, Sall AA, Freedman DO, Staples JE, Althabe F, Baruah K, Mahmud G, Kandun N, Vasconcelos PF, Bino S, Menon KU 2016 Zika virus and microcephaly: Why is this situation a PHEIC? *Lancet* **387** 719-721

Ioos S, Mallet HP, Leparc Goffart I, Gauthier V, Cardoso T, Herida M 2014 Current Zika virus epidemiology and recent epidemics. *Med Mal Infect* **44** 302-307

Kraemer MU, Sinka ME, Duda KA, Mylne A, Shearer FM, Brady OJ, Messina JP, Barker CM, Moore CG, Carvalho RG, Coelho GE, Van Bortel W, Hendrickx G, Schaffner F, Wint GR, Elyazar IR, Teng HJ, Hay SI 2015 The global compendium of Aedes aegypti and Ae. albopictus occurrence. *Sci Data* **2** 150035

Kuno G, Chang GJ 2007 Full-length sequencing and genomic characterization of Bagaza, Kedougou, and Zika viruses. *Arch Virol* **152** 687-696

Lanciotti RS, Kosoy OL, Laven JJ, Velez JO, Lambert AJ, Johnson AJ, Stanfield SM, Duffy MR 2008 Genetic and serologic properties of Zika virus associated with an epidemic, Yap State, Micronesia, 2007. *Emerg Infect Dis* **14** 1232-1239

Ledermann JP, Guillaumot L, Yug L, Saweyog SC, Tided M, Machieng P, Pretrick M, Marfel M, Griggs A, Bel M, Duffy MR, Hancock WT, Ho-Chen T, Powers AM 2014 Aedes hensilli as a potential vector of Chikungunya and Zika viruses. *PLoS Negl Trop Dis* **8** e3188

Li C, Xu D, Ye Q, Hong S, Jiang Y, Liu X, Zhang N, Shi L, Qin CF, Xu Z 2016 Zika Virus Disrupts Neural Progenitor Development and Leads to Microcephaly in Mice. *Cell Stem Cell* **19** 120-126

Marrs C, Olson G, Saade G, Hankins G, Wen T, Patel J, Weaver S 2016 Zika Virus and Pregnancy: A Review of the Literature and Clinical Considerations. *Am J Perinatol* **33** 625-639

Marston HD, Lurie N, Borio LL, Fauci AS 2016 Considerations for Developing a Zika Virus Vaccine. *N Engl J Med* **375** 1209-1212

Victora CG, Schuler-Faccini L, Matijasevich A, Ribeiro E, Pessoa A, Barros FC 2016 Microcephaly in Brazil: how to interpret reported numbers? *Lancet* **387** 621-624

Miner JJ, Cao B, Govero J, Smith AM, Fernandez E, Cabrera OH, Garber C, Noll M, Klein RS, Noguchi KK, Mysorekar IU, Diamond MS 2016 Zika Virus Infection during Pregnancy in Mice Causes Placental Damage and Fetal Demise. *Cell* **165** 1081-1091

Mlakar J, Korva M, Tul N, Popović M, Poljšak-Prijatelj M, Mraz J, Kolenc M, Resman Rus K, Vesnaver Vipotnik T, Fabjan Vodušek V, Vizjak A, Pižem J, Petrovec M, Avšič Županc T 2016 Zika Virus Associated with Microcephaly. *N Engl J Med* **374** 951-958

Musso D, Gubler DJ 2016. Zika virus. *Clin Microbiol Rev* **29** 487-524

Musso D, Roche C, Nhan TX, Robin E, Teissier A, Cao-Lormeau VM 2015 Detection of Zika virus in saliva. *J Clin Virol* **68** 53-55

Musso D, Roche C, Robin E, Nhan T, Teissier A, Cao-Lormeau VM 2015 Potential sexual transmission of Zika virus. *Emerg Infect Dis* **21** 359-361

Musso D, Nhan T, Robin E, Roche C, Bierlaire D, Zisou K, Shan Yan A, Cao-Lormeau VM, Broult J 2014 Potential for Zika virus transmission through blood transfusion demonstrated during an outbreak in French Polynesia, November 2013 to February 2014. *Euro Surveill* **19** pii: 20761

Oehler E, Watrin L, Larre P, Leparc-Goffart I, Lastere S, Valour F, Baudouin L, Mallet H, Musso D, Ghawche F 2014 Zika virus infection complicated by Guillain-Barre syndrome--case report, French Polynesia, December 2013. *Euro Surveill* **19** pii: 20720

Oliveira Melo AS, Malinger G, Ximenes R, Szejnfeld PO, Alves Sampaio S, Bispo de Filippis AM 2016 Zika virus intrauterine infection causes fetal brain abnormality and microcephaly: tip of the iceberg? *Ultrasound Obstet Gynecol* **47** 6e7

Pan American Health Organization. Epidemiological update: neurological syndrome, congenital anomalies, and Zika virus infection. Jan 17, 2016. http://www.paho.org/hq/index.php?option=com_content&view=article&id=11599&Itemid=41691 (accessed Feb 9, 2016).

Petersen LR, Jamieson DJ, Powers AM, Honein MA 2016 Zika virus *N Engl J Med* **374** 1552-1563

Pierson TC, Diamond MS 2013 Flaviviruses. In DM Knipe, PM Howley (eds.), *Fields virology*, 6th ed., Lippincott Williams and Wilkins, Philadelphia, p. 747-794

Plourde AR, Bloch EM 2016 A Literature Review of Zika Virus. *Emerg Infect Dis* **22** 1185-1192

Shao Q, Herrlinger S, Yang SL, Lai F, Moore JM, Brindley MA, Chen JF 2016 Zika virus infection disrupts neurovascular development and results in postnatal microcephaly with brain damage. *Development* **143** 4127-4136

Schuler-Faccini L, Ribeiro EM, Feitosa IM, Horovitz

DD, Cavalcanti DP, Pessoa A, Doriqui MJ, Neri JI, Neto JM, Wanderley HY, Cernach M, El-Husny AS, Pone MV, Serao CL, Sanseverino MT; Brazilian Medical Genetics Society–Zika Embryopathy Task Force 2016 Possible association between Zika virus infection and microcephaly – Brazil, 2015. *MMWR Morb Mortal Wkly Rep* **65** 59-62

Simpson DI 1964 Zika virus infection in man. *Trans R Soc Trop Med Hyg* **58** 335-338

Thomas DL, Sharp TM, Torres J, Armstrong PA, Munoz-Jordan J, Ryff KR, Martinez-Quiñones A, Arias-Berríos J, Mayshack M, Garayalde GJ, Saavedra S, Luciano CA, Valencia-Prado M, Waterman S, Rivera-García B 2016 Local transmission of Zika virus – Puerto Rico, November 23, 2015–January 28, 2016. *MMWR Morb Mortal Wkly Rep* **65** 154-158

World Health Organization, 2016 (http://apps .who.int/iris/bitstream/ 10665/204526/ 2/zikasitrep_4Mar2016_eng.pdf) http://doi.org/10.1056/NEJMp1607762

Zanluca C, De Melo VCA, Mosimann ALP, Dos Santos GIV, dos Santos CND, Luz K 2015 First report of autochthonous transmission of Zika virus in Brazil. *Memorias Do Instituto Oswaldo Cruz* **110** 569–572

Zika virus outbreak. Micronesia (Yap). Suspected. ProMed June 27, 2007 [cited 2008 May 8]. Available from http://www.promed.mail.org, archive no.: 20070627.2065.

13

Human hair follicle biomagnetism: potential biochemical correlates

Abraham A. Embi[1] and Benjamin J. Scherlag[2]

[1] 13442 SW 102 Lane Miami, FLA, 33186, USA
[2] Heart Rhythm Institute, University of Oklahoma Health Sciences Center, Oklahoma City, Oklahoma, USA

Correspondence should be addressed to Abraham A. Embi; E-mail: embi21@att.net

Abstract

Background: The S100 protein family is linked to energy transfer in cells of vertebrates at a molecular level. This process involves the electron transfer chain and therefore, as inferred from Faraday's Law, electron movement will induce electromagnetic fields (EMFs). Biological entities emit photoelectrons that can be tracked and visualized by small paramagnetic nano-sized iron particles.
Methods: We have developed an optical microscopic approach for imaging electromagnetic activity of hair follicles utilizing nano-sized iron particles (mean diameter 2000nm) in Prussian Blue Stain solution (PBS Fe 2000).
Results: We found that the human hair follicle emits electromagnetic fields (EMFs) based on metabolic activity within the follicle, which is associated with the activity of selective S-100 proteins.
Conclusions: Our results link the molecular biochemical energy associated with the S100 family of proteins and biomagnetism of human hair follicles.

Introduction

In the first report of EMF measurements made from the human heart, Baule & McFee (1963) used two large coils placed over the chest to cancel ambient magnetic interference. Better resolution and less noise was later achieved by Cohen and his associates who recorded EMFs from the brain (Cohen 1972) and the heart (Cohen & Kaufman 1975, Cohen et al. 1983) using a superconducting quantum interference device (SQUID). We hypothesize that all living matter maintains an intrinsic, electromagnetic homeostatic mechanism at quantum levels, based on biochemical and biophysical processes. From sub-atomic and atomic interactions, there are ordered amplifications manifesting as metabolic activity to maintain homeostasis. Photon-phonon and photon-photon transductions, i.e. piezo-electricity and photoelectricty may be underlying mechanisms to explain bio-electromagnetic order and balanced function in living things. Utilizing the intrinsic paramagnetic properties of fine iron particles, and the imaging characteristics of iron by Prussian Blue stain, a solution was developed which could be applied to human hairs *ex vivo*. Magnetic energy was detected and visualized by the stained, aggregated iron particles applied to the follicle and shaft of human hairs. Since some of the S100 family of proteins involved in energy exchanges have been localized in anatomically discreet areas of the human hair follicle (Mitoma *et al.* 2014), we proceeded to attempt a correlation between such

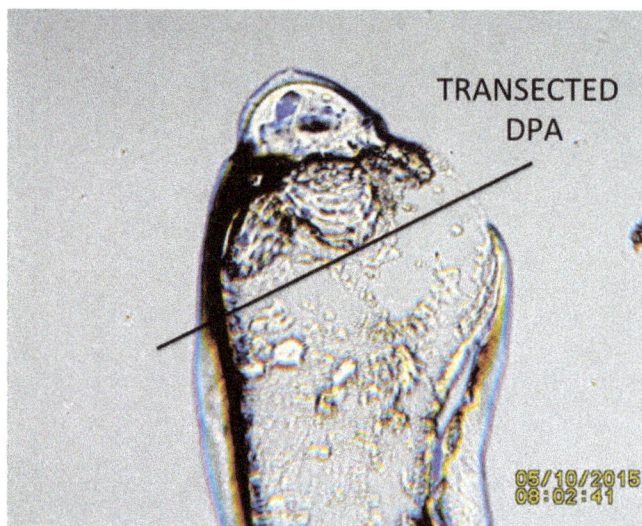

Figure 1. Microphotograph (x14 magnification) showing sebum imprint of human hair follicle (bulb area), depicting the dermal papilla area (DPA). The transected area is shown above the straight line, similarly to Figure 3. The plucked human hair is covered by sebum that is left behind after the hair is removed.

Figure 2. Microphotograph (X10 magnification) of a SSP of PBSFe2000 of human hair dry field. Living matter (hair follicle) is seen triggering crystallization. A depicts the hair bulb; B depicts the crystals adhering to the DP area and C depicts the layered crystallization. Please note that the KFeCN crystals triggered by the EMFs remain in front of the hair follicle which is the putative S100, S4, S6 localization (see Supplementary Video 2).

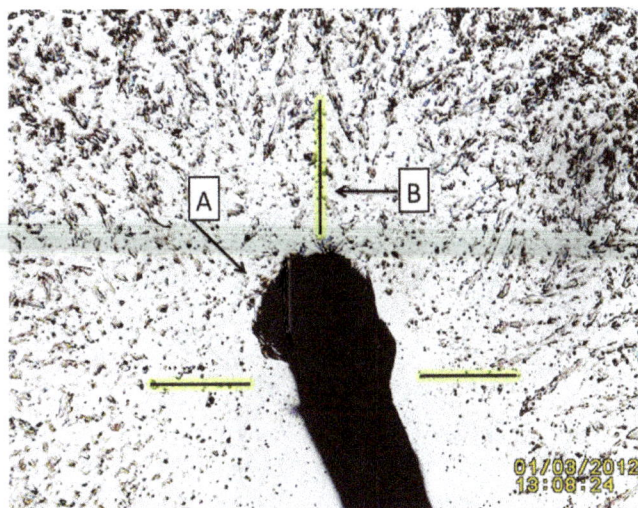

Figure 3. SSP PBSFe2000 after evaporation. Microphotohraph (x10 magnification) showing the human hair follicle post-blade excision of the DP area. Notice the absence of heavy crystallization in front of the hair (facing the evaporation line). This absence correlates with the lack of EMFs strong enough to trigger heavy crystallization. This finding is attributed to the removal of the DP as illustrated in Figure 1.

areas; specifically, the follicular tissue and electromagnetic activity.

Materials and Methods

Preparation of the iron containing solution

A fine iron particle solution was prepared by mixing several grams of powdered iron filings (Edmond Scientific Co., Tonawanda, NY) in 200 cc of deionized water (resistivity, 18.2 MΩ.cm). After standing for several hours the supernatant was carefully decanted for sizing of the iron nano-sized iron particles. The particle size and distribution of the nanoparticles from the supernatant was determined using dynamic light scattering (DLS) and the zeta potential using phase analysis light scattering by a Zeta potential analyzer (ZetaPALS, Brookhaven Instruments Corp, Holtsville, NY). For sizing, 1.5 ml of the solution in de-ionized water was scanned at 25 °C and the values were obtained in nanometers (nm). A similar aliquot of the fine iron particle solution was scanned for 25 runs at 25 °C. for determining zeta potentials. Zeta potential values were displayed as millivolts (mV). Using a transfer pipette, aliquots of the solution containing the iron particles (with a mean particle size of 2000 nm) were combined with Prussian Blue Stain (PBS, 2.5% potassium Ferrocyanide and 2.5% hydrochloric acid).

The single slide preparation (SSP)

Hairs (n = 4) in the Telogen phase were individually placed on a clean slide (size 25 x 75 x1mm). The

freshly plucked human hairs (from the author's scalp) were positioned in the center of the slide. This was facilitated by the inherent stickiness of the hair root. Three drops of the PBSFe2000 solution were placed in the center of the aforementioned clean glass microscopic slide. Care was also taken to cover the root and shaft area and then the liquid was allowed to evaporate. The viewing and event recordings of the evaporation (still pictures or videos) of the slides were done in the normal mode at X10 magnification with a video microscope (Celestron LCD Digital Microscope II model #4341; Torrance ,California, USA).

Transected Hair follicles

Hairs were placed on a clean slide (size 25 x 75 x 1mm). The freshly plucked *ex vivo* hairs were individually positioned in the center of the slides. At this point with the aid of a magnifying glass and a razor blade, the hair follicle was transected at approximately the distal part of the root (Figure 1).

Upon viewing in the microscope (X10 magnification), the transections that did not satisfy our criteria (n=13) of separating the distal bulb from the follicle were discarded. Four samples (successfully transected) (n = 4) were studied by applying 3 drops of the PBSFe2000 solution over the follicle and shaft, mounted on a clean slide and then the liquid was allowed to evaporate. The viewing and event recordings of the evaporation were handled similarly in a series of controls (n = 2), using a non-animate source of EMF (see ancillary experiments in the Results section).

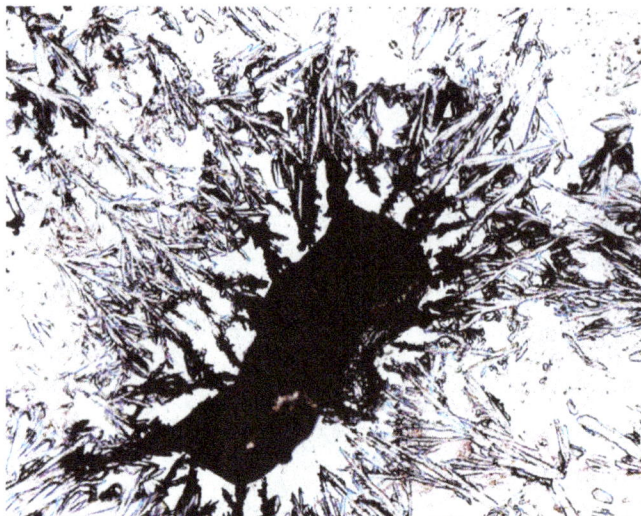

Figure 4. SSP of PBSFe2000 dry field. Microphotograph (x14 magnification) of excised tissue from the human hair follicle shown in Figure 1. Anatomically, this tissue corresponds to the DP area. Notice the heavy crystals attracted to the tissue surface is attributed to the maintenance of metabolism with the presence of S100 proteins and associated EMFs.

Results

In the wet field of the follicles mounted in a SSP, we consistently noticed iron particles circulating around the distal part of the hair follicle (bulb) (n = 4), similarly to our previous findings (Embi *et al.* 2015a, b). The circulating iron particles were interpreted as the presence of EMFs emanating from the hair root. In the dry field after evaporation all control hairs showed layered crystallization of the PBSFe2000. These layers correlated with the EMFs emitted by the follicle. Anatomical areas of the hair such as the hair shaft covered by cuticles expressed weak S100 A2 as well as showing the absence of layered crystallization. In contrast, the distal root area where the dermal papilla (DP) is located has been shown to express S100 A4 and S100 A6 proteins (Mitoma *et al.* 2014). In all our experiments this area consistently triggered heavy concentration of layered crystallization directly adherent to the bulb as seen in Figure 2.

In other experiments, the hair follicle was cut or transected proximal to the DP, thereby separating the DP from the rest of the follicle. When subjected to the same SSP procedures, the rest of the cut follicle (minus the DP) exhibited no crystals adhering to its surface (Figure 3). On the other hand, the isolated DP tissue fragment showed a clear intense crystallization pattern, which adhered to its surface (Figure 4).

Ancillary Experiments
Inanimate magnetized material
When a non-living entity such as a magnet fragment

which is absent of any S100 proteins, was cut to a similar size of a hair follicle and placed in a SSP with PBSFe200, layered crystallization ensued (n = 2). There were no crystals which adhered to the part of the magnet first encountering the evaporation line, which, by its location in this model mimics the excised DP. This ancillary experiment served as a control since non-animate EMFs from the magnet trigger an orderly display of crystallization lines, stopping short of touching the magnetic fragment itself. (Figure 5 and Supplementary Video 1)

Inanimate nonmagnetic material
When a non-living entity such as a dead wood fragment which is absent of any S100 proteins, was cut to a similar size of a hair follicle and placed in a SSP with PBSFe200, layered crystallization failed to ensue. This ancillary experiment served as a negative control. The absence of EMFs is attributed to the failed crystallization. See Figure 6 below and Supplementary Video 2.

Discussion

Major findings
Using nano-sized iron particles (having an average diameter of 2000 nm) mixed with a Prussian blue solution for staining iron, we demonstrated putative images of the EMFs emanating from the DP area of the human hair. Specifically, iron particle aggregates outlined the

Figure 5. Demonstration of EMFs triggering crystallization. Magnet fragment in SSP PBSFe2000 dry field, showing the orderly EMFs emanating from a non-living magnetic source. Although the inanimate magnetized object triggered crystallization there was an absence of heavy crystals adhering to the tip of the magnet (arrow). Also note that the further distance from the magnetic source the wider the gap between the curved crystallization lines (see Supplementary Video 1).

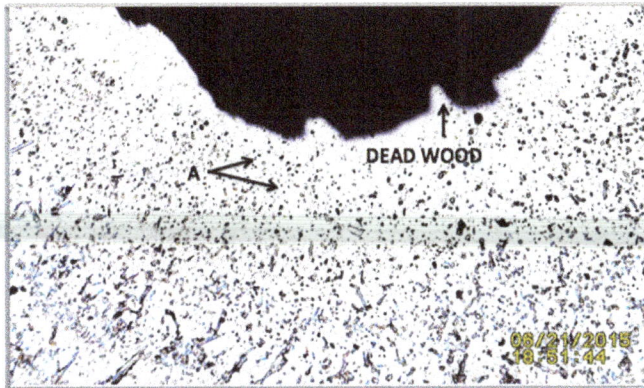

Figure 6. A SSP of a dead wood fragment (toothpick) in PBSFe2000, as a negative control. A depicts iron particles. Notice that an inanimate (dead) material is unable to trigger large crystals formation. This is attributed to the absence of EMFs (see Supplementary Video 2).

anatomical areas where S100 proteins have been located.

Background

A metabolic process common in animals (cellular respiration) involves the electron transport chain. This process consists of electrons transferred along a series of electron donor and receptor compounds coupled to a proton (H^+) gradient across cell membranes. The ensuing charge differential or voltage is used to drive energy production in the form of adenosine triphosphate (ATP). As inferred by Faraday's law, electron movement within cells will induce an electromagnetic field (EMF) emanating from both plant (Scherlag *et al.* 2015) and animal tissues (Embi *et al.* 2015a, b), acting as electrical conductors.

S100 proteins are only expressed in vertebrates. They consist of multiple constituents, which have both intracellular as well as extracellular functions. Furthermore, S100 are involved in aspects of regulation and energy metabolism. Those functions range from energy metabolism to cell proliferation, among others. In regard to energy metabolism, we hypothesize that the follicular metabolism is not only the source of electromagnetic energy but also free electrons, i.e. photoelectrons, which emanate from the follicle and can be tracked by the paramagnetic nano-sized iron particles.

Since our simple approach using nano-sized iron particles mixed with Prussian Blue Stain detects EMFs, is relatively inexpensive and provides an acute procedure, i.e., crystallization within <2 hours, it could easily be extended to various types of cellular research with minimal investment. For example, the S100 protein family has been reported to be present in multiple stages of tumorigenesis and cancer progression (Bresnick *et al.* 2015). Further research, using the

methodology described in the present report, might be useful in evaluating chemotherapeutic drug effects by EMFs changes shown by cells and tissues.

Conclusion

Our results link the molecular biochemical energy associated with the S100 family of proteins and biomagnetism.

References

Baule GM & McFee R 1963 Detection of the magnetic field of the heart. *American Heart Journal* **66** 95-96

Bresnick AR, Weber DJ & Zimmer DB 2015 S100 proteins in cancer. *Nat Rev Cancer* **15** 96-109

Cohen D 1972 Magnetoencephalography: Detection of the Brain's electrical activity with a superconducting magnetometer. *Science* **175** 664-666

Cohen D & Kaufman LA 1975 Magnetic determination of the relationship between the ST segment shift and the injury current produced by coronary occlusion. *Circ Res* **36** 414-424

Cohen D, Savard P, Rifkin RD, Lepeshkin E & Strauss WE 1983 Magnetic measurement of S-T and T-Q segment shifts in humans. Part II; Exercise-induced S-T segment depression. *Circ Res* **53** 274-279

Donato R, Cannon BR, Sorci G, Riuzzi F, Hsu K, Weber DJ & Geczy CL 2013 Functions of S100 proteins. *Curr Mol Med.* **13** 24-57

Embi AA, Jacobson JI, Sahoo K & Scherlag BJ 2015a Demonstration of Inherent Electromagnetic Energy Emanating from Isolated Human Hairs. *Journal of Nature and Science* **1** e55

Embi AA, Jacobson Jl, Sahoo K & Scherlag BJ 2015b Demonstration of Electromagnetic Energy Emanating from Isolated Rodent Whiskers and the Response to Intermittent Vibrations. *Journal of Nature and Science* **1** e52

Mitoma C, Kohda F, Mizote Y, Miake A, Ijichi A, Kawahara S, Kohno M, Sonoyama H, Mitamura Y, Kaku Y, Inoue H, Sasaki Y, Ohno F, Okabe N, Take N, Mizote M, Masuda A & Furue M 2014 Localization of S100A2, S100A4, S100A6, S100A7, and S100P in the human hair follicle. *Fukuoka Igaku Zasshi* **105** 148-156

Scherlag BJ, Huang B, Zhang L, Sahoo K, Towner R, Smith N, Embi AA & Po SS 2015 Imaging the electromagnetic field of plants (Vigna radiata) using iron particles: Qualitative and quantitative correlates. *Journal of Nature and Science* **1** e61

14

Crystal structure and molecular docking studies of octahydrocycloocta[b]pyridine-3-carbonitriles as potential inhibitors against *Mycobacterium tuberculosis*

RA Nagalakshmi[1], J Suresh[1], S Maharani[2] and R Ranjith Kumar[2]

[1]Department of Physics, The Madura College (Autonomous), Madurai, India
[2]Department of Organic Chemistry, School of Chemistry, Madurai Kamaraj University, Madurai, India

Correspondence should be addressed to Janakiraman Suresh; E-mail: ambujasureshj@yahoo.com

Abstract

The compounds 1-benzyl-2-imino-4-p-tolyl-1,2,5,6,7,8,9,10-octahydrocycloocta[b]pyridine-3-carbonitrile (Ia) and 1-benzyl-2-imino-(4-methoxyphenyl-1,2,5,6,7,8,9,10-octahydrocycloocta{b]pyridine-3-carbonitrile (Ib) were synthesized. The crystal structures of the compounds were determined by single crystal X ray diffraction. The compounds $C_{26}H_{27}N_3$ (Ia) and $C_{26}H_{27}N_3O$ (Ib) crystallize in the triclinic system (a = 10.2304(4) Å, b = 10.5655(4) Å, c = 11.8271(4) Å, α = 101.755(2) °, β = 106.934(2) °, γ = 114.071(2) ° and Z = 2 for I(a) and a = 10.2738(4) Å, b = 11.1654(5) Å, c = 11.4162(4) Å α = 98.549(2) °, β = 106.183(2) °, γ = 117.070(2) ° and Z = 2 for I (b)). In both compounds (Ia) and (Ib) the pyridine ring adopts a planar conformation and the cyclooctane ring adopts a twisted boat chair conformation. The synthesized compounds were screened for their antibacterial activities against the enzyme enoyl acyl carrier protein reductase, which is involved in the fatty acid biosynthesis of the mycobacterial cell wall. Both compounds showed good antibacterial activities. The synthesis of the compounds, their structure determination, their conformation, their intra- and intermolecular interactions and docking study results are given.

Introduction

Tuberculosis (TB) is an infectious disease caused by *Mycobacterium tuberculosis* and is the second leading cause of death in developing countries. Primary infection of *M. tuberculosis* takes place within the lungs but can involve any organ (Dietrich & Doherty 2009). Multidrug resistant tuberculosis and extensively drug-resistant tuberculosis have highlighted the deficiency of current standard treatment regimens. The bacterial drug resistance is mainly due to the slow growth, unusual cell envelope and genetic homogeneity (Besra *et al.* 1994). The cell wall of *M. tuberculosis* contains certain unique long chain (C70 to C90), α-alkyl and β-hydroxy fatty acids, referred to as mycolic acids. Its unusual lipid bilayer, formed by the latter, together with glycolipids and lipoprotein makes the cell wall tough, and the drugs normally used for the treatment of the infection, partially effective (Pieters 2008). Hence, there is a need to develop new drug candidates which have anti-mycobacterial activity.

The heterocycles containing nitrogen show good anti-mycobacterial activities (Sriram *et al.* 2006). Many naturally occurring and synthetic compounds containing the pyridine scaffold show important pharmacological properties (Temple *et al.* 1992). Cyanopyridine derivatives possess anti-microbial activities (Dandia *et al.* 1993). As a continuation of our work for developing new drugs for *M. tuberculosis* (Suresh *et al.* 2012), the 1-benzyl-2-imino-4-p-tolyl-1,2,5,6,7,8,9,10-octahydrocycloocta[b]pyridine-3-carbonitrile (Ia) and 1-benzyl-2-imino-(4-methoxyphenyl-1,2,5,6,7,8,9,10-octahydrocycloocta[b] pyridine-3-carbonitrile (Ib) compounds were synthe-

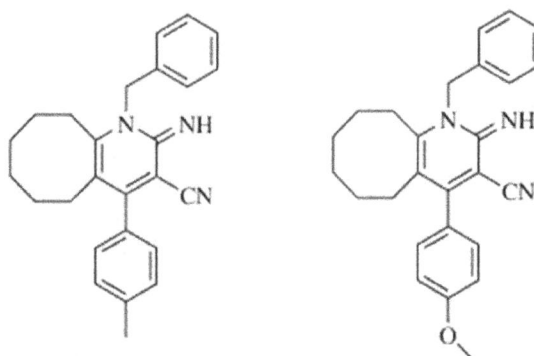

Figure 1. Chemical diagram of the molecule (Ia) (left) and (Ib) (right).

Table 1. The crystal data, intensity data collection, structure solution and structure refinement parameters of compounds (Ia) and (Ib).

Empirical formula	$C_{26} H_{27} N_3$	$C_{26} H_{27} N_3 O$
Formula weight	381.51	397.51
Temperature	293(2) K	293(2) K
Wavelength	0.71073 Å	0.71073 Å
Crystal system /space group	Triclinic/P-1	Triclinic/ P-1
Unit cell dimensions	a = 10.2304(4) Å	a = 10.2738(4) Å
	b = 10.5655(4) Å	b = 11.1654(5) Å
	c = 11.8271(4) Å	c = 11.4162(4) Å
	a = 101.755(2)°	a = 98.549(2)°
	b = 106.934(2)°	b = 106.183(2)°
	g = 114.071(2)°	g = 117.070(2)°
Volume	1037.69(7) Å3	1060.20(7) Å3
Z/ Density (calculated)	2/ 1.221 Mg/m3	2/1.245 Mg/m3
Absorption coefficient	0.072 mm-1	0.077 mm-1
F(000)	408	424
Crystal size	0.21x0.19 x0.18 mm3	0.21 x 0.19 x 0.18 mm3
Theta range for data collection	2.27 to 25.50°.	2.16 to 25.49°.
Limiting indices	-12<=h<=12, -12<=k<=12, -14<=l<=14	-12<=h<=12, -13<=k<=13, -13<=l<=13
Reflections collected	23565	25081
Independent reflections	3865 [R(int) = 0.0272]	3947 [R(int) = 0.0302]
Completeness to theta = 25.50°	100.0 %	99.9 %
Refinement method	Full-matrix least-squares on F2	Full-matrix least-squares on F2
Data / restraints / parameters	3865 / 2 / 268	3947 / 16 / 311
Goodness-of-fit on F2	1.052	1.035
Final R indices [I>2sigma(I)]	R1 = 0.0453, wR2 = 0.1175	R1 = 0.0474, wR2 = 0.1258
R indices (all data)	R1 = 0.0710, wR2 = 0.1335	R1 = 0.0761, wR2 = 0.1463
Extinction coefficient	0.020(3)	0.034(4)
Largest diff. peak and hole	0.275 and -0.177 e.Å-3	0.197 and -0.238 e.Å-3

sized and their structural characteristics were studied using single crystal X-ray diffraction. Furthermore, they were screened for anti-mycobacterial activities using Autodock.

Synthesis

Preparation of compound (Ia)

A mixture of cyclooctanone (1 mmol), 4-methyl benzaldehyde (1 mmol) and malononitrile (1 mmol) were taken in ethanol (10 mL); pTSA (0.5 mmol) was subsequently added. The reaction mixture was heated under reflux for 2-3 h. After completion of the reaction (TLC), the reaction mixture was poured into crushed ice and extracted with ethyl acetate. The excess solvent was removed under vacuum and the residue was sub-

jected to column chromatography using petroleum ether/ethyl acetate mixture (97:3 v/v) as eluent to afford pure product. The pure product was crystallized using ethyl acetate, as solvent. The melting point was 501 K and the yield was 69%.

Preparation of compound (Ib)

A mixture of cyclooctanone (1 mmol), 4-methoxy benzaldehyde (1 mmol) and malononitrile (1 mmol) were taken in ethanol (10 mL); pTSA (0.5 mmol) was subsequently added. The reaction mixture was heated under reflux for 2-3 h. After completion of the reaction (TLC), the reaction mixture was poured into crushed ice and extracted with ethyl acetate. The excess solvent was removed under vacuum and the residue was subjected to column chromatography using petroleum ether/ ethyl acetate mixture (97:3 v/v) as eluent to afford pure product. The pure product was crystallized using ethyl acetate, as solvent. The melting point was 511 K and the yield was 70%.

Structure Determination and Refinement

Single crystals of suitable sizes of both compounds were selected for crystal structure determination. The intensity data of the crystals were collected using a Bruker AXS Kappa APEX II single crystal CCD Diffractometer equipped with graphite-monochromated MoKα radiation (λ=0.71073Å) at room temperature. The data collection, data reduction and absorption corrections were carried out using the APEX2, SAINT-Plus and SADABS programs (Bruker 2004). The trial structures of both the compounds were obtained using SHELXL-97. The structures were refined by the full-matrix least-squares method using SHELXL-97 (Sheldrick 2008). The non-hydrogen atoms were also refined with independent anisotropic displacement parameters using the same program (Sheldrick 2008). The NH hydrogen atoms were located from a difference Fourier map and refined isotropically. The other H atoms were placed in calculated positions and allowed to ride on their carrier atoms with C---H = 0.93 Å (aromatic CH), 0.96 Å (methyl CH_3) or 0.97 Å (methylene CH_2), and N---H = 0.86 Å. Isotropic displacement parameters for H atoms were calculated as $U_{iso} = 1.5U_{eq}(C)$ for CH_3 groups and $U_{iso} = 1.2U_{eq}$ (carrier atom) for all other H atoms. The methoxy substituted molecule was disordered. The carbon atoms C8 and C9, with their corresponding hydrogen atoms of the cyclooctane ring of the compound (Ib) are disordered with occupancy factors 0.33485:0.60186, respectively. The disordered carbon atoms of the cyclooctane ring were refined using SHELXL-97 (Sheldrick 2008).

Docking Studies

Isoniazid (INH) was initially found to be active against tuberculosis in 1952, while its mode of action was still unknown (Bernstein et al. 1952, Fox 1952). It was later shown that INH inhibits the biosynthesis of mycolic acid, that covers the surface of mycobacteria (Takayama et al. 1972). There is strong evidence that this is the pathway that exerts the main effect of INH treatment (Winder 1982, Quemard et al. 1991). Using gene transfer systems for mycobacteria, a gene called InhA was identified as a target for INH (Banerjee et al. 1994). These studies revealed that InhA preferentially catalyzes the NADH-dependent reduction of 2-trans-enoyl-ACP molecules with 16 or more carbons (Dessen et al. 1995, Quemard et al. 1995). This serves as the last step in fatty acid elongation. Despite the increasing numbers of primary and secondary resistance, isoniazid remains one of the key drugs to treat tuberculosis (Cohn et al. 1997). Further studies show that INH can penetrate the mycobacterial envelope through passive diffusion; it then inhibits the InhA enzyme by covalently attaching to NADH within the protein active site (Rozwarski et al. 1998). Additionally, inhibition of InhA blocks mycolic acid biosynthesis, eventually leading to cell death (Vilcheze et al. 2000). The enoyl-acyl carrier protein reductase from Mycobacterium tuberculosis (InhA) is a fundamental target for antituberculosis intervention (Dessen et al. 1995, Rozwarski et al. 1998).

Nowadays, even though the antibacterial effects of isoniazid and triclosan (a newer InhA inhibitor) establish the suitability of the target for drug intervention, these compounds themselves have several limitations, including resistance and suboptimal bioavailability (Wang et al. 2004). Here, we report a new chemically distinct series, 1-benzyl-2-imino-4-(4-methylphenyl)-1, 2,5,6,7,8,9,10-octahydrocycloocta[b]

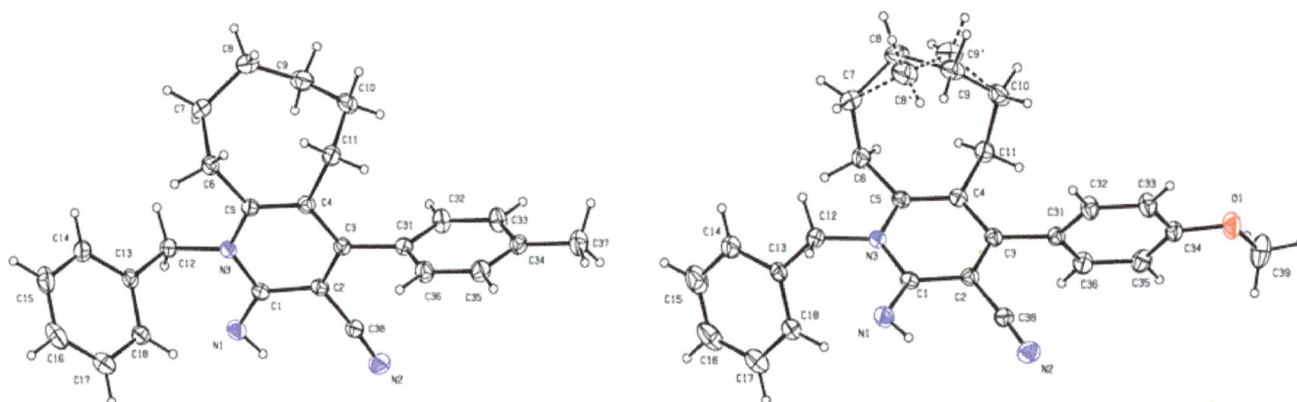

Figure 2. The molecular structure of compounds **(Ia)** (left) and **(Ib)** (right), showing the atom numbering scheme. Displacement ellipsoids are drawn at 20% probability level, using Platon (Spek 2008).

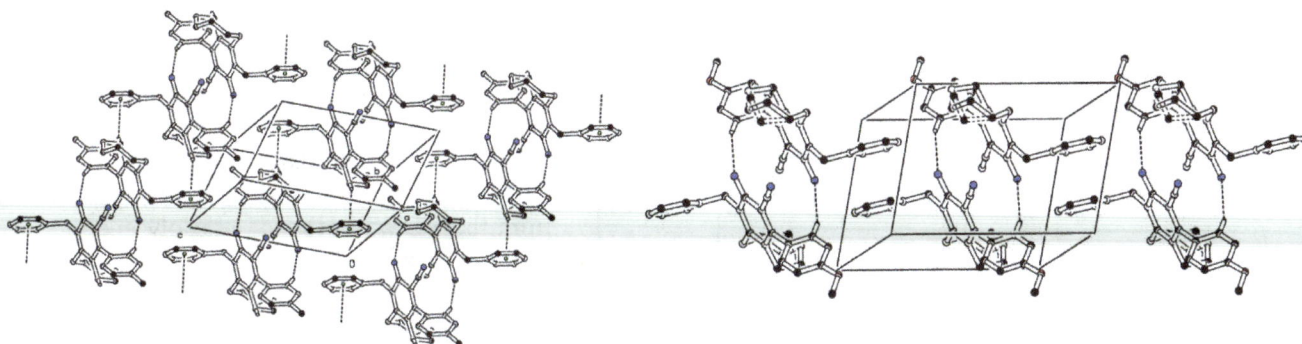

Figure 3. The inversely related molecules form a ring motif; they are further linked through C-H...π (left) and van der walls (right) interactions in the case of compound (Ia) (left) and (Ib) (right).

pyridine-3-carbonitrile, which has anti-mycobacterial activity. We have attempted, aided by a docking approach, to elucidate the extent of specificity of enoyl acyl carrier protein towards synthesized compounds, as anti-tubercular agents. The coordinates of enoyl acyl carrier protein (2NSD) with Nad were then retrieved from the RCSB protein data bank (http://www.pdb.org). The protein structure was cleaned using the whatif online server. Enoyl acyl carrier protein's active site pocket was identified using the online server Computed Atlas of Surface Topography (CASTp). The preparation of the protein and ligand input structures and the definition of the binding sites were carried out under a GRID-based procedure. A rectangular grid box was constructed over the protein with grid points 90X90X90 Å^3 separated by 0.375 Å under the docking procedure. The lowest energy cluster returned by AutoDock for each compound was considered and used for further analysis. Consequently

runs were setup to 100 for each inhibitor. All other parameters were maintained at their default settings. All docking result visualizations are depicted using the PDBsum online server.

Results and Discussion

The molecular structures of compounds (Ia) and (Ib) are shown in Figure 2. In both compounds, the cyclooctane ring adopts a twisted boat-chair conformation. The pyridine ring is planar with an r.m.s. deviation of 0.0041 (1) Å in (Ia) and 0.0085 (1) Å in (Ib). The N1 atom is deviating by -0.0446 (1) Å from the mean plane of the pyridine ring in compound (Ia) and 0.393 (1) Å in compound (Ib). The aromatic ring (C31-C36) and phenyl rings (C13-C18) are planar with an r.m.s. deviation of 0.0059 Å and 0.0070 Å in compound (Ia) and in compound (Ib). The sum of the angles around atom N3 (360°) indicates that the atom N3 is in sp^2 hybridization in both compounds. The nitrile atoms C38 and N2 are displaced from the mean plane of the pyridine ring by 0.0384(1) Å and 0.0612(1) Å, respectively, in compound (Ia) and -0.0090(1) Å and 0.0050(1) Å, in compound (Ib). The methyl benzene ring attached to the pyridine ring is in an (-) anti-clinal conformation with torsion angle C2--C3--C31--C36 = -98.5(2) ° in compound (Ia) and in an (+) anti-clinal conformation with torsion angle C2--C3--C31--C36 = 99.8(2) ° in compound (Ib). The torsion angle C32--C34--C33--C37 = -179.3° (2) indicates the methyl carbon atom attached to the benzene ring is in an (-) anti peri planar conformation in compound (Ia). The plane formed by the cyano group is twisted away from the plane of pyridine ring in both compounds; 50.42 (1)° in compound (Ia) and 36.20 (1)° in compound (Ib).

Crystal Packing

In the crystal structure of the compound (Ia) the atom C32 of the aryl ring is involved in the intermolecular

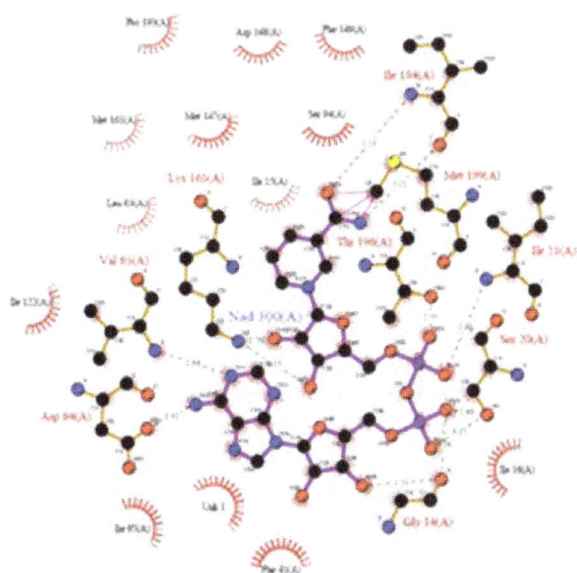

Figure 4. Enoyl acyl carrier protein reductase interactions for compound (Ia).

Table 2. Hydrogen bonds [Å and °] of compound (Ia). Cg1 is the centroid of the (C13-C18) phenyl ring.

D-HA	d(D-H)	...d(HA)	...d(DA)	<DHA
C(32)-H(32)...N(1)[i]	0.93	2.61	3.473(2)	154
C(11)-H(11)...Cg1[ii]	0.97	2.93	3.583(2)	126

Symmetry transformations used to generate equivalent atoms:
(i)   (- x, - y, 1 – z)
(ii) (1 - x, 1 – y, 1 – z)

interaction C32—H32···N1[i] (with the N1 atom) and forms a ring motif $R_2^2(14)$ (Bernstein *et al.* 1995) along the b axis [symmetry code: (i) (- x, - y, 1 – z)]. These adjacent ring motifs are linked together by C-H...π[ii] interactions [symmetry code: (ii) (1 - x, 1 - y, 1 – z)], (Figure 3). Furthermore, in the crystal structure of compound (Ib) the intermolecular interaction C32—H32···N1[i] with the N1 atom forms a ring motif R_2^2 (14) (Bernstein *et al.* 1995) along the b-axis [symmetry code: (i) (1 - x, 1 - y, 1 – z)]. These adjacent ring motifs are linked together by van der Walls interactions (Figure 3).

The C—N distance is longer [3.473 (3) Å] in compound (Ia) than in compound (Ib) [3.465 (3) Å] even though the C—H···N angle [154 (2) °, 154 (17) °] is similar in both structures. This shows that the 4-methoxy substituent forms a stronger hydrogen bond than the methyl substituent. The intermolecular hydrogen bond geometry of compound (Ia) and compound (Ib) is listed in Table 2.

Docking analysis

The target protein structure of 2NSD with Nad was docked with the synthesized compounds using Auto-Dock4 version 2 (Goodsell 1998). The target protein contains two monomers. Only one monomer was selected for docking studies (Figure 4).

Table 3. Hydrogen bonds [Å and °] of compound (Ib)

D-HA	d(D-H)	...d(HA)	...d(DA)	<DHA
C(18)-H(18)...N(1)	0.93	2.61	3.284(3)	130
C(32)-H(32)...N(1)[i]	0.93	2.60	3.464(3)	154

Symmetry transformation used to generate equivalent atoms:
(i)   (1 - x, 1 - y, -z)

The synthesized compounds were found to display good binding affinity to the receptor as shown in Table 3 and in Figure 6, with minimum binding energy. From Table 4, it was found that the stabilized binding interactions of our compounds are due the strong influence from the VDW + Hbond + desolv energy, than the electrostatic interactions.

The hydrophobic pocket of InhA is flexible whereas its hydrophilic pocket appears more rigid for the binding of inhibitors (Punkvang *et al.* 2010). InhA inhibitors may interact with its hydrophilic pocket formed by the backbone of Gly [96], Phe [97] and Met [98]. Hydrophobicity is favourable for the MIC value but its appears to be unfavorable for the activity against *M. tuberculosis*. Hence, there should be a fine balance between the hydrophobic and hydrophilic properties of direct InhA inhibitors, to ensure their effectiveness against both InhA and *M. tuberculosis*. These structural requirements are favorable not only for MIC values but also assist the binding of InhA inhibitors in the hydrophilic InhA pocket. The key structural feature of such InhA inhibitors should be taken. Thus, having a bulky group encompassing suitable hydrophobic and hydrophilic properties could improve the MIC value of direct InhA inhibitors.

In compound (Ia), the nitrogen atom of the pyridine ring is hydrogen bonded to GLY [96], one of the catalytic residues in the InhA active site. The hydrogen bond is formed in the hydrophilic pocket and has hydrophobic interactions with the hydrophobic pocket. Hence it shows good anti-mycobacterial activity.

Rozwarski *et al.* (1999) determined the crystal structure of InhA with a C fatty acyl substrate; they reported that hydrophobic amino acids of the loop are important for substrate binding into the cavity. Of note, the last few carbon atoms of the substrate were shown to interact with Ala [198], Met [199], and Ile [202], which are all hydrophobic. A fatty acid shorter than 16 carbons might not fit well in the enzyme, as there will be missing interactions with Ala [198], Met [199], and Ile [202].

There is strong evidence that the interactions with the three amino acids are important determinants for loop ordering. Our inhibitor (Ib) is able to directly interact with one of these residues (Ala [198]), generating a defined loop structure and having hydrophobic interactions to the important loop residues of InhA, leading to a slow tight binding inhibition. Inhibitor residence time is an important factor for *in vivo* drug activity (Lavie *et al.* 1997, Lewandowicz *et al.* 2003, Tummino *et al.* 2008). Slow onset inhibitors will spend a

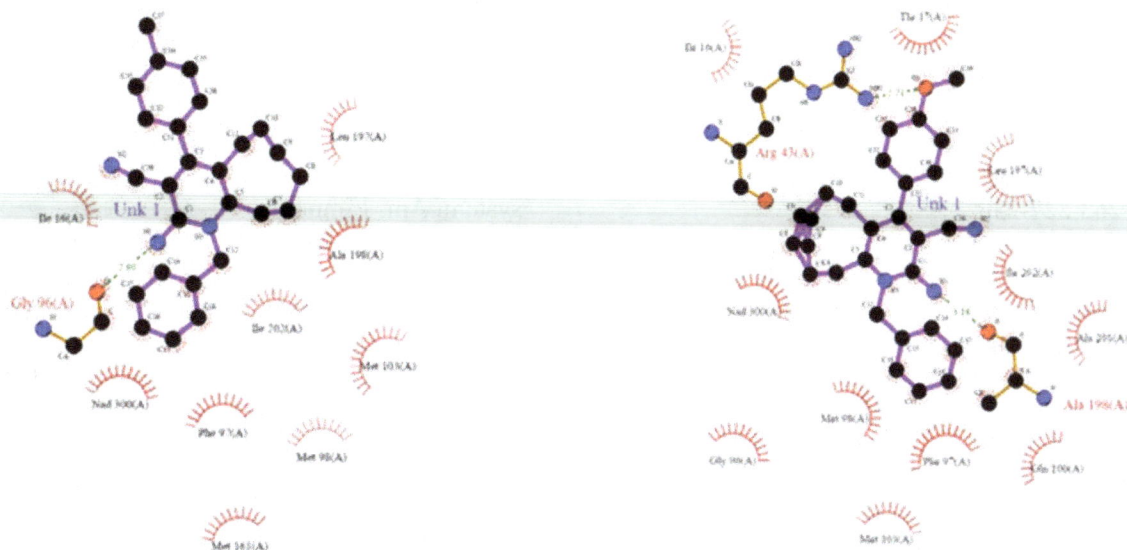

Figure 5. Interactions of compound (Ia) (left) and compound (Ib) (right) with enoyl acyl carrier protein.

longer time bound to their target molecules and can remain bound even at low free drug concentrations. In this regard, our inhibitor (Ib) shows the features of slow onset inhibitors.

Conclusion

For our goal to propose new inhibitors, we synthesized two compounds of pyridine derivatives. The compounds were docked and their binding free binding energies were evaluated. Both compounds show good binding affinity with acceptable free binding energy which indicates that they give good stable complexes. Hence our synthesized ligands show good antitubercular activities. From the docking analysis of the two compounds with the receptor of enoyl acyl carrier protein it was found that (Ib) shows better binding with the receptor, as compared to (Ia).

Supplementary Material

Crystallographic data (excluding structure factors) for

structures of (Ia) and (Ib) reported in this paper have been deposited with the Cambridge Crystallographic data Centre as supplementary publication CCDC 1017700 & CCDC 1017701. Copies of the data can be obtained, free of charge, on application to, CCDC, 12 Union Road, Cambridge, and CB2 1 EZ UK; Email: de-posit@ccdc.cam.uk or at http://www.ccdc.cam.ac.uk/.

Acknowledgements

JS thanks the UGC for the FIST support. JS and RAN thank the management of The Madura College (Autonomous), Madurai for their encouragement and support. RRK thanks University Grants Commission, New Delhi, for funds through Major Research Project F. No. 42-242/2013 (SR). SM thanks the university grants commission, New Delhi for the fellowship under UGC-BSR-JRF meritorious scheme.

Table 4. Binding energy of compounds (Ia) and (Ib).

Ligand	Binding Energy (Kcal/mol)	Intermolecular Energy (Kcal/ mol) (A+B)	VDW + Hbond + desolv energy (Kcal/ mol) (A)	Electrostatic energy (Kcal/ mol) (B)
Ia	-7.83	-9.03	-9.02	-0.01
Ib	-8.67	-10.16	-9.93	-0.23

References

Banerjee A, Dubnau E, Quemard A, Balasubramanian V, Um KS, Wilson T, Collins D, de Lisle G & Jacobs WR Jr 1994 inhA, a gene encoding a target for isoniazid and ethionamide in Mycobacterium tuberculosis. *Science* **263** 227-230

Besra GS & Chatteriee D 1994 Lipids and carbohydrate of Mycobacterium tuberculosis. In *Tuberculosis: Pathogenesis, Protection, and Control*, pp 285-306. Eds BR Blood. Washington, DC: ASM Press

Bernstein J, Lott WA, Steinberg BA & Yale HL 1952 Chemotherapy of experimental tuberculosis. *Am Rev Tuberc* **65** 357-374

Bernstein J, Davis RE, Shimoni L & Chang NL 1995 Patterns in Hydrogen Bonding: Functionality and Graph Set Analysis in Crystals. *Angew Chem Int Ed Engl* **34** 1555-1573

Blanchard JS 1995 Enzymatic characterization of the target for isoniazid in Mycobaterium tuberculosis. *Biochemistry* **34** 8235-8241

Bruker 2004 APEX2 and SAINT. Bruker AXS Inc., Madison, Wisconsin, USA

Cohn DL, Bustreo F & Raviglione MC 1997 Drug-resistant tuberculosis: review of the worldwide situation and the WHO/IUATLD global surveillance project. *Clin Infect Dis* **24** S121-S130

Dandia A, Sehgal V & Singh P 1993 Synthesis of fluorine containing 2-aryl-3-pyrazolyl/pyranyl/isoxazolinyl-indole derivatives as antifungal and antibacterial agents. *Indian J Chem* **32** 1288-1291

Dessen A, Quemard A, Blanchard JS, Jacobs WR & Sacchettini JC 1995 Crystal structure and function of the isoniazid target of Mycobacterium tuberculosis. *Science* **267** 1638-1641

Dietrich J & Doherty TM 2009 Interaction of Mycobacterium tuberculosis with the host: consequences for vaccine development *APMIS* **117** 440-457

Lavie A, Vetter IR, Konrad M, Goody RS, Reinstein J & Schlichting I 1997 Structure of thymidylate kinase reveals the cause behind the limiting step in AZT activation. *Nat Struct Biol* **4** 601-604

Lewandowicz A, Tyler PC, Evans GB, Furneaux RH & Schramm VL 2003 Achieving the ultimate physiological goal in transition state analogue inhibitors for purine nucleoside phosphorylase. *J Biol Chem* **278** 31465-31468

Goodsell DS, Morris GM & Olson AJ 1998 Automated docking of flexible ligands: Applications of autodock *J Mol Recog* **9** 1-5

Pieters J 2008 *Mycobacterium tuberculosis* and the macrophage: Maintaining a Balance. *Cell Host Microbe* **3** 399-407

Punkvang A, Saparpakorn P, Hannongbua S,

Wolschann P, Beyer A & Pungpo P 2010 Investigating the structural basis of arylamides to improve potency against M. tuberculosis strain through molecular dynamics simulations. *Eur J Med Chem* **45** 5585-5593

Rajni & Meena LS 2011 Unique characteristic features of *Mycobacterium tuberculosis* in relation to immune system. Am. J Immunol **7** 1-8

Rozwarski DA, Grant GA, Barton DH, Jacobs WR Jr & Sacchettini JC 1998 Modification of the NADH of the isoniazid target (InhA) from Mycobacterium tuberculosis. *Science* **279** 98-102

Rozwarski DA, Vilchèze C, Sugantino M, Bittman R & Sacchettini JC 1999 Crystal structure of the Mycobacterium tuberculosis enoyl-ACP reductase, InhA, in complex with NAD+ and a C16 fatty acyl substrate. *J Biol Chem* **274** 15582-15589

Sheldrick GM 2008 A short history of *SHELX*. *Acta Cryst* A **64** 112-122

Sriram D, Yogeeswari P & Madhu K 2006 Synthesis and in Vitro Antitubercular Activity of Some 1-[(4-sub) Phenyl]-3-(4-{1-[(Pyridine-4 Carbonyl) Hydrazono] Ethyl}Phenyl)Thiourea. *Bioorg Med Chem Lett* **16** 876-878

Spek AL 2009 Structure validation in chemicalcrystallography. *Acta Cryst* D **65** 148-155

Suresh J, Vishnu Priya R, Sivakumar S & Ranjith Kumar R 2012 Spectral analysis and crystal structure of two substituted spiro acenaphthene structures *J Mol Biochem* **1** 183-188

Takayama K, Wang L & David HL 1972 Effect of isoniazid on the in vivo mycolic acid synthesis, cell growth, and viability of *Mycobacterium tuberculosis*. *Antimicrob Agents Chemother* **2** 29-35

Temple C, Rener GA, Raud WR & Noker PE 1992 Antimitotic agents: structure-activity studies with some pyridine derivatives. *J Med Chem* **35** 3686-3690

Tummino PJ & Copeland RA 2008 Residence time of receptor-ligand complexes and its effect on biological function. *Biochemistry* **47** 5481-5492

Quémard A, Lacave C & Lanéelle G 1991 Isoniazid inhibition of mycolic acid synthesis by cell free extracts of sensitive and resistant strains of *Mycobacterium aurum*. *Antimicrob Agents Chemother* **35** 1035-1039

Quemard A, Sacchettini JC, Dessen A, Vilcheze C, Bittman R, Jacobs WR Jr & Blanchard JS 1995 Enzymatic characterization of the target for isoniazid in *Mycobaterium tuberculosis*. *Biochemistry* **34** 8235-8241

Vilcheze C, Morbidoni HR, Weisbrod TR, Iwamoto H, Kuo M, Sacchettini JC & Jacobs WR Jr 2000 Inactivation of the inhA-encoded fatty acid synthase II (FASII) enoyl-acyl carrier protein reductase induces accumulation of the FASI end products and cell lysis of Mycobacterium smegmatis. *J Bacteriol* **182** 4059-4067

Wang LQ, Falany CN & James MO 2004 Triclosan as a substrate and inhibitor of 3'-phosphoadenosine 5'-phosphosulfate-sulfotransferase and UDP-glucuronosyl transferase in human liver fractions. *Drug Metab Dispos* **32** 1162-1169

Winder FG 1982 Mode of action of the antimycobacte-rial agents. In *The Biology of the Mycobacteria*, vol. 1, pp. 353-438. Edited by C. Ratledge & J. Stanford. London: Academic Press

Winder FG & Collins PB 1970 Inhibition by isoniazid of synthesis of mycolic acids in *Mycobacterium tuberculosis*. *J Gen Microbiol* **63** 41-48

15

Study on the internalization mechanism of the ZEBRA cell penetrating peptide

Roberta Marchione[1], Lavinia Liguori[2], David Laurin[1,3] and Jean-Luc Lenormand[1]

[1]TheREx, TIMC IMAG Laboratory, UMR5525, UJF/CNRS, Joseph Fourier University, 38700 La Tronche, France
[2]SyNaBi, TIMC IMAG Laboratory, UMR5525, UJF/CNRS, Joseph Fourier University, 38700 La Tronche, France
[3]Etablissement Français du Sang Rhône Alpes, La Tronche, F-38701 France

Correspondence should be addressed to Jean-Luc Lenormand; E-mail: jllenormand@chu-grenoble.fr

Abstract

Cell-penetrating peptides (CPPs) represent a noninvasive method for delivering functional biomolecules into living cells. We have recently shown that the Epstein-Barr virus transcriptional factor ZEBRA contains a protein transduction domain, named Z9 or minimal domain (MD). Only few of currently identified CPPs including MD are able to rapidly cross the mammalian cell membrane without being entrapped into endosomal compartments, even when fused to cargo macromolecules. In this work, a series of MD deletion mutants has been engineered and their cellular uptake has been analyzed by confocal microscopy and FACS. We identified a domain MD_{11} (8 amino acids shorter than MD) able to enter mammalian cells via a mainly endocytosis-independent mechanism. All the other generated truncated forms exhibited reduced cellular uptake and penetrated into cells through endocytic mechanisms. These results have highlighted the role of the MD_{11} C-terminal region as essential for efficient cellular entry and endosomal escape and open new perspectives for the use of this CPP as carrier for delivering biologically active macromolecules with therapeutic potential.

Introduction

Many therapeutic targets have been found located within cells but the effective transport of hydrophilic active molecules such as proteins or peptides across the cellular plasma membrane has represented a serious obstacle for many decades. A promising approach that seems to be the solution for overcoming the cellular barrier, has emerged with the discovery of the cell-penetrating peptides (CPPs), also referred to as protein transduction domains. Generally, CPPs are defined as relative short peptides with the ability to gain access to the cell interior and promote the intracellular delivery of conjugated cargoes (Langel 2006). Since the discovery of the first CPP from the HIV TAT protein (Frankel & Pabo 1988, Green & Loewenstein 1988), a variety of transducing peptides has been identified, including both naturally occurring domains and synthetically derived sequences. Well known examples include the penetratin peptide (Derossi et al. 1994), VP22 (Elliott & O'Hare 1997), pVEC (Elmquist et al. 2001), polyarginine (Mitchell et al. 2000), Transportan (Pooga et al. 1998), etc. The common feature of CPPs is their typically high content in basic arginine and ly-

sine residues, leading to a positive net charge of the peptides, which is considered to be crucial for initial membrane interaction through binding to negatively charged phospholipids and glycosaminoglycans (Ziegler 2008). The number of applications using CPPs is increasing, and so far more than 300 studies from in vitro to in vivo have been reported (Heitz et al. 2009). Indeed, the interest for CPPs is mainly due to their low cytotoxicity and to the fact that there is no limitation for the type of cargo. CPPs have been used to improve delivery of cargoes that vary greatly in size and nature, including small molecules, oligonucleotides, plasmid DNA, peptides, proteins, nanoparticles, virus and lipid-based formulations (Zorko & Langel 2005). Despite the similarities among CPPs, the translocation mechanisms may vary considerably. The two main pathways suggested for cellular uptake are direct penetration and endocytosis (Madani 2011). However, the impact of these two mechanisms on the biological function of the transported cargo is different. Contrary to direct penetration, during endocytosis, the cargo can be entrapped and consecutively partially degraded into endosomal compartments, thus leading to the loss of its activity. This aspect represents the main limitation towards the therapeutic use of CPPs as delivery systems for biologically active drugs. When conjugated to cargo molecules, the cellular uptake of most widely used CPPs such as penetratin, TAT, pVEC and transportan is shown to proceed through an endocytic pathway (Eiriksdottir *et al.* 2010, Lundin *et al.* 2008, Saalik *et al.* 2004). In a previous study, we identified a novel cell-penetrating peptide able to cross the cell membranes in an endocytosis-independent mechanism even when fused to cargoes, as shown with eGFP and β-galactosidase reporter proteins (Rothe *et al.* 2010). This CPP derives from the Epstein-Barr virus (EBV) ZEBRA transcription factor. A reductionist study of full-length ZEBRA protein has allowed us to identify the amino acid region (named as Minimal Domain, MD) implicated in cellular uptake (Rothe *et al.* 2010). The region required for internalization spans residues 170-220 from the ZEBRA protein and contains two contiguous domains: a positively charged domain (DNA-binding domain, DBD) and a hydrophobic leucin-rich domain (dimerization domain, DIM). The DBD is believed to mediate cell surface binding of the MD to the negatively charged heparan sulfate proteoglycans while the DIM domain facilitates translocation through the lipid bilayer by hydrophobic interactions (Rothe *et al.* 2010).

In the present study, we aimed at reducing the size and the hydrophobicity of the transduction domain and at describing the amino acid sequence required for its cellular uptake. We produced MD truncations in fusion to the eGFP reporter protein and evaluated the ability of these constructions to translocate through the membrane of HeLa cells. We identified a MD shorter peptide (MD_{11}) able to enter mammalian cells with high efficiency by an endocytosis-independent mechanism. Further trimming of the DIM domain from the MD_{11} peptide led to a decrease in the translocation efficiency and to an alteration of the uptake mechanism. The results presented here reveal the role of the whole DIM domain as necessary for endocytosis-independent cell internalization. This import mechanism is an attractive requisite for developing MD_{11}-mediated uptake of macromolecules in therapeutic applications and strengthens this delivery system compared to most others.

Materials and Methods

Cloning, expression and purification of the MD_x-eGFP fusion proteins

The cloning, the expression and the purification of the free eGFP and the recombinant fusion proteins MD_x-eGFP were performed as described in Rothe and Lenormand (2008). Briefly, the DNA fragments encoding for the MD deletion mutants were generated by PCR and ligated upstream of the 5'-end of eGFP gene into pET15b expression plasmid, bearing a His_6-tag sequence. The generated MD truncations are schematically depicted in Figure 1A. The fusion recombinant proteins were produced in *E. coli* BL21 (DE3) cells by inducing the expression with 0.5mM Isopropyl β-D-1-thiogalactopyranoside (IPTG) at OD_{600nm} of 0.8 for 18h at 16°C. In order to recover the proteins of interest, the bacterial cultures were centrifuged at 5000g for 15 min and the cell pellets were sonicated in 20mM Tris/HCl pH 7.4, 500mM NaCl, 10% glycerol, 2mM DTT, 10mM imidazole (5mL per gram wet pellet) supplemented with a complete protease inhibitor cocktail (Roche). As all the MD_x-eGFP were produced with a hexahistidine tag, the soluble fractions were purified onto nickel sepharose HisGraviTrap columns (GE Healthcare) by gravity-flow and eluted in the same buffer by a stepwise increase (at 100, 175, 250 and 500mM) of imidazole content. MD_x-eGFP proteins were separated onto a 15% SDS-polyacrylamide gel electrophoresis and analyzed by Coomassie blue staining or by Western-blotting using an anti-His tag antibody HRP-coupled (Sigma, 1:10000 dilution). Prior to their use for cellular uptake experiments, purified eGFP and MD_x-eGFP proteins were dialyzed against PBS and 25mM HEPES/KOH pH7.0, 150mM NaCl and 10% glycerol respectively using a MWCO 8000 SpectraPor™ dialysis tubing (Spectrum Laboratories). The yield of the purified His-tagged proteins was quantified by BCA Protein Assay Kit according to the manufacturer's instructions (Pierce).

Helical wheel projections of the peptides were generated using online program available at http://rzlab.ucr.edu/scripts. Primary sequence analysis was performed using EMBOSS bioinformatics programs and the GRAVY index was calculated using ProtParam tool at ExPASy Proteomics Server.

Cell culture

HeLa cells were maintained in DMEM (PAA, GE Healthcare) supplemented with 10% heat-inactivated fetal bovine serum (PAA, GE Healthcare), 100 units/mL penicillin and 50μg/mL streptomycin (Gibco). Cells were cultured at 37°C in a humidified 5% CO2 atmosphere incubator.

Confocal microscopy

HeLa cells (5 x 10^4 cells/well) were seeded onto an 8-well Lab-Tek™ chambered coverglass (Nunc) in complete cell culture media 24h before treatment. After removal of the medium, the cell layers were rinsed twice with DPBS and subsequently exposed at 37°C for 4 hours to 0.3μM of recombinant proteins in fresh

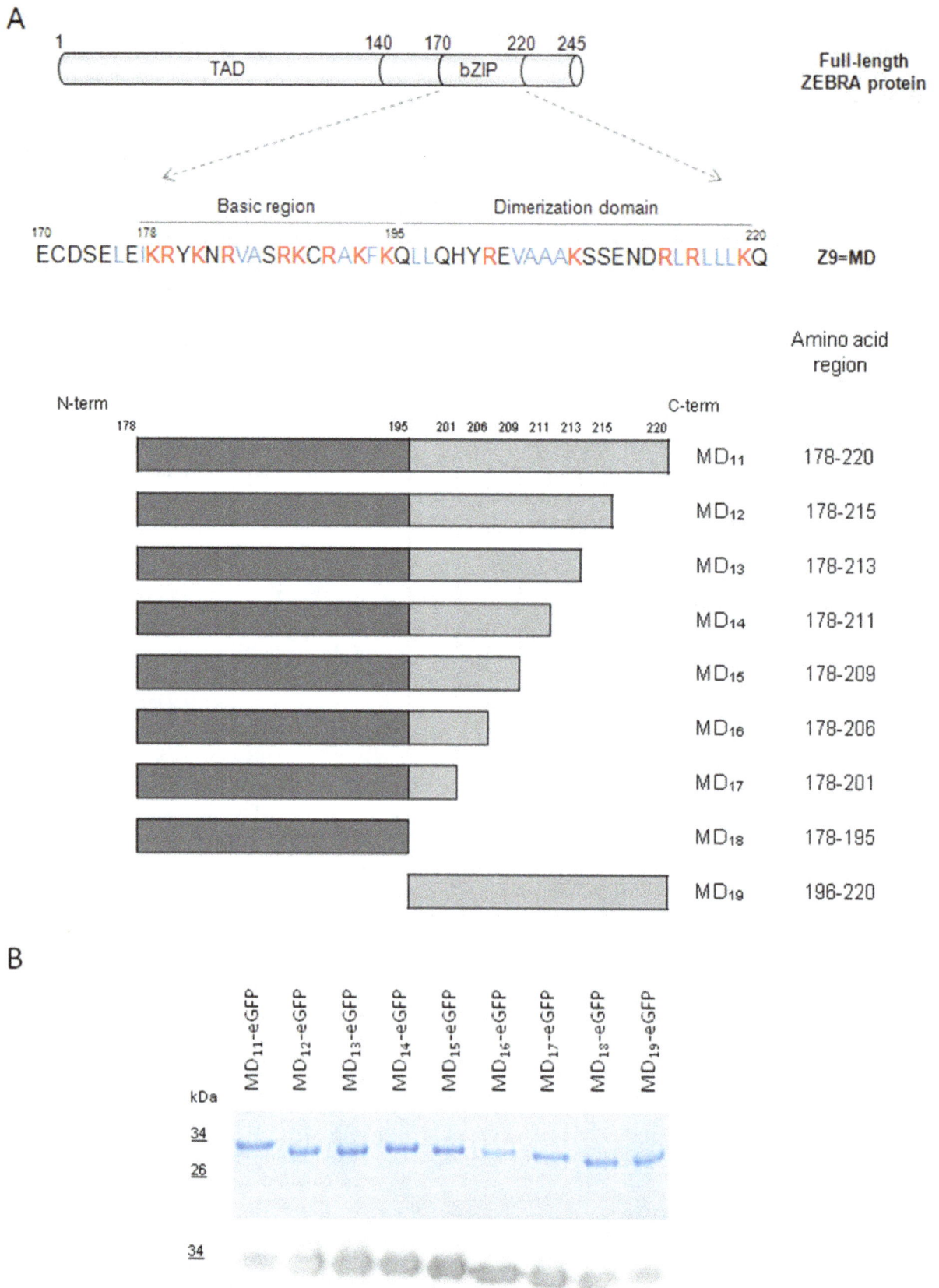

Figure 1. A) Amino-acid sequence of ZEBRA MD cell-penetrating peptide (aa 170-220) and scheme of the deletion mutants. Amino acids are designated with one-letter code and numbered according to the full-length ZEBRA protein sequence. Basic amino acids are shown in red, whereas hydrophobic amino acids in blue. B) Recombinant fusion proteins, containing each peptide fused to the N-term of eGFP, were separated on a 15% SDS-polyacrylamide gel and visualized by Coomassie Blue staining and detected by Western blotting using anti-His antibody.

serum-free medium. Thereafter, the incubation solutions were removed and the cells were washed three times with DPBS. Living cell preparations were observed with a LSM 710 confocal scanning laser microscope (Carl Zeiss, Jena, Germany), using a 63×, NA 1.2, C-apochromat water-immersion objective (Carl Zeiss). The experiment was carried out at 488nm excitation and fluorescence was collected with a 510-560 nm filter. 16 successive optical slices were captured along the cell z axis, with a step of 1μm. For localization study, cells were also fixed for 15 minutes in 4% paraformaldehyde at room temperature and then stained with 10μM Nile Red. The images were acquired with 488nm for eGFP and 633nm for Nile Red excitation and fluorescence was collected between 502 – 541 and 580 – 668nm filters respectively.

Cellular uptake assays

For transduction experiments, 1.5×10^5 HeLa cells/well were cultured on 12-well plates until 70% of confluence. Cells were washed twice with DPBS and incubated with 0.3μM of recombinant proteins in fresh serum-free culture medium for 4 hours at 37°C. After incubation, cells were trypsinized with 0.5% trypsin/EDTA solution (PAA, GE Healthcare) for 10 min at 37°C to remove the extracellular bound proteins. The trypsin was then neutralized by adding complete medium and the cells were washed twice with DPBS.

For endocytic inhibition experiments, cells were first pre-incubated for 30 minutes with different endocytic inhibitors at the following concentrations: 50μg/mL nystatin, 100nM wortmannin and 5mM methyl-β-cyclo-dextrin (Sigma). After pre-incubations with the endocytic inhibitors, the recombinant proteins were added to cell cultures in the presence of drugs and treated as indicated above.

Cell-associated fluorescence was detected using a FACS Canto II BD Biosciences flow cytometer. We used a 488nm laser for excitation and 530/30 bandpass filters for emission. The results are reported as the mean fluorescence intensity from living cells gate of 30000 events recorded and analyzed with the FACSDiva software.

Differences between samples and treatments were evaluated by variance analysis (two-tails, paired values) followed by the least-significant difference test. A difference was considered to be statistically significant with $p < 0.05$.

Preparation and observation of lipid vesicles

Giant unilamellar vesicles (GUVs) were created by sucrose hydration method (Akashi et al. 1996). Three different lipid compositions were used in order to obtain fluid (100% PC), rigid (SM:Chol 50:50 % molar ratio) and semi-fluid (PC:SM:Chol 33:33:33 % molar ratio) GUVs. 50μg of lipid mixtures in chloroform were deposited in an 8-well Lab-Tek™ chambered coverglasses and dried under nitrogen. The dried film was hydrated overnight at 4°C with 25mM HEPES/KOH pH7.4, 250mM sucrose. After re-hydration, 1μg of MD-eGFP was added to GUV solution. The lipid component was labeled with Nile Red dye (10μM), and CSLM images were acquired as described above.

Results

Design and expression of MD analogs

The ZEBRA MD protein transduction domain (amino acids 170-220) consists of 14 basic residues (Lys and Arg) and 15 hydrophobic residues, mainly located at the C-terminal dimerization domain (Figure 1A) (Rothe & Lenormand 2010). To better understand the specific contribution of the different amino acids in the internalization process, we designed a series of iterative MD truncation mutants. An initial deletion was realized by removing the first eight amino acid residues located upstream of the basic region at the N-terminus of the MD, resulting in a sequence named as MD_{11} (Figure 1A). In transduction experiments on HeLa cells, recombinant proteins containing MD_{11} fused to the fluorescent reporter protein eGFP were able to penetrate with the same efficiency as MD (Rothe & Lenormand, unpublished data). Thus, we decided to further optimize this CPP (MD_{11}) by reducing its size and increasing its hydrophilicity. We engineered six truncations by removing from a minimum of five to a maximum of nineteen amino acids from the hydrophobic C-terminus (Figure 1A). In addition, we produced two more truncations containing either only the basic (MD_{18}) or the hydrophobic (MD_{19}) region. We then expressed each peptide fused to the N-terminal end of eGFP. The his-tagged recombinant fusion proteins were produced using an E. coli expression system and purified by nickel affinity chromatography. The purity of the MD_x-eGFP proteins was analyzed by Coomassie blue staining and by western-blotting using an anti-his antibody (Figure 1B). Recombinant proteins were pure at around 85%.

A crystallographic study has revealed that the DNA-bound ZEBRA protein (residues 175-245) is an extended bZIP helix (Petosa et al. 2006). With the use of the helical wheel projections, the sequences of the MD truncations were examined (Figure 2). The wheels for all deletions have both a polar and a hydrophobic character. In each helix, there are two spatially organized regions of polar residues interrupted by a few hydrophobic residues. The exact content in polar and non-polar residues for each truncation is summarized in

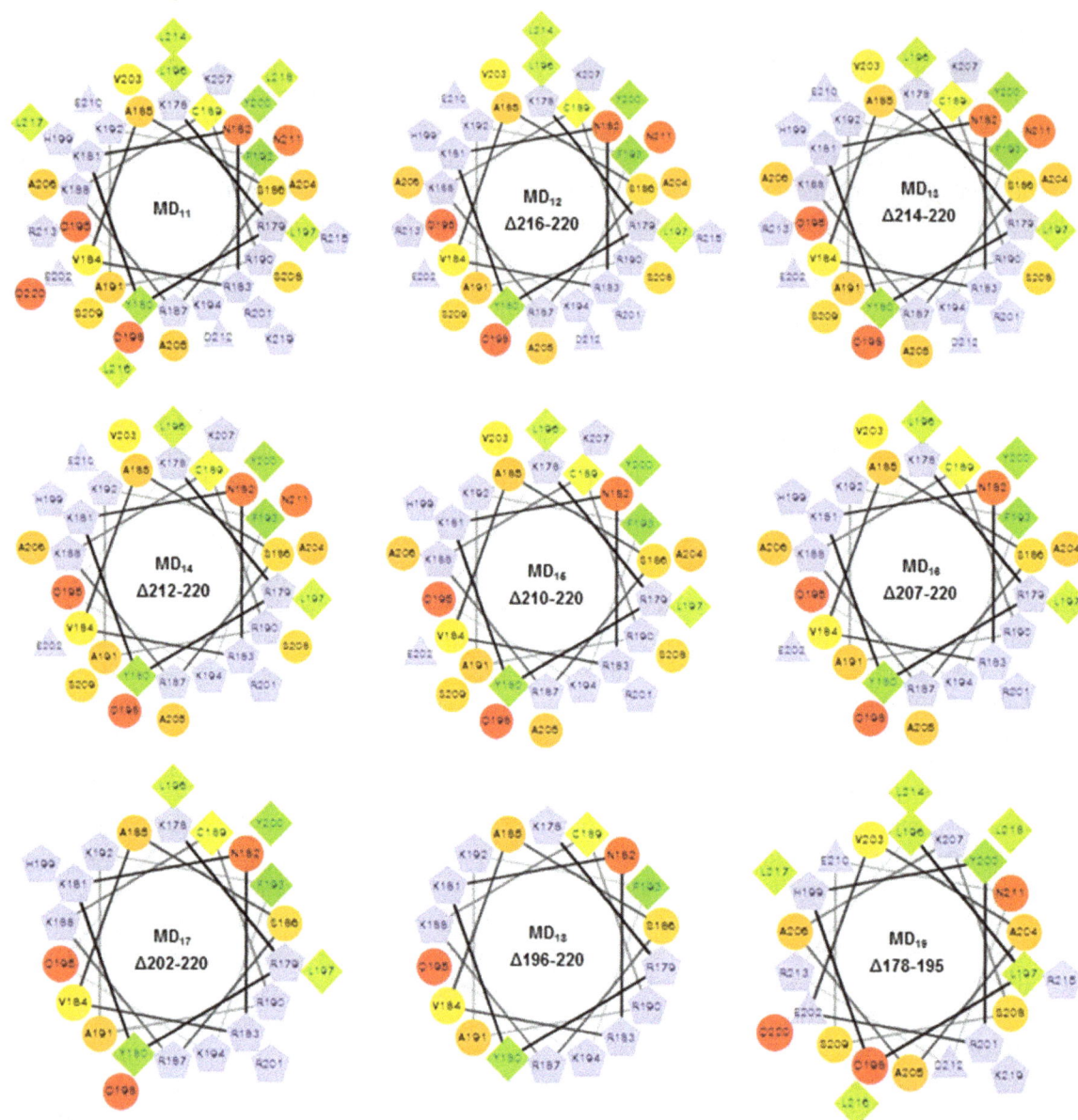

Table 1

Peptide	Number of residues	Net charge at pH 7	Positive charges K,R, H	Negative charges D, E	Aromatic residues Y,F	Polar residues (without net charge) N, Q, S	Non polar residues A, C, F, I, L, V, Y	Grand average of hydropathicity (GRAVY)
MD$_{11}$	43	11,5	15	3	3	8	17	-1.151
MD$_{12}$	38	10,5	14	3	3	7	14	-1.408
MD$_{13}$	36	9,5	13	3	3	7	13	-1.467
MD$_{14}$	34	9,5	12	2	3	7	13	-1.318
MD$_{15}$	32	10,5	12	1	3	6	13	-1.181
MD$_{16}$	29	9,5	11	1	3	4	13	-1.114
MD$_{17}$	24	10,5	11	0	3	4	9	-1.600
MD$_{18}$	18	9	9	0	2	3	6	-1.861
MD$_{19}$	25	2,5	6	3	1	5	11	-0.640

Figure 2. Helical wheels projections of MD$_{11}$ and its truncated analogs. The projections were realized using the online program available at http://rzlab.ucr.edu/scripts. The hydrophilic residues are presented as circles, hydrophobic residues as diamonds, potentially negatively charged as triangles, and potentially positively charged as pentagons. The most hydrophobic residues are presented in green, and the amount of green is decreasing proportionally to the hydrophobicity (with zero hydrophobicity coded as yellow). Hydrophilic residues are coded red, and the amount of red is decreasing proportionally to the hydrophilicity. The charged residues are light blue. The peptide length, the amino acid composition, and the GRAVY index of each peptide are indicated in the Table 1. The parameters were calculated using EMBOSS and ExPASy bioinformatics programs.

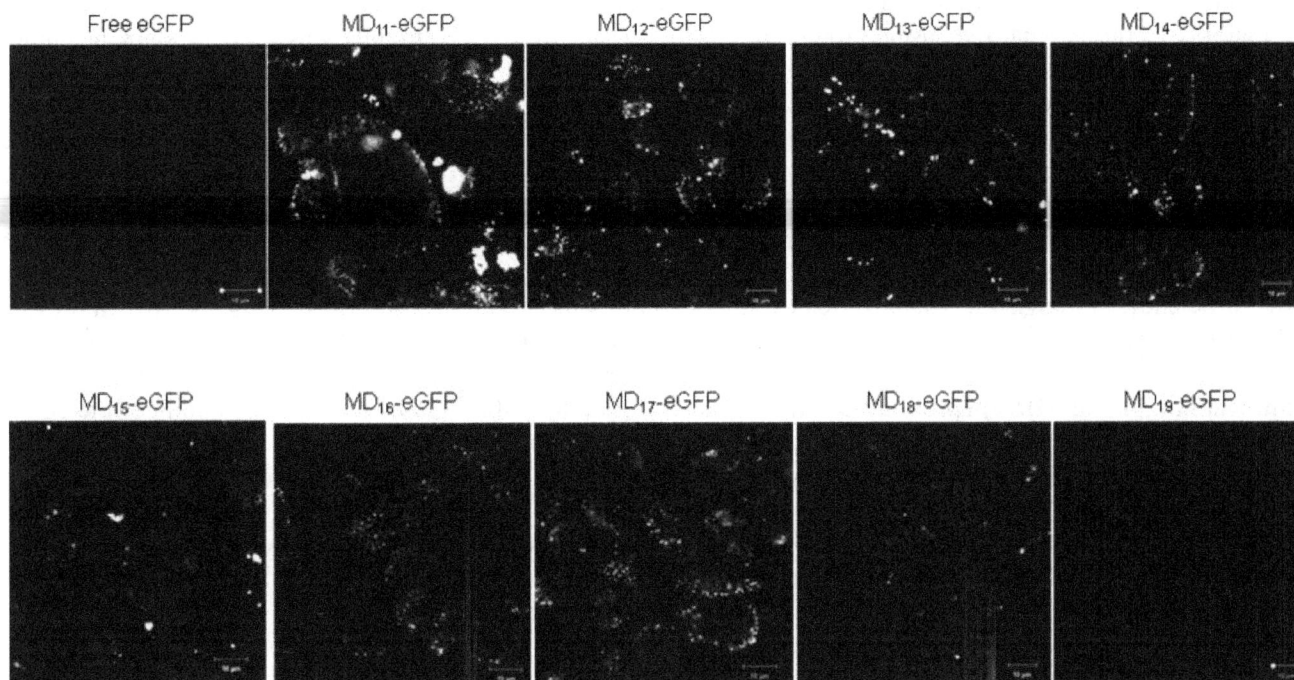

Figure 3. Confocal laser scanning microscopy images of living HeLa cells incubated for 4 hours with 0.3μM of free eGFP and MD_x-eGFP fusion proteins. 16 successive optical slices were captured along the cellular z-axis with a step of 1μm. The presented images correspond to the middle plan of cellular z-axis sectioning.

the Table 1. In all cases, except for MD_{19}, the peptides are net positively charged. The shortening and deletion of the DIM domain affect both charge and hydrophobicity contributions: the non-polar and negative amino acid content is reduced and the hydrophilicity (negative GRAVY value, (Kyte & Doolittle 1982)) of the peptides is increased.

Confocal microscopy

The transduction ability of MD deletions was investigated by confocal microscopy on living HeLa cells after 4 hours treatment with 0.3μM of each recombinant protein according to results reported in Rothe *et al.* (2010). The cellular uptake and the intracellular distribution of the recombinant proteins were checked by direct visualization of the intracellular eGFP fluorescence. As expected, no fluorescence signal was detected on HeLa cells incubated with free eGFP (Figure 3). Except for MD_{19}, Z-stack analysis of treated cells confirmed that MD_{11} and most of its analogs maintained the cell penetrating ability (Figure 3). The degree of intracellular fluorescence accumulation varied between the different MD_x constructions (Figure 3). To better delineate cell contours and evaluate the interaction of the recombinant cell-penetrating proteins with membranes, cells were also fixed and the membrane lipids stained with the fluorescent dye Nile Red (Figure 4). Overlapping green and red images resulted in yellow signals demonstrating the co-localization of

MD_{11}, MD_{14}-MD_{17} eGFP fusion proteins with membrane lipids (Figure 4).

Cellular uptake assays

To quantify the cellular uptake of MD_{11} and its truncated forms, FACS analysis was performed after a four-hour incubation of living HeLa cells with 0.3μM of each fusion protein. The extra-bound fluorescent proteins were removed by trypsin digestion. All peptides were found to penetrate into cells, but the new derived MD_{11} peptides differed remarkably in efficiency of penetration (Figure 5A). No improvement of uptake was observed when the number of MD_{11} amino acid residues was further reduced. Indeed, the maximum efficiency of internalization was observed in HeLa cells incubated with MD_{11}-eGFP.

To evaluate the internalization mechanism of MD_{11} and its truncations, we measured their uptake efficiency in living cells in presence of three endocytosis inhibitors. Nystatin was employed to inhibit the caveolae-mediated endocytosis, wortmannin to inhibit macropinocytosis, and MβCD to disrupt the import through lipid rafts. The presence of nystatin and wortmannin did not interfere with MD_{11}-eGFP internalization (Figure 5B), indicating no participation of the receptor-mediated endocytosis in its uptake process. Incubation with MβCD caused a 30% decreased uptake of MD_{11}-eGFP. Thus, the endocytotic pathway contributed only partially to the MD_{11}-eGFP cellular uptake under the applied conditions but did not account for the majority of the internalized fusion protein. In contrast, a strong decrease in the uptake of MD$_{12-19}$ truncations was observed when the three analyzed endocytosis pathways were inhibited, indicating that the main internalization routes of these peptides were both receptor- and lipid raft- mediated endocytosis. Thus, MD$_{11}$ is the shortest CPP derived from the EBV ZEBRA transcription factor with the highest internalization efficiency through a mainly endocytosis-independent mechanism.

Confocal imaging of MD$_{11}$ peptide-lipid interaction

To further examine the interactions of MD$_{11}$ with lipids, we incubated the fusion protein MD$_{11}$-eGFP with lipid vesicles of different composition and rigidity. Kinetics of protein-lipid interactions, membrane rigidity and changes in membrane morphology can be easily evaluated by microscopy imaging because of the giant vesicle's size. Fluid vesicles were prepared using phosphatidylcholine (PC 100%), rigid and semi-fluid vesicles were obtained by mixing sphingomyelin and cholesterol (SM, Chol 50:50% molar ratio) and PC,

SM, Chol (33:33:33% molar ratio) respectively. The Nile Red dye (2μg/ml) was used to stain the lipids. MD$_{11}$-eGFP fusion protein was incubated with vesicles for 1, 2.5, 3.5, 4.5 and 24 hours at 37°C, and the protein-lipid interaction was monitored by confocal laser scanning microscopy. All the vesicles were homogenously stained with the red dye. We noticed a time-dependent accumulation of green fluorescent signal (eGFP) surrounding the vesicle membranes in all the tested conditions. By overlapping the two fluorescent signals (red from the Nile red and green from the eGFP), it is possible to evaluate the interaction of peptide-lipid that results in a yellow staining of the vesicle membranes. After 3.5 h incubation, three different patterns of co-localization were observed (Figure 6, "merge" panel). A strong co-localization of MD$_{11}$ peptide was recorded on semi-fluid and rigid vesicles containing both SM and Chol (Figure 6B and 6C). Any interaction was observed between MD$_{11}$ and PC fluid vesicle membranes (Figure 6A), even after 24 hours (data not shown). Curiously, a heterogeneous interaction of MD$_{11}$ in restricted clusters of semi-fluid vesi-

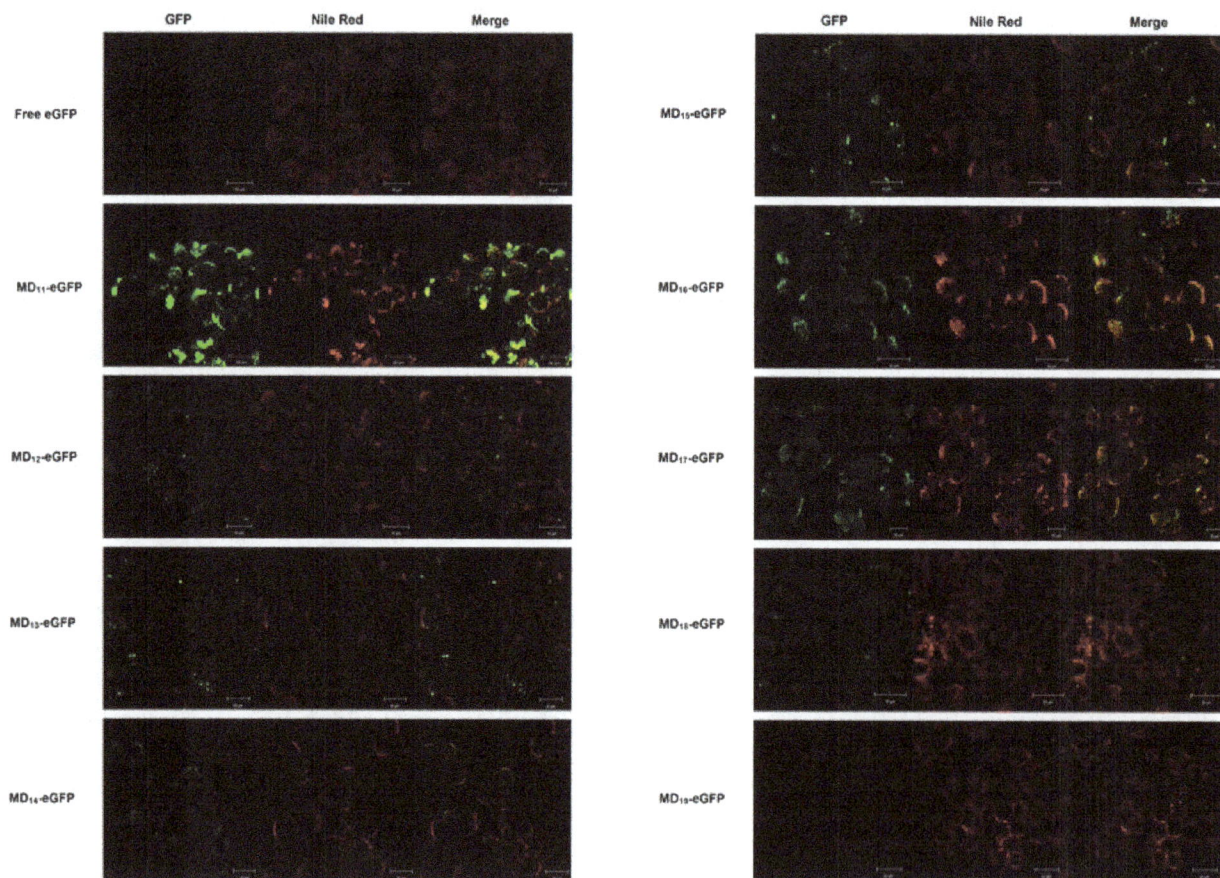

Figure 4. Intracellular localization of MD$_x$-eGFP in HeLa cells. Confocal microscopy images of fixed HeLa cells after a 4-hour incubation with 0.3μM of free eGFP and MD$_x$-eGFP fusion proteins. MD$_x$-eGFP signals are shown in green and the cellular lipid component in red. The showed images correspond to the central plane of cellular z-axis sectioning.

A

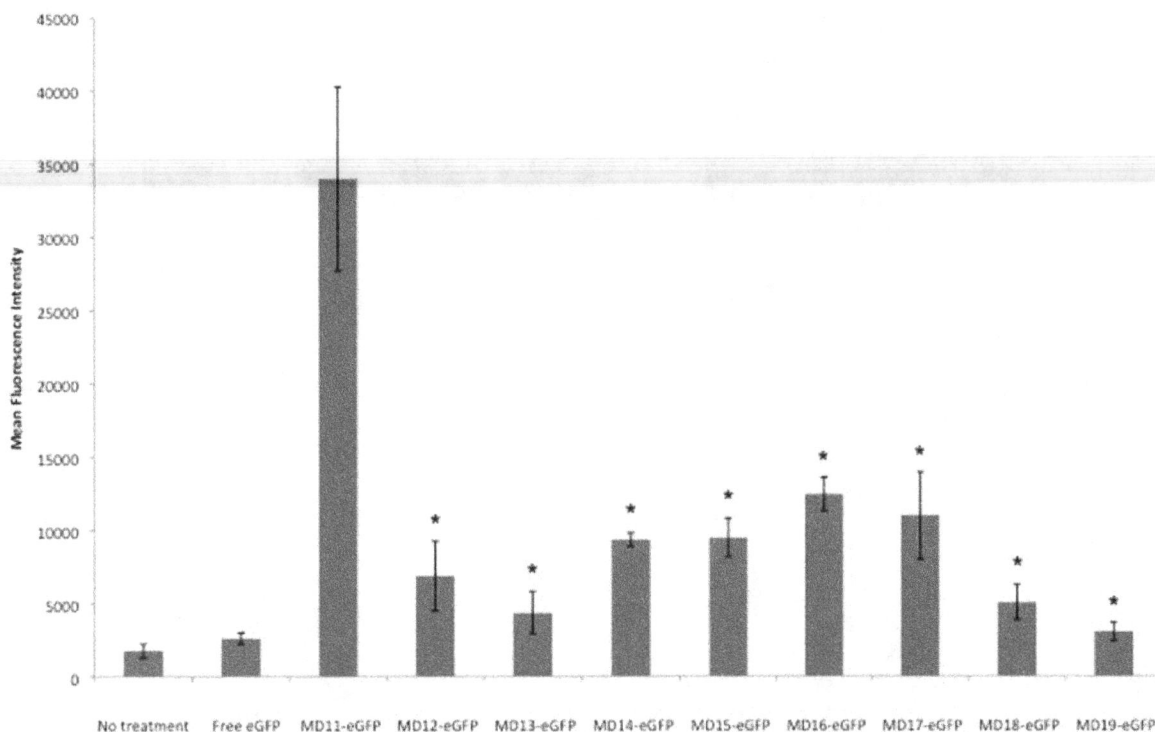

B

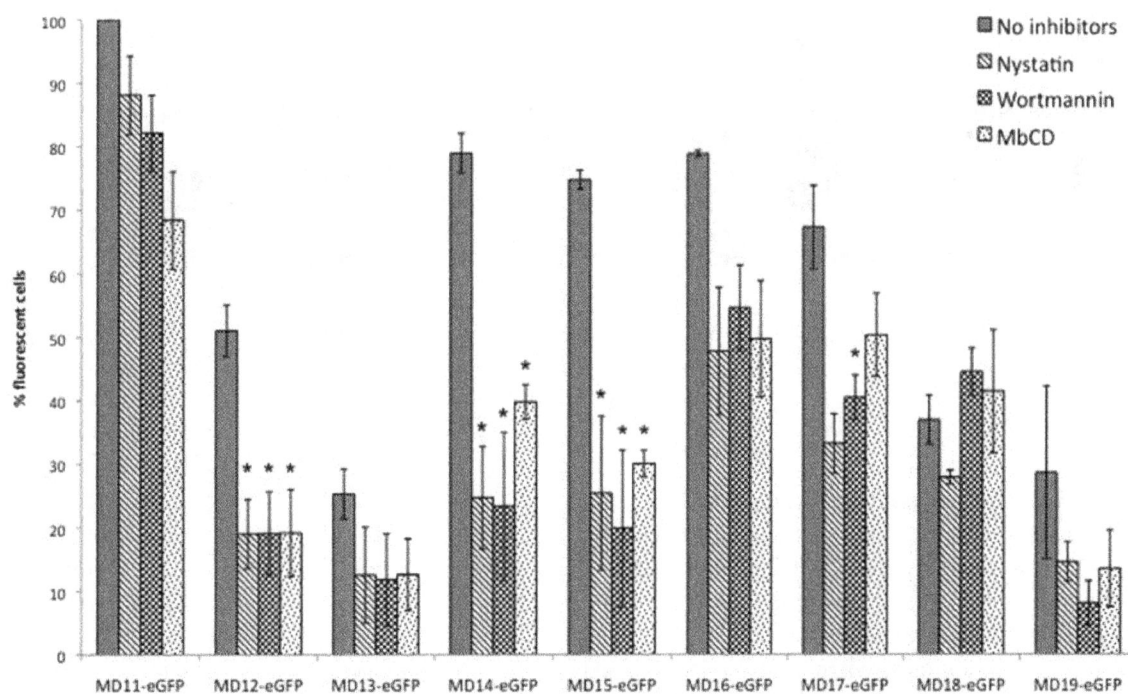

Figure 5 FACS cellular uptake of MD_{11} and its truncated analogs. A) HeLa cells were incubated with 0.3µM of different peptides fused to eGFP for 4h at 37°C. The data are expressed as mean fluorescence intensity (MFI) of eGFP positive cells and the values indicated are the mean±s.d. of four independent experiments. *p < 0.05 MD_{11}-eGFP versus each $MD_{(12-19)}$-eGFP
B) Uptake of MD_{11} and its truncated analogs in presence of endocytosis inhibitors. HeLa cells were pre-treated with 50µg/mL nystatin, 0.1µM wortmannin or 5mM MβCD for 30min at 37°C. 0.3µM of recombinant proteins were incubated with cells for 4h at 37°C. The data are expressed as % of eGFP positive cells and the values indicated are the mean±s.d. of two independent experiments. All fluorescence values have been normalized assuming as 100% the number of fluorescent cells incubated with the MD_{11}-eGFP without inhibitors. *p < 0.05, $MD_{(11-19)}$-eGFP without inhibitors versus the same MD in presence of different inhibitors.

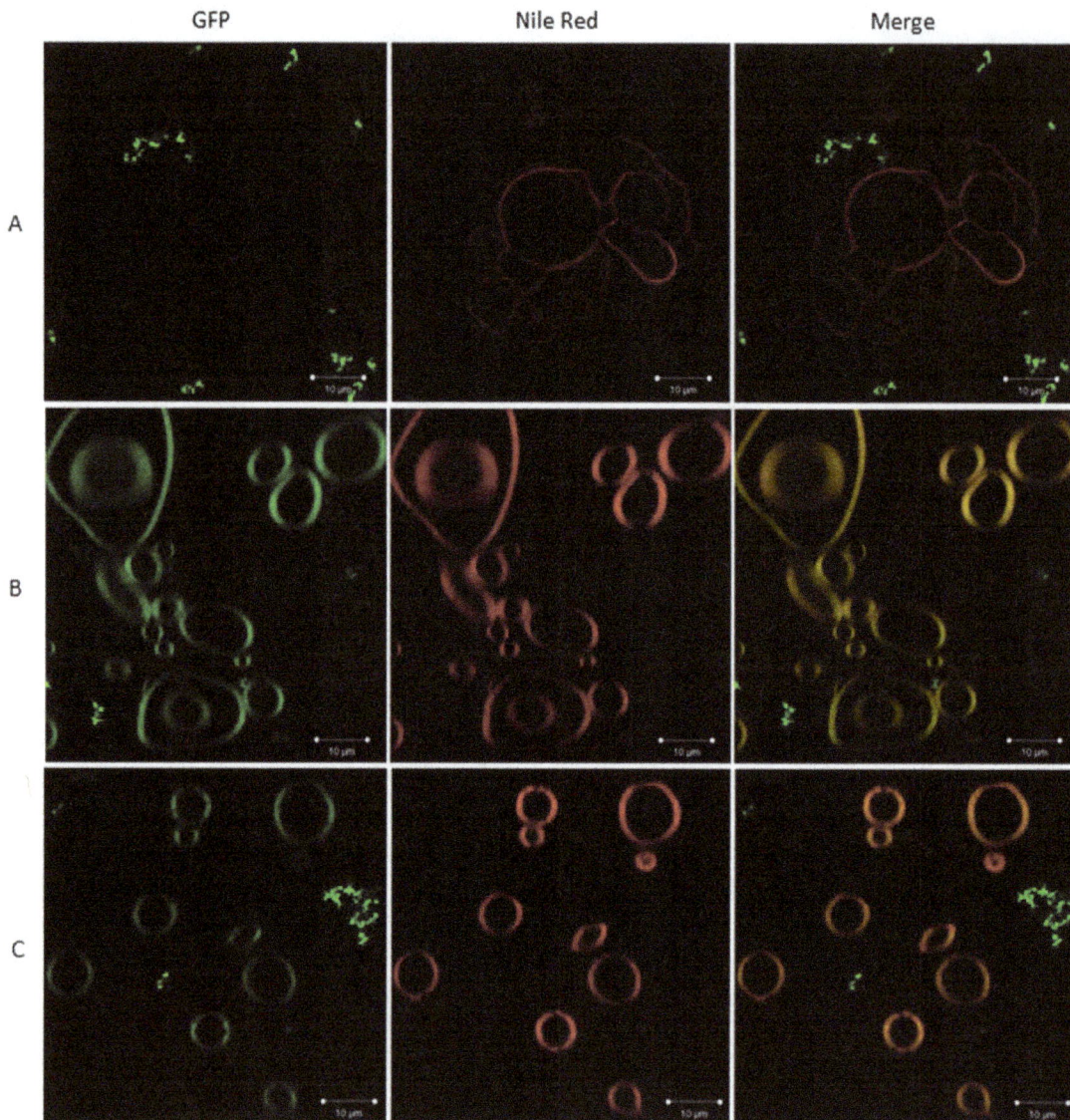

Figure 6. Confocal imaging of MD_{11}eGFP-GUV interaction after 3.5h treatment. MD_{11}-eGFP was incubated with A) fluid (100% PC), B) rigid (SM:Chol 50:50 % molar ratio) and C) semi-fluid (PC:SM:Chol 33:33:33% molar ratio) GUVs. MD_{11}-eGFP signal is shown in green and the lipids in red. The co-localization of the peptide on lipid membranes is shown in the merge panel.

cles was observed after one hour incubation (data not shown).

Discussion

Increasing efforts are currently being made for developing CPPs as delivery systems for therapeutic macromolecules. In this context, the identification of their cellular uptake mechanisms is essential for optimization of appropriate strategies. The majority of CPPs is internalized via endocytosis when coupled to cargo molecules (Eiriksdottir *et al.* 2010, Lundin *et al.* 2008, Saalik *et al.* 2004). This aspect represents a great limit because payload molecules can be entrapped and degraded into endosomal vesicles, with the consequent loss of their biological activity. Due to a potential degradation of the CPP-coupled molecule, this internalization mechanism does not represent a method of choice for the cellular transfer of therapeutic cargoes. There-

fore, identification of new CPPs endowed with direct translocation mechanism is necessary. Recent protein complementation assays have been developed to confirm the cellular uptake of cargoes mediated by CPPs and the endosomal escape of biologically active molecules (Milech *et al.* 2015).

We have recently shown that EBV ZEBRA protein contains a 51 amino acid- sequence (residues 170-220, MD) able to transduce across mammalian cellular membranes by an endocytosis-independent mechanism (Rothe *et al.* 2010). MD peptide contains two regions: the basic DBD and the adjacent leucin-rich DIM domain. Previous results indicated that the presence of both domains could be essential for MD cellular uptake (Rothe *et al.* 2010). Furthermore, enzymatic activity has been detected even in cells transduced with MD fused to high-molecular weight protein, such as β-galactosidase (116kDa) (Rothe *et al.* 2010). This evidence prompts us toward further investigations to explore the employment of MD as carrier of therapeutic macromolecules. The identification of a shorter MD peptide able to enter cells by an endocytosis-independent mechanism is suitable for developing efficient drug delivery strategies. For this purpose, different MD truncations have been designed with the aim to reduce the carrier size and preserve its mechanism of internalization. The direct translocation of hydrophilic cargoes to intracellular compartments represents a great advantage for therapeutic applications as they can be directly delivered into cells in an active state without being entrapped and degraded in endosomes/lysosomes. As the presence of a cargo may influence the internalization process, MD peptides were fused to the eGFP reporter protein and their cellular uptake was investigated by confocal microscopy and FACS analysis.

Starting from an eight amino acid shorter peptide (MD_{11}), six different peptides were produced by reducing the C-terminal end, and two were generated by isolation of DBD (MD_{18}) and the DIM (MD_{19}) domains. All DIM-deleted peptides possess protein transduction properties, even though the highest efficiency has been observed in cells incubated with the original MD_{11}-eGFP protein. Unlike the DIM domain (MD_{19}), the peptide containing only DBD domain (MD_{18}) can pass through the cell membranes, indicating that the cationic charges may play an important role in the uptake mechanism. To quantify the peptide cellular uptake, FACS analysis was performed. In agreement with qualitative data obtained by confocal microscopy, size reduction of the DIM domain induces a decrease in translocation efficiency. Fluorescence values normalized to the highest recorded signal of MD_{11} show a decrease of about 70% for all DIM-deleted peptides and the absence of internalization for MD_{19}-eGFP (DIM domain only).

We can first explain these results considering that the decrease in length of DIM domain could entail structural modifications of MD_{11} helix affecting its penetrating activity. Secondly, as shown for TAT peptide (Hoyer et al. 2012), MD_{11} could penetrate into cells in dimer form and the reduction of the DIM sequence might prevent this process inducing a decrease in the internalization efficiency.

Curiously, the loss in uptake is not directly linked to the number of removed amino acids. In fact, we observe that the deletion of five or seven residues (MD_{12} and MD_{13} respectively) induces a lower cytoplasmic accumulation than the uptake mediated by MD_{16} and MD_{17}. The analysis of the primary sequences shows that, similarly to MD_{11}, both MD_{16} and MD_{17} possess a repeated motif at C-terminal end. This motif is composed of two or three successive hydrophobic residues (Leu or Ala) followed by a basic (Lys) and/or a polar (Gln or Ser) residue. Probably, the presence of this motif is involved in the interaction with hydrophobic membrane lipids and may regulate the cellular uptake. These pieces of evidence highlight that the length of the MD helix and its polar character play a key role in the internalization process.

To check the impact of chain length reduction on internalization mechanism, we investigated the uptake of MD_{11} and its derived peptides in presence of endocytosis inhibitors. Inhibition of caveolae-mediated endocytosis and macropinocytosis does not affect MD_{11}-eGFP uptake, whereas the internalization of all MD_{11} deleted forms is strongly impaired. This loss of endocytosis-independent mechanism is also suggested by the presence of vesicles-like structures in cytoplasm of HeLa target cells incubated with MD_{16} and MD_{17} fusion proteins (Figure 4). These results demonstrate that (i) the MD_{11}-mediated cellular import is neither based on caveolae-mediated endocytosis nor on macropinocytosis and (ii) the deletion of DIM domain induces a switch from a non-endocytic to an endocytic pathway. Future optimizations of MD_{11}-derived CPPs need to take into account that all the residues of the DIM domain are required to guarantee the endocytosis-independent internalization.

To elucidate the roles of lipids in MD_{11}-membrane interactions, we used fluid (PC), rigid (SM, Chol) and semi-fluid (PC, SM, Chol) vesicles. We show that MD_{11} peptide strongly interacts with sphingomyelin and cholesterol, but not with phosphatidylcoline. These evidences well correlate with FACS data on living cells indicating that the depletion of cholesterol by MβCD treatment can have an impact on the MD_{11} entry. Since SM and cholesterol are both involved in lipid raft formation (Simons et al. 1997), we suggest that MD_{11} may interact with lipid-raft membrane domains during internalization. However, any significant accumulation in vesicles' lumen has been detected after MD_{11}-eGFP incubation. Our findings are in agreement with results on other CPPs such as R12-HA, Pep-1 and MPG, that are known to enter cells in endocytic-independent manner, and are not able to translocate and accumulate in artificial vesicles (Deshayes et al. 2004, Henriques et al. 2004, Hirose et al. 2012). These studies have established that direct translocation is strongly dependent on the presence of a negative membrane potential, and is modulated by the lipid composition, the membrane curvature and the local peptide concentration (Henriques et al. 2004, Hirose et al. 2012, Terrone et al. 2006). Furthermore, artificial membranes represent a simplified system compared to biological membrane, and a combination of different methods is necessary to fully clarify the MD_{11} entry mechanism.

Recent studies have reinforced the notion that a characteristic for direct translocation of CPPs is the presence of a hydrophobic moiety and have proved that this process occurs at specific locations of the plasma membrane (Hirose *et al.* 2012, Palm-Apergi *et al.* 2012). The attachment of a hydrophobic peptide tag deriving from human influenza hemagglutinin greatly accelerates the direct penetration of dodeca-arginine (R12) cationic peptide (Hirose *et al.* 2012). The hydrophobicity of the coupled peptide stimulates dynamic morphological alterations in the plasma membrane, which allow the permeation through the lipid bilayer (Hirose *et al.* 2012). Since the DIM domain of the MD_{11} is hydrophobic, we believe that it acts in a similar way and can stimulate the internalization in an endocytosis-independent manner. It was previously demonstrated that the amino acid Leu217 located into the DIM domain was required for the homodimerization of ZEBRA and may avoid heterodimirezation of ZEBRA with other bZIP proteins (Petosa *et al.* 2006). The decrease of cellular uptake with our deletion mutant MD_{12}-eGFP indicates that the removal of Leu217 amino acids may impact first, on the dimerization of ZEBRA and then, on the transduction capacities of the mutant. Other CPPs such as Antennapedia (Antp) or 30Kc19 have a dimerization propensity at the plasma membrane for stimulating the internalization process (Derossi 1994, Park *et al.* 2014). Like these CPPs, the hydrophobic environment of the DIM domain is important for the cellular uptake. According to these pieces of evidence, we believe that the charged DBD domain is involved in the initial interaction with the bilayer while the hydrophobic domain is crucial for its insertion.

In conclusion, with the present study we demonstrate that the MD_{11} C-terminal DIM domain contributes to the internalization but doesn't act as a driving force in this process and that it regulates the MD_{11} endosomal escape. We recently provided the first proof of concept that it can be used as carrier of therapeutic active molecules both in mammalian cells (Marchione *et al.* 2015) and in yeast cells (Marchione *et al.* 2014), demonstrating its promising potential for the development of new therapeutic strategies.

Competing interests

The authors declare that they have no competing interests.

Acknowledgements

We thank Yves Usson (Platforme IBiSA, Imagerie Sciences du Vivant, TIMC IMAG Laboratory, France) for the confocal microscopy facility. We thank Romy Rothe Walther for MD_{11} cloning and Aline Thomas for scientific discussions.

References

Akashi K, Miyata H, Itoh H & Kinosita K Jr. 1996 Preparation of giant liposomes in physiological conditions and their characterization under an optical microscope. *Biophys J* **71** 3242-3250

Derossi D, Joliot AH, Chassaing G & Prochiantz A 1994 The third helix of the Antennapedia homeodomain translocates through biological membranes. *J Biol Chem* **269** 10444-10450

Deshayes S, Gerbal-Chaloin S, Morris MC, Aldrian-Herrada G, Charnet P, Divita G & Heitz F 2004 On the mechanism of non-endosomial peptide-mediated cellular delivery of nucleic acids. *BBA* **1667** 141-147

Eiriksdottir E, Mager I, Lehto T, El Andaloussi SH & Langel U 2010 Cellular internalization kinetics of (luciferin-)cell-penetrating peptide conjugates. *Bioconjugate Chem* **21** 1662-1672

Elliott G & O'Hare P 1997 Intercellular trafficking and protein delivery by a herpesvirus structural protein. *Cell* **88** 223-233

Elmquist A, Lindgren M, Bartfai T & Langel U 2001 VE-cadherin-derived cell-penetrating peptide, pVEC, with carrier functions. *Exp Cell Res* **269** 237-244

Frankel AD & Pabo CO 1988 Cellular uptake of the tat protein from human immunodeficiency virus. *Cell* **55** 1189-1193

Green M & Loewenstein PM 1988 Autonomous functional domains of chemically synthesized human immunodeficiency virus tat trans-activator protein. *Cell* **55** 1179-1188

Heitz F, Morris MC & Divita G 2009 Twenty years of cell-penetrating peptides: from molecular mechanisms to therapeutics *Br J Clin Pharmacol* **157** 195-206

Henriques ST & Castanho MA 2004 Consequences of nonlytic membrane perturbation to the translocation of the cell penetrating peptide pep-1 in lipidic vesicles. *Biochem* **43** 9716-9724

Hirose H, Takeuchi T, Osakada H, Pujals S, Katayama S, Nakase I, Kobayashi S, Haraguchi T & Futaki S 2012 Transient focal membrane deformation induced by arginine-rich peptides leads to their direct penetration into cells. *Mol Ther* **20** 984-993

Hoyer J, Schatzschneider U, Schulz-Siegmund M & Neundorf I 2012 Dimerization of a cell-penetrating peptide leads to enhanced cellular uptake and drug delivery. *Beilstein J Org Chem* **8** 1788-1797

Kyte J & Doolittle RF 1982 A simple method for displaying the hydropathic character of a protein. *J Mol Biol* **157** 105-132

Langel U 2006, in: E.L.U. CRC Press/Taylor & Francis (Ed.) Handbook of Cell-Penetrating Peptides, Boca Raton, London, New York.

Lundin P, Johansson H, Guterstam P, Holm T, Hansen M, Langel U & El Andaloussi SH 2008 Distinct uptake routes of cell-penetrating peptide conjugates. *Bioconjugate Chem* **19** 2535-2542

Madani F, Lindberg S, Langel U, Futaki S & Graslund A 2011 Mechanisms of cellular uptake of cell-penetrating peptides. *J Biophys* **2011** 414729

Marchione R, Daydé D, Lenormand JL & Cornet M 2014 ZEBRA cell-penetrating peptide as an efficient delivery system in Candida albicans. *Biotechnol J* **9** 1088-1094

Marchione R, Laurin D, Liguori L, Leibovitch MP, Leibovitch SA & Lenormand JL 2015 MD11-mediated delivery of recombinant eIF3f induces melanoma and colorectal carcinoma cell death. *Mol Ther Methods Clin Dev* **2** 14056

Milech N, Longville BA, Cunningham PT, Scobie MN, Bogdawa HM, Winslow S, Anastasas M, Connor T, Ong F, Stone SR, Kerfoot M, Heinrich T, Kroeger KM, Tan YF, Hoffmann K, Thomas WR, Watt PM & Hopkins RM 2015 GFP-complementation assay to detect functional CPP and protein delivery into living cells. *Sci Rep* **16** 18329

Mitchell DJ, Kim DT, Steinman L, Fathman CG & Rothbard JB 2000 Polyarginine enters cells more efficiently than other polycationic homopolymers. *J Pept Res* **56** 318-325

Palm-Apergi C, Lonn P & Dowdy SF 2012 Do cell-penetrating peptides actually "penetrate" cellular membranes? *Mol Ther* **20** 695-697

Park HH, Sohn Y, Yeo JW, Park JH, Lee HJ, Ryu J, Rhee WJ & Park TH 2014 Dimerization of 30Kc19 protein in the presence of amphiphilic moiety and importance of Cys-57 during cell penetration. *Biotechnol J* **9** 1582-1593

Petosa C, Morand P, Baudin F, Moulin M, Artero JB & Muller CW 2006 Structural basis of lytic cycle activation by the Epstein-Barr virus ZEBRA protein. *Mol Cell* **21** 565-572

Pooga M, Hallbrink M, Zorko M & Langel U 1998 Cell penetration by transportan. *FASEB J* **12** 67-77

Rothe R & Lenormand JL 2008 Expression and purification of ZEBRA fusion proteins and applications for the delivery of macromolecules into mammalian cells. *Curr Protoc Protein Sci* **18** 18.11

Rothe R, Liguori L, Villegas-Mendez A, Marques B, Grunwald D, Drouet E & Lenormand JL 2010 Characterization of the cell-penetrating properties of the Epstein-Barr virus ZEBRA trans-activator. *J Biol Chem* **285** 20224-20233

Säälik P, Elmquist A, Hansen M, Padari K, Saar K, Viht K, Langel U & Pooga M 2004 Protein cargo delivery properties of cell-penetrating peptides. A comparative study. *Bioconjugate Chem* **15** 1246-1253

Säälik P, Niinep A, Pae J, Hansen M, Lubenets D, Langel Ü & Pooga M 2011 Penetration without cells: membrane translocation of cell-penetrating peptides in the model giant plasma membrane vesicles. *J Control Release* **153** 117-125

Simons K & Ikonen E 1997 Functional rafts in cell membranes. *Nature* **387** 569-572

Terrone D, Sang SL, Roudaia L & Silvius JR 2003 Penetratin and related cell-penetrating cationic peptides can translocate across lipid bilayers in the presence of a transbilayer potential. *Biochem* **42** 13787-13799

Ziegler A 2008 Thermodynamic studies and binding mechanisms of cell-penetrating peptides with lipids and glycosaminoglycans. *Adv Drug Deliv Rev* **60** 580-597

Zorko M & Langel U 2005 Cell-penetrating peptides: mechanism and kinetics of cargo delivery. *Adv Drug Deliv Rev* **57** 529-545

16

SOCS, inflammation and metabolism

Kyoko Inagaki-Ohara[1] and Akihiko Yoshimura[2]

[1] Research Institute, National Center for Global Health and Medicine (NCGM), 1-21-1, Toyama Shinjuku, Tokyo, Japan

[2] Department of Microbiology and Immunology, Keio University School of Medicine, 35 Shinanomachi, Shinjyuku, Tokyo, 160-8582, Japan

Correspondence should be addressed to Kyoko Inagaki-Ohara; Email: kyoinagaki@ri.ncgm.go.jp

Abstract

Obesity is characterized by the development of low-grade chronic inflammation, which is a contributing factor in defective energy metabolism. A hallmark of metabolic dysregulation, obesity is a life-style disease that contributes to diabetes, hypertension, and dyslipidemia. Further, recent studies warn that obesity can be a risk factor for certain cancers and exacerbates infectious diseases. This association is called the "metabolic domino". Suppressor of cytokine signaling (SOCS) proteins are negative feedback regulators of cytokine and hormone signaling mediated by the JAK-STAT signaling pathway. SOCS proteins regulate cell-cell communication through JAK-STAT-dependent cytokines and signaling by Toll-like receptors (TLRs) and they may be influenced by dietary factors such as fatty acids and glucose. In this review, we focus on the role of the JAK-STAT-SOCS signaling cascade in metabolic disorder and obesity-related diseases.

Introduction

Human society has survived facing starvation, predation and infection. Advances in sciences including biology, medicine, agriculture, and engineering, have enriched lives. However, we now confront metabolic syndrome comprising disorders such as diabetes, cardiovascular disease, inflammation, and cancer. Obesity is the underlying cause of metabolic syndrome, which is a pandemic disease attributed to the changing global food system, wherein processed foods are heavily marketed, readily available, and affordable.

Because of their influence on metabolic syndrome, members of the suppressor of cytokine signaling (SOCS) family of proteins are the focus of intensive and numerous studies. Cytokine inducible SH2-protein (CIS)/SOCS proteins inhibit the activation of the JAK-STAT pathway and regulate signaling by interleukins (ILs), interferons (IFNs), members of the tumor necrosis factors (TNF) superfamily, growth factors, and hormones (Endo et al. 1997, Yoshimura et al. 2007). SOCS proteins regulate immune responses such as infection, inflammation and allergy, leukocyte homeostasis, and cell growth as well as metabolic processes such as glucose turnover (Howard & Flier 2006). Our literature search identified over 500 publications regarding the relationship between the SOCS family and metabolic syndrome (Figure 1, left). Among SOCS proteins, SOCS3 is potently involved in the progression of obesity and diabetes. These diseases are risk factor for cancers, infection, stroke, myocardial infraction, ulcers, infertility, and gallstones. Among a series of diseases, SOCS3 associates with obesity-related cancers (Figure 1, right). Recently, obesity has attracted great attention, because it is linked to the pathogenesis of certain cancers, including those of the colon, esophagus, breast, stomach, and pancreas (Wolin et al. 2010). Because patients with obesity-associated cancers experience higher mortality and are more resistant to chemotherapy, increased research efforts in this area are urgently required. In this review, our main focus is on the underlying mechanisms of metabolic dysregulation of the SOCS signaling pathway.

Structure and function of SOCS proteins

The genes that encode components of the JAK-STAT signaling pathway are transcriptionally regulated by its own SOCS family members (Inagaki-Ohara et al. 2013, Yasukawa et al. 2000). The structures of CIS/SOCS proteins are similar and include a central Src-homology 2 (SH2) domain that varies in length with limited homology in their N-terminal regions and a SOCS box motif in their C-terminal domains (Figure 2). The SOCS box interacts with Elongins B and C and

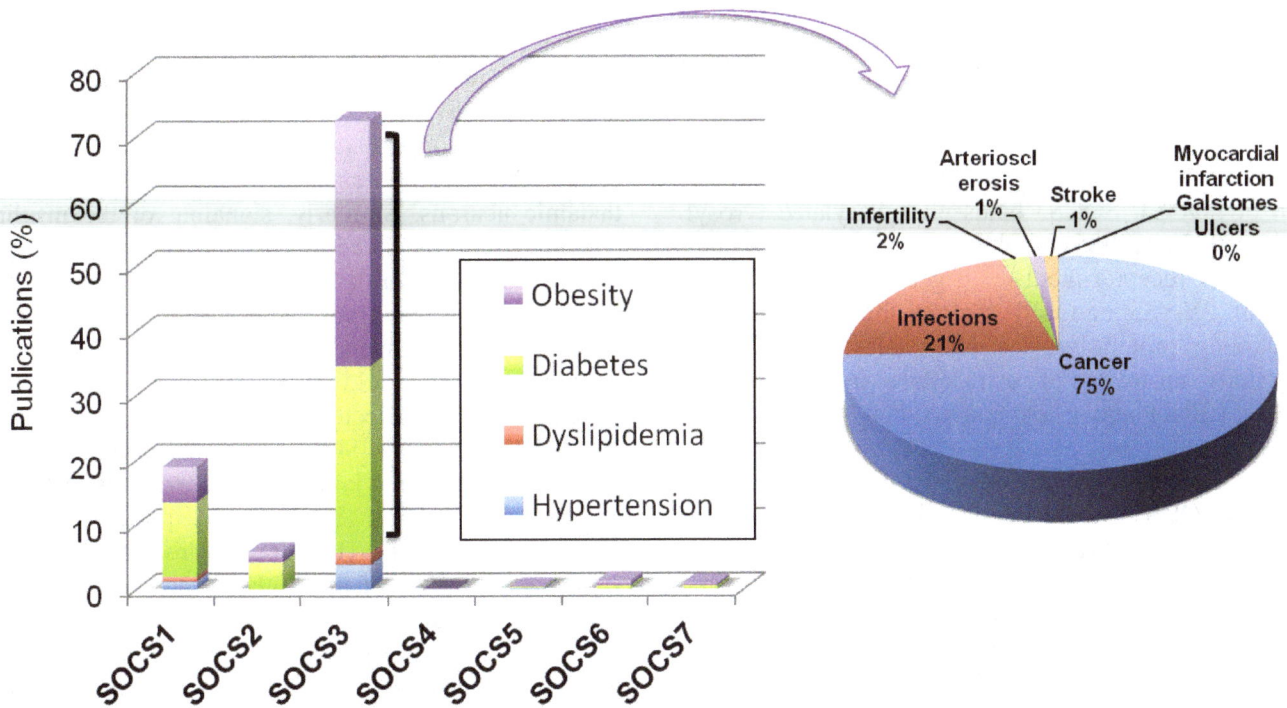

Figure 1. SOCS3 is a critical molecule in the development of metabolic syndrome and obesity-associated diseases. Each bar indicates the percentage of publications regarding metabolic syndrome on each SOCS family member according to a search of PubMed. 100 x (each SOCS)/(SOCS1-7) (left). Each pie chart shows the percentage of publications on obesity and diabetes-associated diseases involved in alteration of SOCS3 expression (right).

Cullin 5 to catalyze the ubiquitination of bound signaling protein, the RING-finger-domain-only protein RBX2 (which recruits E2 ubiquitin–transferase) (Figure 2) (Yoshimura *et al.* 2012). CIS/SOCS family proteins as well as other SOCS-box-containing molecules, likely act as E2 ubiquitin ligases. Because SOCS molecules bind to certain tyrosine-phosphorylated proteins, including Mal (toll-like receptor signaling) and IRS1/2 (insulin receptor substrate signaling) (Yoshimura *et al.* 2012), these targets may be ubiquitinated by SOCS. Unlike other SOCS proteins, SOCS1 and SOCS3 include a unique KIR domain, which is required for inhibition of JAK tyrosine kinase activity (Yasukawa *et al.* 1999) (Figure 2). The KIR domain of SOCS3 may function as a pseudosubstrate (Kershaw *et al.* 2013, Yasukawa *et al.* 1999) as well as a direct substrate of the ubiquitin-proteasome system (Piessevaux *et al.* 2008).

The role of SOCS proteins in leptin and insulin signaling

Obesity is characterized by chronic low-grade systemic and local inflammation. Infiltrating macrophages produce IL-1β and TNF-α, and T cells release IFN-γ as well as TNF-α. These pro-inflammatory cytokines are toxic for the pancreatic β-cells. Diabetes is caused when insulin production by β-cells is deficient (type 1

diabetes; T1D) or when cells that express the insulin receptor (IR) cannot respond to physiological concentrations of insulin (type 2 diabetes; T2D). Evidence indicates that genetic and environmental factors cause T1D (Bluestone *et al.* 2010). In contrast, T2D is caused by life-style (e.g. high-fat diet [HFD] and insufficient exercise) and accounts for more than 95% of patients with diabetes. T2D leads to alterations of glucose and lipid metabolism associated with insulin resistance (Lebrun & Van Obberghen 2008). Cytokines accelerate resistance to leptin and insulin in patients with T2D (Suchy *et al.* 2013). SOCS proteins regulate cytokine signaling and play important roles in pathophysiological processes leading to diabetes and obesity-associated diseases as well (Tanti *et al.* 2012).

Leptin Signaling
Leptin (product of ob gene), an adipocyte-derived hormone, binds to its receptor (ObR) in the hypothalamus to decrease food consumption and increase energy expenditure (Friedman & Halaas 1998). ObR is synthesized as multiple isoforms (ObRa–ObRf) as follows: four short isoforms with shortened intracellular tails (ObRa,c,d and f), one secreted (ObRe) and one long isoform (ObRb). Leptin belongs structurally to the long-chain helical cytokine family and activates the JAK-STAT signaling pathway and PI3K via ObRb, which is a type I cytokine receptor, similar to gp130

(Al-Qassab *et al.* 2009, La Cava & Matarese 2004). ObRb lacks intrinsic enzymatic activity and is activated by phosphorylation of its Cterminus by autophosphorylated JAK2, which is bound to the SH2 domain of STAT3 (Howard & Flier 2006). In the nucleus, STAT3 mediates gene transcription, including that of SOCS3. SOCS3 binds phosphorylated tyrosyl residue 985 (PY-985) of ObRb and to JAK2 to attenuate leptin receptor-mediated signaling. Phosphorylation of PY-985 of ObRb recruits the SH2 domain-containing protein-tyrosine phosphatase SHP-2 that functions upstream of extracellular-signal-regulated kinase (ERK) and c-fos transcription (Banks *et al.* 2000).

Leptin-stimulated signaling via STAT3 rapidly induces SOCS3, which inhibits signaling through the leptin receptor. Further studies of gene-targeted mice reveal that leptin's action specific to the central nervous system is sufficient to regulate body weight, food consumption, energy expenditure, glucose metabolism, and behavior (Gautron & Elmquist 2011). However, leptin's anorexigenic effects are suppressed in obese individuals and animals with HFD-induced obesity, despite elevated levels of serum leptin. This pathological condition is termed "leptin resistance" (Gautron & Elmquist 2011). These observations led to the proposal that SOCS3 is a potential mediator of leptin resistance. Further, peripheral administration of leptin to ob/ob mice specifically induces SOCS3 mRNA in regions of the hypothalamus that are important for regulating feeding behavior (Bjorbaek *et al.* 1998). Mice lacking SOCS3 from cells of the entire brain or only from proopiomelanocortin (POMC) neurons, which are leptin target neurons present in the arcuate nucleus of the hypothalamus, are resistant to HFD-induced obesity (Kievit *et al.* 2006, Mori *et al.* 2004). In addition, studies of SOCS3 transgenic mice show that overexpression of SOCS3 in POMC neuron but not in ObRb neurons is sufficient to impart leptin resistance and obesity mediated by antagonizing signaling through phosphorylated STAT3 and mTOR-S6K (Reed *et al.* 2010). In the periphery, leptin inhibits insulin secretion and expression of preproinsulin mRNA in the pancreatic β- cells. SOCS3 expressed by β-cells is involved in leptin-mediated inhibition of preproinsulin synthesis (Mori *et al.* 2007). Furthermore, leptin transactivates STAT3/STAT5b and the promoter of the preproinsulin 1 gene, and SOCS3 inhibits the activities of both promoters (Lebrun & Van Obberghen 2008), suggesting that SOCS3 inhibits directly the JAK-STAT signaling pathway as well as downstream signaling when the pathway is activated by leptin.

Insulin signaling

Insulin is secreted by the pancreas and stimulates the uptake of glucose and nutrients into peripheral target tissues. Like leptin, insulin reduces body weight and food intake, and regulates the expression of genes encoding neuropeptides as well as the activity of hypothalamic neurons. Similarly, signaling of the insulin receptor (IR) is mediated by phosphorylation of tyrosyl residues. Unlike ObRb, the IR has intrinsic tyrosine kinase activity that upon ligand binding, autophosphorylates the transmembrane domain of its β-subunit on tyrosyl residues Y1158, Y1162, and Y1163 to recruit downstream effector proteins, including IR substrates (IRS) 1 and 2. SOCS1 and SOCS3 bind to the IR. SOCS1 phosphorylated on its C-terminus (Y1158, Y1162 and Y1163), SOCS3, and protein-tyrosine phosphatase 1B (PTP-1B) suppress insulin and leptin signaling via different molecular mechanisms (Suchy *et al.* 2013) (Figure 3).

Increased expression of SOCS1 and SOCS3 induces insulin resistance in a variety models of obesity and diabetes via inhibition of JAK-STAT signal-

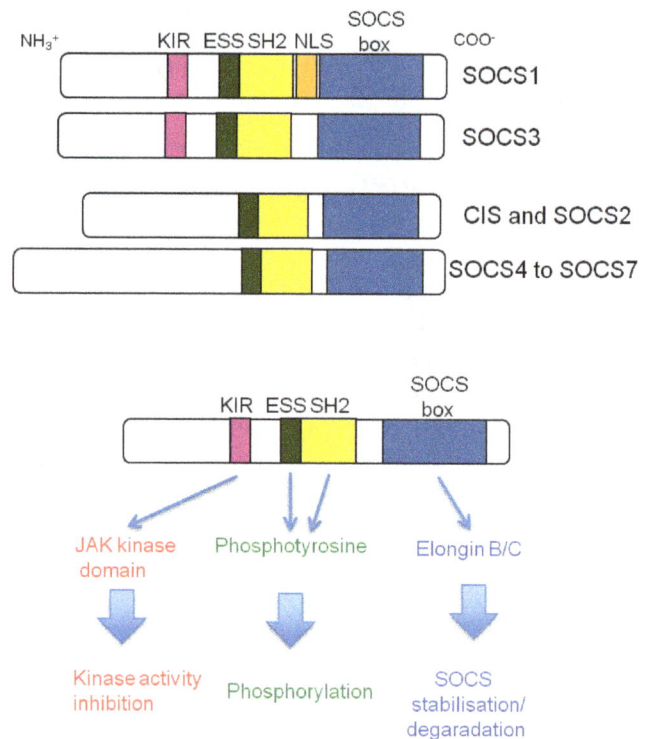

Figure 2. The structure and function of SOCS proteins. The SOCS family consists of eight members. All eight members share a central SH2 domain, extended SH2 domain (ESS), and a C-terminal SOCS box. In addition, SOCS1 and SOCS3 possess a kinase inhibitory region (KIR) that serves as a pseudo-substrate for JAKs that inhibits JAK function. Only SOCS1 contains a nuclear localization signal. A diagram of the extended interactions of SOCS with target proteins. The SOCS box interacts with several ubiquitinating machinery enzymes, i.e. Elongins B and C.

ing, competition for the binding of the IRS1 or targeting the degradation of IRS1 (Howard & Flier 2006, Yoshimura *et al.* 2007). However, SOCS1/RAG2-deficient mice, which survive to adulthood and gain body weight similar to wild-type mice, exhibit inflammation of adipose tissue, accompanied by elevated levels of leptin, TNF-α, and CD68 (a macrophage marker) in adipose tissue (Emanuelli *et al.* 2008). They also exhibit increased transcription of lipogenic genes in the liver, such as Srebp1c and Fas (Emanuelli *et al.* 2008). Furthermore, SOCS1 over-expression alone is

insufficient to block total IFN-γ activity in pancreatic islets (Zaitseva *et al.* 2009). Interestingly, SOCS1-deficient neonatal mice exhibit drastic hypoglycemia and hypoinsulinemia, develop multiorgan inflammatory disease, and die before weaning (Jamieson *et al.* 2005). In contrast, SOCS3 haplo-deficient mice and mice with SOCS3-deficiency in the brain are resistant to HFD-induced obesity and are insulin resistance (Howard *et al.* 2004, Mori *et al.* 2004).

Inhibition of the expression of SOCS3 alone or together with SOCS1 and SOCS3, using antisense oli-

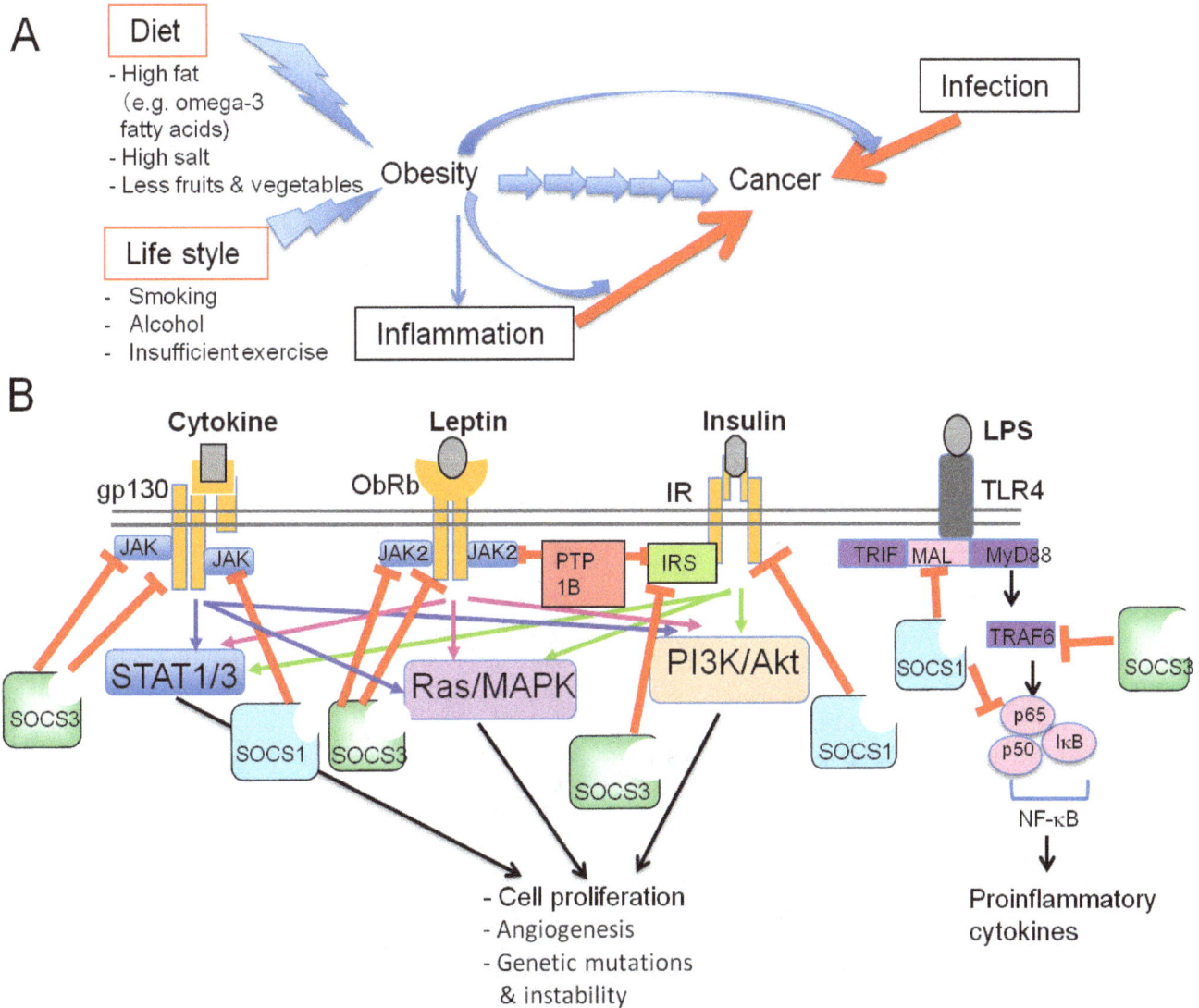

Figure 3. Multiple signaling pathways activated in obesity-associated diseases. A) Diet and lifestyle are critical factors for the development and regulation of obesity. Obesity is characterized by chronic low-grade systemic and local inflammation. Further, obesity accelerates inflammation- and infection-associated cancers through the receptors for cytokines, leptin, insulin, and bacterial components such as LPS via multiple signaling pathways including JAK-STAT, Ras-MAPK, PI3KAkt, and TLR. ObRb, which is analogous to gp130, lacks intrinsic kinase activity as illustrated in B). B) Binding of leptin to ObRb induces autophosphorylation and activation of noncovalently associated JAK2, which in turn, leads to phosphorylation of highly conserved tyrosyl residues in the intracellular domain of ObRb that recruit STAT3. STAT3 activation induces SOCS3 expression. Unlike gp130 and ObRb, IR possesses intrinsic tyrosine kinase activity that recruits and phosphorylates effector IRS molecules that, in turn, recruit adaptor molecules and mediate downstream signaling through PI3K and ERK. PTP1B inhibits leptin and insulin signaling by dephosphorylating JAK2 and IR, respectively.

gonucleotides in the livers of db/db mice, suppresses the expression of lipogenic genes (Ueki et al. 2004). Hepatic SOCS3 mediates insulin resistance; however, aged hepatocyte-specific SOCS3-deficient mice exhibit reduced insulin signaling in muscle, although insulin sensitivity in the liver is enhanced (Torisu et al. 2007). These studies suggest that lack of SOCS3 in the liver promotes systemic insulin resistance mediated by STAT3 activation induced by inflammatory factors produced from the liver. SOCS1-deficiency alone does not prevent HFD-induced obesity and insulin resistance. Considering the role of SOCS3 as a suppressor of IL-1β, IFN-γ, and TNF-α signaling in pancreatic β-cells, the SOCS3 pathway may prevent the onset of T2D as well as T1D (Bruun et al. 2009).

Several factors, such as IL-6, leptin, TNF-α, and infection with hepatitis C virus (HCV) regulate SOCS3 expression. For example, insulin increases the rate of SOCS3 expression in adipose tissue, liver, and muscle tissues (Emanuelli et al. 2001). Double deletions of the gene encoding SOCS3 and PTP-1B in the brain cells, compared with deleting them individually, improves sensitivity to insulin in an additive manner; however, the SOCS3 deletion contributes more significantly, which accounts for the low level of insulinemia detected in the double mutants (Briancon et al. 2010). This strategy is considered an important example of targeted therapy of T2D and obesity.

Function of SOCS proteins in Toll-like receptor (TLR) signaling

The SOCS proteins, in particular SOCS1 and SOCS3, inhibit TLR signaling as well as the JAK-STAT signaling pathway. Chronic inflammatory signaling is a key factor in the development of obesity, causes peripheral insulin, and accelerates leptin resistance. TLRs are a class of receptors that have key roles within the innate immune system by activating proinflammatory signaling cascades upon recognition of microbial and viral products (Medzhitov 2001). In particular, TLR4 contributes to the development of insulin resistance through its activation by an increased number of exogenous ligands, such as dietary fatty acids and lipopolysaccharide (LPS). Activation of the TLR4 signaling cascade induces the production of proinflammatory cytokines, chemokines, and reactive oxygen species (ROS), which are all effectors of innate immunity. TLR4 is expressed by many cells in insulin target tissues, including pancreatic β-cells as well as liver, skeletal muscle (Kim & Sears 2010). Therefore, TLR4 activation may suppress insulin action through proinflammatory cytokine signaling, ROS generation, and producing insulin-desensitizing factors.

The daily diet plays an important role in the development of obesity and diabetes, and the types and caloric content of meals are significant contributors to the inflammatory response. For example, consumption of foods high in carbohydrates and fat induce inflammation and increase in the LPS concentration in the plasma. Consumption of these foods, but not those with high-fiber content, elevates SOCS3 expression in circulating mononuclear cells (Ghanim et al. 2009). Long-chain polyunsaturated omega-3 fatty acids such as DHA and EPA antagonize TLR4 activation by saturated fatty acids and LPS (Lee et al. 2001, 2003, Shi et al. 2006). Consumption of sweets and cream increase the levels of TNF-α, IL-1β, and SOCS3 but not those of SOCS1 (Deopurkar et al. 2010). Further, cream enhances the post-meal spike in the concentration of circulating LPS concentration in contrast to drinking a beverage containing sugar. Moreover, SOCS3 signaling and NF-κB activation in circulating mononuclear cells are remarkably elevated when cream and soft drinks are consumed. These results suggest that the level of SOCS1 and SOCS3 induced by consuming certain kinds of food may differ, although SOCS proteins modulate insulin resistance.

Obesity-related diseases

Obesity-induced inflammation is an important contributor to the development of pathologies such as T2D, atherosclerosis, liver disease, infections, and some types of cancer (Gregor & Hotamisligil 2011, Wolin et al. 2010). In particular, cancer and infection are highly associated with the regulation of SOCS protein expression in obesity-associated diseases (Figure 1, right).

Cancer
Persistent inflammation increases cancer risk, which is driven by genetic alterations that cause inflammation and neoplasia. STATs and NF-κB are key coordinators of innate immunity and inflammation and are executors of tumor promoters (Inagaki-Ohara et al. 2013). Twenty percent of cancers can be attributed to obesity (Wolin et al. 2010), and increased cancer-related mortality (Reeves et al. 2007) attracts great attention as a global health problem. The number of people with diabetes has increased and is predicted to rise to 552 million by 2030 (Wild et al. 2004), suggesting that cancer risk will rise proportionately.

White adipose tissue (WAT) performs multiple functions other than to store lipids. The increase in the mass of WAT during the progression of obesity elevates the production of adipokines such as leptin, IL - 6, TNF-α, causing chronic mild inflammation. Diabetic patients are at significantly higher risk for common cancers including those of the breast, gastrointestinal (GI) (esophageal, colorectal) tract, liver, pancreas, urinary tract, and female reproductive tissues (Xu *et al.* 2014).

Breast cancer

Approximately 60% of breast cancers are hormone-dependent. The hormonal changes that occur post-menopause are attributed to a specific metabolic state that represents a greater risk for breast cancer. These changes are considered indispensable for a more effective therapy (Maccio & Madeddu 2011). Among adipokines, leptin is the most intensively studied regarding its metabolism and role in obesityrelated carcinogenesis, because leptin induces physiological responses in peripheral tissues other than those involved in neuronal activity. Increased expression of leptin and ObRb in human grade-III invasive breast tumors is associated with shorter time to tumor recurrence and mortality (Garofalo *et al.* 2006, Maccio *et al.* 2010).

Tumor cells derived from MMTV-Wnt1 mice, a widely used model of mammary tumors, show decreased growth in ob/ob mice (leptin-deficient) compared with diet-induced obese mice with an intact leptin signaling pathway (Zheng *et al.* 2011). This result suggests that leptin signaling plays an essential role in the growth and survival of tumors induced by MMTV-Wnt1. Tumor initiating stem cells express high levels of ObR to promote tumorigenesis caused by STAT3 activation and by inducing pluripotencyassociated transcription factors such as Oct4 and Sox2 (Feldman *et al.* 2012). In patients with obesity-related breast cancer, the JAK2-STAT3 pathway is activated and SOCS3 is down-regulated (Santillan-Benitez *et al.* 2014). These effects of leptin are mediated through a set of responses of ObRb-positive tumor cells, including a cancer stem cell population that expresses the ObRb. These findings suggest that leptin affects tumor initiation and progression through STAT3 activation.

Gastrointestinal (GI) cancer

The relative risk of GI cancer in obese individuals is highly prevalent among patients with obesityassociated Barrett's esophagus and colon cancer (Wolin *et al.* 2010). The combination of adenomatous polyposis coli (Apc) and db/db mice enhances Apc-driven tumorigenesis of the small intestine and induces gastric and colonic tumors. In contrast, db/db mice do not develop GI neoplasia (Gravaghi *et al.* 2008). Recently, gastric cancer has emerged as an obesity-associated cancer. Normal stomach tissues spontaneously express leptin and ObRb, and their expression levels increase during carcinogenesis (Bado *et al.* 1998, Inagaki-Ohara *et al.*

2014), and induce autocrine signaling (Hoda *et al.* 2007), suggesting that the stomach is more susceptible to ObR signaling than other tissues. Approximately 90% of gastric cancers are gastric adenocarcinomas (GCA), which are further categorized as distal or noncardia GCA and proximal or cardia GCA. Interestingly, cardia GCA is associated with obesity (Cho *et al.* 2012, O'Doherty *et al.* 2012), suggesting that the unique localization of the leptin-ObR signaling pathway to cells of the GI tract functions predominantly in the early phase of human GC and serves as a biomarker. Therefore, targeting this pathway may be invaluable for treatment.

Hepatic cancer

Leptin's oncogenic role, including its capability to enhance tumor invasiveness and migration of hepatocellular carcinoma (HCC) cells, may be antagonized by adiponectin in HCC through suppression of STAT3 and Akt phosphorylation when SOCS3 is up-regulated (Sharma *et al.* 2010). Studies conducted in vitro reveal that the SOCS3 binding site is essential for the interaction with IRS1 and IRS2, whereas the affinity of SOCS1 for a domain within the catalytic loop is crucial for IRS2 (Ueki *et al.* 2004). SOCS1 and SOCS3 bind to the IR in cells, and their overexpression impairs the phosphorylation of IRS1 and IRS2 stimulated by insulin. Insulin resistance is caused by IL-6 due to suppression of tyrosine phosphorylation of IRS-1 through the induction of SOCS3 in murine primary hepatocytes and human hepatocarcinoma cells (Wunderlich *et al.* 2013). Indeed, IL-6-deficient mice are resistant to diethylnitrosamine (DEN)-induced carcinogenesis when fed a HFD (Park *et al.* 2010). Recently, Shimizu *et al.* (2011) showed that several signaling pathways including insulin/IGF-1/PI3K/Akt, ERK, JNK, and STAT3 are important in DEN-induced liver carcinogenesis in db/db mice. SOCS3 expression is reduced in the levers of HCC patients, which is supported by findings that mice with hepatocyte-specific deletion of SOCS3 are resistant to concanavalin A-induced hepatocarcinogenesis (Ogata *et al.* 2006).

Collectively, these studies reveal that leptin exerts its actions centrally and provides beneficial effects to peripheral organs. However, although such peripheral functions of leptin exist, chronic JAK-STAT3- SOCS3 pathway activity in obesity is mainly derived from other signals that, in contrast, act in the periphery as well as the CNS. Further, crosstalk between leptin and IGF-1 significantly increases the proliferation as well as invasion and migration of breast cancer cells, suggesting that the cooperation of several signaling pathways is required to induce obesity-associated carcinogenesis.

Infection

Hepatitis C virus

Overwhelming epidemiological evidence indicates that persistent infection with HCV and HBV is a major risk factor for the development of HCC (Koike et al. 2008). In transgenic mice carrying the gene encoding the HCV core protein (PA28 gamma (+/+) Core Tg), the HCV core protein induces hyperexpression of TNF-α (Miyamoto et al. 2007), and HCV infection causes insulin resistance and T2D, which is sufficient to impair insulin signaling in vitro through the SOCS protein activation and the consequent decrease in IRS-1 expression (Pascarella et al. 2011).

Chronic HCV infection of humans is treated with a combination of Peginterferon (a long acting IFN) and Ribavirin (guanosine analog that inhibits RNA synthesis). These drugs contribute to decreased sensitivity to interferon, which is inhibited by SOCS3 (del Campo et al. 2010). Interestingly, SOCS1-deficient mice show hyperglycemia but die before reaching three weeks of age due to enhanced IFN-γ signaling (Starr et al. 1998). Human embryonic stem cell-derived hepatocytes (hESC-Heps) are capable of supporting the full HCV life cycle and viral infection to a lesser extent compared with the HUh 7 hepatocarcinoma cell line that produces IL-29, a type III IFN, upon stimulation by HCV infection (Zhou et al. 2014). Considering that the level of viral infection and replication in hESC-Heps is increased by addition of JAK inhibitor I that modulates signaling through the JAK-STAT pathway as well as the downstream response to IFN, "tunable" hESC-Heps may serve as a platform for the development of anti- HCV drugs.

Influenza

The association between obesity and influenza was first reported during the 2009 influenza A (H1N1) pandemic (Louie et al. 2011, Morgan et al. 2010). Obese volunteers infected with influenza A (H1N1) show a reduced ability to produce type I IFN in response to the TLR3 ligand, delayed proinflammatory responses, and increased basal expression of SOCS3 but not SOCS1 (Teran-Cabanillas et al. 2014). Dietinduced obese mice exhibit similar responses. In these mice, the number of influenza virus-specific CD8+- memory T cells in the lung, an important cell population for protection against the subsequent virus exposure is decreased, but expression of SOCS1 and SOCS3 is increased (Karlsson et al. 2010). These results suggest that new vaccine strategies are required for obese individuals, because the standard vaccine that induces the proliferation of memory T cells may be less effective in an obese population.

Intestinal microbiota

It is important to understand that the output of signaling pathways that are activated by certain bacteria may protect against intestinal inflammation or obesity. Members of the phyla Bacteroidetes and Firmicutes dominate the gut microbiome, and an increased ratio of Firmicutes to Bacteroidetes is implicated in the development of adult obesity (Eckburg et al. 2005, Ley et al. 2006, Turnbaugh et al. 2006). Obesity is characterized by chronic low-grade inflammation with reduced GI barrier function involving a variety factors and inflammatory mediators (Chakraborty et al. 2010). "Metabolic endotoxemia" can initiate obesity and insulin resistance through gastrointestinal bacteria- triggered SOCS3 signaling. High-fat and high-fructose diets alter the composition of the gut microbiota and the permeability of the gut, which increase the proliferation of enterobacterial species and the levels of circulating LPS (Cani & Delzenne 2009, Kim & Sears 2010). Germ-free mice or mice treated with antibiotics specific for gram-negative bacteria do not exhibit HFD -induced insulin resistance or other metabolic abnormalities associated with obesity (Backhed et al. 2007, Cani et al. 2008). The presence of body fat-inducing gut microbiota may be associated with hypothalamic signs of SOCS3-mediated leptin resistance (Schele et al. 2013). Further, the strong correlation between increased LPS concentrations and HFD-induced endotoxemia, an important component of obesity-associated inflammation in obese patients, is consistent with enhanced expression of TLR4 and NF-κB in circulating mononuclear cells (Cani et al. 2008, Ghanim et al. 2009). Deletion of Tlr4 from the myeloid cells protects mice against HFD-induced inflammation, adipose macrophage infiltration, and insulin resistance (Saberi et al. 2009). TLR5 binds to bacterial flagellin, and mice lacking this receptor display hyperphagia and develop the characteristic features of metabolic syndrome, including hyperlipidemia and insulin resistance (Vijay-Kumar et al. 2010). These metabolic changes correlate with changes in the composition of the gut microbiota, and transfer of the gut microbiota from TLR5-deficient mice to wild-type germ-free mice confers many features of metabolic syndrome to the latter.

Emerging evidence suggest that the GI tract is the origin of inflammation in HFD-induced obesity as well as adipose tissue. A HFD promotes inflammation in the GI tract, which is considered a potential source of inflammation associated with HFD-induced obesity. Adult germ-free (GF) mice have less body fat and do not become obese when they are fed a HFD, and replacing the microbiota of adult GF mice with the microbiota harvested from conventional mice increases in body fat (Backhed et al. 2004, Backhed et al. 2007).

These results indicate that nonpathogenic enteric bacteria in healthy individuals may play a key role in diet-induced obesity.

Mice devoid of PTP1B are resistant to diet-induced obesity (Bence *et al.* 2006, Elchebly *et al.* 1999, Owen *et al.* 2012). Furthermore, knockdown of PTP1B expression in the RAW264.7 macrophage cell line increases the production of IL-6, TNF-α, and IFN-β in response to a variety of TLR ligands, indicating that PTP1B can act as a negative regulator of TLR4-signaling in macrophages (Xu *et al.* 2008). Myeloid cell-specific PTP1B knockout (LysM-PTP1B) mice resist LPS-induced endotoxemia and hepatic damage associated with decreased TNF-α expression and show an increase in basal and LPS-induced IL-10 production associated with enhanced STAT3 activation (Grant *et al.* 2014). These findings suggest that myeloid PTP1B is a previously unrecognized inhibitor of STAT3/IL-10 mediated signaling and may serve as a target for treating inflammation and diabetes in obese patients.

Concluding remarks

Over the past decade, SOCS proteins have been clearly shown to inhibit the leptin and insulin signaling pathways in vitro and in vivo and to influence energy balance and glucose homeostasis. These discoveries highlight the importance of SOCS for regulating the mechanisms of onset and development of diseases involved in metabolic dysfunction in central and peripheral tissues through the JAK-STAT signaling cascade as well as through crosstalk with other mediators (Figure 3). In the future, SOCS proteins will likely provide therapeutic targets for T2D and obesity-associated cancer, although investigations should take into account the induction of SOCS proteins by diverse cytokines and cell types.

Acknowledgements

We would thank to Dr. Takao Shimizu (NCGM, Director General), Dr. Satoshi Matsumoto (Yakult Honsha Co., Ltd.) and Professor Masao Mitsuyama (Kyoto Univer-sity) for steadfast encouraging us. K. I-O is supported by Grants-in-Aid for Scientific Research (C) from the Ministry of Education, Culture, Sports, Science and Technology of Japan (26461391).

Author Contributions

K. I-O. wrote the manuscript and A. Y. provided comments.

References

Al-Qassab H, Smith MA, Irvine EE, Guillermet-Guibert J, Claret M, Choudhury AI, Selman C, Piipari K, Clements M, Lingard S, Chandarana K, Bell JD, Barsh GS, Smith AJ, Batterham RL, Ashford ML, Vanhaesebroeck B & Withers DJ 2009 Dominant role of the p110beta isoform of PI3K over p110alpha in energy homeostasis regulation by POMC and AgRP neurons. *Cell Metab* **10** 343-354

Bäckhed F, Ding H, Wang T, Hooper LV, Koh GY, Nagy A, Semenkovich CF & Gordon JI 2004 The gut microbiota as an environmental factor that regulates fat storage. *Proc Natl Acad Sci U S A* **101** 15718-15723

Backhed F, Manchester JK, Semenkovich CF & Gordon JI 2007 Mechanisms underlying the resistance to diet-induced obesity in germ-free mice. *Proc Natl Acad Sci U S A,* **104** 979-984

Bado A, Levasseur S, Attoub S, Kermorgant S, Laigneau JP, Bortoluzzi MN, Moizo L, Lehy T, Guerre-Millo M, Le Marchand-Brustel Y & Lewin MJ 1998 The stomach is a source of leptin. *Nature* **394** 790-793

Banks AS, Davis SM, Bates SH & Myers MG Jr 2000 Activation of downstream signals by the long form of the leptin receptor. *J Biol Chem* **275** 14563-14572

Bence KK, Delibegovic M, Xue B, Gorgun CZ, Hotamisligil GS, Neel BG & Kahn BB 2006 Neuronal PTP1B regulates body weight, adiposity and leptin action. *Nat Med* **12** 917-924

Bjorbaek C, Elmquist JK, Frantz JD, Shoelson SE & Flier JS 1998. Identification of SOCS-3 as a potential mediator of central leptin resistance. *Mol Cell* **1** 619-625

Bluestone JA, Herold K & Eisenbarth G 2010 Genetics, pathogenesis and clinical interventions in type 1 diabetes. *Nature* **464** 1293-1300

Briancon N, McNay DE, Maratos-Flier E & Flier JS 2010 Combined neural inactivation of suppressor of cytokine signaling-3 and protein-tyrosine phosphatase-1B reveals additive, synergistic, and factor specific roles in the regulation of body energy balance. *Diabetes* **59** 3074-3084

Bruun C, Heding PE, Rønn SG, Frobøse H, Rhodes CJ, Mandrup-Poulsen T & Billestrup N 2009 Suppressor of cytokine signalling-3 inhibits Tumor necrosis factor-alpha induced apoptosis and signalling in beta cells. *Mol Cell Endocrinol* **311** 32-38

Cani PD, Bibiloni R, Knauf C, Waget A, Neyrinck AM, Delzenne NM & Burcelin R 2008 Changes in gut microbiota control metabolic endotoxemia-induced inflammation in high-fat diet-induced obesity and diabetes in mice. *Diabetes* **57** 1470-1481

Cani PD & Delzenne NM 2009 The role of the gut microbiota in energy metabolism and metabolic disease. *Curr Pharm Des* **15** 1546-1558

Chakraborty S, Zawieja S, Wang W, Zawieja DC & Muthuchamy M 2010. Lymphatic system: a vital link between metabolic syndrome and inflammation. *Ann N Y Acad Sci* **1207** E94-102

Cho Y, Lee DH, Oh HS, Seo JY, Lee DH, Kim N, Jeong SH, Kim JW, Hwang JH, Park YS, Lee SH, Shin CM, Jo HJ, Jung HC, Yoon YB & Song IS 2012 Higher prevalence of obesity in gastric cardia adenocarcinoma compared to gastric noncardia adenocarcinoma. *Dig Dis Sci* **57** 2687-2692

del Campo JA, Lopez RA & Romero-Gomez M 2010 Insulin resistance and response to antiviral therapy in chronic hepatitis C: mechanisms and management. *Dig Dis* **28** 285-293

Deopurkar R, Ghanim H, Friedman J, Abuaysheh S, Sia CL, Mohanty P, Viswanathan P, Chaudhuri A & Dandona P 2010 Differential effects of cream, glucose, and orange juice on inflammation, endotoxin, and the expression of Toll-like receptor-4 and suppressor of cytokine signaling-3. *Diabetes Care* **33** 991-997

Eckburg PB1, Bik EM, Bernstein CN, Purdom E, Dethlefsen L, Sargent M, Gill SR, Nelson KE & Relman DA 2005 Diversity of the human intestinal microbial flora *Science* **308** 1635-1638

Elchebly M, Payette P, Michaliszyn E, Cromlish W, Collins S, Loy AL, Normandin D, Cheng A, Himms-Hagen J, Chan CC, Ramachandran C, Gresser MJ, Tremblay ML & Kennedy BP 1999 Increased insulin sensitivity and obesity resistance in mice lacking the protein tyrosine phosphatase-1B gene. *Science* **283** 1544-1548

Emanuelli B, Macotela Y, Boucher J & Ronald Kahn C 2008 SOCS-1 deficiency does not prevent diet-induced insulin resistance. *Biochem Biophys Res Commun* **377** 447-452

Emanuelli B, Peraldi P, Filloux C, Chavey C, Freidinger K, Hilton DJ, Hotamisligil GS & Van Obberghen E 2001 SOCS-3 inhibits insulin signaling and is up-regulated in response to tumor necrosis factor-alpha in the adipose tissue of obese mice. *J Biol Chem* **276** 47944-47949

Endo TA, Masuhara M, Yokouchi M, Suzuki R, Sakamoto H, Mitsui K, Matsumoto A, Tanimura S, Ohtsubo M, Misawa H, Miyazaki T, Leonor N, Taniguchi T, Fujita T, Kanakura Y, Komiya S & Yoshimura A 1997 A new protein containing an SH2 domain that inhibits JAK kinases. *Nature* **387** 921-924

Feldman DE, Chen C, Punj V, Tsukamoto H & Machida K 2012 Pluripotency factor-mediated expression of the leptin receptor (OB-R) links obesity to oncogenesis through tumor-initiating stem cells. *Proc Natl Acad Sci U S A* **109** 829-834

Friedman JM & Halaas JL 1998 Leptin and the regulation of body weight in mammals. *Nature* **395** 763-770

Garofalo C, Koda M, Cascio S, Sulkowska M, Kanczuga-Koda L, Golaszewska J, Russo A, Sulkowski S & Surmacz E 2006 Increased expression of leptin and the leptin receptor as a marker of breast cancer progression: possible role of obesity-related stimuli. *Clin Cancer Res* **12** 1447-1453

Gautron L & Elmquist JK 2011 Sixteen years and counting: an update on leptin in energy balance. *J Clin Invest* **121** 2087-2093

Ghanim H, Abuaysheh S, Sia CL, Korzeniewski K, Chaudhuri A, Fernandez-Real JM & Dandona P 2009 Increase in plasma endotoxin concentrations and the expression of Toll-like receptors and suppressor of cytokine signaling-3 in mononuclear cells after a high-fat, high-carbohydrate meal: implications for insulin resistance. *Diabetes Care* **32** 2281-2287

Grant L, Shearer KD, Czopek A, Lees EK, Owen C, Agouni A, Workman J, Martin-Granados C, Forrester JV, Wilson HM, Mody N & Delibegovic M 2014 Myeloid-cell protein tyrosine phosphatase-1B deficiency in mice protects against high fat diet and lipopolysaccharide-induced inflammation, hyperinsulinemia, and endotoxemia through an IL-10 STAT3-dependent mechanism. *Diabetes* **63** 456-470

Gravaghi C, Bo J, Laperle KM, Quimby F, Kucherlapati R, Edelmann W & Lamprecht SA 2008 Obesity enhances gastrointestinal tumorigenesis in Apc-mutant mice. *Int J Obes (Lond)* **32** 1716-1719

Gregor MF & Hotamisligil GS 2011 Inflammatory mechanisms in obesity. *Annu Rev Immunol* **29** 415-445

Hoda MR, Keely SJ, Bertelsen LS, Junger WG, Dharmasena D & Barrett KE 2007 Leptin acts as a mitogenic and antiapoptotic factor for colonic cancer cells. *Br J Surg* **94** 346-354

Howard JK, Cave BJ, Oksanen LJ, Tzameli I, Bjorbaek C & Flier JS 2004 Enhanced leptin sensitivity and attenuation of diet-induced obesity in mice with haploinsufficiency of Socs3. *Nat Med* **10** 734-738

Howard JK & Flier JS 2006 Attenuation of leptin and insulin signaling by SOCS proteins. *Trends Endocrinol Metab* **17** 365-371

Inagaki-Ohara K, Kondo T, Ito M & Yoshimura A 2013 SOCS, inflammation, and cancer. *JAKSTAT* **2** e24053

Inagaki-Ohara K, Mayuzumi H, Kato S, Minokoshi Y, Otsubo T, Kawamura YI, Dohi T, Matsuzaki G & Yoshimura A 2014 Enhancement of leptin receptor signaling by SOCS3 deficiency induces development of gastric tumors in mice. *Oncogene* **33** 74-84

Jamieson E, Chong MM, Steinberg GR, Jovanovska V, Fam BC, Bullen DV, Chen Y, Kemp BE, Proietto J, Kay TW & Andrikopoulos S 2005 Socs1 deficiency enhances hepatic insulin signaling. *J Biol Chem* **280** 31516-31521

Karlsson EA, Sheridan PA & Beck MA 2010 Diet-induced obesity in mice reduces the maintenance of influenza-specific CD8+ memory T cells. *J Nutr* **140** 1691-1697

Kershaw NJ, Murphy JM, Liau NP, Varghese LN, Laktyushin A, Whitlock EL, Lucet IS, Nicola NA & Babon JJ 2013 SOCS3 binds specific receptor-JAK complexes to control cytokine signaling by direct kinase inhibition. *Nat Struct Mol Biol* **20** 469-476

Kievit P, Howard JK, Badman MK, Balthasar N, Coppari R, Mori H, Lee CE, Elmquist JK, Yoshimura A & Flier JS 2006 Enhanced leptin sensitivity and improved glucose homeostasis in mice lacking suppressor of cytokine signaling-3 in POMC-expressing cells. *Cell Metab* **4** 123-132

Kim JJ & Sears DD 2010 TLR4 and Insulin Resistance. *Gastroenterol Res Pract* 212563

Koike K, Tsutsumi T, Miyoshi H, Shinzawa S, Shintani Y, Fujie H, Yotsuyanagi H & Moriya K 2008 Molecular basis for the synergy between alcohol and hepatitis C virus in hepatocarcinogenesis. *J Gastroenterol Hepatol* **23** S87-91

La Cava A & Matarese G 2004. The weight of leptin in immunity *Nat Rev Immunol* **4** 371-379

Lebrun P & Van Obberghen E 2008 SOCS proteins causing trouble in insulin action. *Acta Physiol (Oxf)* **192** 29-36

Lee JY, Plakidas A, Lee WH, Heikkinen A, Chanmugam P, Bray G & Hwang DH 2003 Differential modulation of Toll-like receptors by fatty acids: preferential inhibition by n-3 polyunsaturated fatty acids. *J Lipid Res* **44** 479-486

Lee JY, Sohn KH, Rhee SH & Hwang D 2001 Saturated fatty acids, but not unsaturated fatty acids, induce the expression of cyclooxygenase-2 mediated through Toll-like receptor 4. *J Biol Chem* **276** 16683-16689

Ley RE, Turnbaugh PJ, Klein S & Gordon JI 2006 Microbial ecology: human gut microbes associated with obesity. *Nature* **444** 1022-1023

Louie JK, Acosta M, Samuel MC, Schechter R, Vugia DJ, Harriman K, Matyas BT & California Pandemic (H1N1) Working Group 2011 A novel risk factor for a novel virus: obesity and 2009 pandemic influenza A (H1N1). *Clin Infect Dis* **52** 301-312

Maccio A & Madeddu C 2011 Obesity, inflammation, and postmenopausal breast cancer: therapeutic implications. *ScientificWorldJournal* **11** 2020-2036

Macciò A, Madeddu C, Gramignano G, Mulas C, Floris C, Massa D, Astara G, Chessa P & Mantovani G 2010 Correlation of body mass index and leptin with tumor size and stage of disease in hormone-dependent postmenopausal breast cancer: preliminary results and therapeutic implications. *J Mol Med (Berl)* **88** 677-686

Medzhitov R 2001 Toll-like receptors and innate immunity. *Nat Rev Immunol* **1** 135-145

Miyamoto H, Moriishi K, Moriya K, Murata S, Tanaka K, Suzuki T, Miyamura T, Koike K & Matsuura Y 2007. Involvement of the PA28gamma-dependent pathway in insulin resistance induced by hepatitis C virus core protein. *J Virol* **81** 1727-1735

Morgan OW, Bramley A, Fowlkes A, Freedman DS, Taylor TH, Gargiullo P, Belay B, Jain S, Cox C, Kamimoto L, Fiore A, Finelli L, Olsen SJ & Fry AM 2010 Morbid obesity as a risk factor for hospitalization and death due to 2009 pandemic influenza A(H1N1) disease. *PLoS One* **5** e9694

Mori H, Hanada R, Hanada T, Aki D, Mashima R, Nishinakamura H, Torisu T, Chien KR, Yasukawa H & Yoshimura A 2004 Socs3 deficiency in the brain elevates leptin sensitivity and confers resistance to dietin-duced obesity. *Nat Med* **10** 739-743

Mori H, Shichita T, Yu Q, Yoshida R, Hashimoto M, Okamoto F, Torisu T, Nakaya M, Kobayashi T, Takaesu G & Yoshimura A 2007 Suppression of SOCS3 expression in the pancreatic beta-cell leads to resistance to type 1 diabetes. *Biochem Biophys Res Commun* **359** 952-958

O'Doherty MG, Freedman ND, Hollenbeck AR, Schatzkin A & Abnet CC 2012 A prospective cohort study of obesity and risk of oesophageal and gastric adenocarcinoma in the NIH-AARP Diet and Health Study. *Gut* **61** 1261-1268

Ogata H, Kobayashi T, Chinen T, Takaki H, Sanada T, Minoda Y, Koga K, Takaesu G, Maehara Y, Iida M & Yoshimura A 2006. Deletion of the SOCS3 gene in liver parenchymal cells promotes hepatitis-induced hepatocarcinogenesis. *Gastroenterology* **131** 179-193

Owen C, Czopek A, Agouni A, Grant L, Judson R, Lees EK, Mcilroy GD, Göransson O, Welch A, Bence KK, Kahn BB, Neel BG, Mody N & Delibegović M 2012. Adipocyte-specific protein tyrosine phosphatase 1B deletion increases lipogenesis, adipocyte cell size and is a minor regulator of glucose homeostasis. *PLoS One* **7** e32700

Park EJ, Lee JH, Yu GY, He G, Ali SR, Holzer RG, Osterreicher CH, Takahashi H & Karin M 2010 Dietary and genetic obesity promote liver inflammation and tumorigenesis by enhancing IL-6 and TNF expression. *Cell* **140** 197-208

Pascarella S, Clement S, Guilloux K, Conzelmann S, Penin F & Negro F 2011. Effects of hepatitis C virus on suppressor of cytokine signaling mRNA levels: comparison between different genotypes and core protein sequence analysis. *J Med Virol* **83** 1005-1015

Piessevaux J, Lavens D, Peelman F & Tavernier J 2008 The many faces of the SOCS box. *Cytokine Growth Factor Rev* **19** 371-381

Reed AS, Unger EK, Olofsson LE, Piper ML, Myers MG Jr & Xu AW 2010 Functional role of suppressor of cytokine signaling 3 upregulation in hypothalamic leptin resistance and long-term energy homeostasis. *Diabetes* **59** 894-906

Reeves GK, Pirie K, Beral V, Green J, Spencer E, Bull D & Million Women Study Collaboration 2007 Cancer incidence and mortality in relation to body mass index in the Million Women Study: cohort study. *BMJ* **335** 1134

Saberi M, Woods NB, de Luca C, Schenk S, Lu JC, Bandyopadhyay G, Verma IM & Olefsky JM 2009 Hematopoietic cell-specific deletion of toll-like receptor 4 ameliorates hepatic and adipose tissue insulin resistance in high-fat-fed mice. *Cell Metab* **10** 419-429

Santillan-Benitez JG, Mendieta-Zeron H, Gomez-Olivan LM, Ordonez Quiroz A, Torres-Juarez JJ & Gonzalez-Banales JM 2014 JAK2, STAT3 and SOCS3 gene expression in women with and without breast cancer. *Gene* **547** 70-76

Schele E, Grahnemo, L, Anesten, F, Hallen, A, Backhed F & Jansson JO 2013 The gut microbiota reduces leptin sensitivity and the expression of the obesity-suppressing neuropeptides proglucagon (Gcg) and brain-derived neurotrophic factor (Bdnf) in the central nervous system. *Endocrinology* **154** 3643-3651

Sharma D, Wang J, Fu PP, Sharma S, Nagalingam A, Mells J, Handy J, Page AJ, Cohen C, Anania FA & Saxena NK 2010 Adiponectin antagonizes the oncogenic actions of leptin in hepatocellular carcinogenesis. *Hepatology* **52** 1713-1722

Shi H, Kokoeva MV, Inouye K, Tzameli I, Yin H & Flier JS 2006 TLR4 links innate immunity and fatty acid-induced insulin resistance. *J Clin Invest* **116** 3015-3025

Shimizu M, Sakai H, Shirakami Y, Yasuda Y, Kubota M, Terakura D, Baba A, Ohno T, Hara Y, Tanaka T & Moriwaki H 2011 Preventive effects of (-)-epigallocatechin gallate on diethylnitrosamine-induced liver tumorigenesis in obese and diabetic C57BL/KsJ-db/db Mice. *Cancer Prev Res (Phila)* **4** 396-403

Starr R, Metcalf D, Elefanty AG, Brysha M, Willson TA, Nicola NA, Hilton DJ & Alexander WS 1998 Liver degeneration and lymphoid deficiencies in mice lacking suppressor of cytokine signaling-1. *Proc Natl Acad Sci U S A* **95** 14395-14399

Suchy D, Labuzek K, Machnik G, Kozlowski M & Okopien B 2013 SOCS and diabetes--ups and downs

of a turbulent relationship. *Cell Biochem Funct* **31** 181-195.

Tanti JF, Ceppo F, Jager J & Berthou F 2012 Implication of inflammatory signaling pathways in obesity induced insulin resistance. *Front Endocrinol (Lausanne)* **3** 181

Teran-Cabanillas E, Montalvo-Corral M, Silva-Campa E, Caire-Juvera G, Moya-Camarena SY & Hernandez J 2014 Production of interferon alpha and beta, proinflammatory cytokines and the expression of suppressor of cytokine signaling (SOCS) in obese subjects infected with influenza A/H1N1. *Clin Nutr* **33** 922-926

Torisu T, Sato N, Yoshiga D, Kobayashi T, Yoshioka T, Mori H, Iida M & Yoshimura A 2007 The dual function of hepatic SOCS3 in insulin resistance in vivo. *Genes Cells* **12** 143-154

Turnbaugh PJ, Ley RE, Mahowald MA, Magrini V, Mardis ER & Gordon JI 2006 An obesity associated gut microbiome with increased capacity for energy harvest. *Nature* **444** 1027-1031

Ueki K, Kondo T & Kahn CR 2004 Suppressor of cytokine signaling 1 (SOCS-1) and SOCS-3 cause insulin resistance through inhibition of tyrosine phosphorylation of insulin receptor substrate proteins by discrete mechanisms. *Mol Cell Biol* **24** 5434-5446

Ueki K, Kondo T, Tseng YH & Kahn CR 2004 Central role of suppressors of cytokine signaling proteins in hepatic steatosis, insulin resistance, and the metabolic syndrome in the mouse. *Proc Natl Acad Sci U S A* **101** 10422-10427

Vijay-Kumar M, Aitken JD, Carvalho FA, Cullender TC, Mwangi S, Srinivasan S, Sitaraman SV, Knight R, Ley RE & Gewirtz AT 2010 Metabolic syndrome and altered gut microbiota in mice lacking Toll-like receptor 5. *Science* **328** 228-231

Wild S, Roglic G, Green A, Sicree R & King H 2004 Global prevalence of diabetes: estimates for the year 2000 and projections for 2030. *Diabetes Care* **27** 1047-1053

Wolin KY, Carson K & Colditz GA 2010 Obesity and cancer. *Oncologist* **15** 556-565

Wunderlich CM, Hovelmeyer N & Wunderlich FT 2013 Mechanisms of chronic JAK-STAT3-SOCS3 signaling in obesity. *JAKSTAT* **2** e23878

Xu CX, Zhu HH & Zhu YM 2014 Diabetes and cancer: Associations, mechanisms, and implications for medical practice. *World J Diabetes* **5** 372-380

Xu H, An H, Hou J, Han C, Wang P, Yu Y & Cao X 2008 Phosphatase PTP1B negatively regulates MyD88- and TRIF-dependent proinflammatory cytokine and type I interferon production in TLR-triggered macrophages. *Mol Immunol* **45** 3545-3552

Yasukawa H, Misawa H, Sakamoto H, Masuhara M, Sasaki A, Wakioka T, Ohtsuka S, Imaizumi T, Matsuda T, Ihle JN & Yoshimura A 1999 The JAK-binding protein JAB inhibits Janus tyrosine kinase activity through binding in the activation loop. *EMBO J* **18** 1309-1320

Yasukawa H, Sasaki A & Yoshimura A 2000 Negative regulation of cytokine signaling pathways. *Annu Rev Immunol* **18** 143-164

Yoshimura A, Naka T & Kubo M 2007 SOCS proteins, cytokine signalling and immune regulation. *Nat Rev Immunol* **7** 454-465

Yoshimura A, Suzuki M, Sakaguchi R, Hanada T & Yasukawa H 2012 SOCS, Inflammation, and Autoimmunity. *Front Immunol* **3** 20

Zaitseva II, Hultcrantz M, Sharoyko V, Flodstrom-Tullberg M, Zaitsev SV & Berggren PO 2009 Suppressor of cytokine signaling-1 inhibits caspase activation and protects from cytokine-induced beta cell death. *Cell Mol Life Sci* **66** 3787-3795

Zheng Q, Dunlap SM, Zhu J, Downs-Kelly E, Rich J, Hursting SD, Berger NA & Reizes O 2011 Leptin deficiency suppresses MMTV-Wnt-1 mammary tumor growth in obese mice and abrogates tumor initiating cell survival. *Endocr Relat Cancer* **18** 491-503

Zhou X, Sun P, Lucendo-Villarin B, Angus AG, Szkolnicka D, Cameron K, Farnworth SL, Patel AH & Hay DC 2014 Modulating innate immunity improves hepatitis C virus infection and replication in stem cell-derived hepatocytes. *Stem Cell Reports* **3** 204-214

17

HCMV activation of ERK-MAPK drives a multi-factorial response promoting the survival of infected myeloid progenitors

Verity Kew[1], Mark Wills[1] and Matthew Reeves[2]

[1]Department of Medicine, Addenbrooke's Hospital, Cambridge, UK
[2]UCL Institute of Immunity & Transplantation, Royal Free Hospital, London, UK

Correspondence should be addressed to Matthew Reeves; E-mail: matthew.reeves@ucl.ac.uk

Abstract

Viral binding and entry provides the first trigger of a cell death response and thus how human cytomegalovirus (HCMV) evades this – particularly during latent infection where a very limited pattern of gene expression is observed – is less well understood. It has been demonstrated that the activation of cellular signalling pathways upon virus binding promotes the survival of latently infected cells by the activation of cell encoded anti-apoptotic responses. In CD34+ cells, a major site of HCMV latency, ERK signalling is important for survival and we now show that the activation of this pathway impacts on multiple aspects of cell death pathways. The data illustrate that HCMV infection triggers activation of pro-apoptotic Bak which is then countered through multiple ERK-dependent functions. Specifically, ERK promotes ELK1 mediated transcription of the key survival molecule MCL-1, along with a concomitant decrease of the pro-apoptotic BIM and PUMA proteins. Finally, we show that the elimination of ELK-1 from CD34+ cells results in elevated Bak activation in response to viral infection, resulting in cell death. Taken together, these data begin to shed light on the poly-functional response elicited by HCMV via ERK-MAPK to promote cell survival.

Introduction

Amongst the first host responses activated upon viral infection are cell autonomous immune responses. These functions are ubiquitous to every cell type and are normally triggered by pattern recognition receptors which detect various components of the incoming pathogen (Bieniasz 2004; Everett & Chelbi-Alix 2007; Randow et al. 2013). One aspect of the cell autonomous immune response is the activation of apoptotic and cell death pathways, which will ultimately eliminate both the pathogen and the infected cell (Clem et al. 1991; Everett & McFadden 1999; Levine et al. 1993). During lytic infection, human cytomegalovirus (HCMV) counters this with an arsenal of virally encoded anti-apoptotic functions that counter stress signals induced by viral binding and, subsequently, throughout the course of viral replication (Brune 2010; Goldmacher et al. 1999; Guo et al. 2015; Reeves et al. 2007; Skaletskaya et al. 2001; Stevenson et al. 2014; Terhune et al. 2007). However, a lack of expression of these functions in non-lytic infections raised the question of how cell survival is achieved – particularly during the initial phases of infection.

Current studies hypothesise that viral infection in cells non-permissive for lytic infection requires the generation of a pro-survival phenotype driven by virally induced up-regulation of key cellular anti-apoptotic proteins (Chan et al. 2010; Peppenelli et al. 2016; Reeves et al. 2012; Stevenson et al. 2014). This event relies on the regulation of a number of pro-survival and pro-death signals via the modulation of cellular signalling pathways initiated upon virus binding and entry (Chan et al. 2010; Reeves et al. 2012). It is the outcome of these signalling events which, ultimately, determines the fate of the cell during these very early stages of infection. Once latency is established, there are additional mechanisms activated which would be consistent with the latent virus propagating an anti-apoptotic state (Poole et al. 2011, 2015; Slobedman et al. 2004).

Apoptosis and cell death is an evolutionarily conserved process that is extensively regulated by Bcl-2 homology domain 3 (BH3) proteins (Doerflinger et al. 2015; Horvitz 1999; Puthalakath & Strasser 2002). The precise mechanism of action still remains equivocal although it is clear that the BH3 proteins trigger the activation of key apoptosis effector proteins (e.g. Bak and Bax) which exert their pro-apoptotic function predominantly at mitochondrial membranes (Wei *et al.* 2001; Westphal *et al.* 2014). A number of regulatory mechanisms have been suggested: The indirect activation model posits that Bax/Bak are required to be retained in an inactive form via direct sequestration by anti-apoptotic BCL-2 family members (Willis *et al.* 2005) and that BH3 proteins must engage with these to release the Bax/Bak to initiate apoptosis (Uren *et al.* 2007; Willis *et al.* 2007). The direct activation model hypothesises that BH3-only activators (e.g. PUMA and BIM) bind to Bak/Bax directly to activate them or, alternatively, other BH3 members (e.g. Bad) target BCL-2 members and inhibit their anti-apoptotic function through sequestration (Doerflinger *et al.* 2015; Erlacher *et al.* 2006; Kuwana *et al.* 2002; Villunger *et al.* 2003; Wei *et al.* 2000). The most recent evidence argues that these events are not mutually exclusive and that, instead, both mechanisms of regulation are likely to be active (Llambi *et al.* 2011; Westphal *et al.* 2014) although the strict delineation between activators and sensitisers in the BH3 family may not be entirely valid (Westphal *et al.* 2014).

Our own research interests have centred on the BCL-2-like protein myeloid cell leukaemia-1 (MCL-1). Unlike other BCL-2 proteins the regulation of MCL-1 is dynamic with a turnover of 30 minutes under certain experimental conditions (Adams & Cooper 2007; Warr & Shore 2008) and it is thus considered a highly responsive determinant of haematopoietic cell viability (Perciavalle & Opferman 2013) – the ablation of this protein from progenitor cells of the haematopoietic lineage being lethal (Opferman *et al.* 2005; Perciavalle & Opferman 2013). MCL-1 (along with BCL-XL) has been suggested to block Bak activity to exert an anti-apoptotic function (Willis *et al.* 2005) and is regulated by both the PI3K and ERK-MAPK pathways (Huang *et al.* 2000; Mills *et al.* 2008). Pertinently, MCL-1 is also regulated by HCMV through activation of the ERK and PI3K pathways in a cell type specific manner (Chan *et al.* 2010; Reeves *et al.* 2012). Depletion of MCL-1 from monocytes or THP1 cells renders them unable to prevent virally induced cell death upon infection (Chan *et al.* 2010; Reeves *et al.* 2012). In CD34+ cells, HCMV survival is also associated with MCL-1 (Reeves *et al.* 2012) although the absolute requirement for endogenous MCL-1 in normal CD34+ cells rendered similar genetic analysis through MCL-1 deletion intractable. The inducible survival effect in CD34+ cells required virus binding and was likely dependent on the engagement of glycoprotein B with an unknown receptor. Furthermore, the survival effect induced was

transient (Reeves *et al.* 2012) and thus did not appear to be propagated for as long as was observed in the monocyte model of infection (Chan *et al.* 2010).

ELK-1 is part of a 27 member superfamily of transcription factors that dictate a diverse range of processes including haematopoiesis, differentiation and survival, oncogenesis and inflammation (Sharrocks 2001). ELK-1 is a nuclear phosphoprotein that exhibits bimodal activity: redundant promoter binding via dimers with other Ets transcription factors and specific binding to a subset of genes through an interaction with serum response factor (SRF) (Odrowaz & Sharrocks 2012; Sharrocks 2001). Amongst these genes are MCL-1 and c-Fos, which, characteristically for this subset of genes, are dynamically regulated (Boros *et al.* 2009; Treisman *et al.* 1992). The activation of ELK-1 requires phosphorylation of serine residue 383 promoting the recruitment of transactivating co-factors to promoters in order to drive transcription (Boros *et al.* 2009; Gille *et al.* 1992, 1995). Furthermore, ELK-1 mediated activation of MCL-1 gene expression has been shown to be important for survival in a number of experimental models (Booy *et al.* 2011; Demir *et al.* 2011; Sun *et al.* 2013; Townsend *et al.* 1999) and likely contributes to the role of ELK-1 in cancer.

Herein we wanted to further explore the consequences of HCMV infection on the development of an anti-apoptotic phenotype in CD34+ cells based on our previous studies of MCL-1 and ERK-MAPK signalling. Here we show that the HCMV protection from cell death is concomitant with an inhibition of prolonged Bak activation in response to chemical and viral insult. Secondly, we show that the protective phenotype driven by virus induced ERK-MAPK signalling correlates with the down-regulation of pro-apoptotic BH3 proteins PUMA and BIM and, concomitantly, is dependent on the phosphorylation and thus activation of ELK-1 – a transcription factor important for MCL-1 expression and cell survival. Importantly, knock down of ELK-1 expression is sufficient to abrogate the protective effect elicited by HCMV infection. Taken together, these data suggest that HCMV infection drives a survival phenotype by simultaneously up-regulating MCL-1 proteins levels and reducing the levels of antagonistic interaction partners. We propose that the net effect tips the balance in favour of survival contributing to the successful establishment of latent infection.

Materials and Methods

Virus, cell lines, culture and reagents
The Merlin strain of HCMV was purified from infected human fibroblasts as previously described

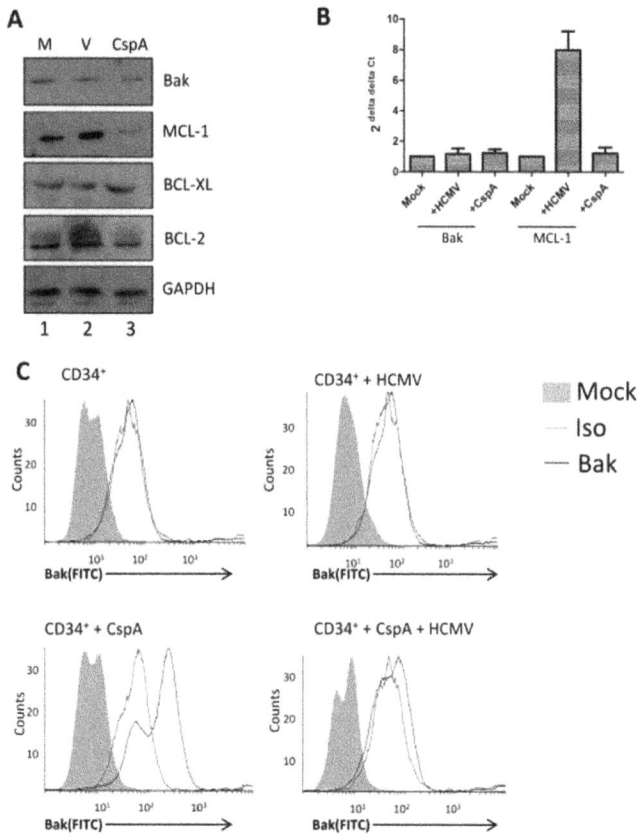

Figure 1. HCMV infection blocks cisplatin A mediated activation of Bak. (A) Western blot analysis of CD34+ cells either mock (M), HCMV infected (V) or cisplatin A treated (CspA) for Bak, MCL-1, BCL-XL, Bcl-2 or GAPDH expression 2 hours post-infection. (B) qRT-PCR analysis of RNA isolated from mock, HCMV infected or cisplatin A treated cells for Bak or MCL-1 expression 2 hours post-infection. Changes in gene expression were identified using GAPDH and $2^{-\Delta\Delta CT}$ method. n=3 (C) CD34+ cells were mock or HCMV infected and then incubated with DMSO or cisplatin A 3 hours post-infection. Cells were then permeabilised and stained with an antibody that specifically recognises an N terminal peptide exposed in activated Bak protein or with an isotype matched control. Fluorescent staining was achieved using a FITC-Goat anti-mouse antibody and then cells analysed by flow cytometry.

(Compton 2000). Primary CD34+ haematopoietic cells (Lonza, Slough, UK) were resuscitated for 24 hours prior to any studies of cell viability in X-vivo-15 (Biowhittaker, Lonza, Slough, UK) media supplemented with 10% human serum. For all subsequent studies, cells were switched into serum-free X-vivo media supplemented with 2mM L-Glutamine. For all infection experiments an MOI of 5 (based on titration in fibroblasts) was used.

Inhibition of the ERK-MAPK pathway was achieved using U0126 (final concentration 1uM; Calbiochem/Millipore, Darmstadt, Germany). The inhibitor was added directly to the culture media with 0.1%

DMSO used as the solvent control 1 hour prior to virus infection.

Nucleic acid isolation, reverse transcription, PCR and Western Blot

Ten micrograms of DNase I treated RNA isolated using RNAeasy spin columns was reverse transcribed using ImpromII RT kit (Promega, Madison, WI) or Quantitect Reverse Transcription Kit (Qiagen, Hilden, Germany). Chromatin Immunoprecipitations were performed as previously described (Kew *et al.* 2014). Briefly, 10^6 cells were fixed and lysed and then subjected to sonication to shear DNA into 200-500bp fragments. DNA was then incubated with anti-ELK1 antibody (Cell Signaling, Danvers, MA; 1:100 dilution), anti-phospho ELK-1 antibody (Cell Signaling, Danvers, MA; 1:100 dilution) or an isotype matched control (mouse IgG1; SIGMA, St Louis, MO). DNA was rescued from the IP complexes and amplified with MCL-1 promoter specific primers.

Gene and promoter specific primers were then used to amplify target sequences by real time PCR using SYBR green amplification kit (Qiagen, Hilden, Germany). MCL-1 (gene): 5'-TGC AGG TGT GTG CTG GAG TAG and 5'-GCT CTT GGC CAC TTG CTT TTC', GAPDH: 5'-GAG TCA ACG GAT TTG GTC GT and 5'-TTG ATT TTG GAG GGA TTC TCG, 18S: 5'- GTA ACC CGT TGA ACC CCA and 5'- CCA TCC AAT CGG TAG TAG CG, Bak: 5'-GCC CAG GAC ACA GAG GAG GTT TTC and 5'-AAA GTG GCC CAA CAG AAC CAC ACC, Bim: 5'-CAC AAA CCC CAA GTC CTC CTT and 5'-TTC AGC CTG CCT CAT GGA A, Puma: 5'-ACG ACC TCA ACG CAC AGT ACG and 5'-TGG GTA AGG GCA GGA GTC C, UL138: 5'- GAG CTG TAC GGG GAG TAC GA and 5'- AGC TGC ACT GGG AAG ACA CT; MCL-1 (promoter): 5'-TAG GTG CCG TGC GCA ACC CT and 5'-ACT GGA AGG AAG CGG AAG TGA GAA (Booy *et al.* 2011).

For Western Blot, 10^5 cells were lysed in Laemmli buffer and subjected to SDS-PAGE electrophoresis. Following transfer, blots were incubated with anti-MCL-1 (Cell Signaling, Danvers, MA; 1:500), anti-ELK-1 or anti-phospho-ELK-1 (phosphor-serine 383; Cell signaling, Danvers, MA; both 1:750), anti-Bak (#06-536, Millipore, Darmstadt, Germany; 1:500), anti-PUMA (Santa Cruz, CA, 1:200) anti-actin (Abcam, Cambridge, UK; 1:1000) or anti-GAPDH (Abcam, Cambridge, UK; 1:2000) for 1 hour followed by detection with the appropriate HRP-conjugated secondary antibody. Specific bands were visualized by ECL detection (Amersham, Horsham, UK).

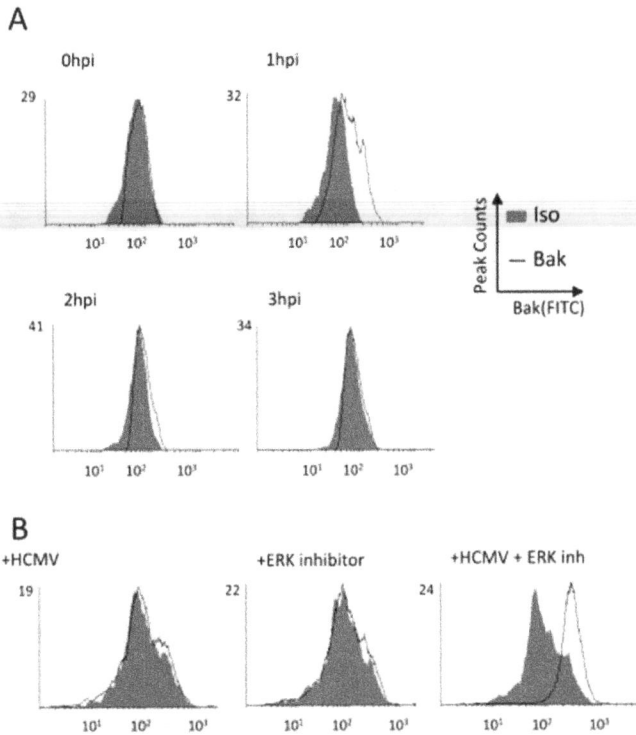

Figure 2. HCMV infection induces a transient activation of Bak which is not reversed when ERK responses are inhibited. (A) CD34+ cells were infected with HCMV and then at times 0 to 3 hours post infection cells were permeabilised and stained for evidence of Bak activation by flow cytometry. (B-C) CD34+ cells were either pre-treated with an ERK inhibitor (B) or subjected to ELK1 or control siRNA knockdown (C) and then either mock of HCMV infected. Cells were then permeabilised and stained for evidence of Bak activation by flow cytometry.

siRNA knockdown

CD34+ cells were transfected with Silencer® ELK-1 siRNAs or Silencer® negative control (Thermofisher, Waltham, MA) using Viromer Green transfection reagent as described by the manufacturer (lipocalyx/ Cambridge Biosciences, UK). CD34+ cells were transfected in bulk in X-vivo serum free media and then plated at the required density for downstream analyses at 48 hours post-transfection. Prior to plating, the viromer:siRNA mix was removed from the cell culture 4 hours post-transfection by pelleting the CD34+ cells at 300g followed by resuspension in fresh X-vivo 15 media.

Cell Death assay and Bak activation

Induction of cell death was achieved using the chemotherapeutic drug cisplatin A (Rosenberg et al. 1965). Cisplatin A (25uM-250uM; SIGMA-Aldrich, St Louis, MO) or 0.1% DMSO (mock) control was added for 4 hours to trigger the apoptotic pathway. Since the effects of cisplatin A induced cell death are not evident

until at least 12-24 hours post treatment (Barry et al. 1990) cell viability was performed 21 hours after cisplatin A addition. To determine levels of cell death, cells were stained with a TUNEL detection kit (Roche, Basel, Switzerland) as described by the manufacturer and analysed for apoptotic cell death by immune-fluorescent microscopy.

Alternatively, cell death was induced using the virus as a ligand through blockade of ERK-MAPK signaling (U0126; 1uM, Calbiochem) 1 hour prior to infection with HCMV. 0.1% DMSO was used as solvent control.

Bak activation in CD34+ cells by flow cytometry was measured as previously described (Griffiths et al. 1999). Briefly, 10^5 cells were incubated with rabbit serum for 20 minutes and then with an anti-Bak antibody directed against the N terminal region (#06-536, Millipore, Darmstadt, Germany; 1:50 dilution in PBS) or an isotype matched control for 15 minutes at 4°C. Cells were washed, incubated with goat serum for 20 minutes and then incubated with a goat anti-rabbit FITC conjugated secondary antibody (1:100 dilution in PBS) for 15 minutes at 4°C. Unstained, isotype stained and Bak stained cells were then analysed by flow cytometry.

Assays for latent infection

CD34+ cells were infected with HCMV 48hrs post treatment with siRNAs in X-vivo-15 media for 3 hours. Cells were pelleted, supernatant removed, washed in PBS and then re-suspended in fresh X-vivo media and cultured for 3 days. RNA was then isolated, converted to cDNA and amplified in a UL138 (viral gene) and 18S (cellular gene) specific PCR. Changes in gene expression were calculated using $2^{-\Delta\Delta CT}$ method (UL138 and 18S RNA).

Results

HCMV infection reduces the level of activated Bak in cisplatin A treated cells

Our previous study had shown that HCMV infection could block the induction of CD34+ cell death in response to the chemotherapeutic drug, cisplatin A (Reeves et al. 2012). Cisplatin A-induced degradation of survival molecules and the triggering of cell death are functionally entwined events (Yang et al. 2007). Here we show that in CD34+ cells, cisplatin A promotes the degradation of pro-survival MCL-1, but not the related BCL-XL or Bcl-2 proteins (Figure 1A). As expected, HCMV infection promoted MCL-1 up-regulation but, interestingly, no effect on BCL-2 or BCL-XL was again observed (Figure 1A). A key target of MCL-1 is Bak. However, no impact on the total lev-

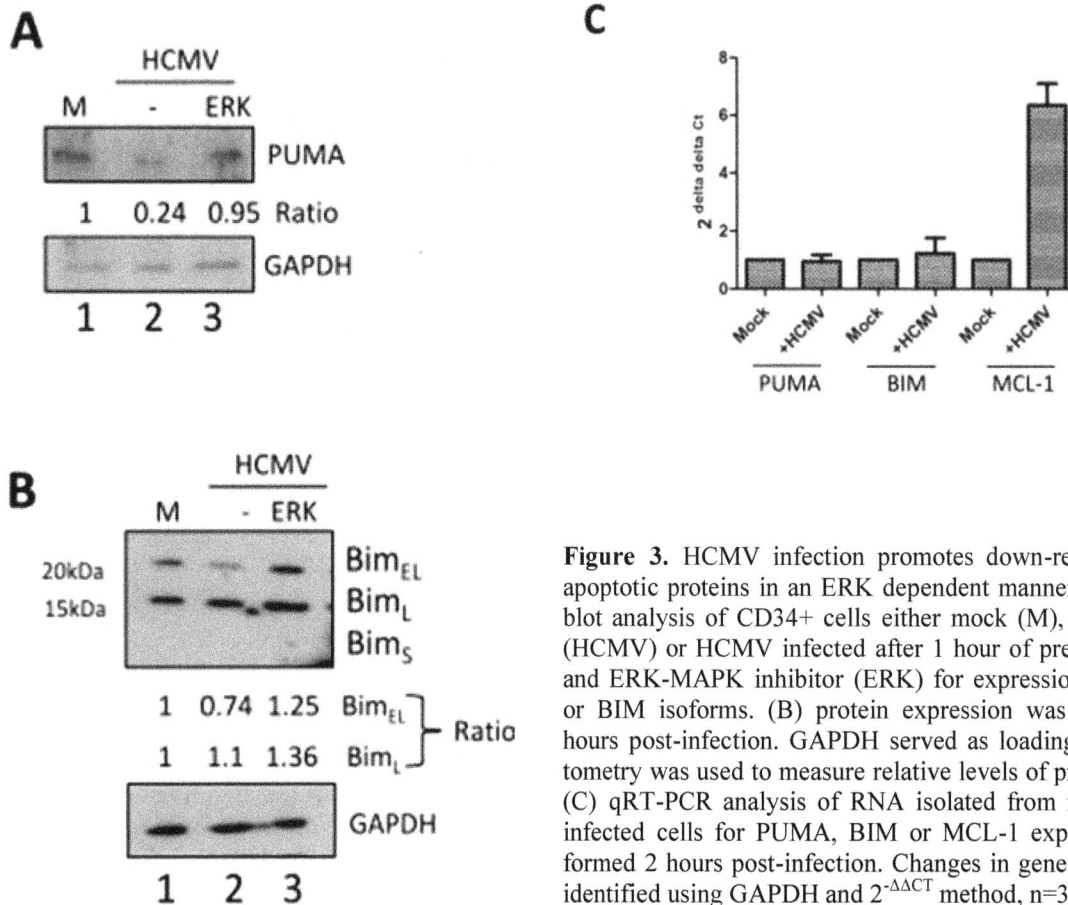

Figure 3. HCMV infection promotes down-regulation of pro-apoptotic proteins in an ERK dependent manner. (A,B) Western blot analysis of CD34+ cells either mock (M), HCMV infected (HCMV) or HCMV infected after 1 hour of pre-incubation with and ERK-MAPK inhibitor (ERK) for expression of PUMA (A) or BIM isoforms. (B) protein expression was performed at 2 hours post-infection. GAPDH served as loading control. Densitometry was used to measure relative levels of protein expression (C) qRT-PCR analysis of RNA isolated from mock or HCMV infected cells for PUMA, BIM or MCL-1 expression was performed 2 hours post-infection. Changes in gene expression were identified using GAPDH and $2^{-\Delta\Delta CT}$ method, n=3.

els of Bak were observed whether treated with cisplatin A or infected with HCMV (Figure 1A). Similarly, no impact on the transcription of Bak was detected at these early times post-infection (Figure 1B). In contrast, HCMV infection was a clear inducer of MCL-1 transcription (Figure 1B). We next investigated the activation of Bak and although levels in the total protein did not change, evidence of increased activation of Bak in response to cisplatin A was detected (Figure 1C). The specific activation of Bak can be detected by flow cytometry utilising the conformational change that exposes Bak epitopes when active (i.e. pro-apoptotic) Bak is released from the inactive complex it is normally sequestered in. As expected, resting CD34+ cells show very little evidence of Bak activation (Figure 1C) consistent with their viability status. In contrast, stimulation of CD34+ cells with cisplatin A resulted in a substantial increase in the level of activated Bak detectable in the cells (Figure 1C). Pertinently, minimal levels of active Bak were detectable in HCMV infected cells at 3hpi. Crucially, pre-infection with HCMV prior to cisplatin A treatment markedly reduced the detection of active Bak (Figure 1C) - consistent with the protective phenotype of HCMV against

cisplatin A. Taken together, these data suggested HCMV engendered a cellular phenotype that was directly antagonistic to cisplatin A induced cell death.

HCMV activation of ERK-MAPK is necessary to prevent prolonged activation of Bak upon infection
To investigate this further we first asked whether Bak activation occurred during viral infection – which would underpin the need to promote a cellular environment that antagonises Bak activation under normal infection conditions. Intriguingly, a time course analysis revealed that Bak was clearly activated in infected cells at 1hpi and that, over time, the level of Bak activation was reversed by 3hpi (Figure 2A).

To explore this further we next exploited the knowledge that inhibition of survival pathways triggered by HCMV promoted the death phenotype (Reeves *et al.* 2012). CD34+ cells were pre-incubated with ERK-MAPK inhibitor for 1 hour prior to infection (Figure 2B). Cells were then stained for Bak activation. As expected, there was little evidence of Bak activation at 3hpi in infected cells or cells treated with ERK inhibitor alone. However, the infection of cells with inhibited ERK-MAPK signalling resulted in a

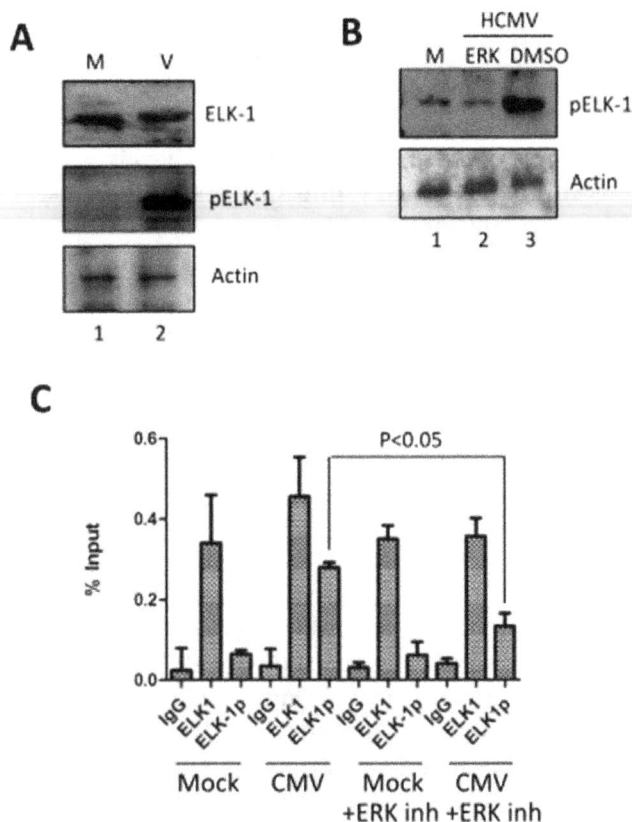

Figure 4. HCMV targets ELK1 for phosphorylation in an ERK dependent manner. (A) Western blot analysis of CD34+ cells, either mock (M) or HCMV infected (V) for ELK1 and ELK1 phosphorylation, 2 hours post-infection. Actin served as loading control. (B) Western blot analysis of CD34+ cells, either mock (M) or HCMV infected (HCMV) cells, with or without prior incubation with ERK-MAPK inhibitor (ERK) or DMSO control for 1 hour at 2 hours post-infection. Actin served as loading control. (C) Chromatin immuno-precipitation assays were performed on CD34+ cells with isotype (IgG), ELK1 or phosphor-ELK1 (ELK-1p) antibodies. ChIPs were performed on CD34+ cells, HCMV infected CD34+ cells or the equivalent but with prior incubation with U0126 ERK inhibitor for 1 hour. Cells were analysed at 2 hours post infection, n=3. Students t-test was used to test for significance at p<0.05.

clear Bak activation phenotype (Figure 2B) and thus clearly pheno-copied the cisplatin effect. Taken together, the data show that HCMV infection activates Bak which it then reverses via the concomitant activation of ERK-MAPK signalling.

HCMV modulates the activity of multiple Bak-interacting functions in protected cells

The activation of Bak in the mitochondrial membrane is elicited through a complicated interplay between effector and inhibitor mechanisms (Llambi et al. 2011). Having already observed changes to the levels

of the MCL-1 regulator we next asked whether any impact on activators of Bak were triggered by HCMV. Two key pro-apoptotic molecules reported to be upstream of Bak in this pathway are the pro-apoptotic PUMA and BIM BH3 proteins (Westphal et al. 2014). Western blot analyses indicated that basal levels of both PUMA and 2 of 3 BIM isoforms (Bim_{EL} and Bim_L) were detectable in uninfected cells. Furthermore, HCMV infected cells analysed at 1hpi suggested that a down-regulation of both PUMA and Bim_{EL} was evident. In contrast no changes in Bim_L were detected (Figure 3A, B). We also noted that the highly pro-apoptotic isoform Bim_S (Marani et al. 2002) was not detectable nor was expression induced upon infection. The down-regulation of both PUMA and BIM_{EL} products was specific to the protein as no effect on RNA levels was observed (Figure 3C). Finally, the reduction in both PUMA and BIM_{EL} levels by HCMV was dependent on ERK-MAPK activity (Figure 3A, B).

Thus HCMV activation of ERK-MAPK was promoting several related effects that would be conducive for cell survival – BIM isoform and PUMA degradation alongside an up-regulation of MCL-1 levels. MCL-1 protein levels within the cell are regulated by multiple processes. As well as increased transcription, MCL-1 is also regulated post-translationally with ERK hypothesised to promote a stabilising phosphorylation event of MCL-1 that antagonises degradation. Thus to investigate the potential contribution of transcription and translation of MCL-1 in response to HCMV, we reasoned we were required to understand the basis of HCMV induced up-regulation of MCL-1 mRNA expression.

Key elements responsible for the regulation of the MCL-1 promoter include binding sites for the ELK family of proteins (Booy et al. 2011). Western blot analysis of infected cells revealed that although HCMV binding did not increase the levels of ELK-1 in the cell, there was an apparent mobility shift which would be indicative of ELK-1 phosphorylation. This was confirmed using a phosphor-ELK1 specific antibody (Figure 4A). We next showed that the phosphorylation, and thus activation of ELK-1, was dependent on ERK-MAPK signalling (Figure 4B). To link this clear activation of ELK-1 upon infection with MCL-1 expression we performed chromatin immuno-precipitation analyses. These revealed that the ELK1 protein was bound to the MCL-1 promoter in CD34+ cells irrespective of whether they were virally infected or not (Figure 4C). Although the data suggested that more ELK1 was bound in virally infected cells, it was the analysis of phosphorylated ELK1 bound to the MCL-1 promoter which exhibited the clearest phenotype. Here, viral infection promoted an increase in the

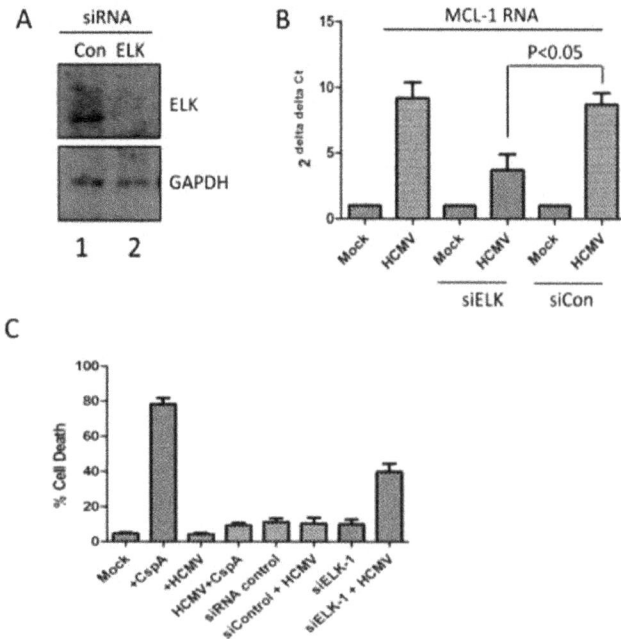

Figure 5. Depletion of ELK1 from CD34+ cells abrogates the HCMV survival response. (A) Western blot analysis of CD34+ cells 48 hours post-transfection with a control (Con), ELK1-specific (ELK) siRNA for ELK1 and GAPDH expression. (B) qRT-PCR analysis of RNA isolated from mock or HCMV infected cells 2 hours post-infection that have first been treated with either mock, control (Scr KD) or ELK1 (ELK KD) siRNAs. Changes in gene expression were identified using GAPDH and $2^{-\Delta\Delta CT}$ method, n=3. Students t-test was used to test for significance at p<0.05. (C) CD34+ cells were either mock or HCMV infected and then, 3 hours post-infection, incubated with cisplatin A or DMSO control. Alternatively, CD34+ cells were transfected with control or ELK-1 specific siRNAs and then either mock or HCMV-infected. In all experiments cell viability was measured 24 hours post-infection. Graph is the average of 2 independent experiments analysed in triplicate.

detectable levels of pELK1 at the MCL-1 promoter and this event was significantly dependent on ERK signalling (Figure 4C).

To test whether ELK-1 was directly involved in the pro-survival phenotype we used siRNA knockdown in CD34+ cells to address this. Delivery of ELK-1 siRNAs significantly reduced ELK-1 protein present in the cell (Figure 5A) but did not result in any overt effects on cell viability over the short time frame of analysis (Figure 5C). Consistent with the ChIP data (Figure 4C), ELK-1 knockdown in the CD34+ cells dramatically impacted on the ability of HCMV to up-regulate MCL-1 RNA expression (Figure 5B). Together with the phenotypic impact on MCL-1 gene expression, there was a clear abrogation of the virally induced survival response (Figure 5C). Although the delivery of ELK-1 siRNAs alone was not deleterious

to the cell, viral infection triggered a marked increase in cell death (Figure 5C) indicating that the elimination of ELK-1 from CD34+ cells renders them more sensitive to HCMV induced cell death.

ELK-1 knockout cells show elevated levels of Bak activation upon HCMV infection resulting in a less efficient establishment of latency

Next, we revisited our studies of Bak activation using our ELK-1 knock down cells (Figure 6A). Unsurprisingly, ELK-1 depletion from cells had little impact on levels of Bak activation. However, infection of ELK-1 knock down cells again resulted in an increased detection of activated Bak (Figure 6A). Although the analysis suggested that not all cells displayed evidence of activated Bak – which may reflect relative efficiency of siRNA knock down – the data clearly suggested that removal of ELK-1 from CD34+ cells promoted Bak activation in response to HCMV infection.

Finally, to assess if this was having any impact on the virus we assessed the ability of HCMV to establish latency (Figure 6B). Control or ELK-1 siRNA treated cells were infected with HCMV and analysed 3 days post-infection for evidence of latent viral gene expression. It was evident that the failure to prevent virally induced cell death manifested with significantly reduced levels of UL138 expression, which would be consistent with a failure to establish a latent infection.

Discussion

In this study we have investigated the phenotype of the pro-survival environment elicited by infection of CD34+ cells with HCMV. This and previous studies provide accumulating evidence that HCMV is required to promote an anti-apoptotic environment through the activation of signalling pathways in non-permissive cells in order to ensure long term survival (Chan *et al.* 2010; Peppenelli *et al.* 2016; Reeves *et al.* 2012; Stevenson *et al.* 2014). These events are likely critical for the pathogenesis and dissemination of virus in monocytes and the establishment of latency in CD34+ cells.

The requirement for a pro-survival signalling cascade is necessitated by the evident induction of pro-death responses which likely represent anti-viral responses to infection. This induction of pro-death responses is assumed to be due to the highly pro-apoptotic phenotype associated with viral infection when the activity of the survival factors is impaired (Chan *et al.* 2010; Reeves *et al.* 2012). Thus HCMV cannot necessarily completely eliminate pro-death signalling but, instead, relies on the concomitant up-regulation of survival factors to counter-balance this. It is of interest that a pan up-regulation of the anti-

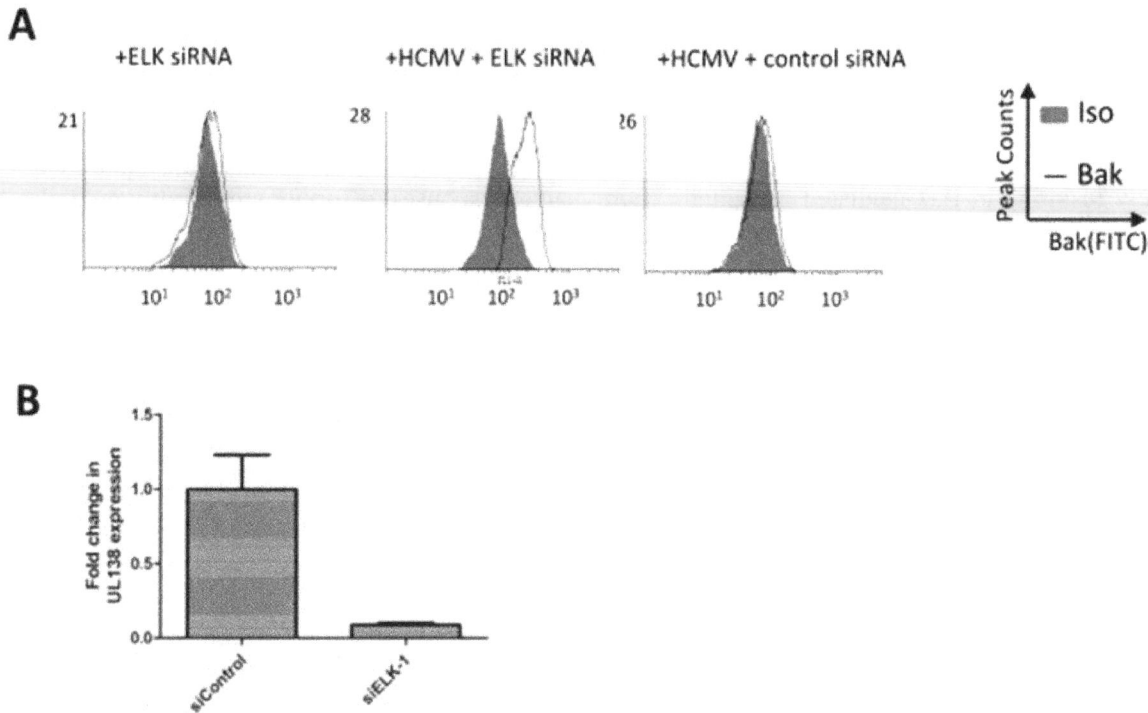

Figure 6. ELK-1 is required for the reversal of Bak activation and the establishment of HCMV latency. (A,B) CD34+ cells were subjected to ELK1 or control siRNA knockdown and then either mock of HCMV-infected. Cells were then either permeabilised and stained for evidence of Bak activation by flow cytometry at 3hpi (A) or analysed 3 days post infection for evidence of viral latent gene expression (B).

apoptotic machinery is not evident in the infected CD34+ cells. An analysis of 3 major players MCL-1, BCL-XL and BCL-2 revealed that only MCL-1 levels were affected by HCMV infection during the initial phases of latent infection in this cell type. However, we cannot dismiss a role for the other proteins completely, as the localisation or binding to target pro-apoptotic proteins of these other BCL-2 family members may be modified to augment the virally induced survival observed. Furthermore, our studies have really focussed on the role of MCL-1 as a transitory regulator during the very initial stages of infection. It is important to note that data from studies performed in the monocyte model have shown that BCL-2 activity becomes increasingly important to maintain the viability of the infected cells (Collins-McMillen *et al.* 2015), suggesting that different molecules are required at different stages of viral infection. Consistent with this, reports in progenitor model systems have suggested that PEA-15 up-regulation via the activity of inter-leukin-10 could play a role in long term latency in CD34+ cells (Poole *et al.* 2011, 2015; Slobedman *et al.* 2004). In this regard, the apparent targeting of the MCL-1 member of the BH3 family during the very early stages could be consistent with the central role that MCL-1 plays in the regulation of apoptosis in haematopoietic cells (Opferman *et al.* 2005) and also

reflective of its more dynamic regulation in the cell (Adams & Cooper 2007). Unlike counterparts (e.g. Bcl-2, Bcl-Xl), MCL-1 has a much shorter half-life (Adams & Cooper 2007). Ablation of MCL-1 results in spontaneous cell death in a number of cell lineages, indicating that, firstly, endogenous levels of other BCL-2 proteins are not sufficient to compensate and secondly, it has a central role in haematopoietic cell survival (Dzhagalov *et al.* 2008; Opferman 2007; Opferman *et al.* 2005). This exquisite sensitivity of primary cells to MCL-1 levels may also involve the complex interaction of MCL-1 with mitochondria (Huang & Yang-Yen 2010; Perciavalle *et al.* 2012). As well as a classical role in apoptosis, MCL-1 has also been shown to regulate ATP biogenesis (Perciavalle *et al.* 2012) which we know, from studies of lytic HCMV infection, has a profound effect on the viability of infected cells (Reeves *et al.* 2007). Whether the impact of MCL-1 on mitochondrial bioenergetics is important here is not clear but, given the relatively short lived nature of the protection in the experimental conditions, it is perhaps unlikely. It is more plausible to have a role either in the MCL-1 mediated protection of monocytes where elevated MCL-1 expression persists for 72 hours (Chan *et al.* 2010) or, it could possibly augment the function of beta 2.7 during lytic HCMV infection.

Cisplatin A mediated MCL-1 degradation is a

trigger for apoptosis (Yang *et al.* 2007) and thus tumours exhibiting resistance to cisplatin A often display elevated levels of MCL-1 that is resistant to drug induced degradation (Michels *et al.* 2014). It is hypothesised that the degradation of MCL-1 exposes Bak to the activity of pro-apoptotic proteins like PUMA and BIM, ultimately resulting in Bak-mediated mitochondrial dysfunction (Letai *et al.* 2002). The precise nature of the inhibitory effect of MCL-1 on Bak is debated: MCL-1 could bind to Bak directly to block activation or could block activation by sequestering pro-apoptotic BH3 proteins like PUMA and BIM or, in fact, perform both functions (Erlacher *et al.* 2006, Letai *et al.* 2002; Marani *et al.* 2002; Willis & Adams 2005).

Another possibility is that although PUMA and BIM are pro-apototic proteins themselves and thus viral induced degradation is a mechanism to eliminate a direct activator, another consequence of the down-regulation of these proteins could be contribute to increasing the level of 'free' MCL-1 in the cell. Thus the up-regulation of MCL-1 and the down-regulation of a binding partner that sequesters it increase the likelihood of an interaction with Bak. Additionally, very recent data suggests that the interaction of MCL-1 with Bak is induced under conditions that promote Bak oligomerisation/activation (Dai *et al.* 2015). Thus, viral infection – an apparent trigger of Bak activation – promotes a concomitant increase in MCL-1 levels which likely counters this. Indeed, it is interesting to note that when Bak activation was analysed at multiple points during the initial infection, we detected evidence of Bak activation at very early times which was quickly reversed.

Many of these observations were dependent on virally induced ERK-MAPK signalling. Viral induced activation of ERK-MAPK signalling and subsequent survival is a recurring theme suggesting that this pathway is a common target (Dai *et al.* 2016; Liu & Cohen 2013; Pleschka 2008; Pontes *et al.* 2015). The multifactorial response driven by ERK may also partially explain the observation that ELK-1 activity, whilst important, was never as detrimental as inhibition of upstream ERK signalling. Despite this, the induction of ELK-1 phophorylation was another virally induced response that was dependent on ERK-MAPK signalling. Again, this had a pro-survival effect, whereby elimination of ELK-1 phosphorylation, and thus activation (Cruzalegui *et al.* 1999), either through inhibition of ERK-MAPK or via a siRNA depletion of ELK-1 was deleterious, specifically upon virus infection. We note that evidence of ELK-1 binding to the MCL-1 promoter prior to any stimulation was observed and, although viral infection provided evidence of increased occupancy of ELK-1, phosphorylation of ELK-1 was

most important for MCL-1 transcription (Booy *et al.* 2011; Cruzalegui *et al.* 1999). The observed occupancy of ELK-1 in an unphosphorylated (presumably inactive) form at the promoter may suggest only a partial involvement of ELK-1 in basal MCL-1 expression in the CD34+ cells but, instead, being important for dynamic responses to further stimuli. It would also be consistent with ELK-1 activation representing a mechanism for rapidly inducing MCL-1 expression (Booy *et al.* 2011; Vickers *et al.* 2004). This stimulus-specific activity would explain the virally induced effects on MCL-1 and, also, the transient nature of the elevated MCL-1 expression (Reeves *et al.* 2012). The virus triggers a burst of ERK-MAPK activity upon entry promoting ELK-1 phosphorylation but once ERK signalling is down-regulated through normal feedback mechanisms (Fritsche-Guenther *et al.* 2011) coupled with a loss of the initial virus binding induced signal as the virus enters the cell, then MCL-1 expression returns to basal levels. The occupancy of inactive or even inhibitory (e.g. p50 homodimers) forms of transcription factors at promoters represents a mechanism for rapid induction of gene expression and has been observed at multiple promoters (Altarejos & Montminy 2011; Baer *et al.* 1998; Herrera *et al.* 1989) including in our own studies of CREB and HCMV reactivation (Kew *et al.* 2014). Finally, although our analyses focused on the regulation of MCL-1 we cannot preclude the possibility that ELK-1 activates multiple responses (Boros *et al.* 2009) that contribute to survival upon viral infection. Indeed, pathogens often utilise central components in pathways, thus it would not be unsurprising to detect further ELK-1 controlled responses that are important.

Prescient to this study is that ERK-MAPK signalling under certain conditions can, for example, be pro-apoptotic (Cagnol & Chambard 2010; Cagnol *et al.* 2006; Lu & Xu 2006) and thus in future studies it will be interesting to determine how HCMV directs the ERK response - possibly via the activation of concomitant pathways - to generate the pro-survival phenotype observed. Put simply, the kinase pathway that implements the final effector function (i.e. ERK-MAPK) is not dictating the outcome alone - that decision is defined by the nature of the signalling milieu activating the ERK-MAPK module upstream and the signalling and molecular context that those pathways are being activated in. Thus an aim of future studies is to address the identity of the death signals that are activated by the host cell in response to viral infection.

What this study illustrates is that viruses are the master regulators of signalling cascades and pathways with an impressive ability to hijack, re-direct or partition them to enhance infection. Understanding these events has greatly improved our understanding of cell biology and shed new light on the mechanisms that govern the activity of cellular functions required in key biological processes. Furthermore, it also illustrates how pleiotropic signalling pathways are modified to generate very specific outputs downstream. Greater understanding of these events and how they contribute to cell survival in the context of pathogen infection also has broader implications on our knowledge regarding the decision a cell makes to live or die.

Authors' contributions

VGK, MRW and MBR performed experiments and MRW and MBR analysed the data and wrote the paper.

Funding Information

This work was funded by a Medical Research Council Fellowship (G:0900466) awarded to MBR. MRW is funded by the Medical Research Council Programme Grants (G:0701279 & MR/K021087/1). The funders had no role in study design, data collection and interpretation, or decision to submit the work for publication.

References

Adams KW & Cooper GM 2007 Rapid turnover of mcl-1 couples translation to cell survival and apoptosis. *J Biol Chem* **282** 6192-6200

Altarejos JY & Montminy M 2011 CREB and the CRTC co-activators: sensors for hormonal and metabolic signals. *Nat Rev Mol Cell Biol* **12** 141-151

Baer M, Dillner A, Schwartz RC, Sedon C, Nedospasov S & Johnson PF 1998 Tumor necrosis factor alpha transcription in macrophages is attenuated by an autocrine factor that preferentially induces NF-kappaB p50. *Mol Cell Biol* **18** 5678-5689

Barry MA, Behnke CA & Eastman A 1990 Activation of programmed cell death (apoptosis) by cisplatin, other anticancer drugs, toxins and hyperthermia. *Biochem Pharmacol* **40** 2353-2362

Bieniasz PD 2004 Intrinsic immunity: a front-line defense against viral attack. Nat Immunol **5** 1109-1115

Booy EP, Henson ES & Gibson SB 2011 Epidermal growth factor regulates Mcl-1 expression through the MAPK-Elk-1 signalling pathway contributing to cell survival in breast cancer. *Oncogene* **30** 2367-2378

Boros J, Donaldson IJ, O'Donnell A, Odrowaz ZA, Zeef L, Lupien M, Meyer CA, Liu XS, Brown M & Sharrocks AD 2009 Elucidation of the ELK1 target gene network reveals a role in the coordinate regulation of core components of the gene regulation machinery. *Genome Res* **19** 1963-1973

Brune W 2010 Inhibition of programmed cell death by cytomegaloviruses. *Virus Res* **157** 144-150

Cagnol S & Chambard JC 2010 ERK and cell death: mechanisms of ERK-induced cell death--apoptosis, autophagy and senescence. *FEBS J* **277** 2-21

Cagnol S, Van Obberghen-Schilling E & Chambard JC 2006 Prolonged activation of ERK1,2 induces FADD-independent caspase 8 activation and cell death. *Apoptosis* **11** 337-346

Chan G, Nogalski MT, Bentz GL, Smith MS, Parmater A & Yurochko AD 2010 PI3K-dependent upregulation of Mcl-1 by human cytomegalovirus is mediated by epidermal growth factor receptor and inhibits apoptosis in short-lived monocytes. *J Immunol* **184** 3213-3222

Clem RJ, Fechheimer M & Miller LK 1991 Prevention of apoptosis by a baculovirus gene during infection of insect cells. *Science* **254** 1388-1390

Collins-McMillen D, Kim JH, Nogalski MT, Stevenson EV, Chan GC, Caskey JR, Cieply SJ & Yurochko AD 2015 Human Cytomegalovirus Promotes Survival of Infected Monocytes via a Distinct Temporal Regulation of Cellular Bcl-2 Family Proteins. *J Virol* **90** 2356-2371

Compton T 2000 Analysis of Cytomegalovirus Ligands, Receptors and the Entry Pathway. *Methods Mol Med* **33** 53-65

Cruzalegui FH, Cano E & Treisman R 1999 ERK activation induces phosphorylation of Elk-1 at multiple S/T-P motifs to high stoichiometry. *Oncogene* **18** 7948-7957

Dai H, Ding H, Meng XW, Peterson KL, Schneider PA, Karp JE & Kaufmann SH 2015 Constitutive BAK activation as a determinant of drug sensitivity in malignant lymphohematopoietic cells. *Genes Dev* **29** 2140-2152

Dai M, Feng M, Ye Y, Wu X, Liu D, Liao M & Cao W 2016 Exogenous avian leukosis virus-induced activation of the ERK/AP1 pathway is required for virus replication and correlates with virus-induced tumorigenesis. *Sci Rep* **6** 19226

Demir O, Aysit N, Onder Z, Turkel N, Ozturk G, Sharrocks AD & Kurnaz IA 2011 ETS-domain transcription factor Elk-1 mediates neuronal survival: SMN as a potential target. *Biochim Biophys Acta* **1812** 652-662

Doerflinger M, Glab JA & Puthalakath H 2015 BH3-only proteins: a 20-year stock-take. *FEBS J* **282** 1006-

1016

Dzhagalov I, Dunkle A & He YW 2008 The anti-apoptotic Bcl-2 family member Mcl-1 promotes T lymphocyte survival at multiple stages. *J Immunol* **181** 521-528

Erlacher M, Labi V, Manzl C, Böck G, Tzankov A, Häcker G, Michalak E, Strasser A & Villunger A 2006 Puma cooperates with Bim, the rate-limiting BH3-only protein in cell death during lymphocyte development, in apoptosis induction. *J Exp Med* **203** 2939-2951

Everett H & McFadden G 1999 Apoptosis: an innate immune response to virus infection. *Trends Microbiol* **7** 160-165

Everett RD & Chelbi-Alix MK 2007 PML and PML nuclear bodies: implications in antiviral defence. *Biochimie* **89** 819-830

Erlacher M, Labi V, Manzl C, Böck G, Tzankov A, Häcker G, Michalak E, Strasser A & Villunger A 2006 Puma cooperates with Bim, the rate-limiting BH3-only protein in cell death during lymphocyte development, in apoptosis induction. *J Exp Med* **203** 2939-2951

Fritsche-Guenther R, Witzel F, Sieber A, Herr R, Schmidt N, Braun S, Brummer T, Sers C & Blüthgen N 2011 Strong negative feedback from Erk to Raf confers robustness to MAPK signalling. *Mol Syst Biol* **7** 489

Gille H, Kortenjann M, Thomae O, Moomaw C, Slaughter C, Cobb MH & Shaw PE 1995 ERK phosphorylation potentiates Elk-1-mediated ternary complex formation and transactivation. *EMBO J* **14** 951-962

Gille H, Sharrocks AD & Shaw PE 1992 Phosphorylation of transcription factor p62TCF by MAP kinase stimulates ternary complex formation at c-fos promoter. *Nature* **358** 414-417

Goldmacher VS, Bartle LM, Skaletskaya A, Dionne CA, Kedersha NL, Vater CA, Han JW, Lutz RJ, Watanabe S, Cahir McFarland ED, Kieff ED, Mocarski ES & Chittenden T 1999 A cytomegalovirus-encoded mitochondria-localized inhibitor of apoptosis structurally unrelated to Bcl-2. *Proc Natl Acad Sci U S A* **96** 12536-12541

Griffiths GJ, Dubrez L, Morgan CP, Jones NA, Whitehouse J, Corfe BM, Dive C & Hickman JA 1998 Cell damage-induced conformational changes of the pro-apoptotic protein Bak in vivo precede the onset of apoptosis. *J Cell Biol* **144** 903-914

Guo H, Kaiser WJ & Mocarski ES 2015 Manipulation of apoptosis and necroptosis signaling by herpesviruses. *Med Microbiol Immunol* **204** 439-448

Herrera RE, Shaw PE & Nordheim A 1989 Occupation of the c-fos serum response element in vivo by a multiprotein complex is unaltered by growth factor induction. *Nature* **340** 68-70

Horvitz HR 1999 Genetic control of programmed cell death in the nematode Caenorhabditis elegans. *Cancer Res* **59** 1701s-1706s

Huang CR & Yang-Yen HF 2010 The fast-mobility isoform of mouse Mcl-1 is a mitochondrial matrix-localized protein with attenuated anti-apoptotic activity. *FEBS Lett* **584** 3323-3330

Huang HM, Huang CJ & Yen JJ 2000 Mcl-1 is a common target of stem cell factor and interleukin-5 for apoptosis prevention activity via MEK/MAPK and PI-3K/Akt pathways. *Blood* **96** 1764-1771

Kew VG, Yuan J, Meier J & Reeves MB 2014 Mitogen and stress activated kinases act co-operatively with CREB during the induction of human cytomegalovirus immediate-early gene expression from latency. *PLoS Pathog* **10** e1004195

Kuwana T, Mackey MR, Perkins G, Ellisman MH, Latterich M, Schneiter R, Green DR & Newmeyer DD 2002 Bid, Bax, and lipids cooperate to form supramolecular openings in the outer mitochondrial membrane. *Cell* **111** 331-342

Letai A, Bassik MC, Walensky LD, Sorcinelli MD, Weiler S & Korsmeyer SJ 2002 Distinct BH3 domains either sensitize or activate mitochondrial apoptosis, serving as prototype cancer therapeutics. *Cancer Cell* **2** 183-192

Levine B, Huang Q, Isaacs JT, Reed JC, Griffin DE & Hardwick JM 1993 Conversion of lytic to persistent alphavirus infection by the bcl-2 cellular oncogene. *Nature* **361** 739-742

Liu X & Cohen JI 2013 Inhibition of Bim enhances replication of varicella-zoster virus and delays plaque formation in virus-infected cells. *J Virol* **88** 1381-1388

Llambi F, Moldoveanu T, Tait SW, Bouchier-Hayes L, Temirov J, McCormick LL, Dillon CP & Green DR 2011 A unified model of mammalian BCL-2 protein family interactions at the mitochondria. *Mol Cell* **44** 517-531

Lu Z & Xu S 2006 ERK1/2 MAP kinases in cell survival and apoptosis. *IUBMB Life* **58** 621-631

Marani M, Tenev T, Hancock D, Downward J & Lemoine NR 2002 Identification of novel isoforms of the BH3 domain protein Bim which directly activate Bax to trigger apoptosis. *Mol Cell Biol* **22** 3577-3589

Michels J, Obrist F, Vitale I, Lissa D, Garcia P, Behnam-Motlagh P, Kohno K, Wu GS, Brenner C, Castedo M & Kroemer G 2014 MCL-1 dependency of cisplatin-resistant cancer cells. *Biochem Pharmacol* **92** 55-61

Mills JR, Hippo Y, Robert F, Chen SM, Malina A, Lin CJ, Trojahn U, Wendel HG, Charest A, Bronson RT, Kogan SC, Nadon R, Housman DE, Lowe SW & Pelletier J 2008 mTORC1 promotes survival through translational control of Mcl-1. *Proc Natl Acad Sci U S*

A **105** 10853-10858

Odrowaz Z & Sharrocks AD 2012 ELK1 uses different DNA binding modes to regulate functionally distinct classes of target genes. *PLoS Genet* **8** e1002694

Opferman JT 2007 Life and death during hematopoietic differentiation. *Curr Opin Immunol* **19** 497-502

Opferman JT, Iwasaki H, Ong CC, Suh H, Mizuno S, Akashi K & Korsmeyer SJ 2005 Obligate role of anti-apoptotic MCL-1 in the survival of hematopoietic stem cells. *Science* 307 1101-1104

Peppenelli MA, Arend KC, Cojohari O, Moorman NJ & Chan GC 2016 Human Cytomegalovirus Stimulates the Synthesis of Select Akt-Dependent Antiapoptotic Proteins during Viral Entry To Promote Survival of Infected Monocytes. *J Virol* **90** 3138-3147

Perciavalle RM & Opferman JT 2013 Delving deeper: MCL-1's contributions to normal and cancer biology. *Trends Cell Biol* **23** 22-29

Perciavalle RM, Stewart DP, Koss B, Lynch J, Milasta S, Bathina M, Temirov J, Cleland MM, Pelletier S, Schuetz JD, Youle RJ, Green DR & Opferman JT 2012 Anti-apoptotic MCL-1 localizes to the mitochondrial matrix and couples mitochondrial fusion to respiration. *Nat Cell Biol* **14** 575-583

Pleschka S 2008 RNA viruses and the mitogenic Raf/MEK/ERK signal transduction cascade. *Biol Chem* **389** 1273-1282

Pontes MS, Van Waesberghe C, Nauwynck H, Verhasselt B & Favoreel HW 2016 Pseudorabies virus glycoprotein gE triggers ERK1/2 phosphorylation and degradation of the pro-apoptotic protein Bim in epithelial cells. *Virus Res* **213** 214-218

Poole E, Lau JC & Sinclair J 2015 Latent infection of myeloid progenitors by human cytomegalovirus protects cells from FAS-mediated apoptosis through the cellular IL-10/PEA-15 pathway. *J Gen Virol* **96** 2355-2359

Poole E, McGregor Dallas SR, Colston J, Joseph RS & Sinclair J 2011 Virally induced changes in cellular microRNAs maintain latency of human cytomegalovirus in CD34[+] progenitors. *J Gen Virol* **92** 1539-1549

Puthalakath H & Strasser A 2002 Keeping killers on a tight leash: transcriptional and post-translational control of the pro-apoptotic activity of BH3-only proteins. *Cell Death Differ* **9** 505-512

Randow F, MacMicking JD & James LC 2013 Cellular self-defense: how cell-autonomous immunity protects against pathogens. *Science* **340** 701-716

Reeves MB, Breidenstein A & Compton T 2012 Human cytomegalovirus activation of ERK and myeloid cell leukemia-1 protein correlates with survival of latently infected cells. *Proc Natl Acad Sci U S A* **109** 588-593

Reeves MB, Davies AA, McSharry BP, Wilkinson GW & Sinclair JH 2007 Complex I binding by a virally encoded RNA regulates mitochondria-induced cell death. *Science* **316** 1345-1348

Rosenberg B, Vancamp L & Krigas T 1965 Inhibition of Cell Division in Escherichia coli by Electrolysis Products from a Platinum Electrode. *Nature* **205** 698-699

Sharrocks AD 2001 The ETS-domain transcription factor family. *Nat Rev Mol Cell Biol* **2** 827-837

Skaletskaya A, Bartle LM, Chittenden T, McCormick AL, Mocarski ES & Goldmacher VS 2001 A cytomegalovirus-encoded inhibitor of apoptosis that suppresses caspase-8 activation. *Proc Natl Acad Sci U S A* **98** 7829-7834

Slobedman B, Stern JL, Cunningham AL, Abendroth A, Abate DA & Mocarski ES 2004 Impact of human cytomegalovirus latent infection on myeloid progenitor cell gene expression. *J Virol* **78** 4054-4062

Stevenson EV, Collins-McMillen D, Kim JH, Cieply SJ, Bentz GL & Yurochko AD 2014 HCMV reprogramming of infected monocyte survival and differentiation: a Goldilocks phenomenon. *Viruses* **6** 782-807

Sun NK, Huang SL, Chang TC & Chao CC 2013 Sorafenib induces endometrial carcinoma apoptosis by inhibiting Elk-1-dependent Mcl-1 transcription and inducing Akt/GSK3beta-dependent protein degradation. *J Cell Biochem* **114** 1819-1831

Terhune S, Torigoi E, Moorman N, Silva M, Qian Z, Shenk T & Yu D 2007 Human cytomegalovirus UL38 protein blocks apoptosis. *J Virol* **81** 3109-3123

Townsend KJ, Zhou P, Qian L, Bieszczad CK, Lowrey CH, Yen A & Craig RW 1999 Regulation of MCL1 through a serum response factor/Elk-1-mediated mechanism links expression of a viability-promoting member of the BCL2 family to the induction of hematopoietic cell differentiation. *J Biol Chem* **274** 1801-1813

Treisman R, Marais R & Wynne J 1992 Spatial flexibility in ternary complexes between SRF and its accessory proteins. *EMBO J* **11** 4631-4640

Uren RT, Dewson G, Chen L, Coyne SC, Huang DC, Adams JM & Kluck RM 2007 Mitochondrial permeabilization relies on BH3 ligands engaging multiple prosurvival Bcl-2 relatives, not Bak. *J Cell Biol* **177** 277-287

Vickers ER, Kasza A, Kurnaz IA, Seifert A, Zeef LA, O'donnell A, Hayes A & Sharrocks AD 2004 Ternary complex factor-serum response factor complex-regulated gene activity is required for cellular proliferation and inhibition of apoptotic cell death. *Mol Cell Biol* **24** 10340-10351

Villunger A, Michalak EM, Coultas L, Müllauer F, Böck G, Ausserlechner MJ, Adams JM & Strasser A 2003 p53- and drug-induced apoptotic responses mediated by BH3-only proteins puma and noxa. *Science* **302** 1036-1038

Warr MR & Shore GC 2008 Unique biology of Mcl-1: therapeutic opportunities in cancer. *Curr Mol Med* **8** 138-147

Wei MC, Lindsten T, Mootha VK, Weiler S, Gross A, Ashiya M, Thompson CB & Korsmeyer SJ 2000 tBID, a membrane-targeted death ligand, oligomerizes BAK to release cytochrome c. *Genes Dev* **14** 2060-2071

Wei MC, Zong WX, Cheng EH, Lindsten T, Panoutsakopoulou V, Ross AJ, Roth KA, MacGregor GR, Thompson CB & Korsmeyer SJ 2001 Proapoptotic BAX and BAK: a requisite gateway to mitochondrial dysfunction and death. *Science* **292** 727-730

Westphal D, Kluck RM & Dewson G 2014 Building blocks of the apoptotic pore: how Bax and Bak are activated and oligomerize during apoptosis. *Cell Death Differ* **21** 196-205

Willis SN & Adams JM 2005 Life in the balance: how BH3-only proteins induce apoptosis. *Curr Opin Cell Biol* **17** 617-625

Willis SN, Chen L, Dewson G, Wei A, Naik E, Fletcher JI, Adams JM & Huang DC 2005 Proapoptotic Bak is sequestered by Mcl-1 and Bcl-xL, but not Bcl-2, until displaced by BH3-only proteins. *Genes Dev* **19** 1294-1305

Willis SN, Fletcher JI, Kaufmann T, van Delft MF, Chen L, Czabotar PE, Ierino H, Lee EF, Fairlie WD, Bouillet P, Strasser A, Kluck RM, Adams JM & Huang DC 2007 Apoptosis initiated when BH3 ligands engage multiple Bcl-2 homologs, not Bax or Bak. *Science* **315** 856-859

Yang C, Kaushal V, Shah SV & Kaushal GP 2007 Mcl-1 is downregulated in cisplatin-induced apoptosis, and proteasome inhibitors restore Mcl-1 and promote survival in renal tubular epithelial cells. *Am J Physiol Renal Physiol* **292** F1710-F1717

18

Introducing Thetis: a comprehensive suite for event detection in molecular dynamics

Eleni Picasi[1#], Athanasios Tartas[1#], Vasileios Megalooikonomou[2] and Dimitrios Vlachakis[1,2]

[1]Genetics and Computational Biology Group, Laboratory of Genetics, Department of Biotechnology, Agricultural University of Athens, 75 Iera Odos, 11855, Athens, Greece
[2]Computer Engineering and Informatics Department, School of Engineering, University of Patras, 26500 Patras, Greece
[#]Equal contributions

Correspondence should be addressed to Dimitrios Vlachakis; Email: dimvl@aua.gr

Abstract

A suite of computer programs has been developed under the general name Thetis, for monitoring structural changes during molecular dynamics (MD) simulations on proteins. Conformational analysis includes estimation of structural similarities during the simulation and analysis of the secondary structure with emphasis on helices. In contrast to available freeware dealing with MD snapshots, Thetis can be used on a series of consecutive MD structures, thus allowing a detailed conformational analysis over the time course of the simulation.

Introduction

Molecular dynamics (MD) simulations permit the study of complex, dynamic processes that occur in biological systems (Dalkas *et al.* 2013). This computational method calculates the time dependent behaviour of a molecular system. MD simulations can be applied, for example, for the evaluation of protein stability, conformational changes, protein folding, molecular recognition in proteins, DNA, membranes or complexes and ion transport in biological systems (Kabsh *et al.* 1983). Moreover, molecular dynamics provide the mean of carrying out structure refinement studies (X-ray), determination studies (NMR) or even *in silico* drug design experiments (Papageorgiou *et al.* 2016, Vlachakis *et al.* 2013, 2014).

Given that molecular dynamics trajectories are very complicated functions of the proteins and the environment, comparing different trajectories, even under the same conditions, is not straightforward (Vlachakis *et al.* 2013b). Several methods have been suggested that attempt to set the criteria for the evaluation and the comparison of MD trajectories at different levels of complexity (Vlachakis *et al.* 2013b). The simpler methods are geometry based and make use of the root-mean squared deviations between structures, while the more complicated methods are based on the time variation of the various properties of the system during the MD simulation (Vlachakis *et al.* 2014b). A great number of computer programs that focus on the trajectory analysis are available, whereas software for the actual monitoring of conformational changes over the course of the MD simulation in detail is not available (Vlachakis *et al.* 2015).

In this report, we present a new suite of computer programs for monitoring structural changes during molecular dynamics (MD) simulations on proteins with emphasis on helices (Vlachakis *et al.* 2015). Thetis is freeware and offers extensive possibilities for detailed conformational analysis using a series of consecutive coordinate files (MD structures) that are produced by taking successive time-snapshots throughout the process of molecular dynamics (MD) simulations (Vlachakis *et al.* 2017).

Description of the program

Thetis is written in optimised Basic (Liberty Basic v4.0) and was developed initially on a Pentium 4 machine running Windows. The program has also been tested in various UNIX systems running Linux (i386) or IRIX (SGi) with the aid of freeware emulators. There is a multitude of freeware windows emulators for Linux capable of providing an adequate platform for our programs.

The Windows-based version of Thetis

supports multiple processor systems (hyperthreading, dual core or dual CPU).

The programs are provided in standalone versions, compressed in zip format. All the essential libraries and accompanying sub-routines are provided in compressed zip format too. The full size of the suite does not exceed 5 MB in total size.

Thetis is controlled via simple pre-formatted text files through a set of keywords. Parameters used by the program, such as hydrogen bonding distance cut-offs, residue range for analysis and many more can be fully customized in order to meet the needs of the experiment. Thetis is flexible in the type and quantity of output that it generates and can read protein structure coordinate data in PDB or ENT format.

The program can operate in either of two modes, either reading, analyzing, and visualizing structures on an individual basis or automatically batch-processing sets of structures for the large-scale analysis of multiple proteins.

Because, long MD calculations generate enormous amounts of data (e.g. big matrices of data) that need to be loaded onto the computer's RAM, our programs are capable of operating in two modes:
a) Fast implementation: large computer memory is required (where, each coordinate file is accessed once and all information is stored permanently in the computer's memory).
b) Slow implementation: small memory is adequate (where, each coordinate file is accessed many times, and as a result each batch of information is read when needed and straight afterwards dumped from the computer's memory).

Having identified hydrogen bonds, Thetis proceeds to assign individual residues to a secondary structure type by matching observed hydrogen-bonding patterns to those characteristic of ideal secondary structures. The three main types of helical secondary structure analyzed by the program are the a, pi and 3_{10} helices. Following Kabsch-Sander and/or the Ramachandran definitions, these patterns can be described by simple logical conditions expressed in terms of the hydrogen bond connectivity matrix. As soon as the protein system has been loaded and analysed the most representative calculations that can be done with Thetis are:

•	Recognition of 3_{10}-, α- and pi- helices using as criterion the H bonding pattern of the polypeptide chain. Contrary to the algorithm of DSSP2 (http://swift.cmbi.ru.nl/gv/dssp), Thetis presents analytically the score each residue receives (3, 4 or 5 depending on the type of helix) per entry judging whether it belongs in helical formation or not. This is achieved by checking for repetition of overlapping residue

windows throughout the sequence. This way it becomes possible to monitor the stability of helices during the course of MD simulations. The output for each MD snapshot entry is recorded in a single txt file.
•	Generation of a txt file containing all φ/ψ and ω conformational angles for each amino acid of each MD snapshot entry.
Structural analysis of the α-helical conformation of each residue, according to the criteria of Ramachandran. The α-helical area is restricted to the most favoured regions on the plot as defined in the Procheck program 3, 4 (http://www.biochem.ucl.ac.uk/~roman/procheck/procheck.html). The following parameters are calculated for each residue, of each file-snapshot of the molecular dynamics simulation:

$$\rightarrow \text{ DaHC (Degree of } \alpha\text{-helical Conformation)} = \frac{\sum_{i=1}^{n}(-1)^{a_i} \cdot d_i}{n}$$

$$\rightarrow \text{ RMS of angular distances from the } \alpha\text{-region of Ramachandran} = \sqrt{\frac{\sum_{i=1}^{n} d_i^2}{n}}$$

$$\rightarrow \text{ Standard deviation from mean angular distance} = \sqrt{\frac{\sum_{i=1}^{n}\left(d_i - \bar{d}\right)^2}{n}}$$

Where,

n = the numbers of files in question
d_i = the angular distance of each residue point from the borders of the nominal α-helical Ramachandran area in entry file i
a_i = odd, if the residue point is located within the Ramachandran α-helical area
a_i = even, if the residue point is located outside the Ramachandran α-helical area

•	Ramachandran plots can be custom-drawn for a user-defined set of file entries. The plots can be coloured both in monochrome or colour, according to either the residue IDs or MD snapshot entries. This way, the user can either follow the course of conformational changes for each residue or the overall conformational changes of the protein during the course of the simulation.
•	RMSd of the C-alpha can be generated for a user-defined set of residues and time course of the MD simulation. The difference between two structures of the same protein can be evaluated by measuring the relative distance between the corresponding atoms. When contributions from translation and rotation are

subtracted and the absolute values are scaled by the total number of atoms, these distances become the familiar root-mean-squared deviations (RMSD). Generally, the Ca RMSD from the starting structure is reported for MD simulations. For native simulations (typically at 300 K), the Ca RMSD should vary little over time and should not deviate more than a couple of Angstrom units from the starting structure.

- As soon as the Ca RMS distances between the given residue-set have been calculated, an all-by-all pair-wise C^α RMSD matrix is subsequently constructed in order to compare the various MD structures between them. A color scale from blue to red was used to depict the variation of the C^α RMSD values. In such matrices, clusters of similar structures should appear as squares of approximately the same color, about the diagonal.

Using the above, even the smallest changes that may occur in the α-helical conformation during the course of MD can be detected. All output files produced by the software suite are in common-read TXT file-format, except picture files that are output in versatile .pov file-format.

Thetis is a complete suite for the analysis of the helical conformation of a molecular system over the course of molecular dynamics (MD) simulations. Through Thetis it is now possible to:

- Evaluate the α - / $3/_{10}$ - / pi - helical conformation of a set of multiple consecutive snapshots from molecular dynamics simulations, in the universally established .pdb or .ent file format.

- The evaluation of the phi and psi dihedral angles of each residue from each coordinate file over the full MD course.

- The "mobility" of each residue over the MD course. Here, the term mobility refers to the positional coordinate fluctuations of a given residue during MD, estimated from an interactive Ramachandran plot analysis.

- The α - helical or non α - helical conformation of a given (set of) residue(s) from the "core" region of the Ramachandran plot as defined by Procheck. This calculation can either be performed on a file-by-file basis or over a given set of consecutive snapshot-files outputted from molecular dynamics simulations. The tendency of a given residue to acquire α - helical conformation during the course of a molecular dynamics simulation, is also evaluated by determining the distance of each residue's phi/psi angles from the α – helical "core" region of the Ramachandran plot.

- Graphical representation of the conformational fluctuations of a given set of residues on a Ramachandran plot over the selected course of MD

simulation. Colored data points on the Ramachandran plot can either refer to selected residues or vary over time.

The Thetis suite is available as freeware (open -source software). The program's distribution includes documentation, example scripts, and standalone executables for windows (9x, NT, 2k and XP). All copyrights are held by their by their respective owners, unless specifically noted otherwise. All files provided by the Thetis team have been thoroughly tested and examined for viruses using a set of anti-viral applications. It can be downloaded from http://dimitrislab.com

Installation and Parameterization of Thetis

The Thetis suite comes in a single compressed file in .zip format. For Windows XP® and VISTA® running systems uncompressing and running the executables is automatically supported by the operating system. For earlier versions of windows an uncompressing utility such as WinZIP® or WinRAR® must be pre-installed (links to shareware and freeware versions can be found in our website). All programs are supplied in separate folders accompanied by their parameter files. All downloaded files are provided in RON (read only) format, to prevent alterations of the original files. Users are STRONGLY encouraged to copy and re-attribute the Thetis suite in their current working directories.

MODULE #1: "Helixanalysis"

BINARY: Helixanalysis.exe (binary file)
PARAMETER FILE: elixanalysisparameters.txt (parameter file)
USAGE: <path>/Helixanalysis.exe (param file in same folder)
DESCRIPTION: Determines the helical conformation of a given set of residues within the selected time-frame, and performs a set of statistical calculations.

→ **INPUT FILE FORMATTING** (remarks showing in *italics*):
Load file set: C:\Helix\AlaninTest\
Location of the files to be analysed
Filename conserved part: >AlaninHelix_<
Common filename string among candidate files
Starting serial number of file data set: 100
The first serial number of the input file set
Last serial number of file data set: 130
The last serial number of the input file set
Sampling step: 3
The sampling step factor. A sampling factor of 1 will

include all files in the calculation, same way a sampling factor of 2 will take file1, file3, file5, a sampling factor of 5 will include file1, file5,file10 etc.

 # Chain identifier: > <

Applies to .pdb files incorporating more than one chain. Here the chain of interest may be selected, otherwise (if left blank) all chains will be included in the calculation.

 # Output files for each snapshot seperately (Y/N)?: Y

Output format can be either in the form of a single file containing all calculations (choose N) or output per input file format (depending on the number of input files, choose Y).

 # Accepted bond patterns to the $^3/_{10}$ helixes: 001, 010, 100, 110, 101, 011, #

All hydrogen bonding combinations for $^3/_{10}$ helices. See "Hydrogen-helix scoring" section below.

 # Accepted bond patterns to the α helixes: 0001, 0010, 0100, 1000, 1100, 1010, 1001, 0110, 0101,0011, 1110, 1101, 1011, 0111, #

All hydrogen bonding combinations for α helices.

 # Accepted bond patterns to the p helixes: 10000, 01000, 00100, 00010, 00001, 11000, 10100, 10010, 10001, 01001, 01100, 01010, 00110, 00101, 00011, 11100, 11010, 11001, 10101, 10110, 10011, 01110, 01101, 01011, 00111, 11110 ,11101, 11011, 10111, 01111

All hydrogen bonding combinations for pi helices.

→ OUTPUT FILE FORMATTING:

A. Summary output file.

Filename format:

<path>/conserved filename part_helixR_1-50ST2_sums.txt, where:

conserved filename part is the common filename string among input files.

1 is the number of the first input file (taken from the filename).

50 is the number of the last input file (taken from the filename).

2 is the sampling step selected in the parameter file.

File format:

The following columns will appear in the output .txt file:

AA RN 310 3Su aHel aSu piH pSu Kink Rems, where:

AA: is the serial number of the current residue

RN: 3-letter code of the current residue

310:    H-bonds satisfying a $^3/_{10}$ helix. 000 means no H-bond, while 111 means all H-bonds are present.

3Su: an H will appear here for those residues that satisfy the criteria used in the parameter file.

aHel: H-bonds satisfying an α helix. 0000 means no H-bond, while 1111 means all H-bonds are present.

aSu: an H will appear here for those residues that satisfy the criteria used in the parameter file.

piH: H-bonds satisfying an α helix. 00000 means no H-bond, while 11111 means all H-bonds are present.

pSu: an H will appear here for those residues that satisfy the criteria used in the parameter file.

Kink: Will identify kinks in the polypeptide chain.

Rems: Remarks

B. Per file analysis output file.

Filename format: <path>/conserved filename part_helixR_1.txt ,where conserved filename part is the common filename string among input files. 1 is the number of the file, being analyzed.

File format: The following columns will appear in the output .txt file:

AA 3H-bond filter% aH-bond filter% pH-bond filter%, where:

AA: is the serial number of the current residue

3H-bond: is the mean of the H-bonds satisfying a $^3/_{10}$ helix from all files.

filtro%: % of residues in $^3/_{10}$ helical conformation

aH-bond: is the mean of the H-bonds satisfying an α helix from all files.

filtro%: % of residues in α helical conformation

pH-bond: is the mean of the H-bonds satisfying a pi helix from all files.

filtro%: % of residues in pi helical conformation

C. Per residue analysis output file.

Filename format:

<path>/conserved filename part_helixR_1XXX.txt ,where:

conserved filename part is the common filename string among input files.

1 is the number of the file, being analyzed.

XXX is the three-letter code of the residue being analysed (e.g. ALA)

File format:

The following columns will appear in the output .txt file:

File 310 3Su aHel aSu piH pSu Kink Rems, where:

File: is the serial number of the file, being analysed

310:    H-bonds satisfying a $^3/_{10}$ helix. 000 means no H-bond, while 111 means all H-bonds are present.

3Su: an H will appear here for those residues that satisfy the criteria used in the parameter file.

aHel: H-bonds satisfying an α helix. 0000 means no

H-bond, while 1111 means all H-bonds are present.

aSu: an H will appear here for those residues that satisfy the criteria used in the parameter file.

piH: H-bonds satisfying an α helix. 00000 means no H-bond, while 11111 means all H-bonds are present.

pSu: an H will appear here for those residues that satisfy the criteria used in the parameter file.

Kink: Will identify kinks in the polypeptide chain.

Rems: Remarks

MODULE #2: "Renamer"

BINARY: Renamer.exe (binary file)

PARAMETER FILE: Renamerparameters.txt (parameter file)

USAGE: <path>/ Renamer.exe (param file in same folder)

DESCRIPTION: Batch renaming of a set of files for subsequent analysis with Thetis.

→ **INPUT FILE FORMATTING** (remarks showing in *italics*):

Load file set: C:\Helix\AlaninTest\
Location of the files to be analysed.

Filename conserved part: >AlaninHelix_<
Common filename string among candidate files.

Filename Variable part: >001<
Variable filename string among candidate files.

First File number ID: 001
The serial number of the first file to rename.

Last File number ID: 100
The serial number of the last file to rename.

Leading Zeros: 3
If the variable filename string contains leading zeroes, enter the full length of the numerical part or else enter "N".

Output Filename String: >output_<
A small string that will be added to all filenames upon renaming.

Renumber from: 051
For renumbering the output files, enter the number of the first output file (here 051 instead of 001)

MODULE #3: "Dihedralcalc"

BINARY: Dihedralcalc.exe (binary file)

PARAMETER FILE: Dihedralcalcparameters.txt (parameter file)

USAGE: <path>/ Dihedralcalc.exe (param file in same folder)

DESCRIPTION: Program that calculates the Phi, Psi and Omega dihedral angles for a given residue range from multiple coordinate files.

→ **INPUT FILE FORMATTING** (remarks showing in *italics*):

Load file set: C:\Helix\AlaninTest\
Location of the files to be analysed.

Filename conserved part: >AlaninHelix_<
Common filename string among candidate files.

First File number ID: 001
The serial number of the first file to be analysed.

Population: 999
Number of files to be analysed (here 001,003,...,999).

MODULE #4: "Dihedralstats"

BINARY: Dihedralstats.exe (binary file)

PARAMETER FILE: Dihedralstatsparameters.txt (parameter file)

USAGE: <path>/ Dihedralstats.exe (parameter file in same folder)

DESCRIPTION: Program analyses output from Dihedralcalc and calculates the mean distance from Ramachandran plot.

INFO: The Dihedralstats will analyse the output from Dihedralcalc and will calculate the mean distance from the Ramachandran plot "core α-helical" region. A weight file can be used to generate a score of a-helical conformational significance along the MD course. An indicative multiplication factor is the current energy of the system. So, for example: the Dihedralcalc returns a value or 3 Angstroms for ALA001 at time 10 picoseconds when the system energy is 55000 Kcal/mole and 3 Angstroms for the same residue at time 10 nanoseconds when the system energy has dropped to -25000 Kcal/mole. Since the system's overall energy has dropped the RMSd of that residue should have dropped as well in the 10 ns snapshot. So, ALA001 is conformationally unstable. By applying the formula for a specific snapshot: $\mathbf{RMSd}_{res} * \mathbf{E}_{total}^{-1} = \mathbf{CS}$ **score**, where by CS score, we refer to the Conformational Stability of that specific residue at the given time snapshot.

→ **INPUT FILE FORMATTING** (remarks showing in *italics*):

Load file set: C:\Helix\AlaninTest\

Location of the files to be analysed.
 # Filename conserved part: >AlaninHelix_<
Common filename string among candidate files.
 # First file number ID: 001
The serial number of the first file to be analysed.
 # Population: 999
Number of files to be analysed (here 001,003,...,999).
 # Iterative mode (Y/N): N
Select Y (Yes) to produce separate outputs for each residue involved in the calculation, rather than the mean (default: N).
 # Use weight file (Y/N): N
The user can define in a separate file a residue importance weight factor to be used in a scoring function (default: N).
 # Load weight file: C:\Helix\AlaninTest\weight.txt
Location of the weight file.
 # Weight file order (as in weight file: N reverse: Y): N
The weight factor can be either used in the order found in the weight file or reversed.

MODULE #5: "Ramawalk"

BINARY: Ramawalk.exe   (binary file)
PARAMETER   FILE: Ramawalkparameters.txt (parameter file)
USAGE: <path>/ Ramawalk.exe   (parameter file must be in the same folder)
DESCRIPTION: Program that calculates the distance covered by each residue on a Ramachandran plot during the course of MD simulations. The Ramawalk values are indicative of the residue's tendency to acquire α-helical conformation.

→ **INPUT FILE FORMATTING** (remarks showing in *italics*):
 # Load file set: C:\Helix\AlaninTest\
Location of the files to be analysed.
 # Filename conserved part: >AlaninHelix_<
Common filename string among candidate files.
 # First file number ID: 001
The serial number of the first file to be analysed.
 # Population: 999
Number of files to be analysed (here 001,003,...,999).
 # Use weight file (Y/N): N
The user can define in a separate file a residue importance weight factor to be used in a scoring function (default: N).
 # Load weight file: C:\Helix\AlaninTest\weight.txt
Location of the weight file.
 # Weight file order (as in weight file: N reverse: Y): N
The weight factor can be either used in the order found in the weight file or reversed.

MODULE #6: "Ramaplotter"

BINARY: Ramaplotter.exe   (binary file)
PARAMETER   FILE: Ramaplotterparameters.txt (parameter file)
USAGE: <path>/ Ramaplotter.exe   (param file in same folder)
DESCRIPTION: Program that generates multiple residue or multiple positions (of same residue) on a Ramachandran plot with enhanced display capabilities.

→ **INPUT FILE FORMATTING** (remarks showing in *italics*):
 # Load file set: C:\Helix\AlaninTest\
Location of the files to be analysed.
 # Filename conserved part: >AlaninHelix_<
Common filename string among candidate files.
 # Consecutive files (S) or random (R): S
If the filenames of the files are in consecutive order choose S, otherwise R.
 # First file number ID: 001
The serial number of the first file to be analysed.
 # Last file number ID: 999
The serial number of the last file to be analysed.
 # Sampling step: 1
The sampling step factor. A sampling factor of 1 will include all files in the calculation, same way a sampling factor of 2 will take file1, file3, file5, a sampling factor of 5 will include file1, file5,file10 etc.
 # Determine the filenames of the Random files: 1, 10, 100
If the filenames of the files to be analysed are not in consecutive order give the variable numerical string of each one separated by a comma (,).
 # Chain identifier of Random files: > <
When using Random rather than consecutive filenames the ID chain must be selected.
 # Consecutive residues (S) or random (R): S
If the residues of interest within each file are in consecutive order choose S (ALA001, GLU002, ARG003), otherwise R (ALA001, ARG003, MET009).
 # First residue number: 001
The serial number of the first residue to be analysed.
 # Last residue number: 100
The serial number of the last residue to be analysed.
 # Determine the residue numbers when using: 1, 10, 100
If the residue numbers of the amino-acids to be analysed are not in consecutive order give the numerical string of each one separated by a comma (,).
 # Residue spot size: 1.5

The size of the spot of each residue on the Ramachandran plot (minimum is 0.8).

 # Background color: W
The background color of the Ramachandran plot. Two choices: White (W) and black (B).

 # Palette color: W
For a palette from red à purple enter (P), whereas for a palette from red à blue (B).

 # Scale up color: P
To scale up from purple or blue to red enter (R), whereas to scale up from red to purple or blue enter (P).

 # Coloring manner: T
To color dots depending on time progress enter (T), whereas to color dots per residue enter (R).

 # Coloring manner scale: A
A for absolute coloring and R for relative coloring.

 # Relative coloring lower value: 12
The number of the file or residue that will be set to have the lower value in the Relative coloring mode.

 # Relative coloring higher value: 1280
The number of the file or residue that will be set to have the higher value in the Relative coloring mode.

Conclusions

All in all, Thetis is a versatile, fast and flexible suite of programs mainly designed not only to evaluate the helical protein conformation, but also to provide the scientist with an extended and comprehensive overview-analysis of a molecular system during the course of MD simulations. Thetis is written in BASIC and is compatible with all native Windows® or Windows® – emulator equipped computers.

References

Kabsch W & Sander C 1983 Dictionary of Protein Secondary Structure: Pattern Recognition of Hydrogen -Bonded and Geometrical Features. *Biopolymers* **22** 2577

Dalkas GA, Vlachakis D, Tsagkrasoulis D, Kastania A & Kossida S 2013 State-of-the-art technology in modern computer-aided drug design. *Brief Bioinform,* **14** 745-752

Vlachakis D, Armaos A & Kossida S 2017 Advanced Protein Alignments Based on Sequence, Structure and Hydropathy Profiles; The Paradigm of the Viral Polymerase Enzyme. *Mathematics in Computer Science* **11** 197-208

Vlachakis D, Bencurova E, Papangelopoulos N & Kossida S 2014b Current state-of-the-art molecular dynamics methods and applications. *Adv Protein Chem Struct Biol* **94** 269-313

Vlachakis D, Champeris Tsaniras S, Feidakis C & Kossida S 2013 Molecular modelling study of the 3D structure of the biglycan core protein, using homology modelling techniques. *J Mol Biochem* **2** 85-93

Vlachakis D, Champeris Tsaniras S, Ioannidou K, Papageorgiou L, Baumann M & Kossida S 2014 A series of Notch3 mutations in CADASIL; insights from 3D molecular modelling and evolutionary analyses. *J Mol Biochem* **3** 97-105

Vlachakis D, Fakourelis P, Megalooikonomou V, Makris C & Kossida S 2015 DrugOn: a fully integrated pharmacophore modeling and structure optimization toolkit. *PeerJ* **3** e725

Vlachakis D, Tsagrasoulis D, Megalooikonomou V & Kossida S 2013b Introducing Drugster: a comprehensive and fully integrated drug design, lead and structure optimization toolkit. *Bioinformatics* **29** 126-128

Papageorgiou L, Loukatou S, Sofia K, Maroulis D & Vlachakis D 2016 An updated evolutionary study of Flaviviridae NS3 helicase and NS5 RNA-dependent RNA polymerase reveals novel invariable motifs as potential pharmacological targets. *Mol Biosyst* **12** 2080 -2093

19

Retroviral proteases: correlating substrate recognition with both selected and native inhibitor resistance

Gary S Laco

Roskamp Institute, Sarasota, Florida, USA

Correspondence should be addressed to Gary S Laco; E-mail: gary.laco@gmail.com

Abstract

A diverse group of retroviral proteases were analyzed to correlate mechanisms of substrate recognition with resistance to HIV-1 protease active-site inhibitors. Here it was shown that HIV-1 protease utilized a pathway common to many retroviral proteases, for recognition of mutated Gag/Pol cleavage sites, in order to become resistant to active-site inhibitors. While HIV-1 and HIV-2 resulted from independent cross-species transmissions of simian immunodeficiency virus into humans, HIV-2 has native primary resistance to many HIV-1 protease inhibitors as do many other retroviral proteases. The native multi-drug resistance of those proteases contributed to the lack of treatments for the respective life-long infections. Analysis of interactions between retroviral proteases and Gag/Pol substrates revealed that protease interactions weighted towards cleavage site residues P4-P4' resulted in inhibitor sensitivity, while interactions weighted towards residues P12-P5/P5'-P12' gave inhibitor resistance. In addition, a mechanism was identified for human T-cell leukemia virus type-1 protease that allowed re-weighting of the protease interactions with substrate residues P4-P4' and P12-P5/P5'-P12' using anti-parallel beta-sheets that connected the protease flaps to the substrate-grooves. Those anti-parallel beta-sheets are common to all studied retroviral proteases. The critical role of the retroviral protease substrate-grooves in substrate recognition and inhibitor resistance makes them a potential target.

Introduction

Retroviruses infect a remarkably diverse range of vertebrate species that span fish to humans (Barre-Sinoussi *et al.* 1983, Fodor & Vogt 2002, Gallo *et al.* 1983). All retroviruses encode an aspartic acid protease (PR) in which two identical monomers each contribute a catalytic aspartic acid residue to the active-site of the symmetrical PR dimer (Kohl *et al.* 1988, Wlodawer *et al.* 1989). The retroviral PR is typically expressed as either a Gag-Pro polyprotein, or as a Gag-Pro-Pol polyprotein, due to translational frame shifting near the C-terminus of Gag (Jacks *et al.* 1988). As the virus buds from the host cell, the PR first undergoes autocatalytic processing out of the respective polyprotein, and then cleaves Gag to release the structural proteins that are needed for virion maturation: matrix; capsid; and nucleocapsid (Strickler *et al.* 1989). The PR also cleaves Pol to release the enzymes needed for completion of the viral replication cycle in the next host cell: reverse transcriptase and integrase (Strickler *et al.* 1989). The essential role of the HIV-1 PR in virion maturation and activation of the viral enzymes have made the HIV-1 PR an important target in the treatment of human immunodeficiency virus type-1 (HIV-

1) infections. Clinical HIV-1 PR inhibitors are typically peptidomimetics based in part on native cleavage sites; the only non-peptidomimetic inhibitor in current clinical use is tipranavir, which perhaps as a consequence is associated with more severe adverse effects. The HIV-1 protease inhibitors are comparable in length to a four residue peptide substrate and are typically designed to bind to the HIV-1 PR active-site as a transition state mimetic with a water tetrahedrally coordinated between the inhibitor P1/P1' backbone carbonyl oxygens and the backbone nitrogens of the HIV-1 PR flaps, Figure 1. Novel side groups on the inhibitors increase both van der Waals and H-bond interactions with the HIV-1 PR active-site and that contributes to active-site inhibitors being able to outcompete Gag/Pol substrates for binding to HIV-1 PR (Erickson *et al.* 1990, Miller *et al.* 1989).

HIV-1 PR inhibitor resistance typically begins with the selection of primary mutations in the active-site that can either reduce van der Waals contacts due to shorter side-chains (i.e., Ile54Val), or change the electrostatic interactions (i.e., Asp30Asn), between the HIV-1 PR and inhibitor, Figure 1 (Wensing *et al.* 2015). Those active-site mutations can also decrease the interactions between HIV-1 PR and substrate resi-

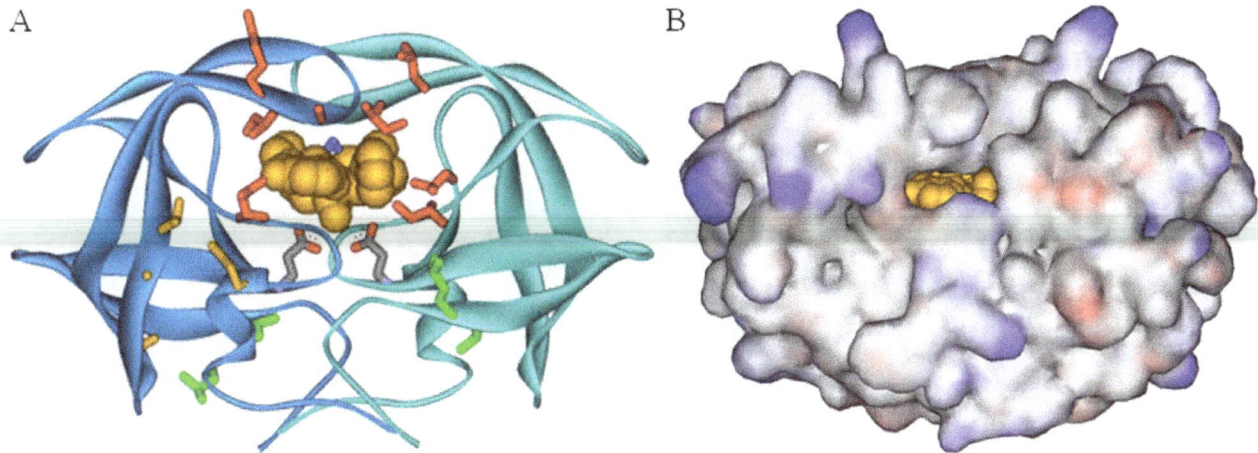

Figure 1. WT HIV-1 PR bound to nelfinavir. A, HIV-1 PR front view with flaps on top, A and B subunits as backbone ribbons, active-site Asp25 (A and B subunits, center) with side chains atoms in: oxygen, red; carbon, grey. Nelfinavir (bronze, CPK rendering) bound in the active-site, hydrogens not shown. Tetrahedrally coordinated water (center, above nelfinavir) with all atoms in blue. Solid colored PR residues that when mutated contributed to either primary, or secondary, inhibitor resistance are shown only once on either the A or B subunits. Primary active-site residues in red, clockwise from center of left subunit: Asp30; Ile47; Met46; Gly48; Ile54; Ile50; Val82; Ile84. Secondary S-groove residues in bronze, clockwise from bottom of left subunit: Ala71; Gly73; Thr74; Asn88. Secondary cleavage-site residues with residue and cleavage site indicted in green, clockwise from bottom of left subunit: Ile93 (P7 PR/RT); Leu90 (P10 PR/RT); Leu10 (P10' p6*/PR); Val11, P11' (p6*/PR). B, HIV-1 PR front view (flaps on top) with electrostatic surface potential (red negative, blue positive) and bound nelfinavir (bronze, CPK rendering).

dues P4-P4' resulting in reduced viral replicative capacity, Figure 2 (Chang & Torbett 2011, Gulnik *et al.* 1995, Kaplan *et al.* 1994, Martinez-Picado *et al.* 1999, Nijhuis *et al.* 1999, Pazhanisamy *et al.* 1996, Prabu-Jeyabalan *et al.* 2004, Schock *et al.* 1996). Next, selection of secondary resistance mutations distant from the active-site (i.e., Ala71Leu, Gly73Thr) restored both HIV-1 PR binding to Gag/Pol cleavage sites and viral replicative capacity while maintaining inhibitor resistance, Figure 1 and 2 (Chang & Torbett 2011, Gulnik *et al.* 1995, Kaplan *et al.* 1994, Martinez-Picado *et al.* 1999, Nijhuis *et al.* 1999, Pazhanisamy *et al.*

1996, Prabu-Jeyabalan *et al.* 2004, Schock *et al.* 1996)

In order to gain insight into the evolution of multi-drug resistant HIV-1 PR (MDR HIV-1 PR) researchers initially focused on understanding HIV-1 PR interaction with the ten primary Gag/Pol polyprotein cleavage sites by using short peptide substrates containing Gag/Pol cleavage site residues P4-P4'. However, a consistent cleavage order for the short peptides could not be determined in part because in order to increase peptide solubility researchers used different assay conditions and peptides of different lengths, including non-native N-and C-terminal residues (Billich

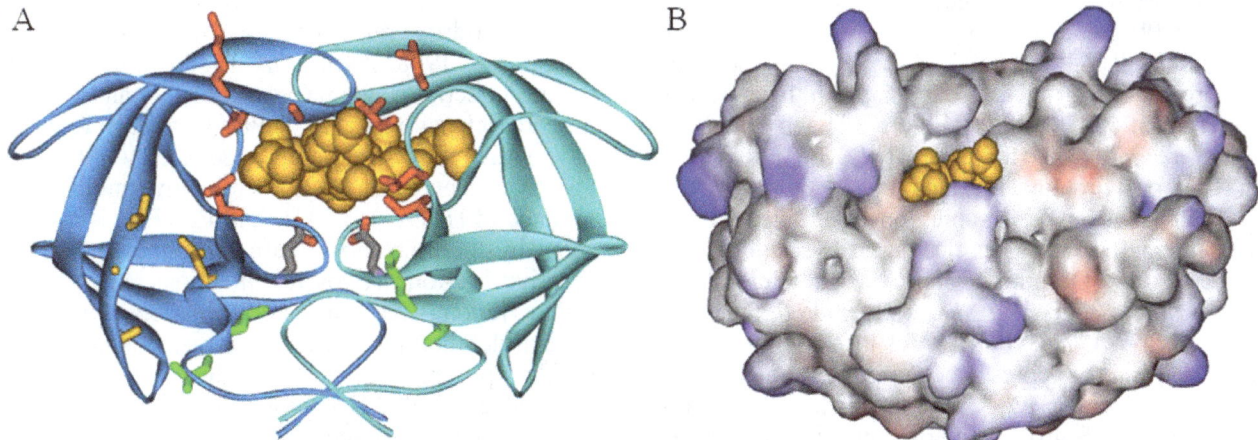

Figure 2. WT HIV-1 PR bound to Gag SP1/NC cleavage site 8-mer. A, HIV-1 PR front view with flaps on top, A and B subunits as backbone ribbons, active-site mutation D25N (A and B subunits, center) side chains atoms: oxygen, red; nitrogen blue; carbon, grey. SP1/NC cleavage site residues P4-P4' (bronze, CPK rendering) bound in the active-site, hydrogens not shown. Tetrahedrally coordinated water not visible (see Fig. 1 A). Solid colored PR residues that when mutated contributed to either primary, or secondary, inhibitor resistance are shown only once on either the A or B subunits. Primary active-site residues in red, clockwise from center of left subunit: Asp30; Ile47; Met46; Gly48; Ile54; Ile50; Val82; Ile84. Secondary S-groove residues in bronze, clockwise from bottom of left subunit: Ala71; Gly73; Thr74; Asn88. Secondary cleavage-site residues with residue and cleavage site indicted in green, clockwise from bottom of left subunit: Ile93 (P7 PR/RT); Leu90 (P10 PR/RT); Leu10 (P10' p6*/PR); Val11, P11' (p6*/PR). B, HIV-1 PR front view (flaps on top) with electrostatic surface potential (red negative, blue positive) and bound SP1/NC cleavage site residues P4-P4' (bronze, CPK rendering).

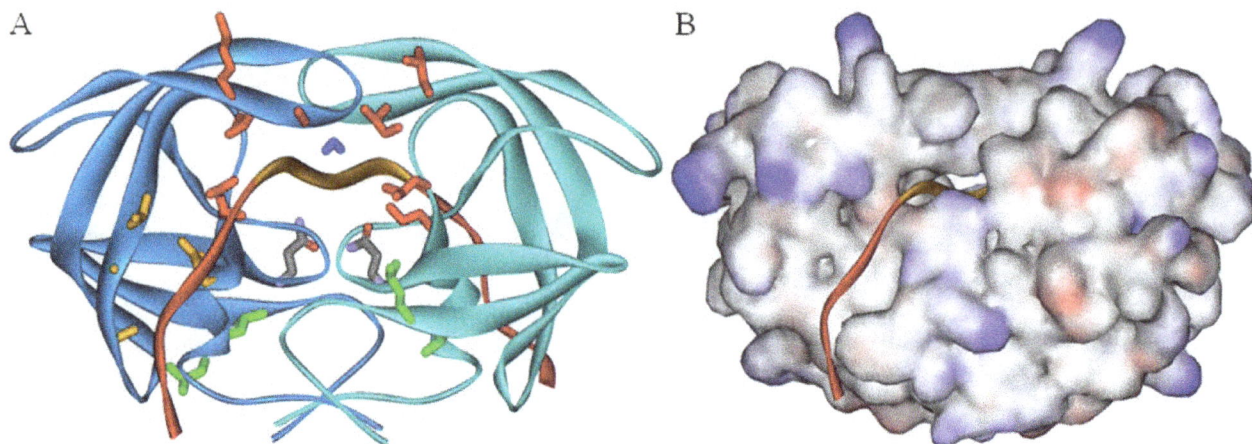

Figure 3. WT HIV-1 PR bound to Gag SP1/NC cleavage site 24-mer. A, HIV-1 PR front view with flaps on top, A and B subunits as backbone ribbons, active-site mutation D25N (center, A and B subunits) side chains atoms: oxygen, red; nitrogen, blue, carbon, grey. SP1/NC cleavage site residues P12-P12' (bronze ribbon |P12-P5|, red ribbon P4-P4') bound in the active-site and S-grooves, hydrogens not shown. Tetrahedrally coordinated water (center, above substrate) with all atoms blue. Solid colored PR residues that when mutated contributed to either primary, or secondary, inhibitor resistance are shown only once on either the A or B subunits. Primary active-site residues in red, clockwise from center of left subunit: Asp30; Ile47; Met46; Gly48; Ile54; Ile50; Val82; Ile84. Secondary S-groove residues in bronze, clockwise from bottom of left subunit: Ala71; Gly73; Thr74; Asn88. Secondary cleavage-site residues with residue and cleavage site indicted in green, clockwise from bottom of left subunit: Ile93 (P7 PR/RT); Leu90 (P10 PR/RT); Leu10 (P10' p6*/PR); Val11, P11' (p6*/PR). B, HIV-1 PR front view with flaps on top and electrostatic surface potential (red negative, blue positive) and bound SP1/NC cleavage site residues P12-P12' (bronze ribbon |P12-P5|, red ribbon P4-P4').

et al. 1988, Darke *et al.* 1988, Kotler *et al.* 1988, Krausslich *et al.* 1989, Tozser *et al.* 1991). It was only when *in vitro* transcribed/translated full length Gag and Gag-Pro-Pol polyproteins and purified full-length Gag polyprotein were used as substrates for HIV-1 PR that a consistent cleavage order was determined (Erickson-Viitanen *et al.* 1989, Pettit *et al.* 2005). These results indicated that cleavage site residues outside of P4-P4' were important for HIV-1 PR recognition of substrates and provided a clue as to how the MDR HIV-1 PR evolved. More recently, it was shown that the HIV-1 PR can also bind the Gag MA/CA cleavage site residues P12-P5/P5'-P12' (|P12-P5|) in the substrate-grooves (S-grooves), one on each face of the symmetric HIV-1 PR dimer, Figure 3 (Laco 2015). In addition, recent NMR studies on HIV-1 PR interaction with Gag polyproteins revealed that residues in the HIV-1 PR S-grooves, as previously defined (Laco 2015), interacted with Gag cleavage site residues outside of P4-P4' (Deshmukh *et al.* 2017). Here those findings were extended *in silico* to include the interaction between WT and MDR HIV-1 PRs and nine Gag/Pol cleavage sites represented as either 8-mers (P4-P4'), or 24-mers (P12-P12'), to evaluate the importance of the S-groove interaction with substrates, while eliminating the inherent variables of *in vitro* peptide cleavage assays. By understanding HIV-1 PR substrate interactions, a strategy can be developed to target HIV-1 that express MDR PR.

In contrast to HIV-1 PR where inhibitor resistance can be selected for *in vitro* and *in vivo*, many other retroviruses express PRs with native resistance to HIV-1 PR active-site inhibitors. For example, whereas

HIV-1 and HIV-2 originated from independent transmissions of simian immunodeficiency virus (SIV) into humans (Hirsch *et al.* 1989, Huet *et al.* 1990, Marx *et al.* 1991, Peeters *et al.* 1989), HIV-2 PR has native primary resistance to many HIV-1 PR clinical inhibitors (Brower *et al.* 2008, Masse *et al.* 2007, Rodes *et al.* 2006, Witvrouw *et al.* 2004). Similarly, while human T-cell leukemia virus type-1 (HTLV-1) resulted from the cross species transmission of simian T-cell leukemia virus (STLV) into humans (Koralnik *et al.* 1994, Voevodin *et al.* 1997), the HTLV-1 PR has native multi-drug resistance to HIV-1 PR inhibitors (Ding *et al.* 1998, Pettit *et al.* 1998). In addition, both equine infectious anemia virus (EIAV) and feline immunodeficiency virus (FIV) PRs have been reported to have native multi-drug resistance to HIV-1 PR inhibitors (Kervinen *et al.* 1998). The native multi-drug resistance of HTLV-1, EIAV, and FIV PRs is a key reason why there are no effective treatments for the respective life-long retroviral infections. In the case of HTLV-1, long-term untreated infections can result in several debilitating neurological diseases, as well as a rapidly progressing and terminal T-cell leukemia (Goncalves *et al.* 2010, Proietti *et al.* 2005). The published structures of the above PRs allowed for the *in silico* analysis of the interactions between the PRs and both substrates and inhibitors, Results (Gustchina *et al.* 1996, Kovalevsky *et al.* 2008, Laco *et al.* 1997, Li *et al.* 2005, Rose *et al.* 1996a). In addition, the generation of three-dimensional (3D) models of the HIV-1 Gag and Pol polyproteins provided insight into the role of cleavage site accessibility on PR recognition of substrates. The analysis revealed the native resistance mechanisms for these PRs, and will be important in the development of effective inhibitors for retroviruses that express PRs with either selected, or native, resistance to active-site inhibitors.

Materials and Methods

Computational chemistry

PR/ligand models were energy minimized prior to calculation of interaction energy scores using Accelrys Discovery Studio (Dassault Systèmes, San Diego, CA) with parameters set to approximate *in vitro* conditions for direct interactions between proteins and ligands as previously described (Laco 2011, 2015). The only explicit water in the PR substrate/inhibitor models was a water tetrahedrally coordinated between either the HIV-1 PR flaps Ile50 (A and B-subunits) backbone nitrogens, or structurally equivalent residues in the HIV-2/SIV-cpz/SIV-sm/HTLV-1/EIAV/FIV PRs, and either the respective substrate residue P1 and P1' backbone carbonyl oxygens or inhibitor oxygens, harmonic restraints were placed on those H-bonded heavy atoms as previously described (Laco 2015). The indicated structure coordinate files were used to build the respective PR substrate/inhibitor models: WT HIV-1 PR bound to the Gag SP1/NC cleavage site was based on 1KJ7.pdb [19] with the addition of the HIV-1 PR HXB2 residues Val3 and Ser37 (Ratner *et al.* 1985), the 24-mer SP1/NC substrate backbone orientation served as the starting point for all PR/substrate models for consistency (Laco 2015). The MDR 3761 HIV-1 PR mutations relative to WT HIV-1 PR [46] were built into 1KJ4.pdb [19] as previously described (Laco 2015). HIV-1 substrates used in the *in silico* studies were based on the HIV-1 HXB2 strain (Ratner *et al.* 1985). The EIAV PR (1FMB.pdb), FIV PR (2FIV.pdb), HTLV-1 PR (2B7F.pdb) were treated the same as HIV-1 PR (Gustchina *et al.* 1996, Kovalevsky *et al.* 2008, Laco *et al.* 1997, Li *et al.* 2005). HIV-2 PR was generated by mutating the SIV-sm PR, as described herein, using the HIV-2 subtype A, isolate BEN, sequence (UniProtKB-P18096), uniprot.org (Consortium 2015). The SIV-cpz and SIV-sm PR models were generated by mutating the published SIV PR structure 1YTJ.pdb (Rose *et al.* 1996a) in Discovery Studio to the amino acid sequences of the SIV-cpz MB66 isolate (UniProtKB-Q1A267) and the SIV-sm S4 F236 isolate (UnitProtKB-P12502) (D'Arc *et al.* 2015), respectively, uniprot.org (Consortium 2015). The HTLV-1 PR was reverse engineered by either mutating two S-groove residues (I85A/T88G) to give HTLV-1 2X PR, or the two S-groove residues (I85A/T88G) and two active-site residues (V56I/A59I) together to give HTLV-1 4X PR. The default Discovery Studio orientations for the mutated HTLV-1 2X and 4X PR side chains were used for the minimizations. Before starting the respective minimizations, the orientations of HTLV-1 2X and 4X PRs native residues, and bound 24-mer substrate residues, were identical to the HTLV-1 PR and bound 24-mer substrate.

HIV-1 Gag and Pol structure-based 3D models

The HIV-1 full-length Gag polyprotein (MA/CA/SP1/NC/SP2/p6) and Pol polyprotein (p6*/PR/RT/RH/IN) 3D models were generated using published structures of Gag and Pol proteins (protein, PDB ID): MA-CA, 1L6N.pdb (Tang *et al.* 2002); CA-SP1, 4XFX.pdb (Gres *et al.* 2015); NC, 1MFS.pdb (Lee *et al.* 1998); p6, 2C55.pdb (Fossen *et al.* 2005); PR monomer, 1Q9P.pdb (Ishima *et al.* 2003); RT, 3T19.pdb (Gomez *et al.* 2011); IN, 4NYF.pdb (Wang *et al.* 2001), while the transframe encoded p6* structure was generated using the Robetta full-chain protein prediction server, http://robetta.bakerlab.org (Ovchinnikov *et al.* 2016). The proteins were bonded together using short random coil linkers containing missing residues, and then the structures were typed with the consistent force field (CFF) and energy minimized using implicit solvent and the same parameters in Discovery Studio as used above for HIV-1 PR (Laco 2015). Note: the relative

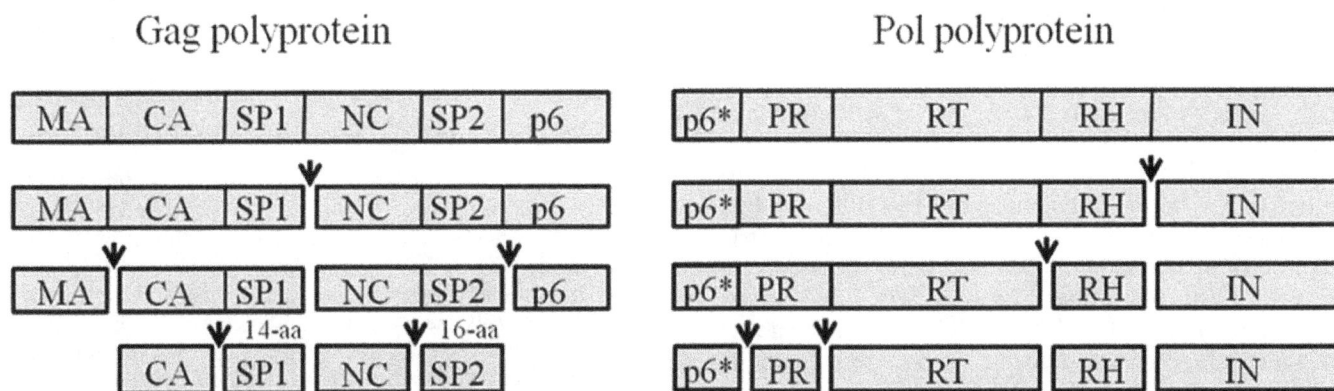

Figure 4. Schematic of HIV-1 Gag and Pol polyproteins and cleavage sites. Top row: interior vertical lines indicate cleavage sites; arrows indicate relative cleavage order from top to bottom for each polyprotein when cleaved in trans by HIV-1 PR (Pettit et al., 2005; Tritch *et al.*, 1991). The length of the CA/SP1 (14-aa) and NC/SP2 (16-aa) cleavage sites C-terminal tails are indicated. The Gag-Pro-Pol polyprotein is composed of Gag with a TF region starting after NC that changes the reading frame to express p6* and the Pol polyprotein (not shown).

orientations of the subdomains within the Gag and Pol polyproteins are in part for illustrative purposes as compared to the more extended orientations found within an immature HIV-1 virion (Bharat *et al.* 2014).

Protein sequence alignments
Protease 3D structures were aligned using PROMALS3D (Pei *et al.* 2008). The following pairs of viral Gag-Pro-Pol polyproteins were aligned using MUSCLE (Multiple Sequence Comparison by Log-Expectation, (Edgar 2004, McWilliam *et al.* 2015): HIV-1 (H2, group M, subtype B, isolate HXB2, Uni-ProtKB-P04585) and SIV-cpz (isolate MB66, Uni-ProtKB-Q1A267); and HIV-2 (subtype A, isolate BEN, UniProtKB-P18096) and SIV-sm (S4, isolate F236, UnitProtKB-P12502), uniprot.org (Consortium 2015).

Statistical analysis
The GraphPad QuickCalcs *t* test calculator and paired *t* test were used for all statistical analyses (http://www.graphpad.com/quickcalcs/ttest1.cfm).

Results

HIV-1 Gag/Pol cleavage site alignments, residues P12-P12'
The HIV-1 PR is essential for virion maturation in the infected cell and for activation of the enzymes required to complete the viral replication cycle in the next host cell (Kohl *et al.* 1988, Le Grice *et al.* 1988). Both processes require HIV-1 PR to cleave the ten primary Gag/Pol polyprotein cleavage sites in the correct order and rate, Figure 4.

How the HIV-1 PR accomplishes this is complicated by the fact that the HIV-1 PR cleavage sites do not have a consensus sequence for residues P4-P4' (Pearl & Taylor 1987). As a result, substrate residues bound by the HIV-1 PR S-grooves (|P12-P5|) were included in an alignment and revealed that even for this expanded cleavage site alignment (P12-P12') there were still no residues at any position conserved in all ten cleavage sites, and for P10' there were no conserved residues, Figure 5.

However, there were two positions where there was conservation at four cleavage sites: P2, Asn (4X); and P1, Phe (4X), Figure 5. The HIV-1 PR like other retroviral PRs, including the FIV PR, can bind peptide substrates in either the N-to C-terminal, or C-to N-terminal, orientation due to the structural symmetry of the PR dimer (Laco *et al.* 1997, Prabu-Jeyabalan *et al.* 2002). When the Gag/Pol cleavage site residues P12-P1 were aligned C-to N-terminal (i.e., P1-P12) with the respective P1'-P12' residues Phe was found at |P1| in three additional cleavage sites for a total of 7X, and His was found at |P12| in two additional cleavage sites for a total of 4X, followed by thirty-four residues/positions where the conservation increased by an additional cleavage site including at

|P10| where six residues were conserved, Figure 5 yellow highlights. It was proposed that the heterogeneity of the Gag/Pol cleavage sites was due to the dual roles of the P12-P12' residues: first as recognition sequences that regulated Gag/Pol polyprotein cleavage order; and second as structural components in the N-and C-termini of the respective Gag/Pol cleavage products (Laco 2015), with the exception of the Gag derived SP1 and SP2 peptides, which are not part of any mature viral protein, Figure 4.

WT HIV-1 PR *in silico* interaction with Gag/Pol cleavage sites
The WT HIV-1 PR interactions with Gag/Pol cleavage sites were analyzed *in silico* using 8-mer and 24-mer peptide substrates in order to explore the role of the WT HIV-1 PR S-grooves in substrate recognition, Table 1 (Laco 2015). This approach avoided the issues associated with *in vitro* peptide assays, which while having good kinetic resolution suffered from variable results (Billich *et al.* 1988, Kotler *et al.* 1988, Krausslich *et al.* 1989, Tozser *et al.* 1991). The WT HIV-1 PR *in silico* interaction energy scores with substrates were compared to the *in vitro* cleavage of the Gag/Pol polyproteins. This was done to determine *in silico* which substrate length (8-mer/24-mer) best matched the published *in vitro* Gag/Pol cleavage order (Pettit *et al.* 2002, 2005). Previous results for WT HIV-1 PR bound to the Gag MA/CA and SP1/NC sites (Laco 2015) were based on structures of HIV-1 PR with substrates from both HIV-1 NL4-3 and HXB2 strains (King *et al.* 2012, Kozisek *et al.* 2007, Prabu-

```
            P12        P4        P4'      P12'
             |          |          |        |
MA/CA    DTGHSNQVSQNY/PIVQNIQGQMVH
CA/SP1   GVGGPGHKARVL/AEAMSQVTNSAT
SP1/NC   AMSQVTNSATIM/MQRGNFRNQRKI
NC/SP2   HQMKDCTERQAN/FLGKIWPSYKGR
SP2/p6   IWPSYKGRPGNF/LQSRPEPTAPPE
NC/TF    HQMKDCTERQAN/FLREDLAFLQGK
p6*/PR   GADRQGTVSFNF/PQVTLWQRPLVT
PR/RT    NLLTQIGCTLNF/PISPIETVPVKL
RT/RH    LEKEPIVGAETF/YVDGAANRETKL
RH/IN    DKLVSAGIRKVL/FLDGIDKAQDEH
```

Figure 5. HIV-1 Gag/Pol cleavage site alignments P12-P12'. HXB2 sequence, / indicates scissile bond. Residue conservation across all sites indicated as follows: bold, 2X; underlined, 3X; red, 4X. The Gag NC/SP2 and NC/TF cleavage sites have identical residues P12-P2'. The translational frame shift used to generate the Gag-Pro-Pol polyprotein occurs between the NC/TF cleavage site residues P2' and P3'. The transframe region contains p6* at the C-terminus. Gray highlighted residues made electrostatic/H-bond interactions with the MDR HIV-1 PR S-groove resistance residue Thr73 (A/B subunits). P10' yellow highlighted residues were conserved when the P12-P1 residues were aligned C-to N-terminal (P1-P12) with the P1'-P12' residues (see Results, not shown).

Table 1. WT HIV-1 PR interaction energy scores for Gag/Pol substrates.

Cleavage Site	HIV-1 PR in vitro Cleavage rate (Order) Gag[1]	HIV-1 PR in vitro (Cleavage Order) Gag-Pro-Pol[2]	WT HIV-1 PR in silico Interaction energy		24-mer Interaction Breakdown	
			8-mer	24-mer	|P12-P5|	P4-P4'
MA/CA	10X↓(3)	(2)	-136	-248	-123	<u>-125</u>
CA/SP1	400X↓(5)	ND	-125	-205	-90	<u>-114</u>
SP1/NC	1 (1)	(1)	-122	-257	<u>-141</u>	-116
NC/SP2	350X↓(4)	ND	-121	-245	<u>-137</u>	-108
SP2/p6	9X↓(2)	ND	-120	-236	<u>-123</u>	-112
P6*/PR		(4)	-126	-226	-106	<u>-120</u>
PR/RT		(5)	-123	-224	<u>-112</u>	<u>-112</u>
RT/RH		(3)	-139*	-255*	-125	<u>-130</u>
RH/IN		(2)	-129*	-254*	<u>-133</u>	-121

WT HIV-1 PR interaction energy in kcal/mol for Gag and Pol cleavage sites as 8-mers and 24-mers, with a breakdown of the 24-mer substrates scores for |P12-P5| and P4-P4' (strongest interactions underlined). [1,2] The *in silico* interaction energy scores were compared to the published *in vitro* cleavage orders for the same sites in the context of Gag and Gag-Pro-Pol polyproteins where Gag-Pro-Pol contained an inactive PR with active HIV-1 PR supplied in trans (Pettit et al., 2002; Pettit et al., 2005). *The 3D context of RT/RH and RH/IN sites were examined in Results.

Jeyabalan *et al.* 2002, Wang *et al.* 2012). Here all substrates used for *in silico* interaction energy scores were based on the HIV-1 HXB2 strain for consistency, Figure 5 (Ratner *et al.* 1985). The WT HIV-1 PR interaction energy scores with nine Gag/Pol cleavage sites were reported alongside the published *in vitro* cleavage order for the same sites when part of Gag/Pol polyproteins, Table 1 (Pettit *et al.* 2002, 2005).

Notably, the scores for the MA/CA and SP1/NC site 8-mers were out of order with the published *in vitro* cleaved Gag/Pol polyproteins, while the 24-mers were in order (excluding RT/RH and RH/IN sites, Table 1). The RT/RH and RH/IN sites 8-mer and 24-mer scores were out of order due to being among the strongest interactions with WT HIV-1 PR (Table 1). Interestingly, the WT HIV-1 PR scores for the SP1/NC site went from one of the weaker 8-mer scores (incorrect order) to the strongest 24-mer score (correct order), Table 1. These results support the importance of the WT HIV-1 PR S-groove contacts with residues |P12-P5| in substrate recognition, Table 1 (Laco 2015). When the interaction energy scores between WT HIV-1 PR and substrates were broken down for substrate residues |P12-P1| versus P4-P4' it was found that four out of the nine cleavage site scores were weighted towards the active-site/P4-P4'. The PR/RT site scores for substrate residues |P12-P1| and P4-P4' were evenly distributed between the S-grooves and active-site, Table 1.

MDR HIV-1 PR *in silico* interaction with Gag/Pol cleavage sites

The evolution of HIV-1 PR multi-drug resistance typically begins with primary resistance mutations selected for in the active-site/flaps (i.e., Ile54Val and Asp30Asn), which reduces PR binding to inhibitors as

well as Gag/Pol substrates (Laco 2015), resulting in inhibitor resistance along with reduced viral replicative capacity, Figure 1 Introduction. Next secondary resistance mutations are selected for in the PR S-grooves (i.e., A71I and G73T) that increased interactions with substrate residues |P12-P5| (Laco 2015), and that restored viral replicative capacity while maintaining inhibitor resistance, Figure 3 Introduction. The combination of HIV-1 PR primary and secondary resistance mutations resulted in multi-drug resistance and treatment failure (Chen *et al.* 1995, Gulnik *et al.* 1995, Schock *et al.* 1996).

Here the interactions between a MDR HIV-1 PR, HIV-1 PR 3761 (Wang *et al.* 2012), which contains seven resistance mutations including the S-groove A71I and G73T mutations (Laco 2015), and nine HIV-1 Gag/Pol cleavage sites was analyzed *in silico* (Table 2). Like the WT HIV-1 PR (Table 1), the MDR HIV-1 PR interactions with the MA/CA and SP1/NC site 8-mers were out of order with the *in vitro* cleaved Gag/Pol polyproteins, while the respective 24-mer scores were in order, Table 2. Similar to the WT HIV-1 PR, the MDR HIV-1 PR interactions with the RT/RH and RH/IN sites were the among the strongest making them out of order with the *in vitro* cleaved Gag/Pol polyproteins, Table 2. When the interactions between the MDR HIV-1 PR and substrates were broken down for substrate residues |P12-P5| and P4-P4', it was found that for eight out of the nine cleavage sites the interaction between the MDR HIV-1 PR and substrate was weighted towards the S-grooves and |P12-P5|. Only the CA/SP1 site was weighted towards the active-site and P4-P4', Table 2. The MDR HIV-1 PR S-groove secondary resistance residue Thr73 made direct electrostatic/H-bond interactions with residues in all nine of the Gag/Pol cleavage sites, see Figure 5

Table 2. MDR HIV-1 PR interaction energy scores for Gag/Pol substrates.

Cleavage Site	HIV-1 PR in vitro Cleavage rate (Order) Gag[1]	HIV-1 PR in vitro Cleavage (Order) Gag-Pro-Pol[2]	MDR HIV-1 PR in silico Interaction energy		24-mer Interaction Breakdown	
			8-mer	24-mer	|P12-P5|	P4-P4'
MA/CA	10X↓(3)	(2)	-126	-248	-132	-116
CA/SP1	400X↓(5)	ND	-115	-205	-101	-104
SP1/NC	1(1)	(1)	-117	-254	-147	-107
NC/SP2	350X↓(4)	ND	-110	-246	-144	-102
SP2/p6	9X↓(2)	ND	-116	-237	-127	-110
P6*/PR		(4)	-117	-244	-131	-113
PR/RT		(5)	-117	-224	-127	-97
RT/RH		(3)	-129*	-259*	-136	-123
RH/IN		(2)	-120*	-256*	-146	-110

MDR HIV-1 PR interaction energy in kcal/mol for Gag and Pol cleavage sites as 8-mers and 24-mers, with a breakdown of the 24-mer substrates scores for |P12-P5| and P4-P4' (strongest interactions underlined). [1,2] The *in silico* interaction energy scores were compared to the published *in vitro* cleavage orders for the same sites in the context of Gag and Gag-Pro-Pol polyproteins where Gag-Pro-Pol contained an inactive PR with active HIV-1 PR supplied in trans (Pettit et al., 2002; Pettit et al., 2005). *The 3D context of the RT/RH and RH/IN sites were examined in Results.

gray highlighted residues.

Interestingly, two incongruous results stand out in Tables 1 and 2. The first is the magnitude of difference in the *in vitro* cleavage rates reported for the CA/SP1 site (400-fold lower) and NC/SP2 site (350-fold lower) relative to the SP1/NC site when part of a Gag polyprotein (Pettit *et al.* 2002), and the 0.04 to 0.20-fold difference for the WT and MDR HIV-1 PRs *in silico* interaction energy scores for the CA/SP1 and NC/SP2 sites 24-mers relative to the SP1/NC site 24-mer (Tables 1 and 2). One explanation is that due to earlier cleavages at the SP1/NC and SP2/p6 sites, the CA/SP1 and NC/SP2 sites only had 14 and 16 residues, respectively, that extended past the C-terminal

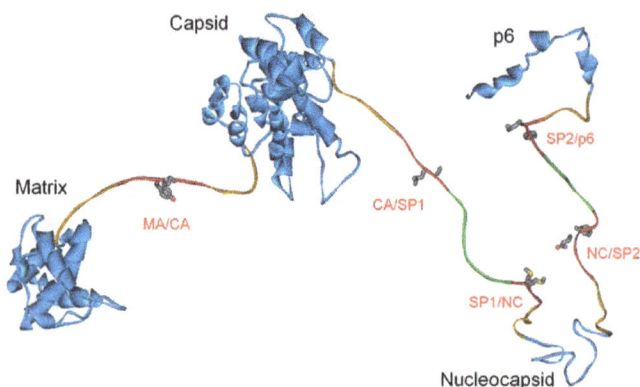

Figure 6. HIV-1 Gag polyprotein 3D model. Protein backbone as a ribbon, individual subdomains in blue with black labels, cleavage site backbone for residues |P12-P5| in bronze, P4-P4' in red. Cleavage site labels in red with residues P1/P1' shown: carbon, grey; oxygen, red; nitrogen, blue; sulfur, yellow. The overlapping portions of the CA/SP1 and SP1/NC, and NC/SP2 and SP2/p6, cleavage sites are green, Fig. 4 and 5. The overall orientation was influenced by illustrative considerations and in not meant to represent the more extended orientation within an immature virion.

side of the scissile bond, Figure 4. This is in contrast to all other Gag/Pol cleavage sites that were stabilized by structured domains on both sides of the scissile bond, Figure 4. The orientations of the CA/SP1 and NC/SP2 site C-terminal tails could hinder HIV-1 PR binding to the respective cleavage site residues P1'-P12' *in vitro*, while *in silico* those potential issues were not taken into account. Complicating the issue is that the context of those sites within the Gag polyprotein could not be analyzed due to the lack of a full-length Gag structure. The second incongruous result from Tables 1 and 2 is that while the RT/RH and RH/IN sites were cleaved slower *in vitro* than the SP1/NC site in the respective Gag/Pol polyproteins, the RT/RH and RH/IN 24-mer substrates had some of the strongest *in silico* interactions with both the WT and MDR HIV-1 PRs (Tables 1 and 2). However, no obvious explanation could be proposed based on either the position of the cleavage sites in the Pol polyprotein, or the primary sequence of the cleavage sites, Figure 4 and 5. As a result, attention was turned to the 3D context of both the Gag CA/SP1 and NC/SP2 sites, and the Pol RT/RH and RH/IN sites. However, while the structures for nearly all of the mature proteins from the HIV-1 Gag/Pol polyproteins have been solved using crystallography and NMR, to our knowledge no full-length structure-based energy minimized 3D models of the HIV-1 Gag and Pol polyproteins have been published.

HIV-1 Gag and Pol polyprotein 3D models

HIV-1 Gag and Pol polyprotein 3D models were built *in silico* using published structures of Gag and Pol derived proteins, while the p6* protein structure was generated using the Robetta full-chain protein prediction server, Materials and methods (Ovchinnikov *et al.* 2016). The individual proteins were connected as required with random coil linkers containing missing

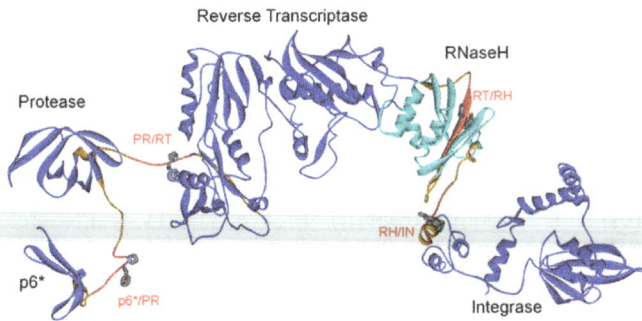

Figure 7. HIV-1 Pol polyprotein 3D model. Protein backbone as a ribbon, individual subdomains in blue with black labels, except for the RNaseH domain in teal. Cleavage site backbone for residues |P12-P5| in bronze, P4-P4' in red. Cleavage site labels in red with residues P1/P1' shown: carbon, grey; oxygen, red; nitrogen, blue. The overall orientation was influenced by illustrative considerations and is not meant to represent the more extended orientation within an immature virion.

residues to generate a full-length Gag polyprotein and Pol polyprotein with an N-terminal p6*, Figure 6 and 7. The Gag polyprotein 3D model revealed that all five cleavage site residues P12-P12' were solvent exposed in extended random coil configurations, Figure 6. As a result, the dramatically reduced *in vitro* cleavage rates for the CA/SP1 and NC/SP2 sites were likely solely to the short C-terminal tails, Figure 4. Those C-terminal tails could be either disordered random coils, or form α-helices (Bharat *et al.* 2014), and in either case interfere with HIV-1 PR binding and cleavage (Table 1). This may explain why the Gag NC/SP2 24-mer scores for both the WT and MDR PRs were stronger than expected based on the *in vitro* Gag cleavage order, in that the high affinity primary sequence may be used to off-set the weaker *in vitro* binding of

the PRs to the short C-terminal tail of the NC/SP2 cleavage site (Tables 1 and 2, Figure 4).

Compared to the Gag cleavage sites, the Pol cleavage sites were not as solvent accessible: the p6*/PR site residues P12-P8, and PR/RT site residues P12-P9, were part of α-helices, however, those structured residues likely contributed only a minor steric hindrance to HIV-1 PR binding since the remaining cleavage site residues were solvent exposed, Figure 7. In contrast, the Pol RT/RH site residues P12-P12' made extensive β-sheet contacts with the RNase H domain, while the RH/IN site residues P12-P6 and P3'-P12' were both part of α-helices that made interactions with RNase H β-sheets and integrase α-helices, respectively, Fig 7. Those finding explain why the Pol RT/RH and RH/IN site 24-mers (P12-P12') had some of the strongest *in silico* interactions with WT and MDR HIV-1 PRs (Tables 1 and 2), while *in vitro* they were reported to be the third and second fastest cleaved sites respectively, in that the high affinity primary sequence of the cleavage sites likely enhanced PR binding to the sites in the context of the Pol polyprotein, Figure 7. The results in Tables 1 and 2 indicated that the *in silico* interaction energy scores between both WT and MDR HIV-1 PRs and the 24-mer Gag/Pol cleavage sites (P12-P12') in general reflected the published *in vitro* cleavage order for sites that were in predominantly solvent exposed regions of the polyproteins, Tables 1 and 2. In contrast, the WT and MDR HIV-1 PRs *in silico* scores with the Gag/Pol cleavage site residues P4-P4' (8-mers) were more out of order with the reported *in vitro* cleavage orders and had lower resolution between the scores for the fast SP1/NC site and the slower CA/SP1, p6*/PR, and PR/

```
                    10/11           20      25    30 33              46 47
HIV-1    PQVTLWQRP LVT IKIGGQL KEALL DTGAD DTV LEEMSLPGRWKPK MI
SIV-cpz  PQITLWQRP ILT VKIGGEI KEALLDTGADDTVIEE IQLEGKWKPKMI

         48 50   54                       7173/74    82 84  88 90 93
HIV-1    GG IGGFI KVRQYDQILIEICGHK AIGT VLVGPTP VNI IGR NLL TQI GCTLNF
SIV-cpz  GGIGGFIKVKQYDNVIIEIQGKKAVGTVLVGPTPVNIIGRNFLTQIGCTLNF


                    10/11           20      25    30 33              46 47
HIV-1    PQVTLWQRP LVT IKIGGQL KEALL DTGAD DTV LEEMSLPGRWKPK MI
HIV-2    PQFSLWKRP VVTAYIED QPVEVLL DTGADDSI VAGIELGDNYTPK IV
SIV-sm   PQFSLWRRP IVTAYIEE QPVEVLL DTGADDSI VAGIELGPNYTPK IV

         48 50   54                       7173/74    82 84  88 90 93
HIV-1    GG IGGFI KVRQYDQILIEICGHK AIGT VLVGPTP VNI IGR NLL TQI GCTLNF
HIV-2    GGIGGFINTKEYKNVEIKVLNKRV RA TIMTGDTP INI FGRNILTALGMSLNL
SIV-sm   GGIGGFINTKEYKDVKIKVLGKVI KGTIMTGDTP INI FGRNLLTAMGMSLNL
```

Figure 8. Structure-based HIV and SIV protease alignments. Top, HIV-1 PR/SIV-cpz PR; bottom, HIV-2 PR/SIV-sm PR with HIV-1 PR included as a resistance residue reference. Gray highlighted residues indicate amino acid identity between aligned PRs. HIV-1 PR residues that when mutated contributed to inhibitor resistance in bold, primary resistance residues indicated by underlined numbers, HIV-2 PR and SIV PR native resistance residues in bold. Active-site Asp25 (D25) italicized. Potential resistance residues in HIV-2 and SIV-sm at position 20 underlined.

RT sites, Tables 1 and 2.

Evolution of HIV-1 and HIV-2 from SIV: Structure-based PR alignments

The evolution of HIV-1 PR multi-drug resistance to active-site inhibitors results in treatment failure (Brown *et al.* 2003, Mocroft *et al.* 2003, Richman *et al.* 2004, Rosenbloom *et al.* 2012). This could represent a novel evolutionary pathway for HIV-1 PR inhibitor resistance. Alternatively, it could be a common pathway used by all retroviral proteases to adapt, during cross-species transmissions (Wolfe *et al.* 2005), to mutations in Gag/Pol cleavage sites selected for in the non-native host. Mutation of the Gag/Pol cleavage sites also occurs during HIV-1 adaptation to a new native host due to cytotoxic T lymphocyte (CTL) selective pressure on viral proteins including Gag/Pol and the respective cleavage sites (Phillips *et al.* 1991, Prince *et al.* 2012, Seibert *et al.* 1995). It is important to note that the HIV-1 ancestral SIV was from chimpanzees/SIV-cpz (Gao *et al.* 1999, Santiago *et al.* 2002), while the HIV- 2 ancestral SIV was from sooty mangabeys/SIV-sm (Lemey *et al.* 2003), since the amino acid sequences differ for both PRs, Figure 8. The SIV-cpz and SIV-sm PRs share 57% identity, Figure 8. It has been reported that both SIV-sm PR and HIV-2 PR had native resistance to clinical HIV-1 PR inhibitors: SIV-sm, amprenavir; HIV-2, amprenavir, atazanavir, nelfinavir, ritonavir, and tipranavir (Brower

et al. 2008, Desbois *et al.* 2008, Rodes *et al.* 2006, Witvrouw *et al.* 2004). The 3D structures for HIV-1 and SIV-cpz PRs, and HIV-2 and SIV-sm PRs, were aligned to determine whether the PRs contained known HIV-1 PR resistance residues, Figure 8. HIV-1 PR had 83% identity with the SIV-cpz PR that had one native resistance residue (Val10), Figure 8. HIV-2 PR had 85% identity with the SIV-sm PR, with HIV-2 PR having seven native resistance residues while SIV-sm PR had six native resistance residues, Figure 8. The HIV-2, SIV-cpz, and SIV-sm PRs native resistance residues when found in HIV-1 PR at the same 3D position contributed to primary and secondary inhibitor resistance, http://hivdb.stanford.edu (Rhee *et al.* 2003, Wensing *et al.* 2015). When the HIV-2 PR and both SIV PRs were analyzed for the presence of S-grooves, all three PRs were found to have S-grooves similar to HIV-1 PR and HTLV-1 PR, Figure 3 and 10 (Laco 2015).

Evolution of HIV-1 and HIV-2 from SIV: Gag/Pol cleavage site alignments

While HIV-2 and SIV PRs were shown to contain native resistance residues, neither of those viruses had been exposed to HIV-1 PR active-site inhibitors during evolution, Figure 8. As a result, we explored how the HIV-1 and HIV-2 Gag/Pol cleavage sites evolved from the respective SIV Gag/Pol cleavage sites, during adaptation to the human host, to see how that may have contributed to the native resistance residues in the PRs.

Gag Cleavage Sites

```
                  P12------P4------P4'-----P12'
MA/CA                      62.5            49.5
HIV-1     DTGHSNQVSQNY/PIVQNIQGQMVH
SIV-cpz   TADGTSTVSRNF/PIVANAQGQMVH
                           87.5            81.2
HIV-2     PTAPPSGKRGNY/PVQQAGGNYVHV
SIV-sm    PTAPPSGRGGNY/PVQQVGGNYVHL

CA/SP1                     100             62.5
HIV-1     GVGGPGHKARVL/AEAMSQVTNSAT
SIV-cpz   GVGGPSHKARVL/AEAMSQAQHSND
                           100             75
HIV-2     GVGGPGQKARLM/AEALKEAMGPSP
SIV-sm    GVGGPGQKARLM/AEALKEALRPDQ

SP1/NC                     12.5            37.5
HIV-1     AMSQVTNSATIM/MQRGNFRNQRKI
SIV-cpz   AMSQAQHSNDAK/RQFKGPKRIVKC
                           50              50
HIV-2     ALKEAMGPSPIP/FAAAQQRKAIRY
SIV-sm    ALKEALRPDQLP/FAAVQQKGQRKT

NC/SP2                     87.5            43.5
HIV-1     HQMKDCTERQAN/FLREDLAFLQGK
SIV-cpz   QMRNCTNERQAN/FFRETLAFQQGK
                           87.5            68.7
HIV-2     HIMANCPERQAG/FFRVGPTGKEAS
SIV-sm    HVMAKCPERQAG/FFRAWPMGKEAP
```

Pol Cleavage Sites*

```
                  P12------P4------P4'-----P12'
P6*/PR                    75/50           25/87.5
HIV-1     GADRQGTVSFNF/PQVTLWQRPLVT
SIV-cpz   GDREQAVSSANF/PQISLWQRPVVT
                          87.5/100    62.5/87.5
HIV-2     DTSQRGDRGLAA/PQFSLWKRPVVT
SIV-sm    ETLQGGDRGFAA/PQFSLWRRPVVT

PR/RT                     100             87.5
HIV-1     NLLTQIGCTLNF/PISPIETVPVKL
SIV-cpz   NILTQIGCTLNF/PISPIETVPVSL
                          87.5            81.2
HIV-2     NILTALGMSLNL/PVAKIEPIKVTL
SIV-sm    NLLTAMGMSLNL/PIAKVEPIKVTL

RT/RH                     100             81.2
HIV-1     LEKEPIVGAETF/YVDGAANRETKL
SIV-cpz   LEQDPIPGAETF/YVDGAANRETKL
                          87.5            75
HIV-2     LVGDPIPGAETF/YTDGSCNRQSKE
SIV-sm    LVKEPIQGAETF/YVDGSCNRQSRE

RH/IN                     100             87.5
HIV-1     DKLVSAGIRKVL/FLDGIDKAQDEH
SIV-cpz   DKLVSSGIRKVL/FLDGIDKAQEEH
                          87.5            100
HIV-2     DHLVSQGIRQVL/FLEKIEPAQEEH
SIV-sm    DHLVSQGIRQVL/FLKKIEPAQEEH
```

Figure 9. HIV and SIV Gag/Pol cleavage site alignments. Cleavage site residues P12-P12' shown, gray highlighted residues indicate identity between aligned sequences. For each cleavage site: top, HIV-1/SIV-cpz; bottom, HIV-2/SIV-sm. Percent identity for P4-P4' shown above cleavage site scissile bond (/); combined percent identity for |P12-P5| shown on the right side above P12'. *The p6*/PR cleave site percent identities were calculated separately for Gag (P12-P4 and P4-P1, left value) and Pol (P1'-P4' and P4'-P12', right value).

Table 3. Protease interaction energy scores with MA/CA 24-mer residues |P12-P5| and P4-P4'.

HIV-1	MDR HIV-1	HIV-2	SIV-cpz	SIV-sm	HTLV-1	EIAV	FIV
-123/-125	-132/-116	-120/-115	-135/-111	-127/-114	-128/-107	-152/-120	-130/-110

HIV-1, MDR HIV-1, HIV-2, SIV-cpz, SIV-sm, HTLV-1, EIAV, FIV PRs interaction energy scores (kcal/mol) with the corresponding MA/CA substrate residues |P12-P5| and P4-P4'.

The HIV-1 and HIV-2 Gag/Pol cleavage site residues (P12-P12') were aligned with the respective cleavage sites from the nearest ancestral SIV based on gag gene sequences, Figure 9 (D'Arc et al. 2015). The HIV-1 and HIV-2 cleavage site residues were divided into two groups (|P12-P5| and P4-P4') with the percent identity to the respective SIV cleavage sites indicated on the right side for |P12-P5|, and above the scissile bond for P4-P4', Figure 9. Note that the p6*/PR site residues P12-P1 were grouped with the Gag cleavage sites, while the residues P1'-P12' were grouped with the Pol cleave sites. HIV-1 had similar identity as HIV-2 with all of the respective SIV Gag cleavage site residues P4-P4' and p6*/PR residues P4-P1 (two-tailed P value = 0.1087), Figure 9. In contrast, HIV-2 had significantly more identity with the SIV-sm Gag cleavage site residues |P12-P5|, than did HIV-1 with the respective SIV-cpz cleavage site residues |P12-P5| (two-tailed P value = 0.0090), Figure 9. The HIV-1 and HIV-2 Pol cleavage site residues P12-P12' were overall highly conserved with the respective SIV Pol cleavage sites, Figure 9. These results indicated that HIV-2 had greater identity with SIV-sm Gag cleavage sites than did HIV-1 with SIV-cpz Gag sites, and the HIV-2 Gag cleavage site identities were more evenly distributed across cleavage site residues P12-P12', Figure 9. Mutation of the solvent accessible Gag cleavage sites, during viral adaptation to the human host, may have been

one of the factors that drove the evolution of HIV-1 PR to weight substrate interactions towards the active-site and more conserved Gag cleavage site residues P4-P4', versus the S-grooves and divergent Gag cleavage site residues |P12-P5|, Figure 9.

Distribution of PR interactions with 24-mer substrates in silico
Based on the analysis of the HIV-1 and HIV-2 Gag/Pol cleavage site evolution from the respective SIVs the following hypothesis was proposed: due to CTL selective pressure in the human host (Phillips et al. 1991, Prince et al. 2012, Seibert et al. 1995), HIV-1 cleavage sites had more residues conserved P4-P4' versus |P12-P5| from SIV-cpz and this drove selection of an HIV-1 PR that had stronger active-site interactions with substrate residues P4-P4'. At the same time, HIV-1 PR S-groove interactions with substrate residues |P12-P5| were weakened in order to: 1) balance HIV-1 PR affinity for substrates to maintain the overall Gag/Pol polyprotein cleavage order required for virion maturation, Table 1; and 2) prevent premature activation of HIV-1 PR during formation of the immature virion (Strickler et al. 1989). One consequence was that HIV-1 PR became sensitive to active-site inhibitors. In contrast, during HIV-2 adaptation to the human host the Gag/Pol cleavage sites had a more balanced and higher conservation of SIV-sm cleavage site residues |P12-P5|

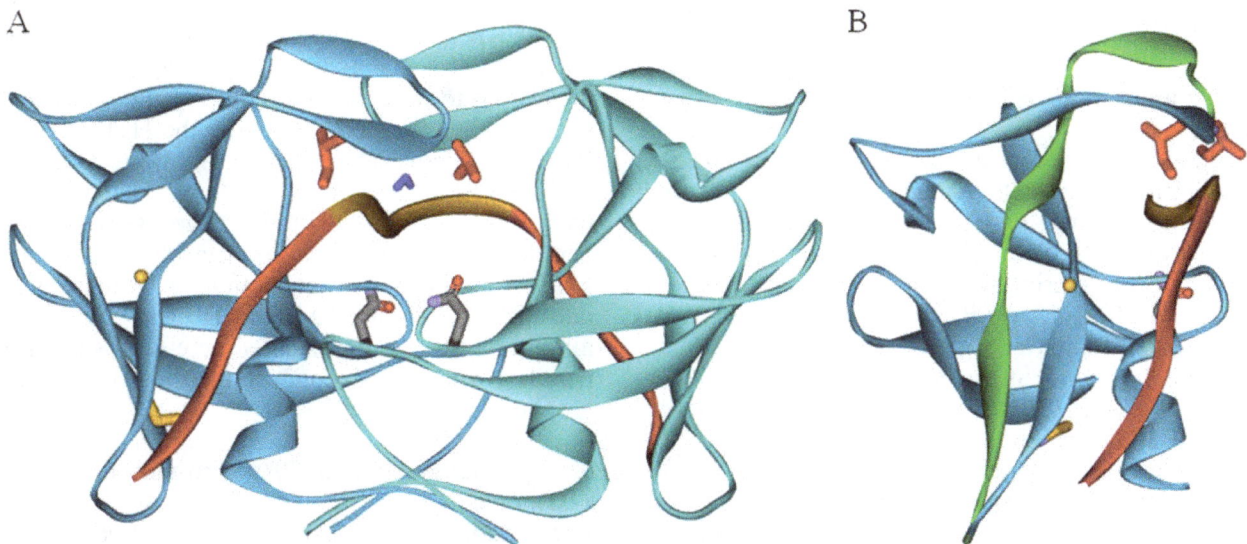

Figure 10. HTLV-1 PR with bound MA/CA 24-mer substrate. A) Front view with flaps on top, A and B subunits as backbone ribbons, active-site mutation D32N center (A and B subunits). Atom colors: oxygen, red; nitrogen, blue; carbon, grey. Tetrahedrally coordinated water with all atoms blue (center, above substrate). Back mutated residues shown starting clockwise bottom left: S-groove I85A and T88G carbons in bronze; active-site V56I and A59I carbons in red. Gag MA/CA substrate backbone |P12-P5| in red; P4-P4' in bronze. B) Side view of A (left subunit) showing the anti-parallel beta sheet with one ribbon colored green starting at the tip of the flap and ending at the bottom of the S-groove, with the other strand of the anti-parallel beta sheet below as a blue ribbon.

Table 4. HTLV-1 PRs interaction energy scores with MA/CA 24-mer and indinavir.

	HTLV-1 PR	HTLV-1 2X PR (S-groove)	HTLV-1 4X PR (S-groove/AC)	
MA/CA 24-mer	-235.9 (128.6/-107.2)	-231.0 (-123.7/-107.3)	-235.6 (-126.4/-109.2)	
Indinavir	-84.1	-84.8	-86.7	

HTLV-1 PR and mutant HTLV-1 PRs interaction energy scores (kcal/mol) with MA/CA 24-mer substrate residues: P12-P12'; |P12-P5|; P4-P4'. Bottom, scores for the same PRs and the active-site inhibitor indinavir.

Figure 9, while HIV-2 PR also acquired the secondary resistance residue Ala73, Figure 8. That mutation could help stabilize HIV-2 PR S-groove interactions with cleavage site residues |P12-P5| and along with the inherited S-groove secondary resistance residue Val71 may have contributed to HIV-2 PR weighting interactions with Gag/Pol cleavage sites more towards the S-grooves and cleavage site residues |P12-P5|. This in turn would allow HIV-2 Gag/Pol substrates to out-compete the binding of active-site inhibitors to HIV-2 PR. In order to test this hypothesis, the interaction energy scores between HIV-2, SIV-cpz, and SIV-sm PRs and the respective MA/CA 24-mer substrates were calculated for |P12-P5| and P4-P4'. WT HIV-1 PR interaction with the MA/CA substrate was weighted towards the active-site and residues P4-P4', while the HIV-2, SIV-cpz, and SIV-sm PRs interactions with the respective MA/CA substrates were all weighted towards the S-grooves and MA/CA substrate residues |P12-P5|, Table 3. The weighting of the HIV-2, SIV-cpz, and SIV-sm PRs interactions towards the S-grooves and substrate residues |P12-P5| correlated with resistance to active-site inhibitors. These findings support the hypothesis that PR/substrate interactions dominated by S-groove binding to cleavage site residues |P12-P5| allows Gag/Pol polyprotein cleavage sites to out-compete the binding of active-site inhibitors to PR (Laco 2015).

In silico analysis of HTLV-1 PR native multi-drug resistance

In contrast to the native primary resistance of HIV-2 and SIV-sm PRs (Results), the HTLV-1 PR has native multi-drug resistance to HIV-1 PR clinical inhibitors (Louis *et al.* 1999, Pettit *et al.* 1998), and contains a total of ten native primary/secondary resistance mutations (Laco 2015, Li *et al.* 2005). HTLV-1 PR has been reported to have an S-groove residue that *in silico* made direct H-bond interactions with substrate residues |P12-P5| (Laco 2015). That HTLV-1 PR S-groove residue (Thr88) aligned in 3D with the WT HIV-1 PR Gly73 that when mutated contributed to HIV-1 PR inhibitor resistance (Laco 2015). Here we tested whether HTLV-1 PR could be reverse engineered *in silico*, in a way opposite to how HIV-1 PR evolved into a MDR HIV-1 PR, to make the HTLV-1 PR inhibitor sensitive while maintaining interactions with substrates. HTLV-1 PR S-groove residues Ile85 and Thr88, equivalent to the HIV-1 PR secondary resistance mutations Ala70Ile and Gly73Thr (Rhee *et al.* 2003, Wensing *et al.* 2015), contributed to interactions

with substrate residues |P12-P5|, Table 4 (Laco 2015). Here those S-groove residues were back mutated to equivalent residues found in WT HIV-1 PR to give the HTLV-1 2X PR (Ile85Ala and Thr88Gly), Figure 10.

The interaction between HTLV-1 2X PR and the MA/CA 24-mer substrate decreased by 4.9 kcal/mol consistent with a loss of S-groove contacts to substrate residues |P12-P5|, Table 4. Next, the HTLV-1 PR active-site residues Val56 and Ala59, equivalent to HIV-1 PR Ile47Val and Ile50Ala primary resistance mutations, were back mutated to Val56Ile and Ala59Ile and along with the S-groove mutations Ile85Ala and Thr88Gly resulted in the HTLV-1 4X PR, Figure 10. The HTLV-1 4X PR had a 2.6 kcal/mol stronger interaction with indinavir while restoring interactions with the MA/CA 24-mer substrate to a level similar to that for the native HTLV-1 PR, Table 4. Interestingly, when the interaction energy scores were broken down for the HTLV-1 4X PR and the MA/CA 24-mer substrate, the interactions were still found to be weighted towards the S-grooves and |P12-P5|, Table 4. Since the HTLV-1 4X PR S-groove interactions with substrate residues |P12-P5| were still the dominant interaction with the MA-CA substrate the potential exists *in vitro* for the MA/CA cleavage site, in the context of the Gag polyprotein, to out-compete active-site inhibitor binding to the HTLV-1 4X PR. These results indicate the potential complexity of reverse engineering the HTLV-1 PR into an inhibitor sensitive PR. It is interesting to note that by acquiring the bulky active-site mutations Val56Ile/Ala59Ile, the HTLV-1 4X PR flaps had to move away from the substrate residues P4-P4' in order to accommodate them. That flap movement was then translated to the S-grooves via the anti-parallel beta-sheets that connect the flaps and S-grooves and increased interactions with substrate residues |P12-P5|, Figure 10 B and 11. The HIV-1 PR and all other retroviral PRs studied here have similar anti-parallel beta-sheets connecting the flaps and S-grooves, Figure 3 B.

Non-primate retroviral PRs with native multi-drug resistance

In order to extend the native resistance findings for the HIV-2 and HTLV-1 PRs, non-primate retroviruses were next examined. EIAV and FIV have both been reported to encode native MDR PRs (Kervinen *et al.* 1998), with structures published for both (Kervinen *et al.* 1998, Laco *et al.* 1997). Structure based alignment of EIAV and FIV PRs with HIV-1 PR revealed that they had 28% and 24% identity with HIV-1 PR, re-

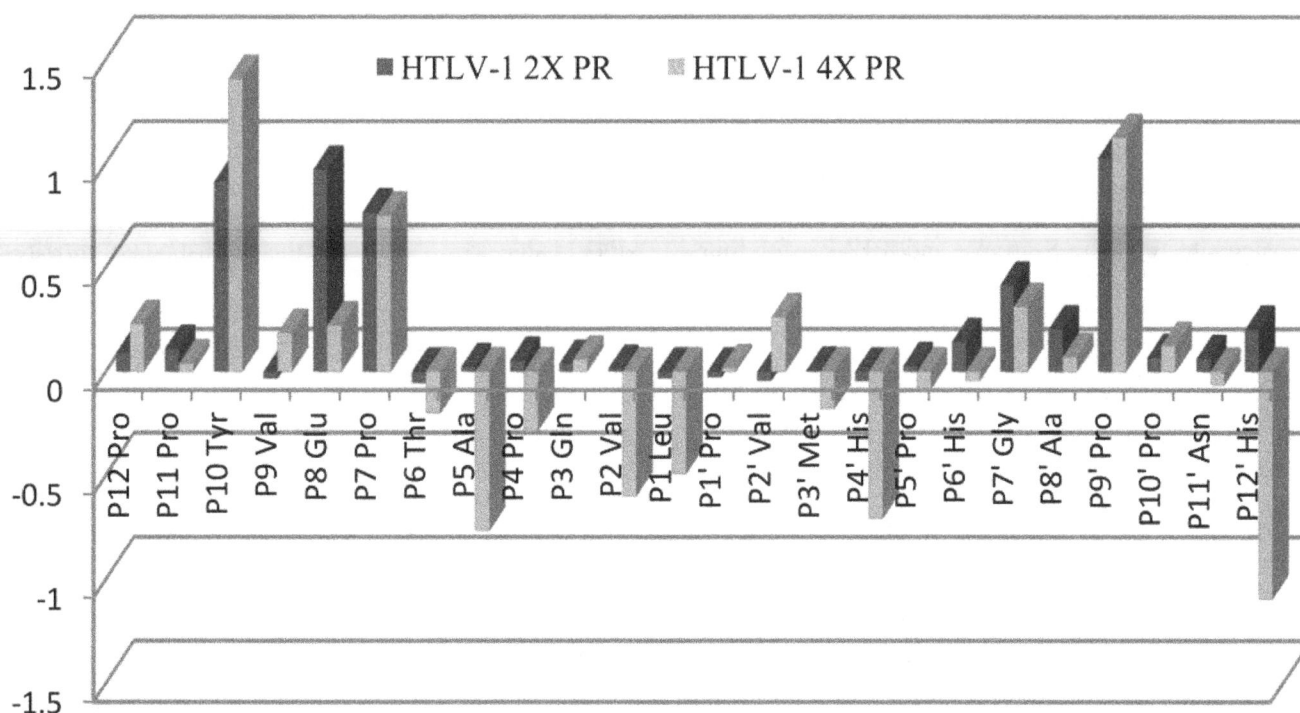

Figure 11. Reverse Engineered HTLV-1 PRs interaction with MA/CA 24-mer on a per residue basis. The reverse engineered HTLV-1 2X PR and 4X PR interaction energy scores, on a per residue basis, with the MA/CA substrate residues P12-P12' (kcal/mol). Scores relative to the native HTLV-1 PR residue interaction energy scores that were set to zero.

spectively, Figure 12. When both PRs were analyzed for native resistance residues, EIAV and FIV PRs were found to each have six residues that when found at the same 3D position in HIV-1 PR contributed to primary and secondary inhibitor resistance, and included S-groove residues, Figure 12 (Rhee *et al.* 2003, Wensing *et al.* 2015).

In the structure of an unliganded MDR HIV-1 PR the flaps were in an open conformation (Yedidi *et al.* 2014), as was the case for two MDR HIV-1 PRs with bound inhibitor (Liu *et al.* 2013a, b). In contrast, the structure of an unliganded WT HIV-1 PR had the flaps in a closed conformation (Wlodawer *et al.* 1989). These published structures demonstrated that the orientation of the flaps correlated with the sensitivity of the respective PRs to active-site inhibitors. Here inhibitor-resistant PRs were shown to have weaker interactions between the flaps/active-site and substrate residues P4-P4', which was compensated for by stronger interactions between the S-grooves and substrate residues |P12-P5|, including the MDR HIV-1, HIV-2, SIV-sm, HTLV-1, EIAV, and FIV PRs (Table 3).

In addition, like HTLV-1 PR, EIAV and FIV PRs had a number of residues that were found at resistance positions in HIV-1 PR, though had not been reported as resistance residues for HIV-1 PR (Rhee *et al.* 2003, Wensing *et al.* 2015). For example, HIV-1 PR Asp30 H-bonds to nelfinavir stabilizing the interaction between WT HIV-1 PR and nelfinavir, while the Asp30Asn mutation results in the loss of that H-bond making the mutant HIV-1 PR resistant to nelfinavir (Kolli *et al.* 2014). The EIAV PR Thr30 aligns with

HIV-1 PR Asp30; it is likely that the shorter Thr30 side chain also does not H-bond to nelfinavir and so explains EIAV PR resistance to nelfinavir, Figure 12 (Kervinen *et al.* 1998). Likewise, the HIV-1 PR Gly48 when mutated to Val contributed to saquinavir resistance due to the loss of a stabilizing H-bond between the HIV-1 PR Gly48Val backbone carbonyl oxygen and saquinavir (Liu *et al.* 2008). Interestingly, the FIV PR Ile57 was found at the equivalent 3D position to HIV-1 PR Gly48 and so may play a similar role as the HIV-1 PR Gly48Val mutation in FIV PR resistance to saquinavir (i.e., RO31-8959), Figure 12 (Lin *et al.* 2000). Next the interaction energy scores were calculated for EIAV PR and FIV PR bound to the respective MA/CA 24-mer substrates. The EIAV and FIV PRs interaction with the respective MA/CA substrates were also weighted towards the S-grooves and substrate residues |P12-P5|, versus the active-site and residues P4-P4' (Table 3). Given that EIAV and FIV have been estimated to be in the respective hosts for millions of years (Cook *et al.* 2013, Pecon-Slattery *et al.* 2008), these findings suggest that there could be an evolutionary trend for retroviral PRs to weight substrate interactions towards the S-grooves and substrate residues |P12-P5| with one consequence being native resistance to active-site inhibitors.

Discussion

We have shown that the WT HIV-1 PR and MDR HIV-1 PR *in silico* interactions with solvent exposed Gag/Pol 24-mer cleavage sites correlated better with the reported *in vitro* cleavage order than did the respective 8-mers, Tables 1 and 2. This demonstrated the importance of the HIV-1 PRs S-groove interactions with substrate residues |P12-P5| in determining the Gag/Pol cleavage order. At the same time, WT HIV-1 PR weighted the interaction with the 24-mer substrates towards the active-site and residues P4-P4' for four cleavage sites and the S-groove and |P12-P5| for four cleavage sites, while the the PR/RT site was equally weighted between the S-grooves and active-site (Table 1). The Gag/Pol cleavage sites in which the WT HIV-1 PR interactions with substrates were weighted towards the active-site/P4-P4' (MA/CA, CA/SP1, p6*/PR, RT/RH) may be responsible for the inhibitor sensitivity of the WT HIV-1 PR, due to active-site inhibitors being able to outcompete those substrates for binding to WT HIV-1 PR (Table 1). In contrast, the MDR HIV-1 PR interactions with 24-mer substrates were weighted towards the S-grooves and residues |P12-P5| for eight out of the nine Gag/Pol cleavage sites, which may be responsible for allowing the Gag/Pol substrates to outcompete inhibitor binding to the MDR HIV-1 PR active-site (Table 2), and correlates with the high level resistance of the MDR HIV-1 PR to active-site inhibitors (Wang *et al.* 2012). The only 24-mer cleavage site where the MDR HIV-1 PR interactions were weighted towards P4-P4' was the Gag CA/SP1 site, the slowest cleaved site *in vitro*, Table 2. Inhibition of cleavage at the CA/SP1 site would still allow virion maturation to proceed under optimal inhibitor levels and in the process release more active PR shifting the inhibitor/PR ratio towards PR. These results are supported by a published *in vitro* MDR HIV-1 PR cleavage assay that used a peptide substrate, that does not bind to the PR S

-grooves, and resulted in a relative 8.4-fold weaker *in vitro* activity for the MDR HIV-1 PR as compared to cleavage at the SP1/NC site in full-length Gag (Laco 2015). This suggests that MDR HIV-1 PR inhibitor resistance may be significantly underestimated *in vitro* when short peptide substrates are used due to the weak binding of the MDR HIV-1 PR to substrate residues P4-P4' (Table 2). In contrast, the published *in vitro* results for the WT HIV-1 PR using the same peptide substrate and full-length Gag demonstrated that the WT HIV-1 PR had only a relative 2-fold weaker *in vitro* activity for the peptide versus Gag because the WT HIV-1 PR was more focused on flap/active-site interactions with substrate residues P4-P4', Table 1 (Laco 2015). This suggests that the sensitivity of WT HIV-1 PR to active-site inhibitors may be modestly overestimated *in vitro* when short peptide substrates are used.

Next we examined whether the inhibitor resistance residues selected for in the MDR HIV-1 PR, including primary active-site mutations and secondary S-groove mutations, were found in closely related retroviruses that had not been exposed to protease inhibitors (Results). This was done in order to determine if the MDR HIV-1 PR resistance evolution was either a novel pathway used solely for inhibitor resistance, or a pathway common to all retroviral proteases in order to adapt to mutated Gag/Pol substrates selected to escape *in vivo* CTL detection (Phillips *et al.* 1991, Prince *et al.* 2012, Seibert *et al.* 1995). When the HIV-1 and HIV-2 PRs were aligned with the evolutionary closest SIV PRs, HIV-1 PR was found to have lost the one native resistance residue found in SIV-cpz PR, while HIV-2 PR gained one native resistance residue in addi-

```
            10/11           20      25   30 33                      46/47
HIV-1   1   PQITLWKRPLVTIRIGGQLKEALLDTGADDTVLEE--MN---LPGKWKPK-MI   47
EIAV    1   VTYNLEKRPTTIVLINDTPLNVLLDTGADTSVLTTAHYNRLKYRGRKYQGTGI   53

            48 50    54                   71 73/74    82 84 88 90 93
HIV-1  48   GGIGGFIKVRQYDQIPVEICGHKAIGTVLVGPTPVNIIGRNLLTQIGCTLNF   99
EIAV   54   GGVGGNVETFS-TPVTIKKKGRHIKTRMLVADIPVTILGRDILQDLGAKLVL  104
```

```
            10/11           20      25   30 33                      46/47
HIV-1   1   --PQITLWKRPLVTIRIGGQLKEALLDTGADDTVLEE--MNLP--GKWKPKMI   47
FIV     4   VGTTTTLEKRPEILIFVNGYPIKFLLDTGADITILNRRDFQVKNSIENGRQNM   56

            48 50    54                   71 73/74    82 84  88 90 93
HIV-1  48   GGIGGFIKVRQYDQIPVEI-----CGHKAIGTVLV---GPTPVNIIGRNLLTQIGCTLN   99
FIV    57   IGVGGGKRGTNYINVHLEIRDENYKTQCIFGNVCVLEDNSLIQPLLGRDNMIKFNIRLV  115
```

Figure 12. Structure-based alignment of HIV-1 PR with EIAV and FIV PRs. Gray highlighted residues indicate identity between aligned sequences. HIV-1 PR residues that when mutated contributed to inhibitor resistance in bold, positions of primary resistance residue indicated by underlined numbers, EIAV and FIV PRs native resistance residues in bold, underlined EIAV and FIV residues indicate potential native resistance residues. HIV-1 PR active-site Asp25 (D25) italicized, as are the active-site residues for EIAV (D25) and FIV (D30).

tion to inheriting the six native resistance residues present in SIV-sm PR, Figure 8. Since neither HIV-1, nor HIV-2, evolved from the respective SIV in the presence of protease inhibitors, the evolution of the corresponding Gag/Pol substrates were examined to see how they could explain the resistance residue inheritance pattern of HIV-1 and HIV-2 PRs, Figure 9. The HIV-1 Gag/Pol cleavage sites had a significant loss in identity with the respective SIV-cpz Gag/Pol cleavage site residues |P12-P5|, and this may have contributed to the WT HIV-1 PR weighting interactions towards the flaps/active-site and the conserved cleavage site residues P4-P4', Figure 9 and Table 1. The HIV-1 PR focus on substrate residues P4-P4' correlated with HIV-1 PR losing the SIV-cpz PR Val10 native resistance residue since Val10 is also the P10' residue in the p6*/PR cleavage site and would be bound in the S-grooves, Figure 3 and 5. In contrast, the HIV-2 Gag/Pol cleavage sites overall retained significantly more identity with the SIV-sm cleavage sites, and non-conserved residues in the HIV-2 Gag/Pol cleavage sites were more evenly distributed across cleavage site residues P12-P12', Figure 9. The net result was the evolution of an HIV-2 PR that weighted S-groove interactions towards substrate residues |P12-P5| due in part to the acquisition of the native S-groove secondary resistance residue Ala73. And so during adaptation to the human host, the HIV-1 PR focused more on substrate residues P4-P4' and became acutely sensitive to active-site inhibitors (Rhee et al. 2003, Wensing et al. 2015). In contrast, the HIV-2 PR weighted S-groove interactions towards substrate residues |P12-P5| as shown for the MA/CA site (Table 3) resulting in native resistance to many HIV-1 PR active-site inhibitors, Results. These findings support the hypothesis that HIV-1 PR used a common retroviral PR pathway to acquire multi-drug resistance. As a result, HIV-1 PR active-site inhibitors can drive the evolution of HIV-1 PR resistance as well as select for adaptive mutations in the Gag/Pol cleavage sites (McKinnon et al. 2011). Transmission of the resulting MDR HIV-1 eliminates PR active-site inhibitors as a treatment option. By targeting the HIV-1 PR S-grooves, along with the active-site, the potential loss of PR active-site inhibitors could be avoided. This approach may be critical given that in 2012 of the estimated 1.5 million HIV-1 infected individuals in the United States and Puerto Rico only 30% had the virus suppressed to < 200 copies/mL (Frieden et al. 2015). Likewise, the HIV-2, HTLV-1, EIAV, and FIV PRs, which all contain S-groove resistance residues, could be targeted with both S-groove inhibitors and active-site inhibitors in order to prevent the evolution of PRs that re-weighted substrate interactions towards P4-P4' in order to escape S-groove inhibitors.

The HIV-1 Gag and Pol polyprotein 3D models revealed the solvent accessibility of the Gag cleavage site residues versus several solvent inaccessible Pol cleavage sites (i.e., RT/RH and RH/IN), Figure 6 and 7. In the cell, soluble proteases degrade cytosolic proteins at low levels to release approximately ten residue long peptides that are then bound by MHC class I receptors for presentation on the cell surface for CTL surveillance (Kourjian et al. 2014, Lazaro et al. 2015, Yewdell et al. 1999, Zervoudi et al. 2013). Due to the solvent accessibility of the Gag cleavage sites, they would be prime targets for those cellular proteases, in contrast to the less accessible Pol cleavage sites, Figure 6 and 7. The net result being that during SIV-cpz cross-species transmission into humans CTL selective pressure likely contributed to the mutation of Gag cleavage sites, while Pol cleavage site sequences were highly conserved, Figure 9. The accessibility of the HIV-1 Gag cleavage sites could be a strategy to direct the cellular immune response towards Gag to minimize mutation of the more structurally sensitive Pol enzymes that typically have ordered N-and C-termini, Figure 7. The remarkable ability of HIV-1 to evade the CTL response may be due in part to the Gag cleavage sites being used as bait for cellular proteases since HIV-1 PR recognizes 24-residues of a cleavage site, with the ability to differentially weight substrate interactions between |P12-P5| and P4-P4' in order to accommodate Gag/Pol cleavage site CTL escape mutations. This substrate recognition strategy may allow HIV-1 PR and other retroviral PRs to minimize the impact of cleavage site CTL escape mutations on PR/substrate interactions and consequently virus maturation (Phillips et al. 1991, Prince et al. 2012, Seibert et al. 1995).

A complicating factor in the proposed HIV-1 PR adaptation to mutations in Gag cleavage sites is the timeline; the CTL response selects for escape mutations in Gag cleavage sites during a round of replication, while the selection of adaptive mutations in HIV-1 PR for the mutated cleavage sites would take place in subsequent rounds of replication. Perhaps HIV-1 PR uses a strategy similar to that shown in silico for the HTLV-1 4X PR, which adapted to bulky cleavage site residues P4-P4' by moving the flaps away, with a concomitant strengthening of the S-groove contacts with residues |P12-P5| via the anti-parallel beta sheets, Tables 4 and 5 and Figure 10. This approach to substrates with mutated residues could allow the HIV-1 PR to get by CTL escape mutations in Gag cleavage sites during the same replication cycle. Then in subsequent rounds of replication adaptive mutations in HIV-1 PR could be selected for. It is interesting to note that HIV-1 PR secondary resistance polymorphisms were found in treatment naive patients, including the S-groove A71V/T and G73C/R mutations (Birk & Sonnerborg 1998, Bossi et al. 1999, Kearney et al. 2008, Kozal et al. 1996, Rose et al. 1996b). The HIV-1 PR secondary resistance polymorphisms could represent an archive of adaptive HIV-1 PR mutations selected in response to CTL escape mutations in Gag cleavage sites over the course of an infection (Birk & Sonnerborg 1998, Bossi et al. 1999, Kearney et al. 2008, Kozal et al. 1996, Servais et al. 2001). Those PR S-groove mutations could then be selected for during subsequent inhibitor treatment and contribute to PR resistance. In untreated chronically infected patients >98% of the latent HIV-1 reservoir contained CTL escape mutations (Deng *et al.* 2015).

The HIV-1 PR may respond to both mutations in substrate residues P4-P4' and active-site inhibitors by reweighting substrate interactions towards the S-grooves, since active-site inhibitors essentially mimic mutated substrate from the perspective of the PR in that both disrupt the processing of the Gag/Pol polyproteins. At the same time, the S-grooves may allow for non-covalent tethering of the HIV-1 PR to the solvent exposed Gag/Pol cleavage sites, even with inhibitor bound in the active-site (Figure 1), and allow for virion maturation at a rate concordant with the inhibitor dissociation rate. Together these findings are directed towards a strategy to combat HIV-1 infections for which current treatments are not effective (Frieden *et al.* 2015), and HTLV-1 infections for which there are no treatments.

Conclusions

Based on the finding presented here and elsewhere (Deshmukh *et al.* 2017, Laco 2015), retroviral PRs interact with the Gag/Pol cleavage sites using both the active-site and S-grooves. The PRs can weight those interactions towards either the active-site and cleavage site residues P4-P4, or S-grooves and residues |P5-P12|, to accommodate cleavage site CTL escape mutations while maintaining Gag/Pol cleavage order. Many retroviral PRs evolved S-groove dominated interactions with substrates, and as a consequence native resistance to active site inhibitors. In contrast, HIV-1 PR evolved a more active-site dominated interaction with substrates that resulted in sensitivity to active-site inhibitors, while retaining the ability to reweight interactions towards the S-grooves and substrate residues |P5-P12| to outcompete active-site inhibitors.

Acknowledgements

This work was supported by the Roskamp Foundation and the Laco Science Institute.

References

Barre-Sinoussi F, Chermann JC, Rey F, Nugeyre MT, Chamaret S, Gruest J, Dauguet C, Axler-Blin C, Vezinet-Brun F, Rouzioux C, Rozenbaum W & Montagnier L 1983 Isolation of a T-lymphotropic retrovirus from a patient at risk for acquired immune deficiency syndrome (AIDS). *Science* **220** 868-871

Bharat TA, Castillo Menendez LR, Hagen WJ, Lux V,

Igonet S, Schorb M, Schur FK, Krausslich HG & Briggs JA 2014 Cryo-electron microscopy of tubular arrays of HIV-1 Gag resolves structures essential for immature virus assembly. *Proc Natl Acad Sci U S A* **111** 8233-8238

Billich S, Knoop MT, Hansen J, Strop P, Sedlacek J, Mertz R & Moelling K 1988 Synthetic peptides as substrates and inhibitors of human immune deficiency virus-1 protease. *J Biol Chem* **263** 17905-17908

Birk M & Sonnerborg A 1998 Variations in HIV-1 pol gene associated with reduced sensitivity to antiretroviral drugs in treatment-naive patients. *Aids* **12** 2369-2375

Bossi P, Mouroux M, Yvon A, Bricaire F, Agut H, Huraux JM, Katlama C & Calvez V 1999 Polymorphism of the human immunodeficiency virus type 1 (HIV-1) protease gene and response of HIV-1-infected patients to a protease inhibitor. *J Clin Microbiol* **37** 2910-2912

Brower ET, Bacha U M, Kawasaki Y & Freire E 2008 Inhibition of HIV-2 protease by HIV-1 protease inhibitors in clinical use. *Chem Biol Drug Des* **71** 298-305

Chang MW & Torbett BE 2011 Accessory mutations maintain stability in drug-resistant HIV-1 protease. *J Mol Biol* **410** 756-760

Chen Z, Li Y, Schock HB, Hall D, Chen E & Kuo LC 1995 Three-dimensional structure of a mutant HIV-1 protease displaying cross-resistance to all protease inhibitors in clinical trials. *J Biol Chem* **270** 21433-21436

Consortium TU 2015 UniProt: a hub for protein information. *Nucleic Acids Res* **43** D204-212

Cook RF, Leroux C & Issel CJ 2013 Equine infectious anemia and equine infectious anemia virus in 2013: a review. *Vet Microbiol* **167** 181-204

D'Arc M, Ayouba A, Esteban A, Learn GH, Boue V, Liegeois F, Etienne L, Tagg N, Leendertz F H, Boesch C, Madinda NF, Robbins MM, Gray M, Cournil A, Ooms M, Letko M, Simon VA, Sharp PM, Hahn BH, Delaporte E, Mpoudi Ngole E & Peeters M 2015 Origin of the HIV-1 group O epidemic in western lowland gorillas. *Proc Natl Acad Sci U S A* **112** E1343-1352

Darke PL, Nutt RF, Brady SF, Garsky VM, Ciccarone TM, Leu CT, Lumma PK, Freidinger RM, Veber DF & Sigal IS 1988 HIV-1 protease specificity of peptide cleavage is sufficient for processing of gag and pol polyproteins. *Biochem Biophys Res Commun* **156** 297-303

Deng K, Pertea M, Rongvaux A, Wang L, Durand C M, Ghiaur G, Lai J, McHugh HL, Hao H, Zhang H, Margolick JB, Gurer C, Murphy A J, Valenzuela DM, Yancopoulos GD, Deeks SG, Strowig T, Kumar P, Siliciano JD, Salzberg SL, Flavell RA, Shan L & Siliciano RF 2015 Broad CTL response is required to clear latent HIV-1 due to dominance of escape mutations. *Nature* **517** 381-385

Desbois D, Roquebert B, Peytavin G, Damond F, Collin G, Benard A, Campa P, Matheron S, Chene G,

Brun-Vezinet F & Descamps D 2008 In vitro phenotypic susceptibility of human immunodeficiency virus type 2 clinical isolates to protease inhibitors. *Antimicrob Agents Chemother* **52** 1545-1548

Deshmukh L, Tugarinov V, Louis JM & Clore GM 2017. Binding kinetics and substrate selectivity in HIV-1 protease-Gag interactions probed at atomic resolution by chemical exchange NMR. *Proc Natl Acad Sci U S A* **114** E9855-E9862

Ding YS, Rich DH & Ikeda RA 1998 Substrates and inhibitors of human T-cell leukemia virus type I protease. *Biochemistry* **37** 17514-17518

Edgar RC 2004 MUSCLE: multiple sequence alignment with high accuracy and high throughput. *Nucleic Acids Res* **32** 1792-1797

Erickson-Viitanen S, Manfredi J, Viitanen P, Tribe D E, Tritch R, Hutchison CA, 3rd, Loeb DD & Swanstrom R 1989 Cleavage of HIV-1 gag polyprotein synthesized in vitro: sequential cleavage by the viral protease. *AIDS Res Hum Retroviruses* **5** 577-591

Erickson J, Neidhart DJ, VanDrie J, Kempf DJ, Wang XC, Norbeck DW, Plattner JJ, Rittenhouse JW, Turon M, Wideburg N & et al. 1990 Design, activity, and 2.8 A crystal structure of a C2 symmetric inhibitor complexed to HIV-1 protease. *Science* **249** 527-533

Fodor SK & Vogt VM 2002 Characterization of the protease of a fish retrovirus, walleye dermal sarcoma virus. *J Virol* **76** 4341-4349

Fossen T, Wray V, Bruns K, Rachmat J, Henklein P, Tessmer U, Maczurek A, Klinger P & Schubert U 2005 Solution structure of the human immunodeficiency virus type 1 p6 protein. *J Biol Chem* **280** 42515-42527

Frieden TR, Foti KE & Mermin J 2015 Applying Public Health Principles to the HIV Epidemic--How Are We Doing? *N Engl J Med* **373** 2281-2287

Gallo RC, Sarin PS, Gelmann EP, Robert-Guroff M, Richardson E, Kalyanaraman VS, Mann D, Sidhu GD, Stahl RE, Zolla-Pazner S, Leibowitch J & Popovic M 1983 Isolation of human T-cell leukemia virus in acquired immune deficiency syndrome (AIDS). *Science* **220** 865-867

Gao F, Bailes E, Robertson DL, Chen Y, Rodenburg CM, Michael SF, Cummins LB, Arthur LO, Peeters M, Shaw GM, Sharp PM & Hahn BH 1999 Origin of HIV-1 in the chimpanzee Pan troglodytes troglodytes. *Nature* **397** 436-441

Gomez R, Jolly SJ, Williams T, Vacca JP, Torrent M, McGaughey G, Lai MT, Felock P, Munshi V, Distefano D, Flynn J, Miller M, Yan Y, Reid J, Sanchez R, Liang Y, Paton B, Wan BL & Anthony N 2011 Design and synthesis of conformationally constrained inhibitors of non-nucleoside reverse transcriptase. *J Med Chem* **54** 7920-7933

Goncalves DU, Proietti FA, Ribas JG, Araujo MG, Pinheiro SR, Guedes AC & Carneiro-Proietti AB 2010 Epidemiology, treatment, and prevention of human T-cell leukemia virus type 1-associated diseases. *Clin Microbiol Rev* **23** 577-589

Gres AT, Kirby KA, KewalRamani VN, Tanner JJ, Pornillos O & Sarafianos SG 2015 X-ray crystal structures of native HIV-1 capsid protein reveal conformational variability. *Science* **349** 99-103

Gulnik SV, Suvorov LI, Liu B, Yu B, Anderson B, Mitsuya H & Erickson JW 1995 Kinetic characterization and cross-resistance patterns of HIV-1 protease mutants selected under drug pressure. *Biochemistry* **34** 9282-9287

Gustchina A, Kervinen J, Powell DJ, Zdanov A, Kay J & Wlodawer A 1996 Structure of equine infectious anemia virus proteinase complexed with an inhibitor. *Protein Sci* **5** 1453-1465

Hirsch VM, Olmsted RA, Murphey-Corb M, Purcell RH & Johnson PR 1989 An African primate lentivirus (SIVsm) closely related to HIV-2. *Nature* **339** 389-392

Huet T, Cheynier R, Meyerhans A, Roelants G & Wain-Hobson S 1990 Genetic organization of a chimpanzee lentivirus related to HIV-1. *Nature* **345** 356-359

Ishima R, Torchia DA, Lynch SM, Gronenborn AM & Louis JM 2003 Solution structure of the mature HIV-1 protease monomer: insight into the tertiary fold and stability of a precursor. *J Biol Chem* **278** 43311-43319

Jacks T, Power MD, Masiarz FR, Luciw PA, Barr PJ & Varmus HE 1988 Characterization of ribosomal frameshifting in HIV-1 gag-pol expression. *Nature* **331** 280-283

Kaplan AH, Michael SF, Wehbie RS, Knigge MF, Paul DA, Everitt L, Kempf DJ, Norbeck DW, Erickson JW & Swanstrom R 1994 Selection of multiple human immunodeficiency virus type 1 variants that encode viral proteases with decreased sensitivity to an inhibitor of the viral protease. *Proc Natl Acad Sci U S A* **91** 5597-5601

Kearney M, Palmer S, Maldarelli F, Shao W, Polis MA, Mican J, Rock-Kress D, Margolick JB, Coffin JM & Mellors JW 2008 Frequent polymorphism at drug resistance sites in HIV-1 protease and reverse transcriptase. *Aids* **22** 497-501

Kervinen J, Lubkowski J, Zdanov A, Bhatt D, Dunn BM, Hui KY, Powell DJ, Kay J, Wlodawer A & Gustchina A 1998 Toward a universal inhibitor of retroviral proteases: comparative analysis of the interactions of LP-130 complexed with proteases from HIV-1, FIV, and EIAV. *Protein Sci* **7** 2314-2323

King NM, Prabu-Jeyabalan M, Bandaranayake RM, Nalam MN, Nalivaika EA, Ozen A, Haliloglu T, Yilmaz NK & Schiffer CA 2012 Extreme entropy-enthalpy compensation in a drug-resistant variant of HIV-1 protease. *ACS Chem Biol* **7** 1536-1546

Kohl NE, Emini EA, Schleif WA, Davis LJ, Heimbach JC, Dixon RA, Scolnick EM & Sigal IS 1988 Active human immunodeficiency virus protease is required for viral infectivity. *Proc Natl Acad Sci U.S.A.* **85** 4686-4690

Kolli M, Ozen A, Kurt-Yilmaz N & Schiffer CA 2014 HIV-1 protease-substrate coevolution in nelfinavir resistance. *J Virol* **88** 7145-7154

Koralnik IJ, Boeri E, Saxinger WC, Monico AL, Fullen J, Gessain A, Guo HG, Gallo RC, Markham P, Kalyanaraman V & et al. 1994 Phylogenetic associations of human and simian T-cell leukemia/ lymphotropic virus type I strains: evidence for interspecies transmission. *J Virol* **68** 2693-2707

Kotler M, Katz RA, Danho W, Leis J & Skalka AM 1988 Synthetic peptides as substrates and inhibitors of a retroviral protease. *Proc Natl Acad Sci U S A* **85** 4185-4189

Kourjian G, Xu Y, Mondesire-Crump I, Shimada M, Gourdain P & Le Gall S 2014 Sequence-specific alterations of epitope production by HIV protease inhibitors. *J Immunol* **192** 3496-3506

Kovalevsky AY, Louis JM, Aniana A, Ghosh AK & Weber IT 2008 Structural evidence for effectiveness of darunavir and two related antiviral inhibitors against HIV-2 protease. *J Mol Biol* **384** 178-192

Kozal MJ, Shah N, Shen N, Yang R, Fucini R, Merigan TC, Richman DD, Morris D, Hubbell E, Chee M & Gingeras TR 1996 Extensive polymorphisms observed in HIV-1 clade B protease gene using high-density oligonucleotide arrays. *Nat Med* **2** 753-759

Kozisek M, Bray J, Rezacova P, Saskova K, Brynda J, Pokorna J, Mammano F, Rulisek L & Konvalinka J 2007 Molecular analysis of the HIV-1 resistance development: enzymatic activities, crystal structures, and thermodynamics of nelfinavir-resistant HIV protease mutants. *J Mol Biol* **374** 1005-1016

Krausslich HG, Ingraham RH, Skoog MT, Wimmer E, Pallai PV & Carter CA 1989 Activity of purified biosynthetic proteinase of human immunodeficiency virus on natural substrates and synthetic peptides. *Proc Natl Acad Sci U.S.A.* **86** 807-811

Laco GS 2011 Evaluation of two models for human topoisomerase I interaction with dsDNA and camptothecin derivatives. *PLoS One* **6** e24314

Laco GS 2015 HIV-1 protease substrate-groove: Role in substrate recognition and inhibitor resistance. *Biochimie* **118** 90-103

Laco GS, Schalk-Hihi C, Lubkowski J, Morris G, Zdanov A, Olson A, Elder J H, Wlodawer A & Gustchina A 1997 Crystal structures of the inactive D30N mutant of feline immunodeficiency virus protease complexed with a substrate and an inhibitor. *Biochemistry* **36** 10696-10708

Lazaro S, Gamarra D & Del Val M 2015 Proteolytic enzymes involved in MHC class I antigen processing: A guerrilla army that partners with the proteasome. *Mol Immunol* **68** 72-76

Le Grice SF, Mills J & Mous J 1988 Active site mutagenesis of the AIDS virus protease and its alleviation by trans complementation. *EMBO J* **7** 2547-2553

Lee BM, De Guzman RN, Turner BG, Tjandra N & Summers MF 1998 Dynamical behavior of the HIV-1 nucleocapsid protein. *J Mol Biol* **279** 633-649

Leigh Brown AJ, Frost SD, Mathews WC, Dawson K, Hellmann NS, Daar ES, Richman DD & Little SJ 2003 Transmission fitness of drug-resistant human immuno-deficiency virus and the prevalence of resistance in the antiretroviral-treated population. *J Infect Dis* **187** 683-686

Lemey P, Pybus OG, Wang B, Saksena NK, Salemi M & Vandamme AM 2003 Tracing the origin and history of the HIV-2 epidemic. *Proc Natl Acad Sci U S A* **100** 6588-6592

Li M, Laco GS, Jaskolski M, Rozycki J, Alexandratos J, Wlodawer A & Gustchina A 2005 Crystal structure of human T cell leukemia virus protease, a novel target for anticancer drug design. *Proc Natl Acad Sci U S A* **102** 18332-18337

Lin YC, Beck Z, Lee T, Le VD, Morris GM, Olson AJ, Wong CH & Elder JH 2000 Alteration of substrate and inhibitor specificity of feline immunodeficiency virus protease. *J Virol* **74** 4710-4720

Liu F, Kovalevsky AY, Tie Y, Ghosh AK, Harrison RW & Weber IT 2008 Effect of flap mutations on structure of HIV-1 protease and inhibition by saquinavir and darunavir. *J Mol Biol* **381** 102-115

Liu Z, Yedidi RS, Wang Y, Dewdney TG, Reiter SJ, Brunzelle JS, Kovari IA & Kovari LC 2013a Crystallographic study of multi-drug resistant HIV-1 protease lopinavir complex: mechanism of drug recognition and resistance. *Biochem Biophys Res Commun* **437** 199-204

Liu Z, Yedidi RS, Wang Y, Dewdney TG, Reiter SJ, Brunzelle JS, Kovari IA & Kovari LC 2013b Insights into the mechanism of drug resistance: X-ray structure analysis of multi-drug resistant HIV-1 protease ritonavir complex. *Biochem Biophys Res Commun* **431** 232-238

Louis JM, Oroszlan S & Tozser J 1999 Stabilization from autoproteolysis and kinetic characterization of the human T-cell leukemia virus type 1 proteinase. *J Biol Chem* **274** 6660-6666

Lv Z, Chu Y & Wang Y 2015 HIV protease inhibitors: a review of molecular selectivity and toxicity. *HIV AIDS (Auckl)* **7** 95-104

Martinez-Picado J, Savara AV, Sutton L & D'Aquila RT 1999 Replicative fitness of protease inhibitor-resistant mutants of human immunodeficiency virus type 1. *J Virol* **73** 3744-3752

Marx PA, Li Y, Lerche NW, Sutjipto S, Gettie A, Yee JA, Brotman BH, Prince AM, Hanson A, Webster RG & et al. 1991 Isolation of a simian immunodeficiency virus related to human immunodeficiency virus type 2 from a west African pet sooty mangabey. *J Virol* **65** 4480-4485

Masse S, Lu X, Dekhtyar T, Lu L, Koev G, Gao F, Mo H, Kempf D, Bernstein B, Hanna GJ & Molla A 2007 In vitro selection and characterization of human immunodeficiency virus type 2 with decreased susceptibility to lopinavir. *Antimicrob Agents Chemother* **51** 3075-3080

McKinnon JE, Delgado R, Pulido F, Shao W, Arribas JR & Mellors JW 2011 Single genome sequencing of HIV-1 gag and protease resistance mutations at virologic failure during the OK04 trial of simplified versus standard maintenance therapy. *Antivir Ther* **16** 725-732

McWilliam H, Li W, Uludag M, Squizzato S, Park Y M, Buso N, Cowley AP & Lopez R 2015 Analysis Tool Web Services from the EMBL-EBI. *Nucleic Acids Res* **41** W597-600

Miller M, Schneider J, Sathyanarayana BK, Toth MV, Marshall GR, Clawson L, Selk L, Kent SB & Wlodawer A 1989 Structure of complex of synthetic HIV-1 protease with a substrate-based inhibitor at 2.3 A resolution. *Science* **246** 1149-1152

Mocroft A, Ruiz L, Reiss P, Ledergerber B, Katlama C, Lazzarin A, Goebel FD, Phillips AN, Clotet B & Lundgren JD 2003 Virological rebound after suppression on highly active antiretroviral therapy. *Aids* **17** 1741-1751

Nijhuis M, Schuurman R, de Jong D, Erickson J, Gustchina E, Albert J, Schipper P, Gulnik S & Boucher CA 1999 Increased fitness of drug resistant HIV-1 protease as a result of acquisition of compensatory mutations during suboptimal therapy. *Aids* **13** 2349-2359

Ovchinnikov S, Kim DE, Wang RY, Liu Y, DiMaio F & Baker D 2016 Improved de novo structure prediction in CASP11 by incorporating coevolution information into Rosetta. *Proteins* **84** 67-75

Pazhanisamy S, Stuver CM, Cullinan AB, Margolin N, Rao BG & Livingston D J 1996 Kinetic characterization of human immunodeficiency virus type-1 protease-resistant variants. *J Biol Chem* **271** 17979-17985

Pearl LH & Taylor WR 1987 Sequence specificity of retroviral proteases. *Nature* **328** 482

Pecon-Slattery J, Troyer J L, Johnson WE & O'Brien SJ 2008 Evolution of feline immunodeficiency virus in Felidae: implications for human health and wildlife ecology. *Vet Immunol Immunopathol* **123** 32-44

Peeters M, Honore C, Huet T, Bedjabaga L, Ossari S, Bussi P, Cooper RW & Delaporte E 1989 Isolation and partial characterization of an HIV-related virus occurring naturally in chimpanzees in Gabon. *Aids* **3** 625-630

Pei J, Kim BH & Grishin NV 2008 PROMALS3D: a tool for multiple protein sequence and structure alignments. *Nucleic Acids Res* **36** 2295-2300

Pettit SC, Henderson GJ, Schiffer CA & Swanstrom R 2002 Replacement of the P1 amino acid of human immunodeficiency virus type 1 Gag processing sites can inhibit or enhance the rate of cleavage by the viral protease. *J Virol* **76** 10226-10233

Pettit SC, Lindquist JN, Kaplan AH & Swanstrom R 2005 Processing sites in the human immunodeficiency virus type 1 (HIV-1) Gag-Pro-Pol precursor are cleaved by the viral protease at different rates. *Retrovirology* **2** 66

Pettit SC, Sanchez R, Smith T, Wehbie R, Derse D & Swanstrom R 1998 HIV type 1 protease inhibitors fail to inhibit HTLV-I Gag processing in infected cells. *AIDS Res. Hum. Retroviruses* **14** 1007-1014

Phillips RE, Rowland-Jones S, Nixon DF, Gotch FM, Edwards JP, Ogunlesi AO, Elvin JG, Rothbard JA, Bangham CR, Rizza CR & et al. 1991 Human immunodeficiency virus genetic variation that can escape cytotoxic T cell recognition. *Nature* **354** 453-459

Prabu-Jeyabalan M, Nalivaika E & Schiffer CA 2002 Substrate shape determines specificity of recognition for HIV-1 protease: analysis of crystal structures of six substrate complexes. *Structure* **10** 369-381

Prabu-Jeyabalan M, Nalivaika EA, King NM & Schiffer CA 2004 Structural basis for coevolution of a human immunodeficiency virus type 1 nucleocapsid-p1 cleavage site with a V82A drug-resistant mutation in viral protease. *J Virol* **78** 12446-12454

Prince JL, Claiborne DT, Carlson JM, Schaefer M, Yu T, Lahki S, Prentice HA, Yue L, Vishwanathan SA, Kilembe W, Goepfert P, Price MA, Gilmour J, Mulenga J, Farmer P, Derdeyn CA, Tang J, Heckerman D, Kaslow RA, Allen SA & Hunter E 2012 Role of transmitted Gag CTL polymorphisms in defining replicative capacity and early HIV-1 pathogenesis. *PLoS Pathog* **8** e1003041

Proietti FA, Carneiro-Proietti AB, Catalan-Soares BC & Murphy EL 2005 Global epidemiology of HTLV-I infection and associated diseases. *Oncogene* **24** 6058-6068

Ratner L, Haseltine W, Patarca R, Livak KJ, Starcich B, Josephs SF, Doran ER, Rafalski JA, Whitehorn EA, Baumeister K & et al. 1985 Complete nucleotide sequence of the AIDS virus, HTLV-III. *Nature* **313** 277-284

Rhee SY, Gonzales MJ, Kantor R, Betts BJ, Ravela J & Shafer RW 2003 Human immunodeficiency virus reverse transcriptase and protease sequence database. *Nucleic Acids Res* **31** 298-303

Richman DD, Morton SC, Wrin T, Hellmann N, Berry S, Shapiro MF & Bozzette SA 2004 The prevalence of antiretroviral drug resistance in the United States. *Aids* **18** 1393-1401

Rodes B, Sheldon J, Toro C, Jimenez V, Alvarez MA & Soriano V 2006 Susceptibility to protease inhibitors in HIV-2 primary isolates from patients failing antiretroviral therapy. *J Antimicrob Chemother* **57** 709-713

Rose RB, Craik CS, Douglas NL & Stroud RM 1996a Three-dimensional structures of HIV-1 and SIV protease product complexes. *Biochemistry* **35** 12933-12944

Rose RE, Gong YF, Greytok JA, Bechtold CM, Terry BJ, Robinson BS, Alam M, Colonno RJ & Lin PF 1996b Human immunodeficiency virus type 1 viral background plays a major role in development of resistance to protease inhibitors. *Proc Natl Acad Sci U S A* **93** 1648-1653

Rosenbloom DI, Hill AL, Rabi SA, Siliciano RF & Nowak MA 2012 Antiretroviral dynamics determines HIV evolution and predicts therapy outcome. *Nat Med* **18** 1378-1385

Santiago ML, Rodenburg CM, Kamenya S, Bibollet-Ruche F, Gao F, Bailes E, Meleth S, Soong SJ, Kilby JM, Moldoveanu Z, Fahey B, Muller MN, Ayouba A,

Nerrienet E, McClure HM, Heeney JL, Pusey AE, Collins DA, Boesch C, Wrangham RW, Goodall J, Sharp PM, Shaw GM & Hahn BH 2002 SIVcpz in wild chimpanzees. *Science* **295** 465

Schock HB, Garsky VM & Kuo LC 1996 Mutational anatomy of an HIV-1 protease variant conferring cross-resistance to protease inhibitors in clinical trials. Compensatory modulations of binding and activity. *J Biol Chem* **271** 31957-31963

Seibert SA, Howell CY, Hughes MK & Hughes AL 1995 Natural selection on the gag, pol, and env genes of human immunodeficiency virus 1 (HIV-1). *Mol Biol Evol* **12** 803-813

Servais J, Lambert C, Fontaine E, Plesseria JM, Robert I, Arendt V, Staub T, Schneider F, Hemmer R, Burtonboy G & Schmit JC 2001 Variant human immunodeficiency virus type 1 proteases and response to combination therapy including a protease inhibitor. *Antimicrob Agents Chemother* **45** 893-900

Strickler JE, Gorniak J, Dayton B, Meek T, Moore M, Magaard V, Malinowski J & Debouck C 1989 Characterization and autoprocessing of precursor and mature forms of human immunodeficiency virus type 1 (HIV 1) protease purified from Escherichia coli. *Proteins* **6** 139-154

Tang C, Ndassa Y & Summers MF 2002 Structure of the N-terminal 283-residue fragment of the immature HIV-1 Gag polyprotein. *Nat. Struct. Biol.* **9** 537-543

Tozser J, Blaha I, Copeland TD, Wondrak EM & Oroszlan S 1991 Comparison of the HIV-1 and HIV-2 proteinases using oligopeptide substrates representing cleavage sites in Gag and Gag-Pol polyproteins. *FEBS Lett* **281** 77-80

Tritch RJ, Cheng YE, Yin FH & Erickson-Viitanen S 1991 Mutagenesis of protease cleavage sites in the human immunodeficiency virus type 1 gag polyprotein. *J Virol* **65** 922-930

Voevodin AF, Johnson BK, Samilchuk EI, Stone GA, Druilhet R, Greer WJ & Gibbs CJ, Jr. 1997 Phylogenetic analysis of simian T-lymphotropic virus Type I (STLV-I) in common chimpanzees (Pan troglodytes): evidence for interspecies transmission of the virus between chimpanzees and humans in Central Africa. *Virology* **238** 212-220

Wang JY, Ling H, Yang W & Craigie R 2001 Structure of a two-domain fragment of HIV-1 integrase: implications for domain organization in the intact protein. *EMBO J* **20** 7333-7343

Wang Y, Dewdney TG, Liu Z, Reiter SJ, Brunzelle J S, Kovari IA & Kovari L C 2012 Higher Desolvation Energy Reduces Molecular Recognition in Multi-Drug Resistant HIV-1 Protease. *Biology (Basel)* **1** 81-93

Wensing AM, Calvez V, Gunthard HF, Johnson VA, Paredes R, Pillay D, Shafer RW & Richman DD 2015 2015 Update of the Drug Resistance Mutations in HIV-1. *Top Antivir Med* **23** 132-141

Witvrouw M, Pannecouque C, Switzer WM, Folks T M, De Clercq E & Heneine W 2004 Susceptibility of HIV-2, SIV and SHIV to various anti-HIV-1 compounds: implications for treatment and postexposure prophylaxis. *Antivir Ther* **9** 57-65

Wlodawer A, Miller M, Jaskolski M, Sathyanarayana BK, Baldwin E, Weber IT, Selk LM, Clawson L, Schneider J & Kent SB 1989 Conserved folding in retroviral proteases: crystal structure of a synthetic HIV-1 protease. *Science* **245** 616-621

Wolfe ND, Heneine W, Carr JK, Garcia AD, Shanmugam V, Tamoufe U, Torimiro JN, Prosser A T, Lebreton M, Mpoudi-Ngole E, McCutchan FE, Birx DL, Folks TM, Burke DS & Switzer WM 2005 Emergence of unique primate T-lymphotropic viruses among central African bushmeat hunters. *Proc Natl Acad Sci U.S.A.* **102** 7994-7999

Yedidi RS, Proteasa G, Martin PD, Liu Z, Vickrey JF, Kovari IA & Kovari LC 2014 A multi-drug resistant HIV-1 protease is resistant to the dimerization inhibitory activity of TLF-PafF. *J Mol Graph Model* **53** 105-111

Yewdell J, Anton L C, Bacik I, Schubert U, Snyder HL & Bennink JR 1999 Generating MHC class I ligands from viral gene products. *Immunol Rev* **172** 97-108

Zervoudi E, Saridakis E, Birtley JR, Seregin SS, Reeves E, Kokkala P, Aldhamen YA, Amalfitano A, Mavridis IM, James E, Georgiadis D & Stratikos E 2013 Rationally designed inhibitor targeting antigen-trimming aminopeptidases enhances antigen presentation and cytotoxic T-cell responses. *Proc Natl Acad Sci U.S.A.* **110** 19890-19895

PERMISSIONS

The contributors of this book come from diverse backgrounds, making this book a truly international effort. This book will bring forth new frontiers with its revolutionizing research information and detailed analysis of the nascent developments around the world.

We would like to thank all the contributing authors for lending their expertise to make the book truly unique. They have played a crucial role in the development of this book. Without their invaluable contributions this book wouldn't have been possible. They have made vital efforts to compile up to date information on the varied aspects of this subject to make this book a valuable addition to the collection of many professionals and students.

This book was conceptualized with the vision of imparting up-to-date information and advanced data in this field. To ensure the same, a matchless editorial board was set up. Every individual on the board went through rigorous rounds of assessment to prove their worth. After which they invested a large part of their time researching and compiling the most relevant data for our readers.

The editorial board has been involved in producing this book since its inception. They have spent rigorous hours researching and exploring the diverse topics which have resulted in the successful publishing of this book. They have passed on their knowledge of decades through this book. To expedite this challenging task, the publisher supported the team at every step. A small team of assistant editors was also appointed to further simplify the editing procedure and attain best results for the readers.

Apart from the editorial board, the designing team has also invested a significant amount of their time in understanding the subject and creating the most relevant covers. They scrutinized every image to scout for the most suitable representation of the subject and create an appropriate cover for the book.

The publishing team has been an ardent support to the editorial, designing and production team. Their endless efforts to recruit the best for this project, has resulted in the accomplishment of this book. They are a veteran in the field of academics and their pool of knowledge is as vast as their experience in printing. Their expertise and guidance has proved useful at every step. Their uncompromising quality standards have made this book an exceptional effort. Their encouragement from time to time has been an inspiration for everyone.

The publisher and the editorial board hope that this book will prove to be a valuable piece of knowledge for researchers, students, practitioners and scholars across the globe.

LIST OF CONTRIBUTORS

Andreas Wagner
Institute of Evolutionary Biology and Environmental Studies, University of Zurich, Zurich, 8057, Switzerland
The Swiss Institute of Bioinformatics, Basel, Switzerland
The Santa Fe Institute, 1399 Hyde Park Road, Santa Fe, New Mexico 87501, USA

Aditya Barve
Institute of Evolutionary Biology and Environmental Studies, University of Zurich, Zurich, 8057, Switzerland
The Swiss Institute of Bioinformatics, Basel, Switzerland

Vardan Andriasyan
Institute of Molecular Life Sciences, University of Zurich, Zurich, 8057, Switzerland

Xiao Wu and Fan Jiang
Department of Pathology and Pathophysiology, School of Basic Medicine, Shandong University, Jinan, Shandong Province, China

Dimitrios Vlachakis, Louis Papageorgiou and Sophia Kossida
Bioinformatics & Medical Informatics Team, Biomedical Research Foundation, Academy of Athens, Athens, Greece

Spyridon Champeris Tsaniras
Department of Physiology, Medical School, University of Patras, Rio, 26504 Patras, Greece

Katerina Ioannidou
School of Electrical and Computer Engineering, National Technical University of Athens, Greece

Marc Baumann
Protein Chemistry/Proteomics Unit, Biomedicum Helsinki, Institute of Biomedicine, University of Helsinki, Finland

Sivan Meyuhas and Mahmoud Huleihel
Department of Microbiology, Immunology and Genetics, Faculty of Health Sciences, Ben-Gurion University of the Negev, Beer-Sheva, Israel

Morad Assali
Department of Orology, Soroka Medical Center, Beer-Sheva, Israel

Mahmoud Huleihil
Academic Institute for Training Arab Teachers (AITAT), Beit Berl College, Israel

RA Nagalakshmi and J Suresh
Department of Physics, The Madura College (Autonomous), Madurai, 625 011, India

S Maharani and R Ranjith Kumar
Department of Organic Chemistry, School of Chemistry, Madurai Kamaraj University, Madurai, 625 021, India

Styliani Loukatou, Ioannis Bassis and Sophia Kossida
Computational Biology & Medicine Group, Biomedical Research Foundation, Academy of Athens, Soranou Efessiou 4, Athens 11527, Greece

Louis Papageorgiou
Computational Biology & Medicine Group, Biomedical Research Foundation, Academy of Athens, Soranou Efessiou 4, Athens 11527, Greece
Department of Informatics and Telecommunications, National and Kapodistrian University of Athens, University Campus, Athens 15784, Greece

Vasileios Megalooikonomou
Department of Computer Engineering and Informatics Faculty, University of Patras, Patras 26500, Greece

Paraskevas Fakourelis
Computational Biology & Medicine Group, Biomedical Research Foundation, Academy of Athens, Soranou Efessiou 4, Athens 11527, Greece
Department of Computer Engineering and Informatics Faculty, University of Patras, Patras 26500, Greece

Arianna Filntisi
Computational Biology & Medicine Group, Biomedical Research Foundation, Academy of Athens, Soranou Efessiou 4, Athens 11527, Greece
School of Electrical and Computer Engineering, National Technical University of Athens, Athens, Greece

Wojciech Makałowski
Institute of Bioinformatics, University of Münster, Niels-Stensen-Straße 14, Münster 48149, Germany

Eleftheria Polychronidou
Computational Biology & Medicine Group, Biomedical Research Foundation, Academy of Athens, Soranou Efessiou 4, Athens 11527, Greece
Department of Informatics, Ionian University, Tsirigoti Square 7, Corfu 49100, Greece

Dimitrios Vlachakis
Computational Biology & Medicine Group, Biomedical Research Foundation, Academy of Athens, Soranou Efessiou 4, Athens 11527, Greece
Department of Computer Engineering and Informatics Faculty, University of Patras, Patras 26500, Greece
Bionetwork ltd. 15234, Chalandri, Athens, Greece

Rajni Bala, Ulfath Saba, Meenakshi Varma, Dyna Susan Thomas, Guruprasad Rao, Kanika Trivedy, Vijayan Kunjupillai and Ravikumar Gopalapillai
Seri-biotech Research Laboratory, Central Silk Board, Kodathi, Carmelaram Post, Kodathi, Bengaluru 560 035, India

Deepak Kumar Sinha
Seri-biotech Research Laboratory, Central Silk Board, Kodathi, Carmelaram Post, Kodathi, Bengaluru 560 035, India
Plant-insect interaction group, ICGEB, New Delhi, India

John B. Mailhes
Department of Obstetrics and Gynecology, Louisiana State University Health Sciences Center, Shreveport, Louisiana, USA 71130

Michael L. Roberts
Division of Endocrinology and Metabolism, Center of Clinical, Experimental Surgery and Translational Research, Biomedical Research Foundation of the Academy of Athens, Athens, Greece

Amalia Sertedaki
Division of Endocrinology, Metabolism and Diabetes, First Department of Pediatrics, National and Kapodistrian University of Athens Medical School, "Aghia Sophia" Children's Hospital, Athens, Greece

Nicolas C. Nicolaides and Evangelia Charmandari
Division of Endocrinology and Metabolism, Center of Clinical, Experimental Surgery and Translational Research, Biomedical Research Foundation of the Academy of Athens, Athens, Greece
Division of Endocrinology, Metabolism and Diabetes, First Department of Pediatrics, National and Kapodistrian University of Athens Medical School, "Aghia Sophia" Children's Hospital, Athens, Greece

Tomoshige Kino
Program in Reproductive and Adult Endocrinology, Eunice Kennedy Shriver National Institute of Child Health and Human Development, National Institutes of Health, Bethesda, Maryland

Eleni Katsantoni
Division of Hematology-Oncology, Basic Research Center, Biomedical Research Foundation of the Academy of Athens, Athens, Greece

Paraskevi Moutsatsou
Department of Clinical Biochemistry, University of Athens Medical School, "Attiko" Hospital, Athens, 12462, Greece

Anna-Maria G. Psarra
Department of Biochemistry and Biotechnology, University of Thessaly, Larissa, 41221, Greece

George P. Chrousos
Division of Endocrinology and Metabolism, Center of Clinical, Experimental Surgery and Translational Research, Biomedical Research Foundation of the Academy of Athens, Athens, Greece
Division of Endocrinology, Metabolism and Diabetes, First Department of Pediatrics, National and Kapodistrian University of Athens Medical School, "Aghia Sophia" Children's Hospital, Athens, Greece
Saudi Diabetes Study Research Group, King Fahd Center for Medical Research, King Abdulaziz University, Jeddah, Saudi Arabia

Sonja Wolff, Balbina García-Reyes, Doris Henne-Bruns, Joachim Bischof and Uwe Knippschild
Department of General and Visceral Surgery, Surgery Center, Ulm University Hospital, Ulm, Germany

David Jelinek, Robert A. Orlando and William S. Garver
Department of Biochemistry and Molecular Biology

Randy A. Heidenreich
Department of Pediatrics, School of Medicine, The University of New Mexico Health Sciences Center, Albuquerque, New Mexico, US

Kathleen KM Glover and Kevin M Coombs
Department of Medical Microbiology, Faculty of Health Sciences, University of Manitoba, Winnipeg, Manitoba R3E 0J6 Canada
Manitoba Centre for Proteomics and Systems Biology, Room 799, 715 McDermot Avenue, Winnipeg, Manitoba, Canada R3E 3P4 Canada

Abraham A. Embi
13442 SW 102 Lane Miami, FLA, 33186, USA

Benjamin J. Scherlag
Heart Rhythm Institute, University of Oklahoma Health Sciences Center, Oklahoma City, Oklahoma, USA

Roberta Marchione and Jean-Luc Lenormand
TheREx, TIMC IMAG Laboratory, UMR5525, UJF/ CNRS, Joseph Fourier University, 38700 La Tronche, France

Lavinia Liguori
SyNaBi, TIMC IMAG Laboratory, UMR5525, UJF/ CNRS, Joseph Fourier University, 38700 La Tronche, France

David Laurin
TheREx, TIMC IMAG Laboratory, UMR5525, UJF/ CNRS, Joseph Fourier University, 38700 La Tronche, France
Etablissement Français du Sang Rhône Alpes, La Tronche, F-38701 France

Kyoko Inagaki-Ohara
Research Institute, National Center for Global Health and Medicine (NCGM), 1-21-1, Toyama Shinjuku, Tokyo, Japan

Akihiko Yoshimura
Department of Microbiology and Immunology, Keio University School of Medicine, 35 Shinanomachi, Shinjyuku, Tokyo, 160-8582, Japan

Verity Kew and Mark Wills
Department of Medicine, Addenbrooke's Hospital, Cambridge, UK

Matthew Reeves
UCL Institute of Immunity & Transplantation, Royal Free Hospital, London, UK

Eleni Picasi and Athanasios Tartas
Genetics and Computational Biology Group, Laboratory of Genetics, Department of Biotechnology, Agricultural University of Athens, 75 Iera Odos, 11855, Athens, Greece

Dimitrios Vlachakis
Genetics and Computational Biology Group, Laboratory of Genetics, Department of Biotechnology, Agricultural University of Athens, 75 Iera Odos, 11855, Athens, Greece
Computer Engineering and Informatics Department, School of Engineering, University of Patras, 26500 Patras, Greece

Gary S Laco
Roskamp Institute, Sarasota, Florida, USA

Index

www.ingramcontent.com/pod-product-compliance
Lightning Source LLC
Chambersburg PA
CBHW080634200326

41458CB00013B/4629